Understanding

Biology

Philip Applewhite
Yale University

Sam Wilson
San Rafael, California

In collaboration with
Richard Roe
Del Mar, California

Holt, Rinehart and Winston

New York • Chicago • San Francisco • Dallas
Montreal • Toronto • London • Sydney

Cover and text design: Nancy Axelrod

Cover illustration: Cross-section of a spruce tree bud (Manfred Kage from Peter Arnold)

Credits for Readings appear on page 429.

Art acknowledgements not under illustrations are given on pages 429-430.

Understanding Biology is based in part on *Modern Biology* © 1973 by Otto and Towle. The Appendix is reprinted from *Modern Biology* © 1977 by Otto and Towle.

Library of Congress Cataloging in Publication Data

Applewhite, Philip.
Understanding Biology.

  Bibliography: p.
  Includes index.
  1. Biology.   I. Wilson, Sam, joint author.   II. Title.
QH308.2.A66   574   77-28305
ISBN 0-03-005641-1

Printed in United States of America

8 9 0 1 2   032   9 8 7 6 5 4 3 2

PREFACE

Biology texts typically begin with a flurry of atoms and molecules—a chemical snow job—rationalized by the claim that in order to understand the macrocosm one must first glimpse the microcosm. For students contemplating a career in the biological sciences the claim may be valid. But, for non-majors, who may be taking their only college course in the life sciences, the early barrage of chemistry is more of an anesthetic than a basis on which to build the rest of biology.

Therefore, in designing this text the authors, a biologist and a professional writer, have chosen evolution and ecology, subjects of general interest, to follow the introductory chapter. The choice has a logical as well as a pragmatic justification. The essentials of evolutionary theory were formulated when the science of biochemistry was still in its dark age. (Urea, the first organic compound to be produced in the laboratory, was synthesized only a couple of decades before Darwin published on natural selection.) Moreover, evolution is the cornerstone of modern biology, a central theme or basic assumption in much of contemporary research.

Following the introduction to the principles of evolution we have included a chapter tracing the evolutionary development of the major groups of organisms. The next chapter, an introduction to ecology, logically follows evolution in that environmental forces are the cutting edge of natural selection. By presenting the basics of ecology and evolution early in the book, a foundation is laid for considering more complex aspects of these highly active fields in later chapters. Of course, chemistry can't be totally ignored, even in preliminary surveys of evolution and ecology. Thus, a brief section on the atomic nature of matter and the conventions of chemical symbolism is presented early in the evolution chapter, in the context of primordial chemical evolution.

Only after the broad scope of biology has been established in the first four chapters do we turn to the biochemical level and begin to describe the magnificent microstructure of living systems that modern biologists have discovered. Subsequent chapters consider cellular structure and function, the flow of life's energy, the genetic continuum maintained by cellular and organismic reproduction, and the complex physiology of higher organisms. The book comes full circle, returning to the organism in its environment, with a consideration of behavior, one of the most active and controversial fields of modern biology.

For those who choose not to approach a survey of biology as we have, the chapters are internally structured so that the order in which they are studied need not follow the order of the book. If an instructor prefers a microcosm approach— that is, cells to organisms to populations, Chapters 5 to 15 could precede Chapters 2 to 4.

Of particular interest to non-science students will be the chapters on human sexual biology and disease and aging. These subjects are dealt with in greater detail than in most other texts of this kind. The brief readings from outside sources which are included at the end of each chapter broaden the text's historical base and enhance its relevance to contemporary socio-biological problems. For example, excerpts from the Scopes "Monkey Trial" supplement the discussion of evolution, an excerpt from James Watson's *The Double Helix* elaborates on the discovery of DNA, and a discussion of the recombinant DNA controversy follows the genetics chapter.

Chapter objectives and summaries, and suggestions for further reading, offer valuable learning aids to the student. In addition, there is an Appendix which includes a glossary and metric tables. A separate study guide is available to assist the student in mastering the course material.

We would like to thank a number of people whose valuable comments and criticisms helped in the development of the text.

WILLIAM D. ALMY, Tarrant County Junior College

MARY LEE BARBER, California State University, Northridge

GWENDOLYN R. BURTON, Community College of Denver, North Campus

CHARLES CURTIN, Richard J. Daley College

STUART J. DEARING, Northern Virginia Community College

JOHN S. EDWARDS, University of Washington

DARREL R. FALK, Syracuse University

WILBUR A. GILBERT, Miami-Dade Community College

MATILDA L. GIRADEAU, Florida Junior College

JAMES E. HALL, Central Piedmont Community College

W. HOLT HARNER, Broward Community College

SADAKO H. HOUGHTEN, Los Angeles Pierce College

WALTER M. JEWSBURY, Long Beach City College

CHARLES O. MATHER, Los Angeles City College

STEVEN B. OPPENHEIMER, California State University, Northridge

CLAUDIO PEREZ, Laredo Junior College

ROBERT L. POPE, El Camino College

SUSAN RIECHERT, University of Tennessee

MARTIN S. ROCHFORD, Fullerton College

STANLEY SALTHE, Brooklyn College

EDWARD SAMUELS, Los Angeles Valley College

HOWARD A. SCHNEIDERMAN, University of California, Irvine

TED SHERRILL, Eastfield College

JOHN F. WALPER, Community College of Philadelphia

SANDRA WINICUR, Indiana University at South Bend

LELAND YEE, Community College of Denver, Red Rocks

THEO ZEMEK, College of DuPage

November 1977
New Haven, Connecticut   P.A.
San Rafael, California   S.W.

iv

CONTENTS

Understanding Biology

CHAPTER 1

The Science
of Life

Objectives

1 Understand why "spontaneous generation" does not work.

2 See how science as a discipline emerged from other studies.

3 Understand the scientific method.

4 Be able to list the principles of biology and understand what they mean.

Humans are curious. We seek explanations for phenomena we can not understand. But, we are sometimes too quick to answer our questions. A few hundred years ago, for example, it was generally believed that certain forms of life arose spontaneously from their environments: eels were thought to emerge from pond mud and flies from rotting meat.

The first blow against this notion of **"spontaneous generation"** was struck by an Italian physician, Francesco Redi, in the seventeenth century. He tested the notion that maggots arose directly from meat by letting meat rot in two sets of jars (Figure 1-1). Some jars were open to the air, hence the meat in them could be exposed to flies. The other jars were covered. In due course maggots appeared only in the uncovered jars.

Subsequent experiments showed that other species of organisms could arise only from their own kind. In the mid-nineteenth century the French biologist, Louis Pasteur (Figure 1-2) ended the fallacy of spontaneous generation by showing that microorganisms could not materialize from the foods that supported their growth; they must somehow be transferred into it (Figure 1-3).

THE HERITAGE OF SCIENCE

The myth of spontaneous generation is at least as old as history. One of the most influential figures in the development of science believed it. Aristotle, a Greek philosopher/scientist of the fourth century, B.C., was one of the first whose thoughts of life were written. (The word **biology,** comes from the Greek *bios,* which means "life," and *logos,* which means "study of" or "science of.") Aristotle was constantly surrounded by students at the Lyceum (a forerunner of modern universities) who recorded his every word. He was an intuitive genius; his theories on life were based on astute observations, but they were supported by

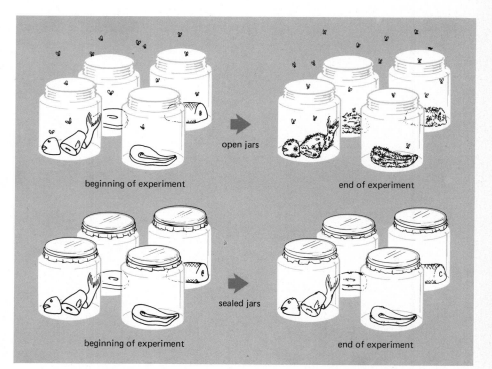

FIGURE 1-1

Francesco Redi's experiments on spontaneous generation.

FIGURE 1-2
Louis Pasteur and his laboratory for biological chemistry in Paris.

little or no experimentation. Unfortunately, he was sometimes wrong, but the real tragedy was not in the errors themselves.

It is a sad fact of human history that the thoughts of Aristotle and a few other ancient scholars went unquestioned for over a millennium. With the decline of the Classical Age and the ascendancy of an authoritarian church, Western scientific thought slid into an age of superstition and faith, a period when dogma replaced observation and original ideas contrary to doctrine were punishable by death. For 1000 years following the fall of the Roman Empire, essentially no distinction was made between science and theology, and bickering over minor points of interpretation of accepted authorities was the extent of permissible intellectual activity in these areas.

With the coming of the Renaissance, scientific thought began to bloom, but escape from the traditions of the Middle Ages was slow. Thomas Aquinas, who taught at the University of Paris in the thirteenth century, was among the first to openly call for more freedom in the analysis of natural phenomena. Yet, 400 years later, when the Italian astronomer, Galileo Galilei dared to assert that the earth

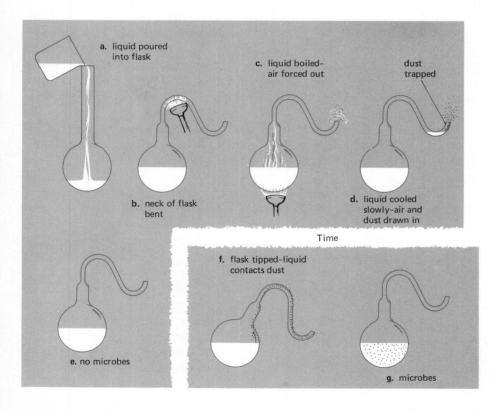

a. liquid poured into flask
b. neck of flask bent
c. liquid boiled—air forced out
d. liquid cooled slowly—air and dust drawn in
dust trapped
Time
e. no microbes
f. flask tipped—liquid contacts dust
g. microbes

FIGURE 1-3
Pasteur's experiment using swan-necked flasks.

revolved about the sun (violating the sacred belief that human beings stood at the center of creation) he was forced to recant under the somber gaze of 10 cardinals and the threat of excommunication. Nevertheless, Galileo carried his search for truth to the grave. In the very year of his death (1642), a man was born who was to expand incredibly human awareness of the universe.

Among other things, Isaac Newton (1642–1727) invented the calculus, formulated the law of gravitation, and discovered some of the basic properties of light. Since Newton's time the physical sciences have flourished, with essentially total freedom to explore the nonliving components of the universe.

Although scientists have long been permitted to examine the properties of inanimate matter, more deeply felt moral and religious questions are leveled at those concerned with the nature of life, particularly human life. Thus, in the mid-nineteenth century, when Charles Darwin proposed that life forms are not static—that humans exist at the crest of a wave of evolutionary change rather than as a separately created, divine species—his ideas sparked a bitter controversy. Even today the teaching of evolution is opposed by fundamentalist religious groups. Nevertheless, evolution is undoubtedly the most significant unifying concept in biology today. The basic tenets of it are considered in the next chapter, and in succeeding chapters the evolutionary theme will repeatedly crop up.

THE "SCIENTIFIC METHOD"

Great advances in the sciences often refute pre-existing beliefs. Perhaps the most important characteristic of a scientist is the ability to let go of previously held ideas in the light of new evidence. Good scientists never reject old ideas, however, merely because they sound strange or foreign. Today, for example, Western doctors are opening their eyes to the fact that **acupuncture,** the ancient Chinese technique for deadening pain, actually works (Figure 1-4). How it works, however, remains a mystery.

A basic tenet of science must be that it is difficult, if not impossible, to know absolute truths. We can only make assumptions based on what our senses perceive, and scientists have come to realize that our senses reveal an incomplete picture of reality. Therefore, through constant reevaluations of ideas, scientists seek closer and closer approximations of truth. It may be that scientists never really prove anything; rather, they disprove alternative explanations so that what is left is the best explanation.

Scientists use logical and orderly procedures of investigation. However, scientific techniques aren't magic formulas that always lead to clear cut answers. On the contrary, the best planned, most carefully done investigations often generate more questions than answers.

Each scientific investigation involves unique problems and approaches to solutions. Hence, generalizations about the **"scientific method"** are of little value when it comes to dealing with specifics. There are certain characteristics that typify most scientific investigations, however. In particular, most research begins with an observation that leads to a question and a review of pertinent research literature. With computers, the retrieval of this information is becoming easier.

Making the observation that people can remember past experiences, for example, one might ask, "What is the mechanism of human memory?" This question is now the subject of intense investigation by scientists throughout the world, and, although there is no generally accepted theory of memory, a great deal of significant data has been accumulated. Thus, to avoid repetition of work already done, a researcher of human memory should first consult the existing literature. The same is true for other fields of research.

FIGURE 1-4
Acupuncture is an ancient Chinese medical technique which involves the insertion of needles.

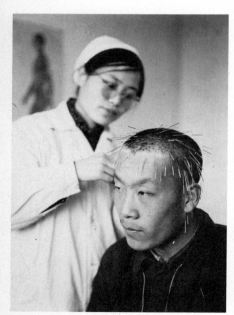

Specialized journals publish results of scientific research. To publish is a recognized obligation of scientists (although some scientists maintain that pressure to "publish or perish" has led to a glut of inconsequential papers). Through their literature, scientists the world over are informed of significant developments and research in progress in their particular fields. Therefore, a library of journals and reference books is essential to a research center. Equally important to scientists are opportunities to personally interact with other scientists having similar interests. The politics of nationalism sometimes interrupt the exchange of ideas in areas related to national security, but communication within the scientific community is essentially world wide.

If available information fails to answer questions, researchers proceed by experimentation. At this point a tentative solution, or **hypothesis** (it might also be called a hunch or an educated guess) is proposed. Then an experiment is designed in which the hypothesis will either be supported or contradicted. In common usage the term "theory" denotes a tentative statement. When scientists speak of **theories,** however, they refer only to statements that have been extensively tested and always found to be true. Hence, an hypothesis becomes a theory only after experimental verification. Theories, too, are subject to change.

Control is essential in experimentation. If, for example, a plant biologist is interested in finding the effect of light intensity on the growth rate of corn, experiments probably would be conducted utilizing several isolated plots of soil. The characteristics of all plots (temperature, moisture, soil composition, and all other factors except light intensity) must be maintained as nearly identical as possible. Only light may differ, and it must be carefully monitored. Maintaining control is commonly the most difficult part of scientific research, particularly in experiments involving organisms in their natural environments.

In considering the preceeding example, one might conclude that experimentation comprises the major portion of all scientific research. But this isn't always so. For instance, in biological oceanography (a field that considers organisms as interacting populations in the marine environment), the bulk of a research program is commonly observational. The techniques of sampling oceanic populations (making net tows at depths of several thousand meters, for instance) are easily disrupted by the whims of nature (Figure 1-5). Thus, it may take years of collecting observational data before an accurate picture of a marine environment is established. In biological oceanography, as in other fields of biology, the observational and experimental phases of the scientific method commonly overlap, since many experiments are directed toward finding accurate methods of observation.

Every phase of an experiment—the way it was planned, the equipment involved, the conditions under which it was conducted, and the results—must be recorded accurately. This record may be in the form of notes, drawings, tables, graphs, calculations, or some combination of these. If large amounts of data have been accumulated, a computer can be utilized to process it.

Research does not end with the elegant array of data spewed from a computer, however, and it should not be thought that anything processed by a computer is automatically true. As the computer maxim says, "Garbage in, Garbage out." By contemplating the results of their observations and experiments, as well as the work and the ideas of fellow scientists, researchers evaluate the worth of the original hypothesis. If experimental evidence doesn't support it, then revisions must be made. Further experimentation may be called for.

THE STATE OF THE LIFE SCIENCES

The most basic scientific questions are as yet unanswered. To answer the question, "What is life?," for example, it is possible to list a number of properties. But we feel there must be an essence that still remains hidden from our knowledge. We do know a great deal about living organisms, however, and it appears that we have the potential to know much more. The following sections summarize some important principles of which we are now aware.

Organisms are centers through which mass and energy flow

One of the results of the modern expansion of scientific awareness has been a merging of traditional study areas. Biology, for example, can no longer be considered apart from chemistry and physics. Scientists now perceive life to involve complex biochemical systems, which act according to the universal laws by which all matter and energy respond.

Plants and animals are accumulations of mass (chemical compounds) that absorb and utilize energy for the continuation of life processes. Indeed, all living organisms require an input of energy in order to live. In the early part of this century, an amazing relationship between mass and energy was discovered. Through the thoughts of Albert Einstein and a few other physicists of vision came the revelation that energy and mass are equivalent. They are related by the simple mathematical statement, $E = mc^2$, in which $E =$ energy, $m =$ mass, and $c =$ the speed of light (approximately 186,000 miles per second or 3×10^8 meters per second).

The "theory of relativity" marked such a significant turning point in the history of science, that most of us are still capable of comprehending it at only a superficial level. A few years before his death, Einstein, himself, stated he was not certain if he was on the right track in his attempt to explain physical reality. Yet, his ideas were given awesome support with the explosion of the first nuclear bomb, a device designed to convert a small amount of mass (a few kilograms) into a devastating amount of energy.

Life's energy originates in the sun through a process basically the same as the explosion of a hydrogen bomb. The primary distinction between plants and animals lies in the manner in which energy is received. Plants absorb energy directly from the sun, (as solar radiation) while animals depend on energy stored by plants and animals (in biochemical compounds). An animal must digest either some of the mass of plants or other animals for its energy.

Part of the energy taken in by plants and animals goes toward renewal of physical structures. The tendency for organized systems to become disorganized is a universal law, **the second law of thermodynamics,** which life does not violate. The biochemical components of life inevitably deteriorate and must be replaced. Hence, plants take in simple nutrients from the soil and atmosphere and (utilizing energy from the sun) organize them into new components. Animals obtain the energy necessary to maintain their bodies by rearranging the molecular structure of biochemical compounds formed by other living things. Ultimately, however, decomposition prevails; organisms age and die.

Organisms grow

Individual life is a losing battle against decomposition. The word *individual* must be stressed in this sentence, since life, as passed from generation to generation through reproduction, is remarkable tenacious. But, early in life, organisms are not only capable of maintaining their structures—they grow.

Growth isn't unique to living systems. An icicle grows as water trickles over its surface and freezes. It is a simple matter of water changing from a liquid to a solid state. Growth of living organisms is entirely different, however. People don't expand by adding food to their bodies without a change in chemical composition. In utilizing nutrients, organisms reorganize chemical structures, forming new interrelating compounds. As a result of these biochemical activities, substances quite unlike those composing nonliving nutrients are incorporated into the makeup of organisms. Thus, living things don't accumulate their substance; they organize it.

Organisms have a unique chemical nature

Living systems utilize energy to organize such complex compounds as proteins, nucleic acids, carbohydrates and fats. Together, these substances of life make up **protoplasm.** Only living organisms produce protoplasm.

Analyses of the protoplasm of numerous plants and animals have revealed basic similarities; all living cells are mostly water, for instance. We are 70% water, but it's what we do with the other 30% that counts! The protoplasm of each species has certain unique properties, and there are subtle differences between individuals of the same species.

Furthermore, the components of protoplasm constantly change. Organisms respond to changes in their surroundings with appropriate shifts in chemical constitution. In reaction to a fearful situation, for example, the adrenal glands (situated near the kidneys) respond by oozing adrenalin into the bloodstream. As it circulates throughout the body minute quantities of this powerful hormone raise an individual to a "fight or flight" state of heart-pounding tension.

Organisms have a cellular structure

Regardless of size and complexity, all living beings are composed of cells—one cell, a few cells, or billions of them. (Humans are composed of nearly 10^{15} cells.) Cells differ in size and shape (from spherical bacteria less than a millionth of a meter in diameter, to strand-like nerve cells over a meter in length), but they all have certain things in common. All cells are bound by membranes, for example, which control what passes into or out of the contained protoplasm. Also, every cell contains a complete copy of the organism's hereditary "code." In other words, it is theoretically possible for a duplicate of a complex organism to be generated from each of its component cells. Hereditary information is contained in complex chemical units, **genes,** which are capable of self-duplication.

Organisms reproduce

Organismic reproduction occurs in many ways. For bacteria and other single-celled species it simply involves dividing in two. Some plants grow from a portion of a root or a stem (Figure 1-5). For many plants and animals reproduction is sexual; it involves formation of special reproductive cells with an endowment of half the genes of a parent. With the merging of two reproductive cells, a new individual is conceived. A human being develops from a mass of living material smaller than a pinhead, which is formed by the union of a **sperm** and an **egg**—the male and female reproductive cells, respectively.

The forms of black bears, sugar maples, and human beings are predictable. Furthermore, certain patterns of behavior are inherited by all members of a given species. (Human babies are born knowing how to nurse, for example.) Organisms resemble their parents. Size, form, and behavioral characteristics are inherited.

FIGURE 1-5
Certain species of plants and animals are able to reproduce a whole organism from only one part.

Genes mutate

As previously mentioned, each of an organism's cells contains a copy of that individual's hereditary "code," which is carried by self-duplicating genes made up of DNA (deoxyribonucleic acid). DNA is the only carrier of genetic information. But, genetic duplication is not always perfect. X-rays, viruses, certain strong chemicals, and various other disturbances can cause changes in genetic structures (**mutations**). If mutations occur in the processes by which reproductive cells are formed, an offspring may develop with characteristics unlike either of its parents or any other members of its species. Usually mutations are detrimental, resulting in physical malformations, death, or disrupted biochemical processes. Occasionally, however, mutation makes an individual better adapted to life than any of the rest of its species.

Organisms have a life span

How long can an organism survive? There is great variation in this. Some species survive for only a few days, while others live for many centuries. For any given species, however, an approximate length of life is predetermined. Barring disease or accidental death, the life span of any particular plant or animal is about the same as that of all others of its kind. No horse is known to have lived beyond 50 years, no dog beyond 25 years. The petunia, marigold, and zinnia plants of a flower garden grow, reproduce, and die in a single season. A white oak tree may live five hundred years. A giant redwood of California would still be young at that age; these remarkable trees may live thousands of years (Figure 1-6).

Is a life span an inevitable limitation of the living condition? If an organism can grow and maintain the organization of its substances during most of its life, why can't it live indefinitely? Perhaps it could if certain physical changes could be avoided, but it is not yet clear what to do.

Certain simple organisms are, in a sense, immortal. An individual bacterium may be formed and mature in a half hour or less. After it has matured, it splits into two bacteria, both of which are capable of growth. Thus, bacteria don't die of old age but continue to live as long as conditions are favorable for growth and cell division. The same is true of the amoeba and other one-celled organisms (Figure 1-7).

Continued growth becomes a problem when organisms become larger and many-celled. However, an experiment performed during the first half of this century seemed to show that portions of complex organisms have the potential for indefinite life. On January 17, 1912, Dr. Alexis Carrel removed a small piece of tissue from the heart of a newly hatched chick. He placed the throbbing mass in a solution of plasma (the fluid part of blood), and put it in a chamber. The warmth and humidity in the chamber duplicated the conditions in the chick's body. For some time, the heart tissue continued to "beat" normally. Gradually, however, the beat slowed, and the end of motion seemed imminent. Apparently, waste products that had formed in the active tissue were accumulating and poisoning it. When the tissue was washed with salt solution, beating returned to normal.

Regular replacement of plasma solved the problem, and the mass began to grow. But, as its bulk increased, a new problem arose. Tissue deep in the mass was no longer receiving nourishment. By dividing it, however, this problem was solved. A routine for care of the tissue culture was established in which new plasma was

FIGURE 1-6
Sequoia gigantia, the giant sequoia or California redwood tree.

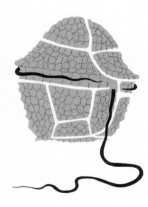

FIGURE 1-7
A one-celled organism (size, about 50 micrometers) called a dinoflagellate protozoan with a whip-like flagella used for movement. Common in fresh water lakes.

added and the mass was divided every forty-eight hours. As part of a normal chick's heart, subject to growth and maturity, it would have died in five to ten years. The experimental culture, however, lived for an incredible thirty-four years.

These results have not been found by other scientists, however. It has been suggested that the culture lived so long only because the plasma added was not pure and in fact contained as contaminants chick heart cells. Even the most exciting experiments sometimes turn out to be misleading, but they still stimulate further necessary research.

Organisms are influenced by their environments

The sensitivity of a living system to its environment is termed **irritability.** The environmental **stimulus** may be light, temperature, moisture, sound, pressure, the presence of a harsh chemical substance, or a source of food. It could also be a threat to survival. The **response,** or reaction of the organism, varies with the nature and intensity of the stimulus.

Each species responds to a given stimulus in a unique way. Certain plants, for example, react to light by growing toward it (Figure 1-8) whereas humans might squint their eyes. Animals are apparently capable of reacting in more complex ways than plants. In higher animals, sight, hearing, taste, touch, and smell are internal responses, or **sensations.** Active responses include fleeing from an enemy, defending oneself in time of danger, hunting for food, and seeking a place in which to settle.

Environmental conditions ultimately define the distributions of populations of organisms. Desert plants and animals, for instance, cannot survive in moist forests. Nor can prairie life survive in marshes. From the arctic to the tropics, and from mountains to valleys, there are certain kinds of organisms that find each environment habitable. Nonliving substances may be influenced by environmental conditions, as is water, which freezes at 0°C and boils at 100°C. But such changes are not of the complexity of biological responses.

FIGURE 1-8
Under moderate light intensities many species of plants are positively phototropic—they grow toward the light. Under high intensities some avoid the light.

Organisms interact with each other

An organism responds not only to its inanimate surroundings, but also to influences by others of its species and/or other forms of life. In order for life to maintain itself, at least two kinds of organisms must exist, with each consuming the chemical output of the other. If plants did not exist and produce oxygen, animals would eventually consume the available supply and die. Similarly, without animals to produce carbon dioxide, plants would soon run out of a supply of this vital substance.

Not all interactions between species are mutually beneficial. Competition, for example, is intrinsic to life. Competition may involve individuals of the same species (intraspecific competition) or two different species (interspecific competition). Male deer in rut flail at each other with their antlers in order to achieve sexual dominance — and a fruit tree competes for water with the grass growing at its base.

The struggle for existence involves many forms of environmental and interorganic relationships. Indeed, as the following chapter shows, these relationships appear to have directed development of the forms of life.

CHAPTER 1: SUMMARY

1 Science as we know it today only gradually emerged from the cultural bounds of society and the realm of philosophy and theology.

2 The "scientific method" is at the heart of experimentation and involves hypothesis, experimentation and theorizing.

3 We cannot precisely define life but we can list several principles of life that apply to all free-living organisms from bacteria to humans.

Suggested Readings

Asimov, Isaac   *A Short History of Biology,* Natural History Press, Garden City, New York, 1965 — An historical survey of significant biological discoveries and the people who made them.

Bronowski, Jacob   *Science and Human Values,* Harper & Row, New York, 1959 — A stimulating discussion of the role of science in modern society.

Butterfield, H.   "The Scientific Revolution," *Scientific American,* September, 1960 — An historical survey of the growth of science since the Renaissance.

Eckehard, Munck   *Biology of the Future,* Franklin Watts, Inc., New York, 1974 — Reviews recent biological developments and considers future possibilities.

Gould, Stephen J.   "On Heroes and Fools in Science," *Natural History,* August-September, 1974 — How some scientists based their discoveries on facts, while others were influenced by superstition.

Wilson, E. Bright, Jr.   *An Introduction to Scientific Research,* McGraw-Hill, New York, 1952 — Applications of the scientific method.

LIFE AND MATTER
René Dubos

Are we yet aware of all the forces influencing life or the ways that organisms respond to their environments? Some scientists think that the characteristics of life are explainable by the laws of chemistry and physics. Nevertheless, there are scientists (scornfully referred to by less mystical peers as "vitalists") who sense a major gap in our understanding of life, particularly the evolutionary process. In the words of the Nobel prize-winning biologist, Albert Szent-Gyorgyi: "In my mind I have never been able to accept fully the idea that adaptation and the harmonious building of complex biological systems, involving simultaneous changes in thousands of genes, are the results of molecular accidents . . . I have always been seeking some higher organizing principle that is leading the living system toward improvement and adaptation . . . Walter B. Cannon, the greatest of American physiologists, often spoke of the 'wisdom of the body.' I doubt whether he could have given a more scientific definition of this 'wisdom.' He probably had in mind some guiding principle, driving life toward harmonious function, toward self-improvement."

The following passage from René Dubos' *The Torch of Life* elucidates a philosophy of science which neither accepts the presence of a guiding principle nor dismisses the possibility.

The work of the past hundred years has established, beyond doubt, that all living things — whatever their size and whether they be men, animals, plants, or microbes — possess many physicochemical characteristics in common. In particular, all of them depend upon the same fundamental reactions for their supply of energy; all synthesize proteins of approximately the same amino acid composition for structural purposes and for enzymatic activity; and all transfer their hereditary endowment from cell to cell through the agency of submicroscopic particles consisting largely of nucleic acids. These physicochemical similarities provide, of course, spectacular confirmatory evidence for the theory of evolution. They greatly increase the likelihood that all living forms studied so far have something common in their origin, that they may indeed all derive from a single point of genesis and have progressively differentiated only in secondary characters. But granted that all forms of life exhibit great chemical similarities, at least two very different working hypotheses can be considered to account for this remarkable unity. One is that the same forces which operate in the inanimate world also act on ordinary matter in such a way as to produce the characters of life; the other is that some unknown principle runs like a continuous thread through all living forms and governs the organization and operation of their physicochemical characteristics.

The first theory is supported by the well-established fact that all living phenomena always go hand in hand with definable reactions which occur according to physicochemical laws identical with those which operate in the inanimate world. This fundamental fact makes it possible to study the living body as if it were an ordinary machine. Indeed, there is no doubt that many of the greatest advances in modern biology, and its applications to medicine, have grown directly out of this theoretical concept.

The proponents of the second hypothesis acknowledge, of course, that all biological phenomena obey known physicochemical laws, but they point out that this fact is not sufficient evidence to prove that life is merely an expression of these laws. Correlation and lack of contradiction with the phenomena of physics and chemistry could be compatible with other theories of life. Furthermore, it appears, at first sight at least, that living processes exhibit many characteristics which are not found in the inanimate world, and there is ground for the assumption that these involve the operation of some directive force as yet unidentified. Such a belief does not imply that living phenomena are outside the rule of natural laws, but rather that science has failed so far to recognize the forces peculiar to the living world, probably because it has focused its attention on the material world.

As appears from this cursory and oversimplified statement, there is one essential question at the basis of all the theoretical arguments concerning the nature of life. The question is to determine whether it is truly a fact that some of the characteristics of living things do not have their counterpart in the nonliving world. It would be useless and unwise to try to guess the answer to this question from the very meager knowledge available at the present time. There are reasons to believe, furthermore, that the problem will be very difficult to formulate in terms of experimental operations and, thus, cannot be solved very soon. What is possible and useful, however, is to try to recognize the properties which are possessed by all living things, and which disappear from them as soon as their life ceases. This can be done objectively, irrespective of one's prejudiced view concerning the origin of life and its essential nature. To describe life as experience manifested through the observable behavior of living things is far less ambitious than attempting to define its ultimate nature. True enough, this humble effort is not likely to provide a philosophical answer to the riddle of existence, but it will certainly help man to learn more of the biological basis of his own nature, and this in turn will help him in formulating, with greater wisdom, a social and ethical philosophy of human life.

A JUG OF WINE, A LOAF OF BREAD, AND YEAST
Sam Wilson

Artists as well as scientists are experimentalists. But, while scientists base their methods on a knowledge of physical processes, artists rely more on intuition. Technology may be born of art or science, and as described in this reading concerning the discovery of anaerobic respiration, science can arise from art.

We have surrounded ourselves with products of industrial technology, the majority of which came about through the creative impetus of science. It was through such scientific breakthroughs as the discovery of the structure of atomic nuclei that such technological developments as the nuclear reactor were made possible.

But, not all technological developments have begun with scientific discoveries. The word "technology," which has come to mean applied science, was derived from the Greek word "technologia," which meant systematic treatment of an art. Among the products of ancient technology were bread and wine, the production of which involves cultures of yeast. Highly refined methods for making bread and wine were practiced long before the functions of yeast were known. In fact, it was not until relatively modern times that yeasts were known to exist, let alone that they are living organisms. Since the techniques of bread- and wine-making were developed more on the basis of intuition and trial and error than on a knowledge of underlying biological processes, early bakers and winemakers were more artists than scientists.

Leaven, which is mentioned in the Bible, was used in early bread-making. It probably consisted of a mixture of yeasts and bacteria maintained in a dough medium. When a batch of dough was prepared for making bread, a small portion was retained for the inoculation of the next batch. In this way the yeast culture was maintained indefinitely.

Wine was produced as early as Neolithic times; an inscription on an Egyptian tomb erected around 2500 B.C. depicts the process. It wasn't until the nineteenth century A.D., however, that scientists began to explain the biology of it. In 1810, Joseph Louis Gay Lussac discovered the chemical reaction by which sugars are broken down to yield alcohol and carbon dioxide:

$$C_6H_{12}O_6 \longrightarrow 2(C_{12}H_5OH) + 2\ CO_2$$

For many years this reaction, fermentation, was thought to be a purely chemical process. But, in the 1850s the correlation was made between fermentation and organismic metabolism. In a series of experiments at the beginning of his brilliant career, Louis Pasteur showed that yeasts were the agents of the fermentation process. He demonstrated that yeasts living naturally on the skin of grapes were responsible for the conversion of sugars to ethyl alcohol and carbon dioxide. In short, alcoholic fermentation is yeast metabolism.

Thus, an ancient artistic technology ultimately led to one of the fundamental discoveries of modern biology.

CHAPTER 2

Introduction to Evolution

Objectives

1 Realize that change is a part of all life.

2 Understand the theories of the formation of the universe.

3 Distinguish between atoms, molecules, and elements.

4 Be aware of the origin of biological molecules and how they formed living organisms.

5 Be able to present the arguments used to support evolution.

6 Understand natural selection.

7 Learn how mutations affect evolution.

8 Learn how the environment affects evolution.

9 Understand how species develop and change.

10 Retrace the evolution of human beings.

BIOLOGICAL EVOLUTION

Change is implicit in existence. Some changes recur periodically, such as the cycle of sleep and wakefulness. Other changes, like the awakening of sexual interests during puberty, happen but once during a lifetime. A little over a hundred years ago a British naturalist showed that life changes over vast spans of time. Forms of life change over the course of many lifetimes.

Evolutionary theory has become the backbone of modern biology, a central theme or a basic assumption in much of contemporary research. The following chapter summarizes the basic ideas of evolutionary theory. References to it appear repeatedly throughout the book.

In 1859, the British naturalist Charles Darwin published *On the Origin of Species by Means of Natural Selection*. In this book he produced compelling evidence for biological evolution. He showed that over many generations new species of plants and animals evolve from preexisting species.

Darwin's publication (a massive work containing an overwhelming number of examples from nature) was needed to gain broad acceptance of the theory of evolution, but the concept had been around for many years. Scientists and natural philosophers before Darwin had remarked upon the similarities between species and proposed that one form of life could gradually change into another. However they did not have the evidence that Darwin presented. Even Darwin's grandfather, Erasmus Darwin, a physician and naturalist, had posed the following question in his writings: ". . . would it be too bold to imagine that all warm-blooded animals have arisen from one living filament . . .?" But, in Erasmus Darwin's time, it was too early for such bold thoughts to catch on.

Since Charles Darwin's time, scientists of many fields have found that long term change is not only a characteristic of life, but that it prevails throughout the universe. In fact, there is now evidence that several billion years ago, changes at the surface of the earth produced life from nonliving matter. The exact sequence leading to the origin of life will probably never be known. With the present knowledge, however, it is possible to theorize a series of steps by which the universe could have evolved and life originated on earth.

UNIVERSAL EVOLUTION

Astronomers have calculated that the universe went through a violent birth (or, perhaps, rebirth) approximately 10 to 20 billion years ago. So much time defies the imagination, but it is possible to draw an analogy between the theoretical age of the universe and an easily conceived span of time, the day. By analogy to the day, the 10 billion year history of the universe can be represented by 24 hours, about 420 million years by each hour, 7 million years by each minute, and 117,000 years by each second. Less than 2000 years have passed since the birth of Christ—the starting point of our method of dating. By analogy, this is but a small fraction of a second in the day representing the age of the universe.

The "big-bang" theory

Some scientists believe that the universe once existed as a single massive body in the void of space. They believe that approximately 10 billion years ago this body exploded, and its fragments were scattered into space. (It is interesting to

FIGURE 2-1
An expanding galaxy similar to our own Milky Way galaxy and containing perhaps as many as 100 billion stars.

speculate where the initial mass came from in the first place.) This concept is appropriately called the **big-bang theory.**

The big bang isn't the only theory of the origin of the universe. The **steady-state** theory which holds that the universe always existed and had no moment of origin has had many followers. However, the big-bang theory is the one which is most widely accepted by scientists.

Today, celestial bodies continue to drift apart, presumably from the force of the big bang (Figure 2-1). Astronomers have measured the rate of spread and estimated the outer limits of the universe. Calculations of universal age are based upon these data.

The early evolution of the earth

According to the big-bang theory the matter of the universe cooled over billions of years following the initial explosion. The debris condensed into galaxies, and within the galaxies, stars, planets, and other solitary units of mass formed.

In its primitive stages (approximately 5 billion years ago), the earth is thought to have been a molten sphere laced with boiling upheavals and fiery explosions. At this time our planet did not have an atmosphere as we know it. Rather, swirling gaseous storms swept around it.

Over many millions of years, the earth cooled and eventually formed a thin crust. Earthquakes and volcanoes rocked it, and gases generated by terrestrial activities accumulated in an atmosphere. Its composition was different, however, than the one surrounding the earth today.

Atoms and the atmosphere

Although today's atmosphere differs from that of primordial ages, it is similar in that it is composed of minute particles hurtling through space. The atmosphere con-

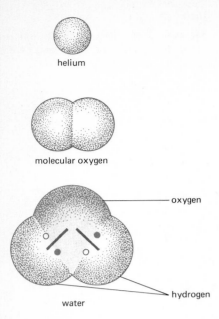

helium

molecular oxygen

oxygen

hydrogen

water

FIGURE 2-2

Some simple chemical forms. Water is a molecule composed of one atom of oxygen and two atoms of hydrogen. Molecular oxygen is composed of two oxygen atoms. Helium is a single atom.

sists of tiny units of matter (some less than a billionth of a meter in diameter to visible pollution particles) hurtling through space. All of the constituents of air are not the same, but they all are single or simple combinations of **atoms,** the building blocks of matter. Ancient Greek philosophers guessed matter's particulate nature; they believed that different elements were composed of unique, indivisible particles. The term atom comes from the Greek, *atomos,* meaning "uncut." More recently, physicists have discovered that atoms are composed of more elementary particles. Atomic substructure and atomic interactions are discussed in Chapter 5.

Atoms may exist singly, or they may be bonded together in structures called **molecules** (Figure 2-2). Atoms of helium, for example, are always unbonded, while atoms of oxygen are almost always bonded. Some molecules are formed from two or more atoms of the same kind. (Molecules of atmospheric oxygen, for instance, consist of two atoms of oxygen). Others are combinations of atoms of different kinds. (Water molecules consist of one atom of oxygen and two of hydrogen). Some of the complex molecules of living organisms are composed of thousands of atoms.

There are about 105 different kinds of atoms, 92 of which occur naturally, (the others are made in the laboratory), and each of these **elements** has a one or two letter designation. The six elements that compose the major portions of living organisms: carbon, hydrogen, oxygen, nitrogen, phosphorous, and sulphur, are respectively indicated by C, H, O, N, P, and S. Other elements, including iron (Fe), copper (Cu), and magnesium (Mg), are present in very small amounts in living systems.

Molecules are abbreviated with the letters representing their constituent atoms and numerical subscripts indicating how many of each kind is included. Molecular oxygen, for example, is represented as O_2. It makes up 21 percent of our present atmosphere. Nitrogen, N_2, is another two-atom molecule, which accounts for 78 percent of the atmosphere. Fractions of a percent of carbon dioxide (CO_2), water (H_2O), helium (He), and a few other gases are also included. The ancient atmosphere, however, probably consisted of ammonia (NH_3), methane (CH_4), hydrogen (H_2), hydrogen cyanide (HCN), and water. Radio astronomers have found these molecules and many others in outer space, suggesting that the same life forming processes that occurred on the earth may be happening elsewhere.

CHEMICAL EVOLUTION

As the ancient atmosphere cooled, water began to condense. Eventually, temperatures fell below the boiling point of water. Rain began to fall, carrying suspended solids from the atmosphere to the earth. Water accumulated in depressions, which became repositories for substances from the atmosphere and from the earth's surface. Eventually, water covered large portions of the earth, forming the ancient seas. Dissolved in the seas were many salts, as well as methane, ammonia, and hydrogen. This mixture was *thermodynamically stable,* that is, these substances did not spontaneously react with each other to form more complex molecules. In order for such reactions to occur, there had to be an input of energy, which may have come from radiation from the sun or electrical discharges from lightning. The added energy was necessary to push the molecules close enough together so that they could react with each other and bind together.

The origin of the "building blocks" of life

In 1953, Stanley Miller and Harold Urey, working at the University of Chicago, performed an experiment in which they used laboratory apparatus designed to simulate conditions in the ancient seas (Figure 2-3). They added water to a large

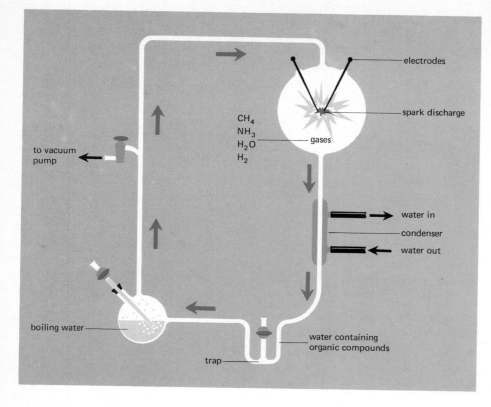

FIGURE 2-3
Apparatus used for the synthesis of amino acids by electric discharge.

flask containing the substances that were believed to have been present in the early waters of the earth. This mixture was then exposed to electrical shock, which was intended to simulate the effects of lightning. At the end of a week, the contents of the flask were found to include several **biochemical compounds,** substances involved in the processes of life. Among them were **amino acids,** the building blocks of protein (Figure 2-4).

The experiments of Miller and Urey implied that some of the basic units of life could have been generated in the ancient oceans. Since 1953, many other experiments of this type have been performed. The results have substantiated the original findings and have indicated that other important biological compounds, including the **purines** and **pyrimidines,** could have been produced in a similar way.

Just as amino acids are joined in a chainlike manner to form protein molecules, so are purines and pyrimidines the building blocks of **nucleic acids.** One kind of nucleic acid, **DNA,** is the constituent of genes, which carries hereditary information in all living organisms. It can perform this function because of a unique characteristic: a DNA molecule can direct its own reproduction. By coding for the arrangement in the proper order of purines and pyrimidines which surround it, a DNA molecule can direct the formation of an exact replica of itself. Later in the book, more detailed information on the structure of DNA will be given, as well as what is known about how it reproduces itself and relays inheritable characteristics from generation to generation. For the time being, however, it is only important to remember two things about DNA. It is the genetic material that carries hereditary information within its structure, and it directs its own reproduction.

The transition from the nonliving to the living

Experiments like those of Miller and Urey have not shown how life was generated from nonliving substances; for example, how the original DNA molecule was produced. But they have indicated how certain components of life could have evolved.

FIGURE 2-4

Many thousands of individual amino acids (2-4a) may be bonded into a single protein molecule. Two atoms of hydrogen and one atom of oxygen (one water molecule) are lost from each amino acid in the process of forming what is called a peptide bond (2-4b). R in the diagrams represents one of twenty possible naturally-occurring side groups of additional atoms. A three-amino-acid protein is shown in Figure 2-4b.

FIGURE 2-5
Fossil trilobites from the Devonian period. Trilobites are members of a large class of extinct marine arthropods whose bodies are divided into three parts by two furrows.

What actually occurred in the transition from nonliving chemicals to living systems remains largely unknown. Nevertheless, approximately three and one-half billion years ago, entities with the characteristics of life existed on earth. Evidence for this ancient presence comes from a direct source—**fossils,** which have preserved the forms of early organisms in rock (Figure 2-5). Fossils have formed as part of the evolution of our planet.

THE GEOLOGICAL TIMETABLE

The face of the earth constantly changes. While life has evolved, land masses have risen above and settled below the seas several times. Mountains have pushed up only to be eroded gradually by winds and rain. Rivers have deepened their channels as they carried more and more land to the sea. Climatic conditions have also changed many times. There have been warm periods and cold periods, periods of heavy rain and periods of drought in all regions of the earth. Changes are still occurring, but most happen so slowly that we are not aware of them.

Sedimentary rock—the earth's timetable

Nature has left traces of these geological changes, as well as of the forms of life that have existed on earth. Layers, or **strata,** of sedimentary rock hold some of these records (Figure 2-6). Rock of this type is formed by **sediment** (debris from erosion and fragments of dead plants and animals) that settles to the bottoms of oceans and other bodies of water. Here it is compressed and eventually solidifies. New layers settle above old ones, leaving a stratified timetable in rock. Analyses of layers of sedimentary rock are journeys through millions of years of geological time.

Geologists have devised several methods for determining the ages of strata. Geologists often use two or more methods to calculate the age of a given stratum of rock. This gives them an idea of the accuracy of their measurements.

FIGURE 2-6
A typical formation of sedimentary rock which has been eroded.

One technique uses the rate of sedimentation. By measuring the rates at which sediment accumulates today and making certain assumptions about past rates, geologists can estimate the age of a stratum. In the open ocean, for example, about one-third of an inch of sediment accumulates in 1000 years. Thus, sediment that is one inch beneath the surface of the ocean floor is approximately 3000 years old.

A more recently developed method of dating utilizes radioactive elements.* Strata of sedimentary rock contain elements that undergo radioactive decay and thereby change into other elements. One such element is uranium. Half of a deposit of uranium will decay to lead in 4.5 billion years. For this reason, uranium is said to have a **half life** of 4.5 billion years. Thus, if uranium and lead are present in a stratum of sedimentary rock, determining the relative amounts of them is a means of calculating the age of the rock.

The fossil record a timetable of life

The geological timetable is of biological significance because it helps us determine the ages of fossils (Figure 2-7). Fossils are formed in two ways. They may be imprints (such as footprints), or they may be replicas of whole organisms, preserved as mineral matter which gradually replaced living substances over the course of millions of years.

Besides traces in sedimentary rock, nature has preserved life forms in other ways. For example, ancient insects have been preserved in **amber,** a fossil plant resin. Remains of ancient animals, including the mammoth (now extinct but related to the modern elephant) have been found with flesh intact in the "deep-freeze" of glacial ice.

The La Brea Tar Pits near Los Angeles have yielded fossil skeletons and other remains of numerous animals of past ages, including extinct antelope, bison, bears, and saber-toothed cats. The tar pits mark the location of a petroleum spring formed many millions of years ago. As the oils in the spring evaporated, a mass of sticky tar was left. It's believed that rain pools accumulated on the surface, and mammals and birds attracted by the water became stuck. Predators apparently attacked the helpless victims but soon found themselves trapped and sinking in the ooze. Today **paleontologists** (researchers of the fossil record) sift through the jumble of bones which remains, trying to reconstruct the extinct animals.

Using the fossil record, paleontologists have pieced together an "evolutionary tree" (Figure 2-8). The first organisms in the sequence of development of life forms, including bacteria, algae, fungi, and protozoans, were present at least 3 billion years ago. Progressively more complex forms of life inhabited the earth in more recent times. Thus, fossils give evidence of evolution.

Comparisons of the structures of modern organisms with fossils of ancient plants and animals have indicated that certain forms of life such as ferns and sharks have survived through millions of years with little change. Other forms, however, have changed over the course of many generations.

As indicated by the fossil record, the changes of evolution occur slowly. Differences that can be perceived only with painstaking observation have occurred over the course of thousands of years. The difference between two related populations of fish which lived 10,000 years apart, for example, may be limited to a single change noted only by a specialist.

* For a discussion of radioactivity, refer to Chapter 4.

era	period	epoch	approximate years since beginning	approximate years duration	conditions and characteristics
cenozoic	quaternary	recent	15,000		moderating climate; glaciers receding
		pleistocene	1,000,000	1,000,000	warm and cold climates; periodic glaciers
	tertiary	pliocene	10,000,000	9,000,000	cold climate; snow building up
		miocene	25,000,000	15,000,000	temperate climate
		oligocene	35,000,000	10,000,000	warm climate
		eocene	55,000,000	20,000,000	very warm climate
		paleocene	70,000,000	15,000,000	very warm climate
mesozoic	cretaceous		120,000,000	50,000,000	warm climate; great swamps dry out; rocky mountains rise
	jurassic		150,000,000	30,000,000	warm climate; extensive lowlands and continental seas
	triassic		180,000,000	30,000,000	warm, dry climate; extensive deserts
paleozoic	permian		240,000,000	60,000,000	variable climate; increased dryness; mountains rising
	pennsylvanian (late carboniferous)		270,000,000	30,000,000	warm, humid climate; extensive swamps; coal age
	mississippian (early carboniferous)		300,000,000	30,000,000	warm, humid climate; shallow inland seas; early coal age
	devonian		350,000,000	50,000,000	land rises; shallow seas and marshes; some arid regions
	silurian		380,000,000	30,000,000	mild climate; great inland seas
	ordovician		440,000,000	60,000,000	mild climate; warm in arctic; most land submerged
	cambrian		500,000,000	60,000,000	mild climate; extensive lowlands and inland seas
(pre-cambrian) proterozoic			1,500,000,000	1,000,000,000	conditions uncertain; first glaciers
(pre-cambrian) archeozoic			3,500,000,000	2,000,000,000	conditions uncertain; origin of life

FIGURE 2-7

The Earth through the ages. This geological timetable shows the sequence and estimated length of the eras, periods, and epochs as defined by geologists.

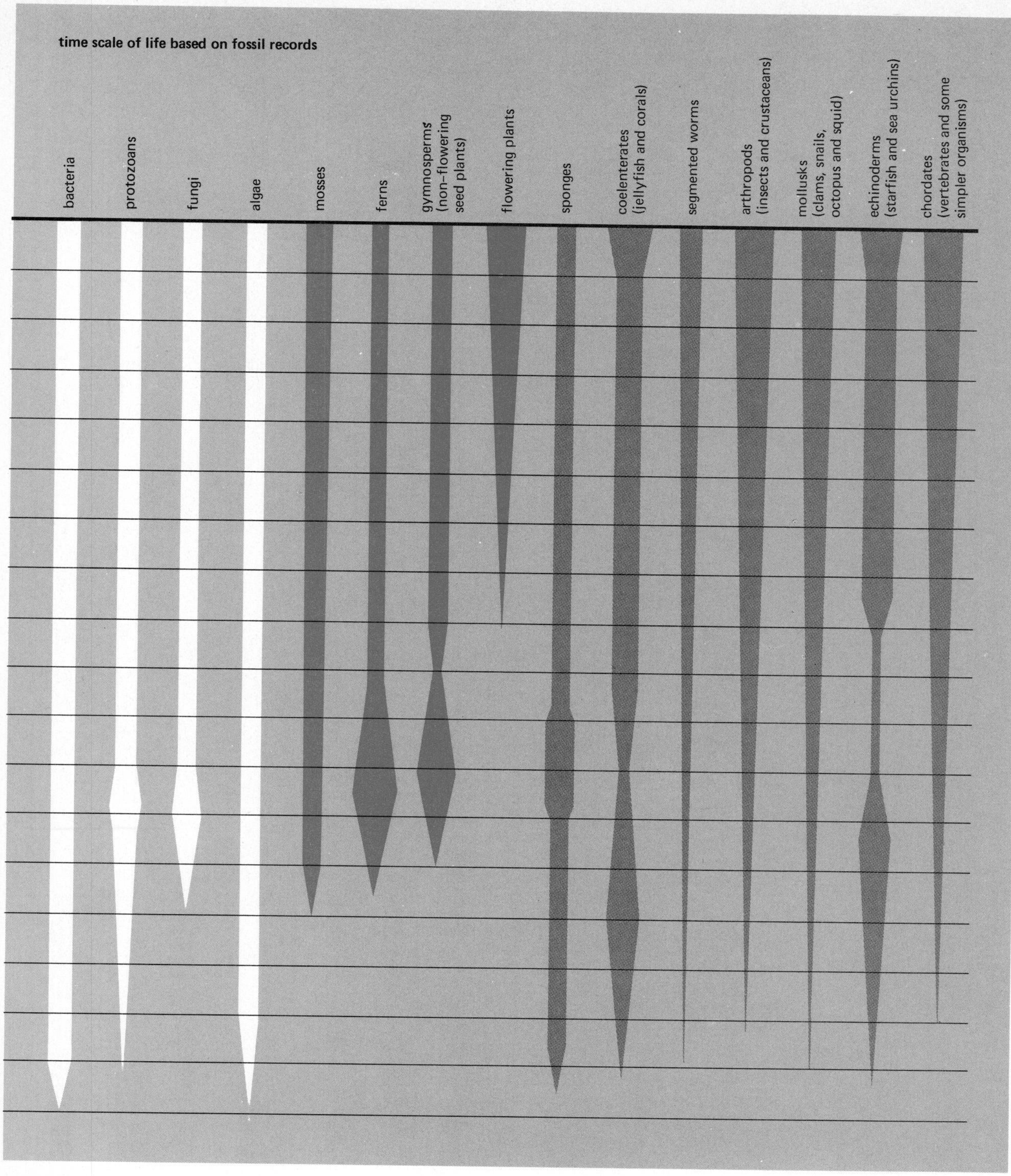
time scale of life based on fossil records
bacteria
protozoans
fungi
algae
mosses
ferns
gymnosperms (non-flowering seed plants)
flowering plants
sponges
coelenterates (jellyfish and corals)
segmented worms
arthropods (insects and crustaceans)
mollusks (clams, snails, octopus and squid)
echinoderms (starfish and sea urchins)
chordates (vertebrates and some simpler organisms)

FIGURE 2-7
The time scale of major groups of organisms shows the approximate time each group is believed to have originated and, to a limited extent, relative increases and decreases in numbers that have occurred.

FIGURE 2-8
A simple evolutionary tree.

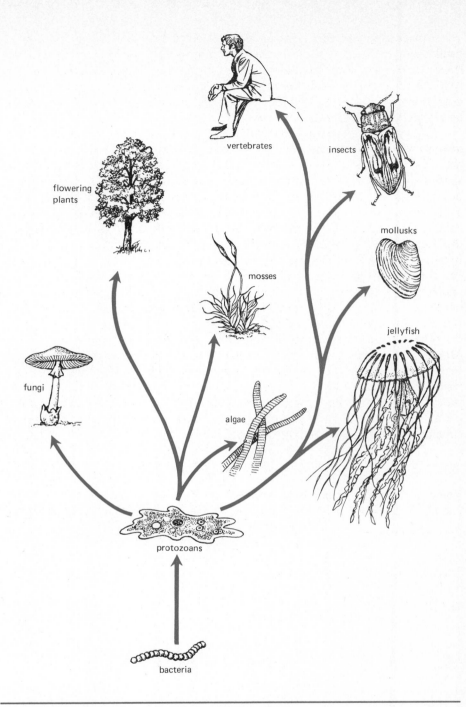

vertebrates

insects

flowering
plants

mollusks

mosses

jellyfish

fungi

algae

protozoans

bacteria

HOMOLOGY—EVIDENCE OF COMMON ANCESTRY

Evidence for evolution isn't limited to the fossil record. Examinations of now-living plants and animals have revealed many recurring physical forms that indicate common ancestries. For example, the bones of a bird's wing, the front leg of a horse, the paddle of a whale, and the human arm are all basically the same in structure. Modifications have apparently occurred over many millions of years to suit different habitats and lifestyles (Figure 2-9). Far back in the history of life, birds, horses, whales, and humans appear to have had a common ancestor. From its primitive appendage, wings, legs, paddles and arms developed.

Similar structures that have developed from the same precursor are termed **homologues,** and there are countless examples of homology throughout the living world. An obvious example in animals is the eye, whose structure is essentially the same in all **vertebrates** (animals with backbones). In some cases, however, homology is masked by extensive modifications. Among flowering plants, leaves, petals, thorns, and tendrils are homologues.

One type of homology gives particularly good evidence for evolution. Most animals have certain body parts that are seemingly useless and often degenerate, although homologous structures are well developed and functional in related species. Such unused structures are said to be **vestigial;** they are remnants of body parts that were of value to an organism's ancestors, but their importance for survival diminished as the environment and habits of the life form changed in the course of evolution. The human appendix, for example, is a small projection from the large intestine without any known function. However, in **mammals** (the class of animals to which humans belong) that eat coarse diets, the appendix is a pouch that serves as an organ of digestion. In rabbits, for example, the appendix is large and functional. Other human vestigial structures include the remains of muscles which can move the ears, wisdom teeth, and the coccyx (or tail vertebrae) which is the vestige of a tail.

The most striking traces of homology have been discovered by **embryologists** (biologists concerned with unborn forms of life and their development). In the first weeks of life, embryonic fish, birds, and mammals are nearly indistinguishable (Figure 2-10). It appears, therefore, that all vertebrates receive the same sets of genes which control early development. Later in the process other genes assume control and each species matures in a unique way, with homologous structures becoming less and less similar.

More evidence for evolution

Some of the best indications of evolution come from comparative studies of biochemistry. Evidence indicates that all organisms produce nucleic acids, especially

FIGURE 2-9

Homologous bones in forelimbs of several vertebrate animals.

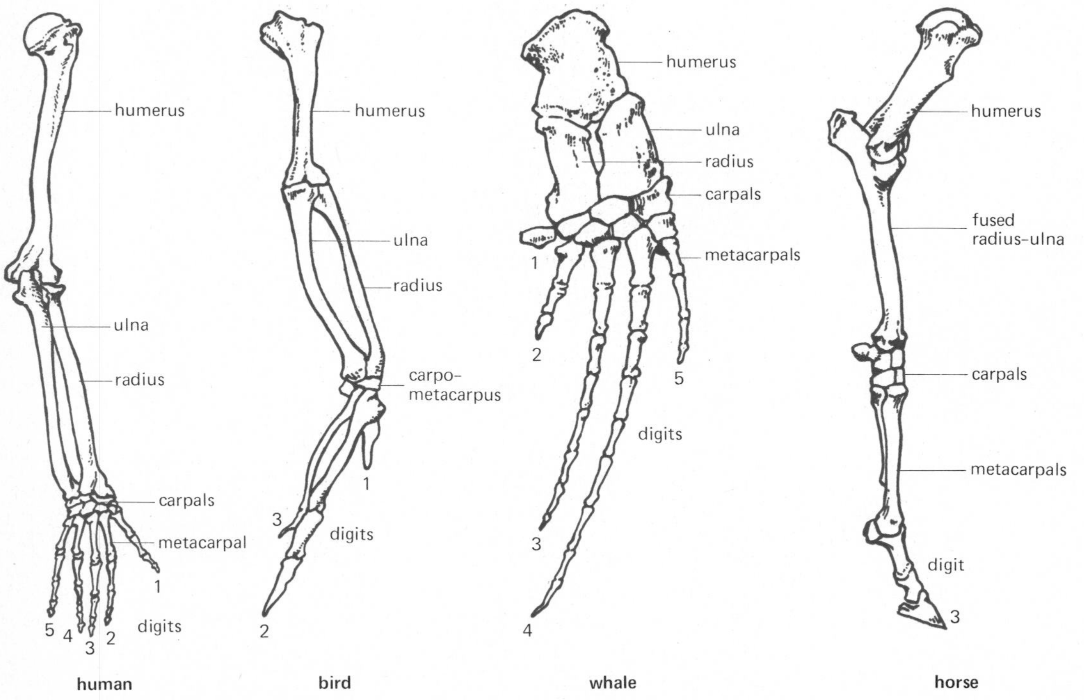

FIGURE 2-10
Different stages in the development of three vertebrate animals. Note their similarities in the very early stages and their differences in the later stages.

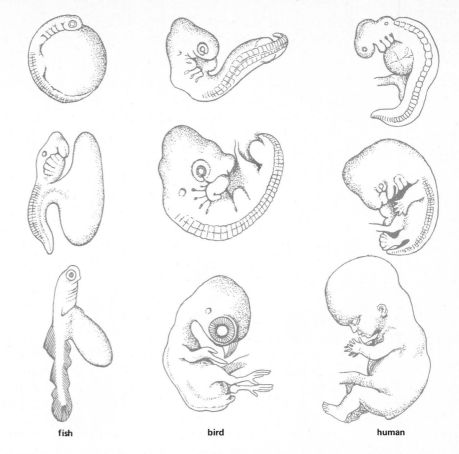

fish   bird   human

the nucleic acid of genes, DNA. Other types of biochemical compounds show similarities from species to species. For example, human diabetics owe their lives to insulin—a protein hormone which facilitates the movement of sugars out of the bloodstream—that is extracted from hogs and cattle. Now insulin can be synthesized in the laboratory by stringing together the proper amino acids in the proper sequence. This ''artificial'' insulin is identical to ''natural'' insulin in every way.

Plant and animal breeding provides further evidence for evolution. Over twenty-five breeds of dogs have been developed from wild, wolflike ancestors, and jungle fowl of India were bred into domestic chickens. Ancient Egyptians created a breed of cat which they considered to be sacred. American Indians of about 7000 years ago developed a hardy breed of domestic corn by crossbreeding species of wild corn with grasses. The history of plant and animal breeding does not prove that similar changes have taken place naturally, but it does strongly point to that possibility. It is a curious fact that, although organisms have been bred for thousands of years, evolution was not accepted until relatively recent years.

LAMARCK'S CONCEPT OF EVOLUTION

A French biologist, Jean Baptiste Lamarck, made one of the early attempts to explain evolution in 1801. His ideas failed to receive experimental verification, but they provide an interesting contrast to Darwin's theory. Lamarck's concept of evolution involved three hypotheses:

The Hypothesis of Need: The production of a new organ or part by a plant or animal results from a need. For example, early ancestors of the snake had legs and short bodies. But with changes in land formations, it became necessary for snakes to walk through narrow places. They began to stretch their bodies and to crawl rather than walk. This line of reasoning formed the basis on which Lamarck formulated his second hypothesis.

The Hypothesis of Use and Disuse: Organs remain active and strong as long as they are used, but disappear gradually with disuse. Lamarck believed that successive generations of snake predecessors continued to stretch and strengthen their bodies. Their legs were used less and less because they interfered with crawling, and they finally disappeared.

The Hypothesis of Inheritance of Acquired Characteristics: All that has been acquired or changed in the structure of individuals during their life is transmitted by heredity to the next generation. Lamarck believed that modern snakes evolved from the forms that had lost their legs through disuse. Thus, they inherited the legless trait from their ancestors.

Lamarck used other examples from nature to explain his hypothetical scheme of evolution. He believed, for example, that the giraffe evolved from a short-legged, short-necked form. When competition for low-growing grasses became too great, the giraffe began to stretch its neck and forelegs in order to reach the leaves of trees. Stretching, according to Lamarck, resulted in succeeding generations with progressively longer necks (Figure 2-11).

Lamarck's hypothesis of inheritance has been tested several times, but each experiment has presented further evidence that acquired traits are not inherited. One researcher, August Weismann, tried cutting off the tails of mice and then mating them. The offspring of the tailless mice had tails. He then cut off the tails of the second generation and mated them. He continued this experiment for twenty generations. The twenty-first generation still had tails the same length as the first.

Of course, cutting off a mouse's tail doesn't improve its ability to survive. In fact, since the operation results in a disrupted sense of balance, it is detrimental. Thus, Weismann's experiment was not a true test of Lamarck's hypothesis. While Weismann demonstrated that a particular acquired characteristic was not passed to offspring, he did not show that needed characteristics could not be acquired and passed on. Nevertheless, no evidence has ever substantiated the hypothesis that organisms can somehow "sense" needed changes, make them, and genetically incorporate them. Scientists can only accept what has been verified.

THE THEORY OF NATURAL SELECTION

According to Darwin, the factors accounting for the development of new species can be summarized as follows:

- All organisms produce more offspring than can survive to maturity.
- Because of overproduction, there is a constant struggle for existence among individuals.
- The individuals of a given species vary.
- The fittest, or the best adapted, individuals of a species survive.
- Surviving organisms transmit variations to offspring.

That the mid-nineteenth century was the right time for the theory of natural selection to come of age is evidenced by the fact that a contemporary of Darwin, Alfred Russel Wallace, independently developed the concept. By a gentleman's agreement, both Darwin and Wallace first presented their ideas at the same time, in 1858, at a meeting of the Linnaean Society in London.

The following sections expand upon the basic points of the theory of natural selection.

FIGURE 2-11
Comparison of theories of Lamarck and
Darwin about the evolutionary process.

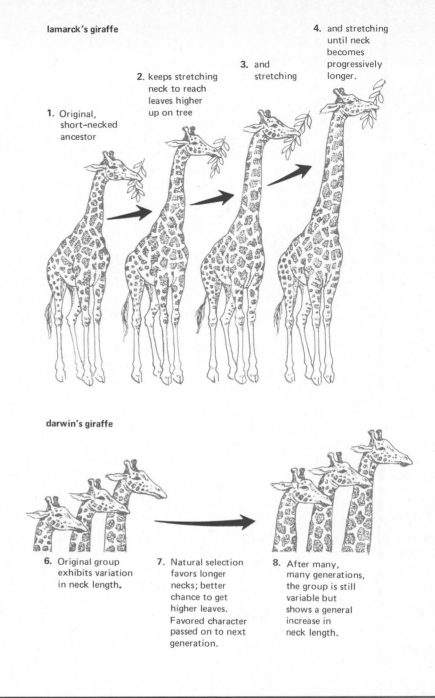

lamarck's giraffe

1. Original, short-necked ancestor

2. keeps stretching neck to reach leaves higher up on tree

3. and stretching

4. and stretching until neck becomes progressively longer.

darwin's giraffe

6. Original group exhibits variation in neck length.

7. Natural selection favors longer necks; better chance to get higher leaves. Favored character passed on to next generation.

8. After many, many generations, the group is still variable but shows a general increase in neck length.

Excessive reproduction

A fern may produce 50 million spores each year. If all the spores resulting from this great fertility matured, in the second year ferns would nearly cover North America. A mustard plant produces about 730,000 seeds annually. If they all took root and matured, in two years they would occupy an area 2,000 times that of the land surface of the earth. The dandelion would do the same in ten years (it has already taken over the lawn of one of the authors). At a single spawning, an oyster may shed 114,000,000 eggs. If all these eggs matured, the ocean would be literally filled with oysters. The elephant is considered to have a slow rate of reproduction. An average elephant lives to be about 100 years old, breeds over a span of 30 to 90 years, and bears about 6 young. If all the young from one pair of elephants survived, in 750 years the descendants would number 19,000,000. There is, however, no such increase.

Struggle for existence

Regardless of the rate of reproduction, only a small fraction of the original number of offspring of most species reach maturity. Each individual seeks food, water, air, warmth, and space, but only a few can obtain these needs in the struggle for survival. This struggle is most intense between members of the same species, because they compete for the same necessities or **ecological niche.**

Variations among individuals

Each individual varies in some respects from other members of its species. Animal breeders take advantage of this fact when they choose those individuals with desirable characteristics. Nurseries are able to produce disease-resistant plants and to control sizes and colors of blooms by careful study and cross-fertilization of particular individuals. In nature variations within species furnish the material for evolution's cutting edge, natural selection.

Natural selection

If, among the trillions of dandelion seeds produced each year, those of a particular parent are most capable of survival in close competition with other plants, they will be most likely to survive until maturity. Those having poorer adaptations will perish by overcrowding. In so severe a struggle, where only relatively few can survive, even slight variations tip the scale in favor of better suited individuals. Darwin termed this **natural selection,** or "survival of the fittest," a phrase which has had an inestimable influence on modern human self-awareness, including economic and political behavior.

Heredity and variation

In general, offspring resemble their parents. If the parents reach maturity because of an unusual genetic variation that gives them special fitness to the environment, those of their descendants that inherit the variation will, in turn, be more likely to survive. If, for example, a monkey has better hearing than other members of its species, it will be able to detect predators more readily, and it will be more likely to survive and reproduce. By the same token, any offspring that inherit unusual hearing capabilities will be better equipped for survival. Since the survivors will reproduce and the ones that perish will not, more sensitive hearing will become common throughout the population over the course of many generations.

While Darwin recognized that inheritable variations occur, he had no knowledge of the mechanisms involved. In fact, in his later writings, Darwin proposed that "pangenes" from all parts of the body merged to form eggs and sperm. Thereby, he suggested acquired characteristics could be passed on to offspring. Thus, Darwin himself was a bit of a Lamarckian.

GENE MUTATIONS AS A CAUSE OF CHANGE

Since Darwin's time, biologists have achieved somewhat of an understanding about how variations of inherited characteristics occur. In part, they are caused by changes, or *mutations,* in the structures of the DNA molecules of genes in reproductive cells. To date *no* research has indicated that "pangenes" or any other mode of input to the reproductive cells exists.

In considering the evolutionary effects of gene mutations it is worth first asking how often they occur.* Genes tend to be stable molecules. The DNA molecules composing them replicate time after time without chemical alterations. Still, mutations do occur, and at rates that geneticists have been able to predict. Genes for seed color in corn, for example, mutate in about 1 out of 2,000 reproductive cells. Hence, since a single stalk of corn produces hundreds of thousands of pollen grains, mutations for seed color occur many times on any given plant. However, only a small fraction of pollen grains fertilize kernels. (Most pollen is scattered far afield by the wind.) Moreover, more than one gene is involved in determining a kernel's color; thus the effect of a mutant gene may be masked. In domestic corn aberrant seed colors seldom develop.

The effects of some gene mutations are so slight that they cause no apparent manifestations. However, the effects of certain mutations are pervasive. For instance, the symptoms of the human genetic disease, phenylketonuria (PKU) include mental retardation, epileptic seizures, and eczema (itchy skin), but it is caused by a single faulty gene.

Variation, environment, and survival

As far as we now know, mutations occur randomly. There is no known mechanism by which mutations occur to the benefit of a species and, in fact, most of them are either insignificant or harmful. It appears to be by chance that genes mutate, giving a giraffe a longer neck, or a rabbit white fur and pale skin. A long neck is of advantage to a giraffe; but albinism is a distinct disadvantage to a rabbit in the wild, rendering it hypersensitive to sunlight and an obvious catch for predators (Figure 2-12). Interactions between organisms and their environments determine whether or not variations favor survival. Thus, environments determine the direction of evolution. If the environmental history of the earth had varied from what it was, we would not look like we do today. We might have evolved further or we might still be jelly fish.

A population of a given species that has long been established in a stable environment is probably well adjusted to the conditions there. In such a population natural selection has eliminated unfit variants, and as long as the environment remains unchanged, it is unlikely that further genetic variation will improve the species. But, if environmental conditions change, or if certain individuals migrate to a new locality, then genetic variations might improve chances of survival and lead to the establishment of a new variety of the species. Possibly a new species could develop.

Migration, variation and environmental change

Members of a migrant population may settle in an area uninhabited by members of the same species, or they may make their new home in an area already colonized by others. If several members of an animal population migrate to a new area, they take with them certain combinations of genes characteristic of the population of which they have been a part. Their arrival in a new area may introduce genetic characteristics that have been absent in an established population. If interbreeding occurs, the offspring will receive sets of genes from both the migrant and established populations. The result can be offspring with a blend of the physical characteristics of the two populations, or possibly, some characteristics absent in both

FIGURE 2-12
An albino rabbit (top) and a brown rabbit (bottom). Albino rabbits are easy marks for predators; hence few of them survive in the habitat of the darker rabbits.

* A discussion of the causes of mutations and how the structure of DNA is affected is given in Chapter 8.

preexisting populations. For example, interbreeding between European explorers and American Indians produced children with skin of an intermediate hue, who were uniquely handsome.

Migrations can lead to environments differing in climate and terrain from original homelands, and changes in environment present new conditions for survival. Thus, the descendents of a migrant segment of a population may evolve along a line totally different from that of the individuals who stay behind.

Ancient migrations of camels, for instance, resulted in several divergent lines of evolution. At one time, camels in various forms existed throughout Asia, Europe, North America, Central America, and South America. They are believed to have originated in North America and spread to Asia via a land bridge that spanned the Bering Strait millions of years ago (Figure 2-13). Migrations extended through North America and Central America to South America. Then, with the coming of the Ice Age, camels completely died out in most areas, resulting in widely separated populations. Today, the Asian camel (with two humps) and the African camel (with one hump) are well adapted for life in the desert. Relatives in South America evolved into quite a different animal—the llama—which lacks a hump but is surefooted and protected with a dense coat of hair, suitable for existence in its cold, rocky environment in the Andes Mountains (Figure 2-14).

Individual variations among the camel populations that originally reached the Andes were probably not very great. But the ones with slightly heavier coats were more likely to survive cold winters and to reproduce the following year. Also, the nimble individuals with shorter legs had better chances to escape predators. The strains that didn't possess these favorable variations eventually died out.

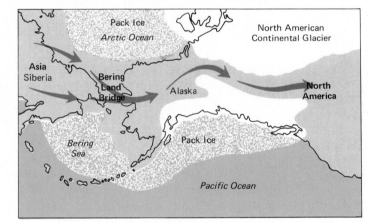

c. 26,000 years ago

FIGURE 2-13

The original human immigrants to North America are believed to be descendants of Siberian hunting parties which crossed a land bridge connecting Siberia with Alaska some 30,000 years ago.

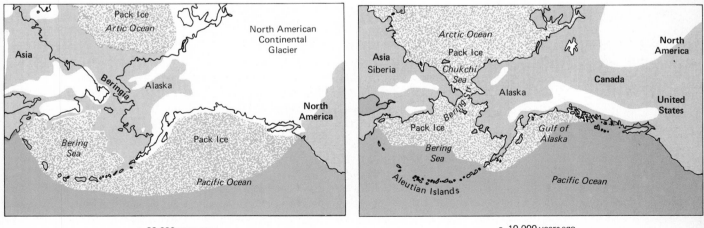

c. 20,000 years ago

c. 10,000 years ago

American Museum of Natural History

FIGURE   2-14
From top to bottom: camel ancestor; Bactrian or Asian camel; South American llama.

Plants don't have means of self-locomotion, but they are capable of far ranging distributions. Carried by wind and water their seeds may spread around the globe. Coconut palms, for instance, have populated widely separated islands because their thickly husked seeds can survive months of floating in salt water. Seeds of certain berries pass unharmed through the digestive tracts of birds, to be widely distributed in excrement.

Protective coloration and environmental change

A moth population in England affords a striking example of the evolutionary impact of environmental change. British scientists documented a change in color of the peppered moths of the Manchester region, which occurred over a span of just fifty years. The findings have become a classic example of the dynamics of evolution.

In the early 1800's the wings of the peppered moth had dark spots on a pale background. They were hardly visible when resting on the light gray bark of trees. Nature had camouflaged the peppered moth, and it would seem that no change in appearance would have been beneficial. But, in 1845, a black variant of the species was captured, and in succeeding years more and more black moths appeared.

A black moth resting on light bark would be easily spotted by predators. However, Manchester was changing and so were its trees. The city was rapidly becoming an industrial center. Smoke poured from factory chimneys, and soot turned light-colored bark nearly black. Consequently black moths, not pale ones, were camouflaged. In the period from 1845 to 1895, the proportion of black moths increased to 99 percent of the population. It is important to remember that the black variant was not produced by the industrial environment. The black moth was a mutant that already existed but now was selected for by the environment.

The change in the peppered moth population interested a certain Professor Kettlewell, who in the fine tradition of British natural history, continued the investigation with a field study. He selected two entirely different areas—a bird reserve in Birmingham (an industrial city similar to Manchester), and the countryside near Dorset (where there was no soot on the trees).

In the bird reserve, Kettlewell released 477 black moths and 137 pale ones. He watched through the day, as birds plucked moths from the trees, and that night he recovered a portion of the remaining moths by attracting them to a light. Forty percent of the black moths but only 19 percent of the pale ones were re-captured.

Near Dorset, Kettlewell released 473 black moths and 496 pale ones. The birds gorged themselves throughout the day. That night only 6 percent of the black moths and 12.5 percent of the pale ones were recaptured.

Like all experiments performed in the field, the observations of Kettlewell were subject to large experimental error. Undoubtedly not all of the moths still alive at the end of the day were attracted to the light, and it is possible that moths of one color were harder to recapture than those of the other color. Nevertheless, comparison of the data from Birmingham and Dorset indicates that natural selection was involved in the color shift of the moths in Manchester. The discovery of **industrial melanism** (as the color change has been termed) in the late nineteenth century gave strong support to the early proponents of evolutionary theory (Figure 2-15). Recent reports from Manchester give a heartening sidelight to this story. As the city has begun to clean up its industrial pollution, there has been a resurgence in the population of pale peppered moths.

FIGURE 2-15
The principle of industrial melanism. In each situation it appears obvious which moth is likely to survive.

Isolation and changes in "gene pools"

Squirrel populations on opposite rims of the Grand Canyon illustrate the evolutionary effect of isolation. On the north rim lives the Kaibab squirrel, with a white tail, and dark belly. The south rim is the home of the Abert squirrel, which appears to be essentially the same except for a gray tail and light belly. The two variants are thought to be direct descendants of the same species, but the fast flowing Colorado River in the rugged depths of the canyon makes interbreeding unlikely. In other words, each population possesses a unique **gene pool** (the collective store of genes of an interbreeding group), which is physically separated from the gene pool of the other population. A mountain range, a dry plain, a desert, or an ocean may act as a physical barrier, preventing plants or animals from interbreeding.

Physical barriers are not the only causes of reproductive isolation, however. Peculiarities in breeding habits may also isolate a gene pool. The habits of the sockeye salmon of the Fraser River are a good example. The Fraser has long been the spawning area of the sockeye, whose life begins far up the river in shallow, cold water. After hatching, fingerlings gradually work their way downstream, finally arriving at the mouth of the river, where they enter the Pacific Ocean. There they live three or four years until maturity. Then a reproductive instinct urges them back to the west coast of Canada and the mouth of the Fraser. Fighting currents and leaping waterfalls, they struggle back to their birthplace—perhaps the very pool in which they hatched. Here they breed and die.

In the Pacific Ocean, sockeye mingle with other types of salmon. However, they never interbreed with other species, only with other sockeyes and always in the Fraser River. Thus, the sockeye salmon has retained an isolated breeding population and consequently remains unique from salmon that breed in other streams.

Other barriers to interbreeding include such factors as differences in the mating seasons of two populations, differences in anatomy that prevent mating, and peculiarities of mating rituals. Within certain groups of otherwise reproductively compatible species of insects, each species has a unique "dance" or other form of ritual that must be performed before sexual union.

Sometimes interbreeding between two distinct populations does occur, but genetic differences result in sterile offspring. The mule (which is the offspring of a cross between a horse and an ass) is such an animal (Figure 2-16). In more extreme cases of genetic incompatibility an embryo forms but dies before birth. Major genetic differences prevent any fertilization whatsoever.

FIGURE 2-16

A mule. If this mule has the power of reasoning he could feel no pride in ancestry nor hope for posterity. He is a sterile hybrid resulting from a cross of different species, the jackass and the mare. He has a strength greater than either parent — a condition known as hybrid vigor.

SPECIES, POPULATIONS, RACES, AND SPECIATION

Up to this point we have used the terms "species" and "population" without strict definitions, but it is worth stating exactly what they mean to biologists. Organisms which are capable of mating and producing fertile offspring are said to be of the same **species,** and an interbreeding group of organisms of the same species constitutes a **population.** Thus, there may be several populations of organisms of the same species which are reproductively isolated.

Members of reproductively isolated populations of the same species may have slightly different physical characteristics. That is, they may be of different **races** or breeds (subspecies). Evolution has produced no better example of this than the various races of human beings (Figure 2-17). With today's opportunities for world-wide travel, social forces are the primary causes of human reproductive isolation. But, it is conceivable that some day there will be only one human race.

The general drift of evolution has not been toward merging of races. Rather it has been toward the development of new species, or **speciation.** New species are forming today, just as they have in past ages. The mechanisms that we have been discussing—variation, migration, environmental change, selection, and isolation—result in speciation.

FIGURE 2-17
Various races of humans.

Speciation in maples

Maple trees are found in many environments of North America, Europe, and Asia. Sixty to seventy species of them thrive in the yards, parks, and forests of the Northern Hemisphere. Characteristics common to all species of maples indicate that they are descendants of the same ancestral stock that lived many years ago.

At some time in the past a variation may have developed in a population of primitive maples that adapted them for life in wetter surroundings than maples had previously occupied. (It might have been a rot-resistant root system.) Other variations resulting in further adaptations to damp environments could have led to the evolution of a new species. The silver maple (or soft maple) appears to be the descendant of such an evolutionary sequence. Today the silver maple is found throughout the eastern part of the United States. It towers to a height of sixty to eighty feet in bottomlands, swamps, stream borders, and floodplains. While it lives best in damp lowlands, the silver maple can also tolerate much drier situations, and for this reason, it is widely planted as a fast growing shade tree. In wet surroundings the silver maple will grow in partial shade, but it requires more exposure to sunlight in drier situations.

Foresters are familiar with a variation of the silver maple that has deeply notched leaves. This cut-leaf maple grows more slowly and seldom reaches the size of a silver maple. It may be a species in the making.

Maples in various forms occupy a wide range of environments in North America. The sugar maple and black maple thrive in the forests of the eastern United States, while the moosewood maple mingles with the pines and hemlock of the northern forest. The striped maple does best in the elevations of eastern mountain ranges. The big leaf maple is limited to a narrow belt in the Pacific coastal area from Alaska to California, where it lives in moist climates in foothills and low mountains.

While each species of maple has unique characteristics, they all have certain things in common. For instance, all maples produce winged seeds and have characteristic patterns of veins in their leaves. Such similarities are manifestations of genes that have not changed as maples evolved.

Adaptive radiation

Adaptive radiation is the branching out of a population, through variation and adaptation, to new and different environments. As variations occur, conditions in the environment determine whether they are favorable or unfavorable.

Many varied members of the deer family live in North America (Figure 2-18). Elk inhabit high mountain forests in the summer and descend in herds to sheltered valleys in the winter. Moose live near lowland lakes and swamps in the summer, feeding on aquatic plants. In the winter, they move to higher ground, to find food and shelter in forests. Caribou live furthest north of all, in spruce forest and barren tundra regions near the arctic circle. It is quite likely that elk, moose, and caribou all evolved from a common ancestral species, whose adaptations to different environments brought on variations of appearance and behavior.

DIVERGENCE AND CONVERGENCE: HOMOLOGY AND ANALOGY

So far we have primarily considered those phenomena that lead to **divergent evolution** (whereby one ancestral group splits into two or more groups, which

FIGURE 2-18
The elk, one of several closely-related types of deer.

become less alike as time passes). But, evolution does not always work that way. Various races of human beings are now tending to merge through interbreeding, and there is another quite different mechanism by which evolution produces similarity rather than dissimilarity.

Organisms having different ancestries but sharing similar environments become similar in form by **convergent evolution.** Tuna fish and dolphins, for example, are superficially alike; both are streamlined and have fins (Figure 2-19). But comparative internal examinations reveal that the obvious similarities are due to common adaptations to life in the sea rather than common direct ancestries. Tunas, like other fish, obtain oxygen through gills and have two-chambered hearts. Dolphins, on the other hand, have lungs, four-chambered hearts, mammary glands, and hair—all characteristics of mammals. Thus, dolphins are more closely related to humans than to tunas. The common environment but different evolutionary ancestries of tunas and dolphins has led to a tragic state of affairs. Schools of dolphins and tuna live in such close contact that the former frequently become entangled in nets set to catch the latter. Because dolphins must periodically surface to breathe, they drown before they can be released.

Earlier in this chapter we discussed **homologous structures** (body parts of related evolutionary heritage, although not necessarily the same function), citing the example of human arms, bird wings, and whale flippers. Homologous structures have become dissimilar as the result of divergent evolution. On the other hand, convergent evolution produces **analogous structures** (body parts similar in function and superficial structure but of different evolutionary heritage). For example, birds and bees both have wings, but they did not develop from a common winged ancestor; hence their wings are analogous rather than homologous.

A comparison of the evolutionary development of the eyes of squid and humans illustrates one of nature's most amazing examples of analogy. As mammals, humans have eyes traceable to fishlike marine ancestors that lived at least 500 million years ago. The eyes of squid also evolved in the marine environment but along a different line. Squid exist at the evolutionary peak of a spineless group of animals called **molluscs** (snails and clams are also molluscs). Thus squid eyes and human eyes developed independently, via different ancestors.

Not all problems of analogy are as straightforward as the preceding one, as illustrated by a reconsideration of the tuna and dolphin example. We have pointed out that the superficial similarities of tunas and dolphins are due primarily to convergence (adaptations to the same environment). Nevertheless, they have indirect evolutionary ties. They are thought to share a distant common ancestor in the precursors of modern jawed fish, **Placodermi.** The last species of Placodermi became extinct about 250 million years ago, but fossils indicate that they had fins and were otherwise similar to the bony fishes (fish with bones as opposed to fish with skeletons of cartilage like sharks) that evolved from them.

Tunas are living examples of bony fishes, but millions of years ago an evolutionary offshoot of this group gave rise to land dwelling vertebrates. Over the course of many generations fins evolved into appendages for walking and **amphibians** (of which frogs and salamanders are now living examples) moved from the marine environment to land. Amphibians gave rise to **reptiles** (which include turtles and crocodiles), and an ancestral reptile, in turn, is thought to have given rise to mammals. As indicated by the lifestyles of many modern amphibians, reptiles and mammals, the adjustment from marine to terrestrial environments was not complete. In fact, in some cases, evolutionary lines led totally back to the sea, and appendages that had been rudimentary legs reverted to fins. Such appears to be the case for dolphins. Thus, although the similarities of the fins of dolphins and tunas have arisen through convergence, there is a roundabout evolutionary link between them. Hence, there are underlying elements of both homology and analogy.

FIGURE 2-19
The tuna (top) and the dolphin (bottom) look somewhat similar but are quite different internally.

HUMAN EVOLUTION

The organs and systems of the human body closely resemble those of other mammals. The similarities between humans and other **primates** (apes and monkeys) are especially striking. Likenesses abound throughout the anatomy: in the skeleton, muscles, teeth, and even in some behavior and facial expressions (Figure 2-20). Similarities are evident in the structure of the heart and blood vessels, lungs, digestive organs, excretory organs, glands, and nearly all other internal organs. The likenesses are even more marked at the chemical level. The digestive enzymes, for instance, are identical in apes and humans. The Rh factor, so important in matching human blood types, was originally discovered in the blood of the rhesus monkey. Such is the evidence that apes and humans evolved from a common ancestor.

Anthropology—the science of humans

Anthropology is the study of the physical and cultural history of humans. Physical anthropologists trace our ancestry as preserved in fossils and compare our physical characteristics with those of other primates. Among the things they consider are brain size and posture.

Unlike other primates, with the near exception of the gibbon ape, humans walk fully erect. This characteristic is reflected in our skeletons. As illustrated in Figure 2-21, the blade of the human pelvis is broad with a flange which projects to the rear to support internal organs. The flange also strengthens the pelvis and serves as an attachment for the large muscles that are used in walking.

Fragments of fossil skulls give clues to the mental capacities of our ancestors. Fossilized skulls are molds of the brains which they once contained; hence skulls indicate changes in total size and relative development of the brain. Comparisons of skulls of human precursors and contemporary humans indicate that the **cerebrum** (the portion of the forebrain which is thought to be the center of memory, learning, and conscious sensation) grew markedly in the course of human evolu-

FIGURE 2-20
Behavioral traits of other primates are amazingly similar to those of humans.

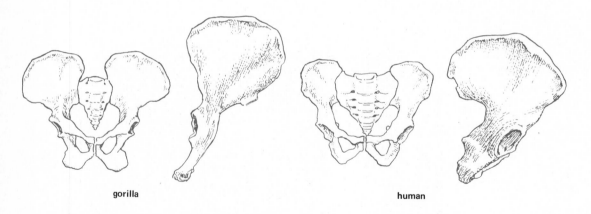

gorilla   human

tion. Relatively little development of the cerebrums of apes occurred in the same period. The brains of apes are smaller than those of humans. They lack a frontal bulge and are encased in heavier skulls (Figure 2-22).

Studies of the environments of primitive ancestors also help to piece together our development. For example, wherever anthropologists find evidence of early humans, they search for fossil remains of food. Charred bones and remains of hearths indicate whether fire was used to cook meat. Anthropologists are particularly interested in evidence of planned and consistent use of tools, since this

FIGURE 2-21
The broad blade of the human pelvis supports the weight of the internal organs. Its posterior projection serves as attachment for the large muscles used in walking. Compare the human pelvis to that of the gorilla.

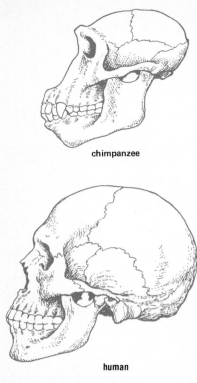

chimpanzee

human

FIGURE 2-22
Comparison of the skulls of a chimpanzee and a human.

FIGURE 2-23
Human hands are well adapted for a firm grip, but are more suitable for manipulation of tools than are those of the ape.

is a sign of the higher intelligence found in humans. Nevertheless, tool use has also been observed in apes and insects.

The use of tools is enhanced by a physical characteristic unique to humans—a thumb that fully opposes the other four fingers (Figure 2-23). Thus our hands are better suited for grasping and manipulating than those of any other primate. Furthermore, our fully erect posture frees our hands for such tool use.

Perhaps the most significant difference between us and other primates is our capacity to use symbols. The ability to communicate with spoken and written languages, and to pass learning from generation to generation has allowed us to undergo a rapid **cultural evolution** not possible in other species. Relatively recent technological developments, including films, television, and communication satellites, have expanded our ability to communicate at a level unattainable only a few decades ago. Because of the slow rate of evolutionary change, it is unlikely that any significant genetic differences exist between humans of today and those of 10,000 years ago. There are, however, enormous cultural differences.

If we were able to trace the history of the primates back perhaps 10 million years, we might find an animal that was the common ancestor of both modern humans and modern apes. Unfortunately, although anthropologists have unearthed a wealth of fossil information about our evolutionary ancestry, the evidence still does not provide a complete picture. Nevertheless, it has been possible to make important inferences about many physical and behavioral characteristics. A fossilized jaw bone, for instance, gives a theoretical indication of whether its possessor had the capacity for speech. The presence of the capacity, of course, does not necessarily mean that a language was used.

African ancestors

In 1924, a worker in a South African quarry noticed a small fossilized skull that had been dislodged by blasting. The fossil was given to a South African medical school, where Professor Raymond Dart recognized its resemblance to human skulls. Dr. Dart named the primate **Australopithecus africanus** (which translates as "southern ape from Africa"). Since then, other bones of this primate have been found and estimated to be between 2.5 and 3 million years old.

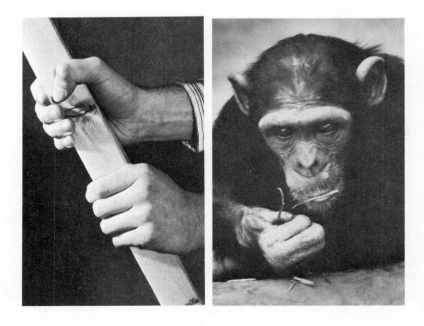

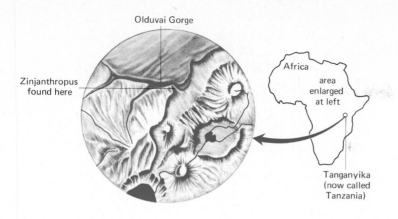

Olduvai Gorge

Zinjanthropus
found here

Africa

area
enlarged
at left

Tanganyika
(now called
Tanzania)

FIGURE   2-24
The site in Africa where *Zinjanthropus* was discovered.

Over the past forty years the search for human ancestors has centered at the Olduvai Gorge in Tanganyika (Figure 2-24). Here the anthropologists Mary and Louis Leakey first found "pebble tools" (chipped stones) thought to have been used for cutting. Then, in 1959, the Leakeys unearthed an ancient camp site, where skull fragments, bones from a foot, fingers, and a lower jaw with well-preserved teeth provided clues about the structure of the primate they named **Zinjanthropus** (Figure 2-25), now called Australopithecus robustus. Radioisotope dating has placed its age at about 1,750,000 years.

The bones of Zinjanthropus and Australopithecus are similar and look much like those of modern apes. The pointed, fanglike teeth of modern apes are not found in these forms, however. Furthermore, the shape of the pelvis and opening for the spinal cord in the skull indicate an upright posture. Anthropologists do not agree as to whether Zinjanthropus and Australopithecus should be catagorized in the ape family or in the family that includes humans, but they do not seem to be on the direct road to humans.

In 1964 Dr. Leakey published an account of the discovery of the remains of another humanlike fossil in the Olduvai Gorge. He named it **Homo habilis,** and dated it as a contemporary of Zinjanthropus. Homo habilis, however, is more humanlike. He seems to have walked and run erect, to have had a well-opposed thumb, and to have eaten both meat and plant foods. Dr. Louis Leakey believed that Homo habilis was the ancestor of modern humans, while Zinjanthropus and Australopithecus were evolutionary dead ends.

In 1972, Richard Leakey (son of the Leakeys) found a skull that he believes pushes the evolutionary heritage of humans back even further than Homo habilis. "Skull 1470" has been dated at around 2.8 million years old, and Leakey believes it is the earliest known fragment of the evolutionary line leading directly to humans (Figure 2-26).

FIGURE   2-25
Dr. L. S. B. Leakey, the eminent British anthropologist who unearthed the remains of one of the earliest known humanlike primates in East Africa.

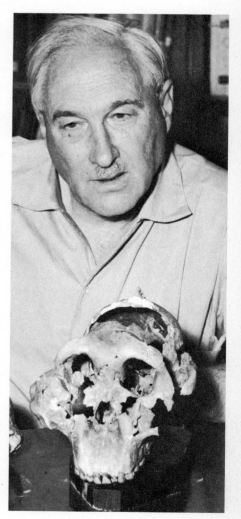

"Java man"

In 1891, a part of a skull, a piece of a jaw, and an upper leg bone were discovered in an excavation on the island of Java. The site was a deposit of sand and gravel left by an early glacier. Similar remains were unearthed in the same general region in 1937, and later in caves near Peking. "Java man," as he was first called, is believed to have walked erect and to be about five and one-half feet tall. He is now called **Pithecanthropus erectus** (or **Homo erectus**). He lived about 500,000 years ago and had a slanting forehead and heavy brow ridges. The skull of Java man indicates that his brain, although only about half the size of a modern human, was more than one-third larger than that of the present-day gorilla. Anthropologists believe that Java man made crude stone weapons and used fire.

FIGURE 2-26

The Leakey view of human evolution, showing two parallel lines of development. (from a drawing by Jay Matternes in R. E. Leakey, "Skull 1470 — New Clue to Earliest Man?" *National Geographic*, June 1973.)

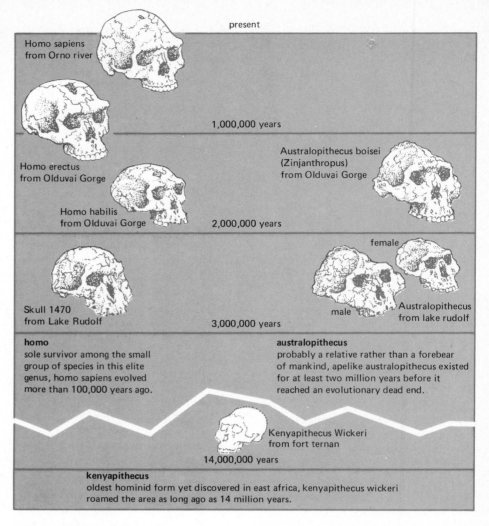

present

Homo sapiens from Orno river

1,000,000 years

Homo erectus from Olduvai Gorge

Australopithecus boisei (Zinjanthropus) from Olduvai Gorge

Homo habilis from Olduvai Gorge

2,000,000 years

female

Skull 1470 from Lake Rudolf

3,000,000 years

male

Australopithecus from lake rudolf

homo
sole survivor among the small group of species in this elite genus, homo sapiens evolved more than 100,000 years ago.

australopithecus
probably a relative rather than a forebear of mankind, apelike australopithecus existed for at least two million years before it reached an evolutionary dead end.

Kenyapithecus Wickeri from fort ternan

14,000,000 years

kenyapithecus
oldest hominid form yet discovered in east africa, kenyapithecus wickeri roamed the area as long ago as 14 million years.

Neanderthals

This race of hominids lived in Europe, Asia Minor, Siberia, and Northern Africa. Neanderthals are believed to have disappeared about 30,000 years ago, near the close of the glacial age. Information about them has been acquired from careful study of almost one hundred skeletons, many of them nearly complete. Neanderthals were about five feet tall, and their bone structure indicates they were powerfully built. Their facial features were coarse; like Pithecanthropus, their foreheads sloped back from heavy brow ridges. Their mouths were large and chins were small.

Neanderthals lived in caves from which they journeyed on hunting expeditions in search of the mammoth, sabertoothed cat, and woolly rhinoceros. Their brains were as large or larger than those of modern humans. Neanderthals used stone tools and weapons, made use of fire, buried their dead with flowers, and lived in groups of families. For reconstructions of these skulls, see Figure 2-27.

Cro-Magnons

Anthropologists classify Cro-Magnons and Neanderthals as subspecies of **Homo sapiens**–modern humans. Cro-Magnons lived in Europe, especially in France and

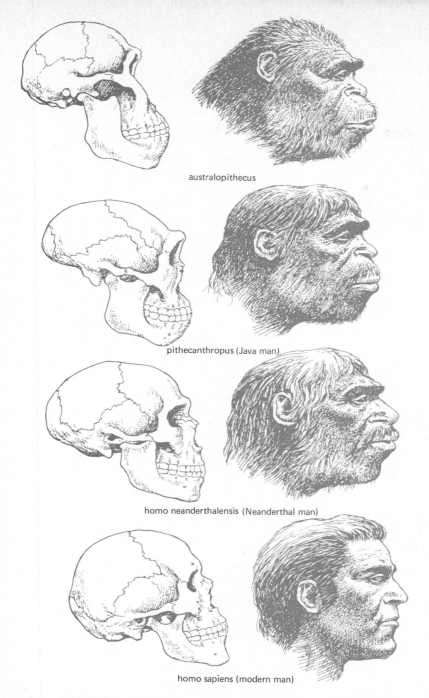

australopithecus

pithecanthropus (Java man)

homo neanderthalensis (Neanderthal man)

homo sapiens (modern man)

FIGURE 2-27
Prehistoric primates which resemble humans and modern man.

Spain, about 300,000 years ago. They had high foreheads and well-developed chins. They lacked the heavy brow ridges of more primitive hominids. Several caves along the southern coast of France have yielded Cro-Magnon weapons of stone and bone as well as skeletons. The walls of these caves bear beautiful drawings of animals of the region.

Many anthropologists believe that the Cro-Magnons and Neanderthals were competitors and that the latter were eventually driven to extinction. There is also evidence, however, of interbreeding.

The present and the future

A comparison of modern and ancient humans shows a steady trend toward "brain over brawn." Few people today could endure the hardships of life in primitive times. Yet we can expect to live three or four times as long as our ancient ancestors. We have learned to mold our existence to make life safer and longer lasting. We have drastically changed environments, checked epidemic diseases and have begun a technology of gene-manipulation.

How should we use our powers? We can see that life need not (indeed, cannot) be static, and perhaps we should take charge of our own evolution. But, the directions in which we should be going are not clear, and even if our goals were clear, it is quite possible that our attempts to achieve them would lead to something totally unexpected. In general, we don't seem to be very sensitive to the implications of our acts. Yet, if one has faith in the potential benefit of change, then is it possible that change can bring an ever increasing sense of living harmony?

CHAPTER 2: SUMMARY

1   Change is a part of all life.

2   The "Big Bang" is the most accepted theory of the origin of the universe.

3   Atoms are elements and atoms form molecules.

4   Life is thought to have originated from the random coming together of atoms to form molecules useful to "life." Why these particular molecules are used in life is not known.

5   There are many arguments and pieces of evidence to suggest evolution is real. Among them are the fossil record, homologous organs and biochemical similarities.

6   Natural selection is a process whereby already existing mutations are selected to survive by the environment.

7   Both the environment and genes cause changes in populations of organisms.

8   Population genetics involves the study of changes in gene pools of populations of organisms and the resulting changes in species.

9   Present day humans and apes are thought to have evolved from a common ancestor but the actual pathways are not yet clear.

Suggested Readings

Boorer, Michael   *Mammals of the World,* Grosset and Dunlap, Inc., New York, 1967 — A survey of the evolution of mammals.

Darwin, Charles R.   *The Origin of Species,* Doubleday & Co., Inc., Garden City, New York, 1960 — Darwin's classic treatise on the survival of the fittest.

Howell, F. Clark and the Editors of *Life*   *Early Man,* Time-Life Books, New York, 1965 — A picture-essay book covering the day-to-day problems and general conditions of life in ancient times.

Luria, S. E.   *Life: The Unfinished Experiment,* Charles Scribner's Sons, New York, 1973 — Predictions for the future of evolution.

Payne, Melvin   The Leakeys of Africa: in Search of Prehistoric Man, *National Geographic,* February, 1965 — How the remains of early humans were found in the Olduvai Gorge in Tanganyika, Africa.

THE SCOPES "MONKEY TRIAL"

"And God created man in his own image,
in the image of God
He created him; male and female
He created them."   *Genesis 1:27*

For fundamentalists, who believe in a literal interpretation of the Bible, this excerpt from Genesis is the last word on the origin of humankind. In spite of overwhelming evidence that we share a common ancestry with the other primates, the fundamentalist view has remained amazingly forceful. In 1973 the California State Board of Education acquiesced to pressure by the fundamentalist lobby and ruled that science texts used in public schools must teach evolution as a speculative theory rather than "scientific dogma." At present, fundamentalists in California are seeking an even stronger ruling, demanding that divine creation be given at least equal time with evolution. They are opposed by an outraged contingent of the state's scientists.

Although the current fundamentalist-evolutionist debate is a matter of interest, for sheer drama it cannot match a confrontation which occurred half a century ago in a sleepy southern town. On March 21, 1925, the Tennessee Legislature passed a strong antievolutionist bill:

Be it Enacted, by the General Assembly of the State of Tennessee, that it shall be unlawful for any teacher in any of the universities, normals, and all other public schools in the State, which are supported in whole or in part by the public school funds of the State, to teach the theory that denies the story of the divine creation of man as taught in the Bible, and to teach instead that man has descended from a lower order of animals.

In order to test the constitutionality of the bill, John Scopes admitted to teaching evolution in his high school biology class. Scopes purposely had himself indicted. The case may have passed unnoticed outside Tennessee had not William Jennings Bryan, a prominent lawyer and national political figure, offered his services to the prosecution. Bryan's move prompted Clarence Darrow, a writer and criminal lawyer, to come to the aid of the defense. The trial proceeded with great fanfare, amidst soft drink stands, in the summer heat of Dayton, Tennessee.

During the trial, a personal battle between Bryan and Darrow soon overshadowed the determination of the guilt or innocence of John Scopes, and as the trial drew to a close, Darrow surprised everyone by calling Bryan as a witness. Following are the closing exchanges of the testimony:

MR. DARROW: Mr. Bryan, do you believe that the first woman was Eve?

MR. BRYAN: Yes.

MR. DARROW: Do you believe she was literally made out of Adam's rib?

MR. BRYAN: I do.

MR. DARROW: Did you ever discover where Cain got his wife?[1]

MR. BRYAN: No, sir; I leave the agnostics to hunt for her.

MR. DARROW: You have never found out?

MR. BRYAN: I have never tried to find.

MR. DARROW: You have never tried to find?

MR. BRYAN: No.

MR. DARROW: The Bible says he got one, doesn't it? Were there other people on the earth at that time?

MR. BRYAN: I cannot say.

MR. DARROW: You cannot say? Did that never enter into your consideration?

Reprinted from the transcript of the Scopes trial.

[1] Genesis 4:17.

MR. BRYAN: Never bothered me.

MR. DARROW: There were no others recorded, but Cain got a wife. That is what the Bible says. Where she came from, you don't know. All right. Does the statement, "The morning and the evening were the first day" and "The morning and the evening were the second day" mean anything to you?[2]

MR. BRYAN: I do not think it means necessarily a twenty-four-hour day.

MR. DARROW: You do not?

MR. BRYAN: No.

MR. DARROW: What do you consider it to be?

MR. BRYAN: I have not attempted to explain it. If you will take the second chapter—let me have the book. [Examining Bible] The fourth verse of the second chapter (Genesis) says: "These are the generations of the heavens and of the earth, when they were created, in the day that the Lord God made the earth and the heavens." The word "day" there in the very next chapter is used to describe a period. I do not see that there is necessity for construing the words, "the evening and the morning" as meaning necessarily a twenty-four hour day: "in the day when the Lord made the Heaven and the earth."

MR. DARROW: Then when the Bible said, for instance, "And God called the firmament Heaven. And the evening and the morning were the second day,"—that does not necessarily mean twenty-four hours?

MR. BRYAN: I do not think it necessarily does.

MR. DARROW: Do you think it does or does not?

MR. BRYAN: I know a great many think so.

MR. DARROW: What do you think:

MR. BRYAN: I do not think it does.

MR. DARROW: You think these were not literal days?

MR. BRYAN: I do not think they were twenty-four-hour days.

MR. DARROW: What do you think about it?

MR. BRYAN: That is my opinion—I do not know that my opinion is better on that subject than those who think it does.

MR. DARROW: Do you not think that?

MR. BRYAN: No. But I think it would be just as easy for the kind of God we believe in to make the earth in six days as in six years or in six million years or in six hundred million years. I do not think it important whether we believe one or the other.

MR. DARROW: Do you think those were literal days?

MR. BRYAN: My impression is they were periods, but I would not attempt to argue as against anybody who wanted to believe in literal days.

MR. DARROW: Have you any idea of the length of the periods?

MR. BRYAN: No, I don't.

MR. DARROW: Do you think the sun was made on the fourth day?

MR. BRYAN: Yes.

MR. DARROW: And they had evening and morning without the sun?

MR. BRYAN: I am simply saying it is a period.

MR. DARROW: They had evening and morning for four periods without the sun, do you think?

MR. BRYAN: I believe in creation, as there told, and if I am not able to explain it, I will accept it.

MR. DARROW: Then you can explain it to suit yourself. And they had the evening and the morning before that time for three days or three periods. All right, that settles it. Now if you call those periods, they may have been a very long time?

MR. BRYAN: They might have been.

MR. DARROW: The creation might have been going on for a very long time?

MR. BRYAN: It might have continued for millions of years.

MR. DARROW: Yes, all right. Do you believe the story of the temptation of Eve by the serpent?

MR. BRYAN: I do.

MR. DARROW: Do you believe that after Eve ate the apple, or gave it to Adam—whichever way it was—God cursed Eve, and at that time decreed that all womankind

---

[2] Genesis 1:5, 8.

thenceforth and forever should suffer the pains of childbirth in the reproduction of the earth?[3]

MR. BRYAN: I believe what it says, and I believe the fact as fully.

MR. DARROW: That is what it says, doesn't it?

MR. BRYAN: Yes.

MR. DARROW: And for that reason, every woman born of woman, who has to carry on the race,—the reason they have childbirth pains is because Eve tempted Adam in the Garden of Eden?

MR. BRYAN: I will believe just what the Bible says. I ask to put that in the language of the Bible, for I prefer that to your language. Read the Bible, and I will answer.

MR. DARROW: All right, I will do that: "And I will put enmity between thee and the woman."[4] That is referring to the serpent?

MR. BRYAN: The serpent.

MR. DARROW (reading): "And between thy seed and her seed; it shall bruise thy head, and thou shalt bruise his heel. Unto the woman He said, I will greatly multiply thy sorrow and thy conception; in sorrow thou shalt bring forth children; and thy desire shall be to thy husband, and he shall rule over thee." That is right, is it?

MR. BRYAN: I accept it as it is.

MR. DARROW: Do you believe that was because Eve tempted Adam to eat the fruit?

MR. BRYAN: I believe it was just what the Bible said.

MR. DARROW: And you believe that is the reason that God made the serpent to go on his belly after he tempted Eve?

MR. BRYAN: I believe the Bible as it is, and I do not permit you to put your language in the place of the language of the Almighty.

You read that Bible and ask me questions, and I will answer them. I will not answer your questions in your language.

MR. DARROW: I will read it to you from the Bible: "And the Lord God said unto the serpent, Because thou hast done this, thou art cursed above all cattle, and above every beast of the field; upon thy belly shalt thou go, and dust shalt thou eat all the days of thy life." Do you think that is why the serpent is compelled to crawl upon its belly?[5]

MR. BRYAN: I believe that.

MR. DARROW: Have you any idea how the snake went before that time?

MR. BRYAN: No, sir.

MR. DARROW: Do you know whether he walked on his tail or not?

MR. BRYAN: No, sir. I have no way to know. [Laughter]

MR. DARROW: Now, you refer to the bow that was put in the heaven after the flood, the rainbow. Do you believe in that?

MR. BRYAN: Read it.

MR. DARROW: All right, Mr. Bryan, I will read it for you.

MR. BRYAN: Your Honor, I think I can shorten this testimony. The only purpose Mr. Darrow has is to slur at the Bible, but I will answer his questions. I will answer it all at once, and I have no objection in the world. I want the world to know that this man, who does not believe in a God, is trying to use a court in Tennessee——

MR. DARROW: I object to that.

MR. BRYAN: To slur at it, and, while it will require time, I am willing to take it.

MR. DARROW: I object to your statement. I am examining you on your fool ideas that no intelligent Christian on earth believes.

[3] Genesis 3.
[4] Genesis 3:15, 16.

[5] Genesis 3:14.

CHAPTER

3

Patterns
of Life

Objectives

1 Learn how organisms are classified by biologists.

2 Be able to distinguish one group of organisms from another.

3 Understand how an organism "works." That is, how they are able to carry on life processes with their particular biological structures.

TAXONOMY

Classification of organisms, **taxonomy,** has long been of concern to biologists. In the eighteenth century the Swedish botanist, Carolus Linnaeus, devised a classification system that in many respects is still in use today. Linnaeus, who knew nothing of evolution, based his taxonomy on structural similarities, whereas the modern system is based on evolutionary relationships. Both approaches lead to approximately the same groupings.

One of Linnaeus' innovations, which has been retained by modern taxonomists, is his system of naming. Linnaeus discarded the common names of organisms and gave them names made up of two Latin words, which referred either to some characteristic of the organism or to the person who named it. Latin was also chosen as the language of biology because it was in common usage among many scientists at the time. His system is also called the binomial system.

Using Latin enables biologists to avoid the confusion of common names. The mountain lion, for example, is also known as the puma, cougar, panther, silver lion, American lion, mountain demon, mountain screamer, king cat, sneak cat, varmint, brown tiger, red tiger, and deer killer. Under the Linnaean system, however, all scientists can easily identify it as *Felis concolor.* The common house cat is *Felis domesticus. Felis onca* is the jaguar and *Felis leo* is the African lion.

Another problem with common names is illustrated in Figure 3-1. But, in spite of the problems with them, they are still the names by which the organisms are best known; even biologists sometimes use them. Thus in our descriptions of organisms, we have used some of the most widely accepted common names. In some cases we will give both the common and Linnaean names.

In Linnaeus' system of naming, the first name refers to the **genus** (group of closely related species) and always begins with a capital letter. The species name follows and usually begins with a small letter. Taxonomic names are either printed in boldface or italics. Remember that organisms of the same species are similar in structural characteristics, and they can mate and produce fertile offspring. Thus, all domestic dogs are of one species, although they may differ in size, color, and shape. All humans belong to the same species.

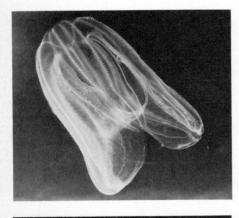

FIGURE 3-1
Common names can be misleading. Although these two animals are known as jellyfish and starfish, neither is a true fish (which have backbones, scales, fins and gills). They are unlike structurally, and they are not even closely related to one another.

GROUPINGS IN THE LINNAEAN SYSTEM

Biological classification begins by dividing all organisms into **kingdoms.** Until recent years biologists considered that there were two kingdoms, the plants and animals. Today there are various opinions on how many kingdoms there should be, but most biologists agree there should be three to five of them. Each kingdom is divided into smaller groups known as **phyla** (singular, **phylum**). Each phylum is divided into **classes,** and a class contains many **orders.** A division of an order is a **family;** a family contains related **genera,** and a genus is composed of more than one **species.** Sometimes individuals of a single species vary slightly, but they can still produce fertile offspring. In this case the species is said to be composed of **sub-species,** or races.

As an example of how organisms are classified into these groupings, consider humankind. The kingdom **Animalia** includes all animals. Most of the animals in the phylum **Chordata** (organisms in this phylum are also referred to as **chordates**), including humans, have backbones and are therefore included in the subphylum **Vertebrata** (vertebrates). The class **Mammalia** (mammals) includes all animals having mammary glands. The order **Primates** includes only a certain group of mammals that stand nearly erect. Monkeys, chimpanzees, and gorillas are among the primates. The family **Hominidae** separates early humanlike forms from the other primates. The genus **Homo** includes all humans, and the species **sapiens** (which means "wise") is the only surviving species of humans on earth.

In the past few decades a great amount of research has focused on simpler forms of life. Some of these organisms are more closely related to one another than to organisms considered to be definitely plant or animal. For example, there are bacteria and algae that engage in photosynthesis like plants, but swim or move like certain animals (Figure 3-2). Traditional classification has provided no place for these "in-between" organisms. Thus, biologists have recognized the need for classifying them in a different way. Some systems, such as the five-kingdom scheme, place bacteria and the blue-green algae (the most primitive form of algae) in a kingdom called **Monera**. These single-celled organisms have one striking characteristic in common—they lack a well-defined cell nucleus. Rather than being contained in these nuclei, their hereditary material is dispersed throughout their cells.

Protista is the second kingdom in the five kingdom scheme. One-celled organisms with membrane-bound nuclei and certain algae are placed in this kingdom. A third kingdom is the **Fungi,** which includes yeasts and molds. The remaining kingdoms are the **Plantae** (plants) and **Animalia** (animals).

The five-kingdom scheme is not the only one proposed by modern biologists, nor is it by any means perfect. It is impossible to say that one system is right and another wrong. We are simply choosing one system as a convenience in organizing the following study of life forms. Some of the similarities and differences among the organisms in the five-kingdom scheme are presented in Table 3-1. Biologists

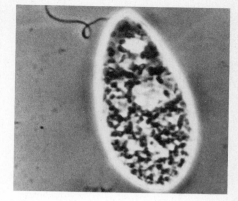

FIGURE 3-2

A photograph of *Euglena*, a green single-celled alga about 50 micrometers long.

TABLE 3-1   CHARACTERISTICS OF THE FIVE KINGDOMS

CHARACTERISTICS	MONERA	PROTISTA	KINGDOMS FUNGI	PLANTAE	ANIMALIA
1. Nuclear membrane?	No	Yes	Yes	Yes	Yes
2. Mitochondria bodies? (present for energy production in the cells)	No	Yes	Yes	Yes	Yes
3. Ability to photosynthesize?	Some do	Some do	No	Yes	No
4. Motility? (ability to move)	Some have it, some do not	Some have it, some do not	Primarily non-motile	Primarily non-motile	Yes
5. Form	One-celled	One-celled	Molds, Multicellular and yeasts single celled	Multicellular except for some algae	Multicellular

believe that the Protista evolved from the Monera and the plants, animals, and fungi evolved in separate lines from the Protista. With any good evolutionary scheme, it is possible to make general statements about an organism just by knowing where it falls in the scheme, even if nothing else about it is known. For example, if we know that a biologist is studying a mold, we know immediately that what is being studied is a multicelled, nonphotosynthetic organism, probably not motile but having a membrane around the cell nucleus.

In the remainder of this chapter we will look at some of the features of selected organisms of the five kingdoms.

KINGDOM MONERA

This kingdom includes the simplest of living organisms, the bacteria and blue-green algae. Many biologists believe that bacteria were the first forms of life on earth. Long before there were green plants capable of photosynthesis, certain primitive bacteria may have utilized energy from iron, sulfur, and nitrogen compounds instead of from the sun. Later, when green plants began building up stores of organic compounds, other kinds of bacteria began using them as a food supply.

Bacteria have survived through the ages and have increased in numbers until today they are the most abundant form of life. Too small to be seen by the naked eye, they live almost everywhere. They thrive in air, water, soil, and the bodies of plants and animals. In fact, any environment that can support life in any form will have its population of bacteria.

Bacterial structure

Compared with most other organisms, bacteria are extremely small. The average length is only a few micrometers — several thousand of them could be placed on the period at the end of this sentence. Bacterial cells are among the simplest cells of the living world.

Some bacterial cells are surrounded by a **slime layer** (Figure 3-3), a gelatinous coat that protects them from unfavorable environmental conditions. Slime

FIGURE 3-3
The general structure of a bacterial cell.

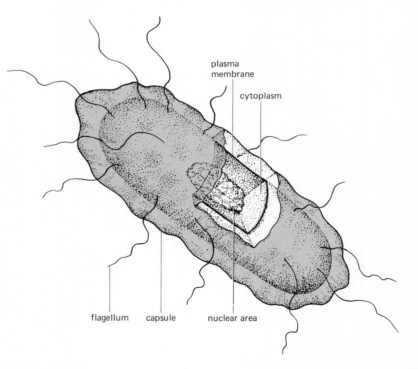

plasma membrane

cytoplasm

flagellum   capsule   nuclear area

layers protect infectious species of bacteria from the defenses of the host. For example, strains of the pneumonia-causing bacterium which have a slime layer are highly virulent, while those that lack one are low in virulence or even noninfectious.

Beneath the slime layer is a **cell wall** that gives a bacterium its characteristic shape. A thin, flexible **plasma membrane** lies just beneath the cell wall. The plasma membrane controls the passage of materials into and out of the cellular contents, the **cytoplasm.** Bacteria lack a well-defined nucleus, but they contain the hereditary compound DNA.

Many bacteria contain granules of stored food and other materials dispersed through the cytoplasm. **Mitochondria,** the centers of **respiration** (the process by which energy is made available for life processes) in other cells, are lacking in bacteria. Respiratory enzymes, usually found in mitochondria, are concentrated on the inner membrane of a bacterium. The oxygen (O_2) necessary for respiration and the carbon dioxide (CO_2) which is given off diffuse through the plasma membrane.

Some bacteria are equipped with threadlike whips, or **flagella** (singular, flagellum) that extend from the plasma membrane and propel bacteria through their fluid environments. In some bacteria, flagella are found all around the cell (Figure 3-4). They are strands of protein molecules performing the same function as proteins in the muscles in higher animals.

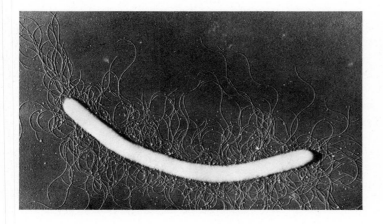

FIGURE 3-4

The flagella of *Proteus vulgaris* (a bacterium) show clearly in this electron micrograph. Size of organism, about two micrometers long.

Bacterial nutrition

Some bacteria cannot synthesize the substances they require and must therefore live in contact with preformed organic matter. They are of great importance to us because they are the principle decomposers of dead organisms.

Most foods would remain edible for many months or even years if bacteria were not present or if they could not reproduce. Hence, we have developed various ways of preserving foods. Since bacteria require moisture, one method of preservation is by drying. Heating above the boiling point kills bacteria (as in the pasteurization of milk), and cooling to near or below the freezing point slows or stops their rate of reproduction. Freezing doesn't necessarily kill bacteria; it may merely cause them to revert to inactive forms, known as **endospores,** which can become active again after a thaw.

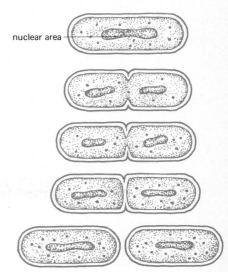

nuclear area

Bacterial reproduction

Bacteria reproduce by undergoing **transverse binary fission;** in simple language, they divide in half (Figure 3-5). When conditions for growth are ideal, bacteria reproduce at an amazing rate. A cell may be formed by a division of a "mother" cell, grow to maturity, and start dividing again in a period of only twenty minutes. Reproduction by fission is **asexual;** it does not involve the combining of hereditary material from two individuals.

FIGURE 3-5
Bacterial reproduction by transverse binary fission.

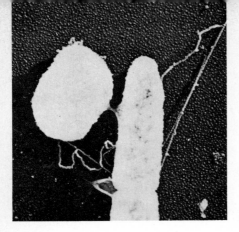

FIGURE 3-6
An electron micrograph of sexual reproduction in bacteria. Note the cytoplasmic bridge joining the two cells.

FIGURE 3-7
Four common blue-green algae as seen through a microscope.

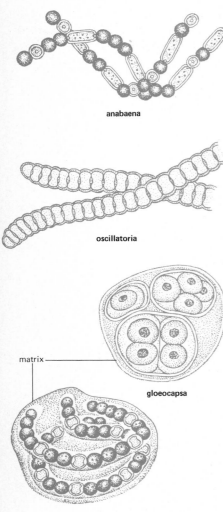

anabaena

oscillatoria

matrix

gloeocapsa

nostoc

Bacteria can reproduce by fission indefinitely, but certain species also reproduce sexually; hereditary material from two individuals combines. For instance, a species of bacteria which normally lives in human intestines, *Escherichia coli* (commonly abbreviated as *E. coli*) exists as two mating types. When "male" and "female" strains of *E. coli* are mixed in a fluid suspension, cells of the opposite sex come together, and a cytoplasmic bridge joins them (Figure 3-6). A "male" cell injects its DNA into a "female" cell. Soon afterward the "male," lacking genetic materials, may die; the cell that had been "female" is now a "male."

Remember that when organisms reproduce sexually, both parents contribute genetic material. The resulting new genetic combinations may produce offspring that differ from both parents. These offspring may be more favorably adapted to their environments and have better chances of survival. Thus, sexual reproduction is an important factor in evolution.

Blue-green algae

Like other species of algae, blue-green algae possess chlorophyll and are capable of photosynthesis. However, they are fundamentally different from other species of algae in that they do not have well-defined nuclei. Like bacteria their hereditary material is in direct contact with the other cellular contents and not enveloped by a membrane. Thus, blue-green algae are classified in the kingdom Monera rather than in the plant kingdom with the rest of the algae.

There are several species of blue-green algae. Some are filamentous chains of cells, while others form slimy matrices in which the cells are embedded (Figure 3-7). You can find blue-green algae in almost every roadside ditch, pond, and stream. Some species live on wet soil and rocks and even thrive in hot springs with a temperature as high as 185° F.

KINGDOM PROTISTA

The members of the kingdom **Protista** are known as protists or protozoans. They are complex single-celled organisms. In the course of evolutionary history protists are thought to have evolved from ancestors resembling the modern Monera. Perhaps a billion years ago the hereditary material within the forerunners of protists became surrounded with a membrane, forming a well-defined nucleus. The nuclear membrane is a partial barrier, allowing certain compounds to pass through, but preventing passage of others. From its protected position within the nuclear membrane, DNA directs processes in the rest of the cell.

Protozoans are fascinating subjects of microscopic study. For single-celled organisms they are amazingly complex. Some have gaping mouths and dart about by means of tiny, hairlike projections that act like oars. Others move by waving long, whiplike flagella. Most of the species live alone, but some live together as part of a colony.

The amoeba—a false-footed protozoan

The genus **Amoeba** includes several species of protozoans (Figure 3-8). An amoeba might be described as "animated jelly." On first seeing it, you might mistake it for a nonliving particle. But this tiny blob of grayish protoplasm moves of its own accord, feeds, reproduces, and performs all the other life processes.

Amoebas can be collected by taking slime from the bottoms of streams and ponds and from the surface of leaves of aquatic plants.

The protoplasm of an amoeba is surrounded by a thin, flexible plasma membrane (Figure 3-8). If you find an active animal, you will notice that the cytoplasm has a constant flowing motion. This streaming cytoplasm presses against the plasma membrane and produces projections called false feet, or pseudopodia (singular, pseudopod). A pseudopod starts as a bulge at any point on the surface of the organism and enlarges as part of the mass of the amoeba flows into it. Soon, another pseudopod may form, and the flow will change to a new direction. The old pseudopod gradually disappears as the cytoplasm flows back to the cell mass. This type of locomotion is called amoeboid movement.

The oxygen necessary to maintain the life of the amoeba diffuses through the cell membrane from the surrounding water. Most of the carbon dioxide and soluble wastes of protein metabolism, such as ammonia, pass out through the cell membrane in the same way. Intake of oxygen and elimination of wastes is a passive process by which the materials involved flow from regions of higher concentration to regions of lower concentration. However, during the process the amoeba takes in excessive amounts of water which must be actively eliminated. If the amoeba did not have a method of ridding itself of this water, it would swell till it burst. Within an amoeba excess water accumulates to form a **contractile vacuole** that contracts when it reaches a critical size and discharges the water through a temporary break in the cell membrane. Hence, the amoeba maintains a nearly constant internal water concentration.

Amoebas have no mouths, but they are capable of ingesting quite large bits of food. When an amoeba comes into contact with an algal cell or another protozoan that is acceptable as food, it engulfs it by surrounding it with pseudopodia. Part of the membrane of the amoeba now becomes the lining of a **food vacuole** inside the cytoplasm. A new membrane forms quickly on the surface at the point at which the food particle entered the cell. Digestion is accomplished by enzymes, formed by the cytoplasm, that pass into the vacuole and act on the food substances. Digested substances pass into the surrounding cytoplasm, and undigested particles are eliminated from the vacuole through temporary breaks in the plasma membrane.

Amoebas respond to conditions around them. Although they lack eyes, they are sensitive to light and seek areas of darkness or dim light. Amoebas have no nervous systems, but they react to movement around them. They tend to move away from large foreign objects, and they are attracted to food substances. Chemicals the foods give off act as stimuli to the amoeba.

In the presence of abundant food and good environmental conditions, an amoeba doubles its volume in a day or two. When it reaches approximately twice its original size it reproduces by cell division (Figure 3-9). Prior to cell division, the nucleus divides, and the two "daughter" nuclei move to opposite ends of the cell. The cytoplasm then constricts near the center, and the two halves pull apart, forming two cells. Each new daughter cell is capable of independent life and growth.

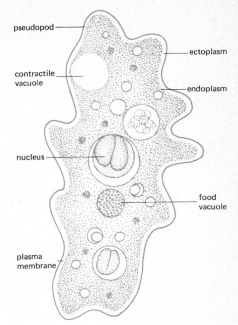

FIGURE 3-8
The structure of an amoeba. Size, about 500 micrometers. This protist assumes many different shapes as it moves.

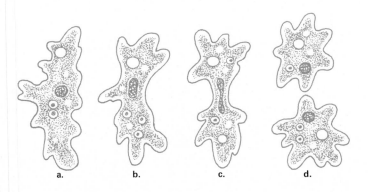

a.   b.   c.   d.

FIGURE 3-9
The amoeba reproduces by means of simple cell division.

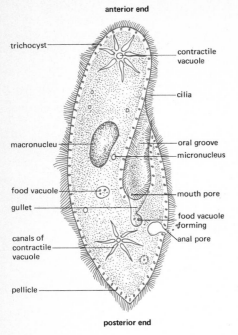

anterior end

trichocyst

contractile
vacuole

cilia

macronucleu

oral groove

micronucleus

food vacuole

mouth pore

gullet

food vacuole
forming

canals of
contractile
vacuole

anal pore

pellicle

posterior end

FIGURE 3-10
The structure of the paramecium. Size,
about 180 micrometers.

The paramecium—a complex protozoan

Various species of the genus **Paramecium** live in quiet or stagnant ponds where
scums form. They can be cultured in the laboratory by collecting submerged pond
weeds and keeping them in a jar with pond water. The jar should be set aside in a
warm place for a few days. As the weeds decay, a scum forms on the surface. Large
numbers of paramecia may be found in this scum.

The paramecium is shaped like a slipper (Figure 3-10). Unlike the amoeba,
the paramecium's shape is fairly rigid because it is surrounded by a thickened outer
membrane, the **pellicle.** However, the paramecium can bend somewhat as it
swims around objects it meets.

Paramecia move by means of hairlike threads called **cilia** that project through
the cell membrane and pellicle. The cilia are arranged in rows and lash back and
forth like tiny oars. They cover the entire cell, but are most easily seen along the
edges. Cilia can beat either forward or backward, thus moving the cell in either
direction or causing it to turn.

A striking feature of the paramecium is a deep **oral groove** along one side
of the cell. This depressed area is lined with long cilia that cause the paramecium
to rotate as it swims. The paramecium has a definite front (**anterior**) end which is
rounded, and a more pointed rear (**posterior**) end—a perfect design in stream-
lining. The oral groove runs from the anterior end toward the posterior end. The
action of the cilia lining the oral groove and the forward movement of the paramecium force food particles into the **mouth pore,** which opens into a funnel-like
tube, the **gullet.** Bacteria and other food particles forced into the gullet enter the
cytoplasm within a food vacuole. When the food vacuole reaches a certain size,
it breaks away from the gullet, and a new one begins to form. Digestion occurs as
food vacuoles drift in the flow of cytoplasm within the cell. Undigested food passes
through an opening in the pellicle, the **anal pore,** which is located near the
posterior part of the cell. Excess water is eliminated through contractile vacuoles
that are located at both ends of the cell.

Paramecia posess organelles of defense called **trichocysts,** which are small
shafts with a barb attached that can be discharged explosively in a fraction of a
second. Trichocysts are used in defense and to capture prey.

Paramecia normally reproduce asexually by fission, through a process
basically the same as fission by amoebas (Figure 3-11a). However, after several
months of cell divisions, paramecia lose vitality and die, unless they undergo a
sexual process known as **conjugation.** During conjugation two paramecia unite at
their oral grooves and exchange genetic material (Figure 3-11b). A paramecium
contains two nuclei: a large **macronucleus,** which regulates the normal activity

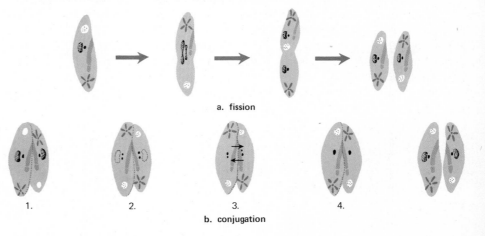

a. fission

FIGURE 3-11
Shown here are the two types of reproduction characteristic of paramecia: (a)
fission and (b) conjugation.

1.   2.   3.   4.

b. conjugation

of the cell, and a small **micronucleus,** which is involved in conjugation. Prior to conjugation the macronucleus disintegrates and the micronucleus divides. Portions of the micronuclei from two paramecia are exchanged during conjugation.

The Euglena—a photosynthesizing protozoan

Several species of the genus **Euglena** live in fresh-water ponds and streams. Under the microscope, a euglena appears oval- or pear-shaped. The anterior end is rounded, and the posterior end is usually pointed (Figure 3-12). The euglena swims by means of a flagellum attached to its anterior end. The flagellum—nearly as long as the one-celled body—rotates, pulling the organism rapidly through the water.

The internal features of a euglena include a combination of plant and animal characteristics. A gullet opening at the anterior end of the cell leads to an enlarged **reservoir.** Since the euglena has never been seen to ingest food, the gullet probably serves only as an attachment of the flagellum. A contractile vacuole close to the reservoir discharges water at regular intervals. Near the gullet is a prominent red **eyespot.** The area adjacent to this tiny bit of specialized protoplasm is sensitive to light and serves to direct the euglena to bright areas in its habitat. The most striking characteristic of the euglena is the presence of numerous oval **chloroplasts** (chlorophyll-containing bodies) Most species of euglena carry on photosynthesis and, in addition, gain nutrition by absorbing dissolved organic matter from the water in which they live. The fact that the euglena can live with or without preformed organic matter was a source of endless debate among those who recognized only two kingdoms of organisms (plants and animals) as to which kingdom the euglena belonged. The problem is easily avoided by adopting a taxonomic scheme that includes at least one more kingdom—the protists.

KINGDOM FUNGI

The fungi (composed of molds and yeasts) are considered to be protists in some classification schemes and simple plants in others. In this discussion they will be described as a kingdom in themselves. Among the fungi are microscopic species, such as yeasts, as well as larger organisms, such as mushrooms.

Like animals, fungi lack chlorophyll and, cannot produce their organic food requirements. They must utilize organic compounds produced by other organisms. Certain fungi are **saprophytes,** living on the remains of dead plants and animals. Others are **parasites,** taking their nourishment from the tissues of still-living hosts.

In obtaining nutrition for themselves, certain fungi cause enormous economic loss to humankind by spoiling food. Others are agents of plant diseases, and still others are parasites of animals, including humans. The well-known scourge of the locker room, athletes foot, is a fungus infection. But, not all species of fungi are harmful; some are even beneficial. Among these are the species that produce the antibiotics used in combating bacterial diseases. Fungi are also used in brewing beer, making bread and cheese, and decomposing dead animals.

Molds

The term **mold** refers to several kinds of fungi that grow on food, wood, paper, leather, cloth, and other organic substances (Figure 3-13). Molds thrive best in warm, moist environments, but certain species survive at temperatures near

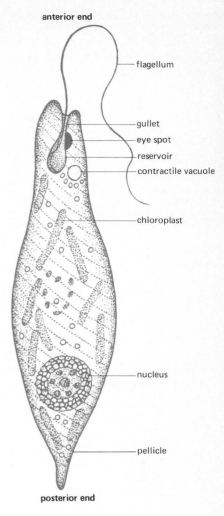

anterior end

flagellum
gullet
eye spot
reservoir
contractile vacuole
chloroplast
nucleus
pellicle

posterior end

FIGURE 3-12

The structure of a *Euglena*. Size, about 50 micrometers. Contrast this protozoan with a paramecium and an amoeba.

Carolina Biological Supply Company

FIGURE 3-13
A cottonlike mass of a water mold, *Sapro-legnia,* on a seed.

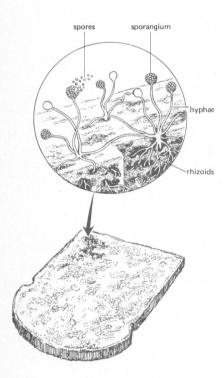

FIGURE 3-14
The structure of bread mold. Note the rootlike rhizoids extending into the bread.

FIGURE 3-15
Sexual reproduction in bread mold requires two physiologically different mating strains.

56

freezing. This makes them a problem in cold-storage plants and in home refrigerators.

If you moisten a piece of bread, expose it to the air for several minutes, and place it in a covered dish in a warm, dark place, bread mold (*Rhizopus nigricans*) is almost certain to appear within a few days. It starts as a microscopic spore that germinates on the surface of the bread, forming a branching mass of silver, tubular **hyphae** that eventually permeate the bread as rhizoids (Figure 3-14).

After a few days of growth, black knobs appear among the hyphae. These are **sporangia,** and they contain spores. More than fifty thousand spores may develop in a single sporangium. When a sporangium matures, its thin outer wall disintegrates, releasing the spores. These are dispersed by air currents, and if they lodge on a suitable food supply they germinate and start the cycle anew.

Sexual reproduction also occurs in bread mold by a form of conjugation. Although *Rhizopus* hyphae look alike, there are two different mating strains that can be designated as plus and minus. During sexual reproduction, the hyphae form short side branches (Figure 3-15). If the tip of a branch of a plus strain contacts the tip of a branch of a minus strain, conjugation occurs. Cross walls form a short distance behind the tips of the side branches, cutting off the terminal cells that become **gametes** (reproductive cells). Dissolving of the walls at the tips of the end branches allows the two gametes to fuse, forming a **zygote.** When conditions are favorable for growth the zygote germinates, and new hyphae emerge.

Following a warm spring rain, mushrooms may be found poking through the ground in orchards, fields, and woodlands. Although they are present during the rest of the year, mushrooms usually exist in a form that is undetected by the casual observer—diffuse, silvery hyphae, which thread their way through soil or decaying wood. A mushroom may live many years, gradually penetrating more and more of the host. Digestive enzymes secreted by the hyphae break down organic substances in the host, changing them to compounds that can be absorbed and used as nourishment. At certain seasons, especially in the spring and fall, small knobs develop on the hyphae just below the ground. These develop into the familiar spore-bearing structure recognized as a mushroom (Figure 3-16).

A mature mushroom consists of a stalk, or **stipe,** which supports an umbrella-shaped **cap.** While pushing up through the soil, the cap is folded downward around the stipe. After forcing its way through the soil, the cap opens, leaving a ring, or **annulus,** around the stipe at the point where the cap and stipe were joined.

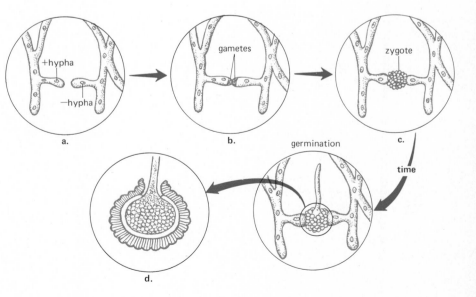

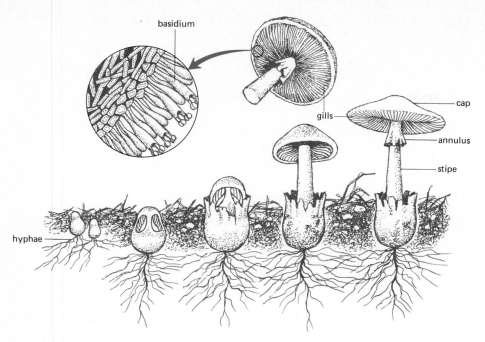

FIGURE 3-16
The development of a mushroom. The inset is an enlargement of a gill.

Most mushrooms contain numerous platelike gills on the undersurface of the cap, radiating out from the stipe like the spokes of a wheel. On the outside of each gill are hundreds of spore-producing **basidia.** A single mushroom may produce as many as ten billion spores, which drop from the basidia when mature and are carried by the wind. If a spore lands in a suitable environment (most of them do not) a new mushroom begins development.

Certain mushrooms are among the world's greatest delicacies, but others are extremely poisonous. They contain **neurotoxins** (chemicals that disrupt the functions of the central nervous system). The most toxic of mushrooms can cause death within a matter of hours.

Yeasts

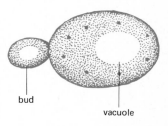

Yeasts are the simplest in structure of the fungi. Each yeast cell is oval or spherical and is bounded by a membrane and thin cell wall. Under ideal conditions, which include a source of nourishment, warmth, and oxygen, yeast cells reproduce rapidly by **budding.** A bud starts as a small knob pushing out from the surface of a mother cell (Figure 3-17). As the bud enlarges, the nucleus of the mother cell divides. One nucleus moves into the bud; the other remains in the mother cell. The base of the bud remains attached and, in turn, produces another bud, forming a chain of yeast cells. Additional buds may also develop on the original mother cell.

The manner in which yeasts utilize sugar in their life processes has made them of use to humans in several ways, including the brewing of beer and bread-baking. By **fermentation** yeasts split sugar molecules, thereby releasing energy for their life processes. Alcohol and carbon dioxide are by-products of fermentation. Various strains of yeast are cultured for industrial use. For example, baker's yeast grows rapidly in dough. As it grows, it forms bubbles of carbon dioxide that expand during baking, causing the dough to rise. The small amount of alcohol produced is driven off by heat. The growth of baker's yeast is inhibited at an alcohol concentration of from 3 to 5 percent, but brewer's yeasts and wine yeasts tolerate alcohol up to a concentration of 14 percent.

FIGURE 3-17
Budding in yeasts. Under ideal growth conditions, a bud develops into a mature yeast cell in about 30 minutes. Size of each yeast, about 50 micrometers.

KINGDOM PLANTAE

Plants bridge the gap between the nonliving and the living worlds. By trapping light energy and utilizing it for the synthesis of organic compounds from simple, inorganic substances they initiate the biosphere's web of life. While all plants have an **autotrophic** (self-feeding) mode of life, they have evolved diverse physical forms as they have adapted to the various sunlit environments of the earth's surface.

The algae

Most people tend to overlook the algae. They pass them off as pond scums or seaweeds and often link them with water pollution. To a biologist, however, they are a fascinating group occurring in a vast assortment of sizes, shapes, and colors.

Most people are not aware, either, of the importance of algae in an aquatic environment. Here, algae are primary food producers and water aerators, supplying oxygen for other forms of life. In a natural aquatic environment algae thrive in close association with other aquatic plants and animals. Either reduction or rapid increase in the algal population upsets this balance. As algae die out, fish and other aquatic animal populations decline with them. The environment is no longer productive. On the other hand, an algal population may increase to the point of choking a lake or pond and converting the water to a death trap for fish and other animals when the plant matter decays and ferments. Humans have produced countless problems of these sorts in aquatic environments by destroying algae or causing conditions that result in their excessive growth.

Relative to other types of plants, such as mosses, ferns, and seed plants, algae are simple organisms. They lack specialized tissues for conducting water, and they also lack such plant organs as roots, stems, and leaves. However, from the standpoint of reproductive processes and biochemical capabilities, they are quite complex. Further, they must certainly be considered as highly successful forms of life, for they have dominated aquatic environments for over one billion years.

Algae occur in a variety of forms: some as we saw are simple enough to be classified as Monera, some are classified as Protista. Many are one-celled and can be seen only with a microscope. These solitary cells may float in water, settle to the bottom, or swim about with lashing flagella. Other algae, including the giant kelps of the ocean, form plant bodies more than one hundred feet long. Regardless of size, however, the vegetative body of an alga, termed a **thallus** (plural, thalli), lacks the specialized tissues and organs possessed by land plants.

In many species of algae, the cells group to form cell colonies. Often, the cells are attached in chainlike groups called **filaments.** Many of the larger algae, including the seaweeds, have broad, ribbonlike thalli composed of thousands of cells.

There are several phyla of algae, but a general idea of their structure and function can be gained by a consideration of species of two of them, **Chlorophyta** ("green algae") and **Phaeophyta** ("brown algae").

PROTOCOCCUS, A LAND-DWELLING GREEN ALGA   **Protococcus** is one of the most common green algae, but is an exception in that it does not live in water. It grows on the moist trunks of trees, unpainted wooden structures, and damp stones. During dry weather it is seldom seen, but in wet weather it is very

FIGURE   3-18

Cells of the green alga *Protococcus* occur singly or in colonies of two or more. This alga forms the green film on rocks and wood boards. Size, about 20 micrometers.

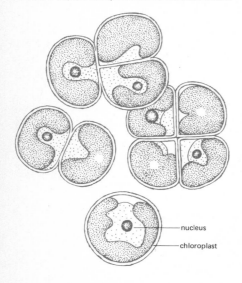

nucleus

chloroplast

evident. On trees it tends to grow on the north side because bark is more moist on the shaded side.

The cells of *Protococcus* are spherical or somewhat oval (Figure 3-18). Each contains a nucleus and a single, large chloroplast. The cells are so small that many thousands cover only a few square inches of bark. They may be carried from tree to tree by birds and insects as well as by the wind during dry weather. *Protococcus* reproduces by fission; it is incapable of sexual reproduction.

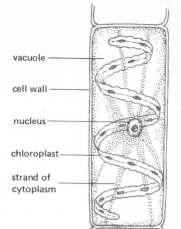

SPIROGYRA, A FILAMENTOUS GREEN ALGA   Almost any pond or quiet pool will have bright green masses of threadlike **Spirogyra** during the spring and fall months. Filaments of *Spirogyra* range in length from a few inches to a foot or more. Under a microscope they appear as a series of transparent cells, arranged end to end, like box cars in a train (Figure 3-19). Each cell in a strand of *Spirogyra* has one or more spiral chloroplasts that wind from one end of the cell to the other.

Spirogyra reproduces in two ways. Asexual reproduction occurs when the cells undergo fission, thus adding length to a filament. Sexual reproduction is by conjugation, which occurs when weather conditions are unfavorable for growth by fission. You are likely to see it in specimens collected early in summer, at the end of the spring growing season, and again in fall, with the approach of cold weather.

Conjugation involves two filaments of *Spirogyra* that line up parallel to each other. A small knob grows out from each cell (Figure 3-19). Each knob continues to grow until it touches the knob of the cell across from it. The tips of the knobs dissolve, forming a passageway between the two cells. Subsequently, the contents of one of the cells flows through the passageway into the interior of the other cell, forming a zygote. Interestingly the contents of all the cells of one filament flow into the cells of the other filament, resulting in one filament of empty cells and another containing rows of zygotes.

Soon after conjugation, the zygotes form thick protective walls, separate from the attached cells, and undergo a rest period. The thick walls protect them from heat, cold, and dryness. In this form *Spirogyra* can survive a cold winter or a summer drought. When conditions are again favorable for growth, cell divisions generate new filaments.

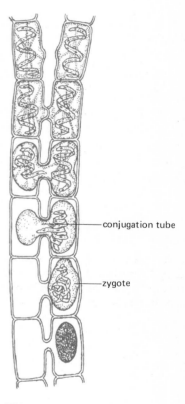

THE BROWN ALGAE   The brown algae, which are commonly known as "seaweeds," are most abundant in cold ocean waters, where they grow from attachment points on rocks near the shore. The brown algae occur in many forms and sizes, all of which are multicellular. Their vegetative bodies may be slender filaments or ribbonlike thalli, over thirty two meters in length. In some respects, the thalli of brown algae resemble the bodies of higher plants.

A rootlike **holdfast** anchors the thallus to a rock or other object in the water, and leaflike blades project from the thallus. Air-filled bladders on certain species serve as floats, giving them buoyancy. Cells of brown algae contain chlorophyll as well as other pigments, including a golden-brown pigment known as **fucoxanthin,** which usually masks the colors of the other pigments.

Fucus, or rockweed, is a brown alga that is common in the cold waters of the northern Temperate Zone. It can be seen lying in vats along rocky coasts when the tide is out (Figure 3-20). The forked thalli, one to three feet in length, float toward the surface at high tide.

Kelps are the largest brown algae. They are common along the coast of western North America. A kelp plant is composed of several flat blades attached by stalks to a holdfast that adheres to a rock. The thalli, which may be over one hundred feet in length, are suspended by air bladders. Skin diving in a kelp bed is an excursion into a beautiful labyrinth of fronds gently swaying with the passing waves.

FIGURE   3-19
Filaments of the green alga *Spirogyra*, showing configuration. The detailed structure of a vegetative cell can be seen above.

The bryophytes—primitive land plants

Evolutionary development of plants in the transition from aquatic to terrestrial life seems to have been along two lines. One line led to the bryophytes, the phylum that includes the mosses and liverworts. The other line resulted in the tracheophytes, or vascular plants, which include ferns and seed-producing plants.

Although the bryophytes were among the first land plants, they have never developed more than primitive adaptations to land environments. For this reason, they cannot compete with seed plants for space. One of the limitations of the bryophytes is their lack of a root system. A large land plant must have large roots that penetrate the soil to reach subsurface water and anchor the aerial plant body. Bryophytes lack true roots. Their absorbing structures are small and reach only a short distance into the soil. Therefore, they can anchor only a small plant body. The lack of conducting channels, or vascular tissues, in the stems of bryophytes limits water conduction to a short distance, thus further restricting the size to which they can grow.

The Mosses In spite of their structural limitations, the bryophytes, especially the mosses, are widely distributed in the vegetation of the earth. They occupy land environments from the Arctic to the Antarctic. In a way the small size of the mosses is an advantage, allowing them to live in harsh environments that larger plants cannot occupy. In the treeless, windswept tundra of the Arctic, for instance, mosses are quite common plants. In more favorable environments, they grow among ferns and seed plants, which provide the shade and moisture that certain mosses require for survival.

If you closely examine a clump of moss, you will find that it is a compact mass of tiny plants. Each plant has a slender stem, usually a few centimeters long, with numerous delicate leaves encircling it. If you carefully remove a plant from the soil, you will find a cluster of hairlike projections growing from the base of the stem.

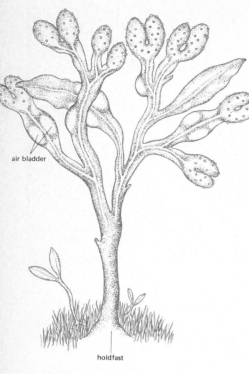

FIGURE 3-20
Fucus, a brown alga that lives along the Atlantic Coast. Inhabiting the intertidal zone, it lies exposed at low tide, but is submerged at high tide.

FIGURE 3-21
Life cycle of the moss. Note that alternation of generations occurs. After fertilization, a sporophyte grows out of, and is parasitic on, the female gametophyte. Spores form at the top of the sporophyte and fall to the ground. These germinate and grow into a new generation of gametophytes which reproduce sexually.

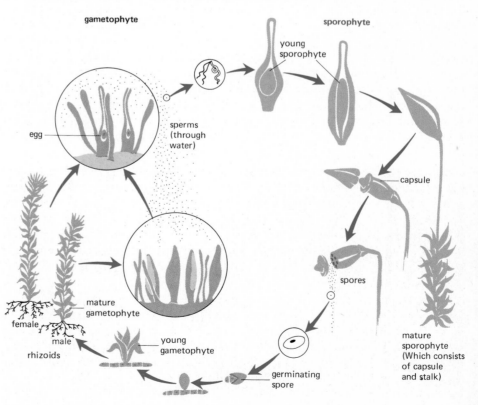

These are not roots but are **rhizoids** that absorb water and minerals and anchor the plant body. Rhizoids function similarly to roots, but they are cellular filaments, while roots are plant organs composed of specialized tissues.

As illustrated in Figure 3-21, the life cycle of a moss is complex. It involves an asexual, spore-producing stage, the **sporophyte,** which forms a sexual, gamete-producing stage, the **gametophyte.** Sexual reproductive organs develop at the tips of the leafy stems of gametophyte moss plants. Depending on the species of moss, male and female organs may be borne on the same plant or on different plants. In order for fertilization to occur, a certain amount of moisture must be present for the sperm to swim from the male organ to the female organ (a thin film of dew is generally sufficient). The union of a sperm and egg produces a zygote.

Fertilization starts the sporophyte stage of the life cycle of a moss. Remaining within the female organ, the sporophyte begins as a slender stalk that grows up from the leafy stem of the gametophyte. The top of the stalk swells, forming a capsule that contains numerous spores. When the spores are ripe, the top of the capsule falls off, and the spores are released, to be distributed by the wind. Upon landing in a suitable environment, a spore germinates, initiating a new gametophyte stage of the cycle.

The Liverworts   Less familiar than mosses are their relatives, the liverworts. These small plants grow in wet places, often along the banks of streams. They have thin, leathery leaves laid flat against the ground, anchored by numerous rhizoids on the lower side. One of the common liverworts, **Marchantia,** resembles a green tongue (Figure 3-22). The life cycle of a liverwort is similar to that of the mosses.

FIGURE 3-22
Marchantia, a type of liverwort.

The ferns

Ferns and their relatives were most numerous three hundred million years ago, when flowering plants had not yet developed. The marshy land and warm climate that prevailed at that time were ideal for the growth of ferns and other early tracheophytes. Ferns were not limited to isolated clumps and patches, as they are today; they were the dominant plants. Tree ferns from thirty to forty feet in height formed dense forests. Smaller species, much like those that exist today, grew in a dense undercover. At present, tree ferns grow only in a few tropical areas, including Hawaii and Puerto Rico. They are the remainders of the once-flourishing age of ferns (Figure 3-23).

FIGURE 3-23
Tropical fern forest in Hawaii. These trees are small versions of huge tree ferns that grew abundantly some 300 million years ago.

FIGURE 3-24

Life cycle of the fern. The gametophyte, which is so small that it is rarely noticed, develops on the ground.

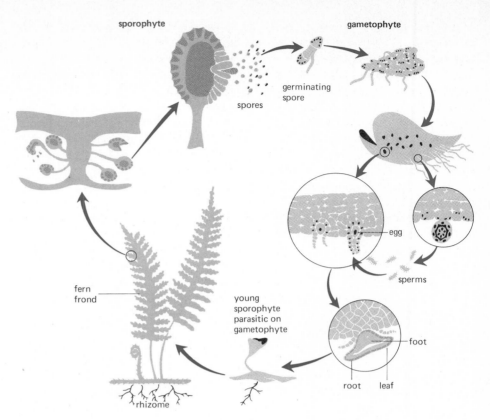

Today we are reaping the benefits of the age of tree ferns. When they thrived, layers of plant remains accumulated in the swampy areas where they grew. Later, movements of the earth compressed these layers, transforming them into coal.

Like mosses, ferns have complex life cycles which involve sporophyte and gametophyte stages. Unlike mosses, however, the predominant stage in the fern life cycle is the sporophyte rather than the gametophyte (Figure 3-24).

In all ferns but tree ferns, the stem of the sporophyte is underground, growing horizontally just below the surface (in tree ferns the stem is vertical and above ground). This underground stem, or **rhizome,** contains vascular tissues that are similar to those of seed plants. Clusters of roots grow from the rhizome and spread through the soil, anchoring the plant and absorbing water and minerals. Fronds arise from the rhizome, and, when the fronds mature, certain of them produce spores on their undersides. The spores are released to the air, and those that land on moist soil germinate, thus initiating the gametophyte generation. At its maximum size, the gametophyte is a flat, heart-shaped structure, rarely more than a centimeter in width. Rhizoids develop on its underside, along with male and female sex organs. When the sperm are mature, they swim to the female sex organs (as with the mosses a film of water from dew or rain is necessary) and fertilize the eggs. The union of a sperm and egg results in a zygote, which is the first cell of the sporophyte stage. It grows by cell division as the gametophyte withers and disappears.

The seed plants

In addition to the ferns, the phylum Tracheophyta includes the seed-producing plants. A seed is an embryo plant covered with one or more protective seed coats.

Food stored in the seed nourishes the young plant until the early period of growth has occurred. In a sense, a seed is a packaged plant, ready for delivery. A seed may be blown through the air, float on water, or be carried on the fur of an animal; it may lie dormant for many months. When moisture and temperature conditions are favorable, the seed coat softens, and the young plant begins to grow by sending a root downward into the soil and a shoot upward. Reproduction through the production and dispersal of seeds is highly efficient and is one of the reasons that seed plants have gained dominance in land environments.

There are two classes of seed plants, the **gymnosperms** and **angiosperms.** The gymnosperms are distinguished from angiosperms on the basis of seed development. The name gymnosperm means "naked seed" and refers to the fact that the seeds develop in exposed positions on the upper surfaces of the scales of cones. Angiosperms, on the other hand, develop within the protective wall of the female sex organ, the **ovary** (Figure 3-25).

FIGURE 3-25

The main difference between the seed of an angiosperm, like the bean, and that of a gymnosperm, like the pine, is that while they are developing, the bean seed is enclosed in a ripened ovary (or fruit), whereas the pine seed lies exposed on a cone scale.

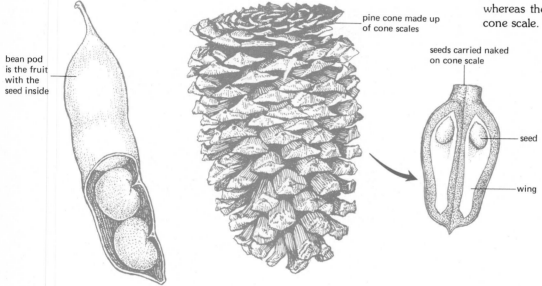

bean pod is the fruit with the seed inside

pine cone made up of cone scales

seeds carried naked on cone scale

seed

wing

Gymnosperms: About 750 species of gymnosperms are living today. These species are the survivors of a much larger plant population that flourished in earlier geological periods. Living gymnosperms are classified in four orders, three of which are represented by only a few remaining species. By far, the most prolific order of gymnosperms is **Coniferales,** which includes such trees as the pines, spruces, firs, and cedars. Their common name, **conifer,** refers to the woody cones in which seeds are borne.

Among all trees, conifers hold the record for height, trunk diameter, and age. A giant redwood (*Sequoia gigantea*), in the Calaveras Grove in California, towers three hundred feet above the ground. This tree is over thirty feet in diameter and is estimated to be more than four thousand years old. Even this trunk diameter and age are surpassed by a cypress tree, the Big Tree of Tule, growing about two hundred miles south of Mexico City. It is fifty feet in diameter and is estimated to be over five thousand years old.

Angiosperms: All flowering plants are classified as angiosperms. These most highly evolved plants represent a long line of development that probably branched from the gymnosperm line. Angiosperms are thought to have developed

FIGURE 3-26

Specialized tissues of a seed plant.

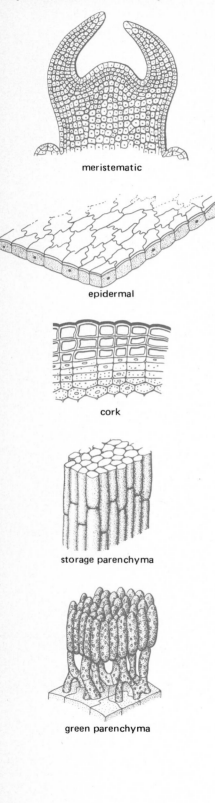

meristematic

epidermal

cork

storage parenchyma

green parenchyma

about a hundred million years ago and evolved rapidly to replace older forms of vegetation.

There are several reasons for the success of angiosperms. Genetic variations have produced numerous forms of them, including trees, shrubs, climbing plants, and floating plants. Because of this variety of form, they can grow and reproduce in a greater variety of environments than any other group of plants. Their efficient mode of reproduction has also given the angiosperms an evolutionary advantage over other plants. Unlike mosses and ferns, angiosperms do not require the presence of water for fertilization, and they have evolved a wealth of means of seed dispersal. Some seeds are carried to new lands by the wind (such as the tufted seeds of the dandelion); others float (such as the coconut); and still others travel on the fur or within the guts of animals (such as the burdock and blackberry). In their myriad of forms, seeds are the most diverse objects for the dispersal of plants yet evolved.

Tissues and Organs of Seed Plants: In the course of evolution, cellular specialization has led from complexity at the cellular level (as in the protists) to the high degree of specialization of certain cells to form tissues that perform specific functions in higher plants and animals. In the plant kingdom the seed plants represent the highest degree of cellular specialization.

The various organs of seed plants perform their processes with great efficiency because of the specialized tissues of which they are formed. The following are some of those tissues (Figure 3-26).

1   **Meristematic tissue** is composed of small, thin-walled cells that are capable of unlimited reproduction. During the growing season, meristematic cells divide almost continuously. As they mature, cells of the meristematic tissue form the permanent tissues of plant organs. Plants differ from animals in that growth occurs at isolated areas of meristematic tissue rather than throughout the plant body. Meristematic tissue occurs in the tips of roots, in the buds of stems, and just beneath the bark in the **cambium layer.**

2   **Epidermal tissue** is a covering layer, usually one cell thick, on the surfaces of roots, stems, and leaves. Epidermal cells reduce water loss and protect inner tissues from injury. The epidermis of a young root is specialized for absorbing water.

3   **Cork** is a covering tissue on the surfaces of woody roots and stems. Cork cells are usually short-lived but remain as a protective, waterproof covering, usually many cells in thickness after they die.

4   **Parenchyma tissue** is composed of thin-walled cells resembling those of meristematic tissue. Parenchyma cells occur in flower petals, leaves, and various regions of roots and stems. **Green parenchyma** tissue contains chlorophyll and is the center of photosynthesis in leaves and young stems.

5   **Sclerenchyma** is a strengthening tissue occurring in various regions of roots, stems, and certain leaves. Cells of sclerenchyma are elongated, thick-walled fibers. They are usually short-lived, but their thick walls remain as supporting structures.

6   **Vascular tissues** are composed of elongated, tubular cells that serve as channels of conduction. Most end walls of the cells in vascular tissue disappear, along with the protoplasm, leaving continuous channels that may be several feet in length. Vascular tissues are paths of transport for water and minerals upward from the roots and food compounds downward from their place of formation in the leaves.

7 **Xylem** is a complex tissue, commonly known as wood. It is composed of vascular tissue and parenchyma, and it forms a prominent region in roots and stems. Xylem conducts water and minerals upward to the leaves.

8 **Phloem** is also complex tissue composed of several simple tissues, including vascular tissue, sclerenchyma, and parenchyma. In a woody stem, phloem is a bark tissue. Phloem transports food produced in the leaves to the rest of the plant and roots.

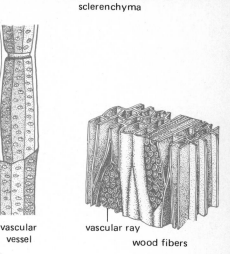

sclerenchyma

vascular vessel

vascular ray

wood fibers

Organs may be composed of several types of tissue. Each organ is highly developed for performing certain activities. The root, stem, and leaf of a seed plant are **vegetative organs.** They perform all the processes necessary for life except the formation of seeds.

The **root** anchors the plant in the ground. It penetrates the soil, absorbs water and minerals, and conducts these to the stem for delivery to the leaves (Figure 3-27). Roots of certain species store food substances and return them to the plant as needed. The large roots of carrots, for instance, serve this function.

The **stem** produces the leaves and displays them to light. The stem is a busy thoroughfare, for it conducts water and minerals upward and transports food that has been produced in the leaves downward (Figure 3-28). Like the root, the stem often serves as a place in which to store food.

The **leaf** is the center of much of the plant's activity. In most plants, it is the chief organ of photosynthesis. It also exchanges gases with the atmosphere in the processes of respiration and photosynthesis (Figure 3-29).

FIGURE 3-27

Major regions of cellular development in a root tip. The thimble-shaped root cap protects the delicate meristematic region from injury by soil particles.

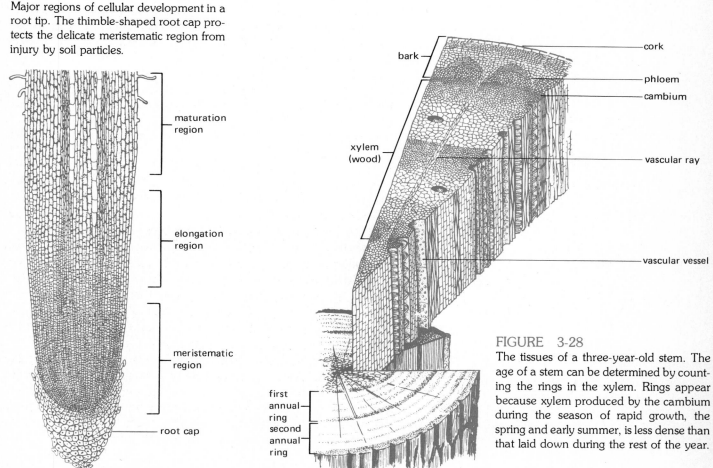

maturation region

elongation region

meristematic region

root cap

bark

xylem (wood)

cork

phloem

cambium

vascular ray

vascular vessel

first annual ring

second annual ring

FIGURE 3-28

The tissues of a three-year-old stem. The age of a stem can be determined by counting the rings in the xylem. Rings appear because xylem produced by the cambium during the season of rapid growth, the spring and early summer, is less dense than that laid down during the rest of the year.

FIGURE 3-29
A cross-section of a leaf.

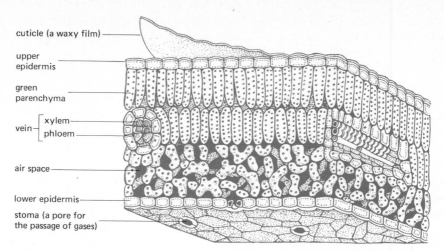

cuticle (a waxy film)
upper epidermis
green parenchyma
vein — { xylem / phloem
air space
lower epidermis
stoma (a pore for the passage of gases)

After a period of growth, a plant usually reproduces. In angiosperms, flowers are organs specialized for sexual reproduction. A portion of the flower develops into a fruit, which contains the seeds. The sequence from flower to fruit and seed is a complex, highly evolved process, which is considered in chapter 11.

KINGDOM ANIMALIA

As the course of evolution has wrought increasing complexity within the plant kingdom, animals have also been evolving more sophisticated forms. All organisms in the animal kingdom are multicellular, but most of them are **invertebrates,** having no backbone, as contrasted to the **vertebrates,** which have backbones. Evolution of the animals probably began with a protist ancestor from which invertebrates developed. Vertebrates later developed from the invertebrates. In tracing the evolutionary development of the animals we will begin with the least complex of the existing phyla.

The sponges

The sponges are a sort of evolutionary "dead ends." They are thought to have evolved from ancestral protists hundreds of millions of years ago, but in subsequent years they have remained a relatively unspecialized group of animals. Most sponges are marine, although there are a few freshwater species. They occur in a variety of shapes and colors (Figure 3-30). They may live singly or in colonies

FIGURE 3-30
Various sponges. The appearance of sponges may depend upon the material composing the skeleton. (Left, Walter Dawn; right, American Museum of Natural History.)

so massed together that they form an encrusting layer over the surface of a rock. Individual sponges vary in size from less than a centimeter to two meters in diameter. When you first look at a living sponge, you may conclude, as Aristotle did, that it is an interesting plant. But, sponges are not capable of synthesizing organic nutrients; they obtain food from their environment in a subtle way.

Most animals are capable of moving about in search of food, but sponges are **sessile;** they are permanently attached at their bases. Therefore, they must draw their food (which consists of single-celled algae, protists, bacteria, and organic particles) to them. Certain cells within sponges, **collar cells,** possess flagella that set up the currents that draw water inward through many small **incurrent pores** (Figure 3-31). As food particles enter, they are caught by the collars and digested. Water from which food has been filtered passes out through a larger **excurrent pore.**

From the collar cells, digested food is absorbed by cells known as **amoebocytes,** which resemble free-living amoebas. The amoebocytes wander throughout the jellylike substance in which the collar cells are embedded, transporting digested food and oxygen. The amoebocytes also carry wastes and carbon dioxide to the collar cells for disposal.

The body of a sponge is held together by a protective outer layer of **epidermal cells,** as well as internal, noncellular structures, called **spicules.** These are hard, pointed objects that are secreted by living cells. Spicules may consist of silicon, calcium carbonate, or a tough but flexible protein called **spongin.**

Sponges reproduce sexually, but they also may reproduce asexually in two ways. They may form **buds,** which are groups of cells that enlarge and live attached to the parent for a time, then break off and live independently. Or, during periods of freezing temperatures or drought, groups of cell masses surrounded by a heavy coat of organic matter form and break off from the parent sponge. These are called **gemmules.** They consist of little groups of amoebocytes and a few spicules. When favorable conditions return, each gemmule has the capacity to grow into another sponge. Reproduction by gemmules is usually characteristic of fresh-water species.

Sponges reproduce sexually by developing eggs and sperms. The sperms are shed into the water and enter another sponge through the incurrent pores. They are taken into the cytoplasm of the collar cells and are then transferred to the egg by the wandering amoebocytes. The fertilized egg develops into a flagellated larva that escapes from the sponge and, after swimming for a while, settles down and grows into a young sponge.

Sponges have amazing powers of regeneration. Because of this, sponge growers are able to increase the number of sponges by cutting them in pieces, which are placed in growing beds. Each piece develops into a separate sponge.

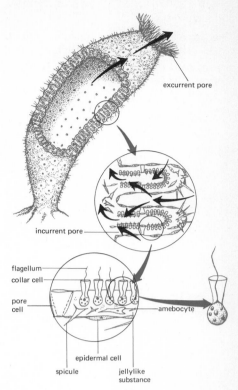

FIGURE 3-31
Water is continually drawn into the sponge by the flagella of the collar cells. It passes through small incurrent pores into the cavity of the sponge and flows out through the large excurrent pore. The black arrows indicate the direction of water flow.

Coelenterata

Like sponges the majority of coelenterates are marine, but the characteristics of this phylum can be seen in **Hydra,** a genus of common fresh-water coelenterates, which live in quiet ponds. The baglike body of the hydra consists of two cell layers separated by a jellylike material called **mesoglea.** The outside layer is the **ectoderm;** the inner layer is the **endoderm** (Figure 3-32). Hydras attach themselves to rocks or water plants by means of a sticky secretion from the cells of their **basal disks.**

The body of a hydra has a single opening that is surrounded by **tentacles** bearing stinging cells, each of which contains a structure called a **nematocyst.** When a small animal comes in contact with one of the tentacles of a hydra, many

FIGURE 3-32

The external and internal structures of the hydra can be seen in this diagram. The animal has two layers of cells with a jelly-like material between.

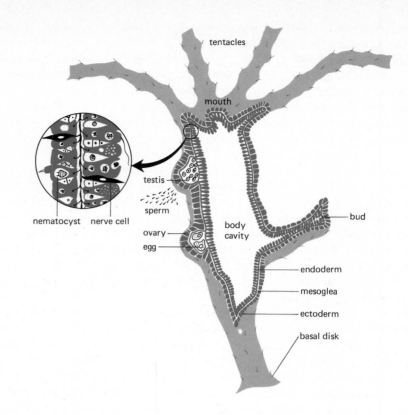

FIGURE 3-33

The photographs below show the hydra using its stinging cells to paralyze its prey; then the hydra's tentacles push the food into the body cavity. The nerve net of the hydra—which lies between the ectoderm and endoderm—transmits impulses and allows coordination of the hydra's activities. (See illustration at bottom of page)

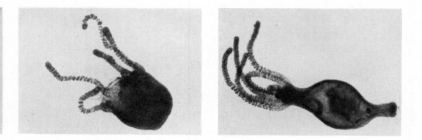

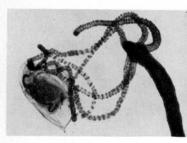

nematocysts are explosively discharged, and they pierce the victim's body with tiny, hollow barbs. Since each barb is attached to the tentacle by a thin thread, the combined effect of many threads prevents the escape of the hapless victim (Figure 3-33). At the base of each barb is a small poison sac that discharges its contents

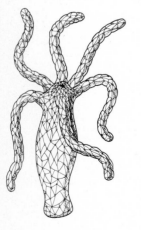

through the hollow barb and into the prey, thus paralyzing it. Once the prey has been paralyzed, the tentacles bend inward and push it through the circular mouth and into the body cavity of the hydra. Specialized endodermal cells that line this space function in digestion and absorption.

The food-getting reaction of the hydra shows a marked advance over the more primitive sponge. The tentacles of the hydra are coordinated in catching food and pushing it into the mouth. Also, if you touch a tentacle with a needle, the body and all the tentacles contract suddenly. The stimulus to one tentacle travels throughout the body via a series of nerve cells. This **nerve net** is next to the mesoglea. The contraction itself is accomplished by the shortening of slender fibers that lie in the ectoderm. These fibers are similar to the muscle cells of higher animals.

Like the sponge, the hydra reproduces asexually by forming buds. A bud appears first as a knob growing out from the side of the adult, as shown in Figure

68

3-33. Later, this knob develops tentacles, and after a period of growth, it separates from the parent and lives independently.

Sexual reproduction by the hydra usually occurs in autumn. Eggs are produced along the body wall in little swellings, called **ovaries;** motile sperm cells are formed in similar structures called **testes.** The egg is fertilized in the ovary, and the zygote grows into a spherical, many-celled structure with a hard, protective cover. In this form it leaves the parent and passes through a period of dormancy before maturing into the adult form.

TWO BODY FORMS   So far in our consideration of the coelenterates we have looked at only the hydra, but there are other, quite different members of this phylum. The body form of a coelenterate may be one of two types. The **polyp** form is well illustrated by the hydra, with its tubular body that has a basal disk at one end and tentacles at the other. The most familiar coelenterate, the jellyfish, has a different body plan. The bell-shaped, free-swimming form of the jellyfish is called a **medusa** (Figure 3-34). A medusa swims by taking water into the cavity of the bell, then forcibly ejecting it. This jet propulsion produces a jerky movement through the water.

The life cycles of some coelenterates involve both polyp and medusa phases. Among these is the common jellyfish, **Aurelia.** The sexually-reproducing medusa of *Aurelia* has a scalloped margin from which tentacles hang. The male medusa sheds sperms into the sea, which fertilize eggs within the body cavity of a female medusa. Within the female, the zygotes are protected for a short time by folds of tissue surrounding the mouth. After leaving the medusa, the young *Aurelia* attach themselves to a rock or other substrate and develop into the polyp form. As illustrated in Figure 3-35, a polyp ultimately reproduces asexually by budding off to form several medusae, completing the cycle.

FIGURE 3-34
A coelenterate with a medusa form.

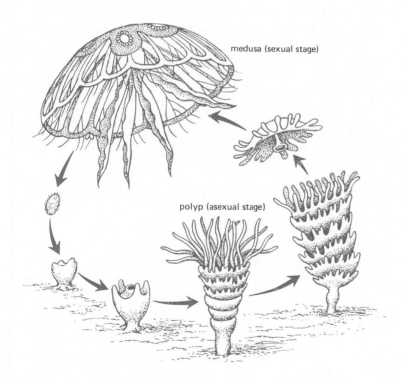

medusa (sexual stage)

polyp (asexual stage)

FIGURE 3-35
Life cycle of *Aurelia*.

OTHER COELENTERATES   The immense coral reefs of tropical seas are feats of architecture achieved by millions of tiny flowerlike polyps, each of which is smaller than a centimeter long. Most coral polyps live in colonies and build skeletons of lime, which they extract from the sea. The skeletons of neighboring polyps are firmly cemented together. When one animal dies, its skeleton remains and serves for the attachment of another polyp.

Physalia, better known as the Portuguese man-of-war, is a purple coelenterate whose nematocysts are powerful enough to stun a human. This organism is actually a colony of polyps. A single large polyp, from which the others are suspended, acts as a float for the entire colony. Some of the polyps of the colony are specialized for stinging, others for digestion, and still others for producing reproductive cells. Although the Portuguese man-of-war lives mainly in tropical waters, it has been known to be swept up into the Gulf Stream, in which it may drift as far as the English coast.

Symmetry

At this point, it is worth taking time out to consider the shapes of organisms. The form of an organism is referred to as its **symmetry.** The amoeba, for example, has no definite shape, and is said to be **asymmetrical.** It can move in any direction with its pseudopods. Certain other protists, such as radiolarians, possess **spherical symmetry.** Like a basketball they can be divided into two equal parts by any plane passing through the diameter of their bodies (Figure 3-36a). Organisms with spherical symmetry often lack an efficient method of locomotion and usually float on or near the surface of water.

For most organisms, however, gravitational orientation plays a role. Consider, for example, the sea anemone. This coelenterate, which is similar in structure to a hydra, has tentacles at one end that are well suited for reaching around in its environment. The sea anemone has **radial symmetry** (Figure 3-36b). It can be divided into two equal parts by any plane that passes through the central axis. Some sponges and most coelenterates are radially symmetrical.

Actively moving organisms are better adapted for their way of life by having **bilateral symmetry,** which means, literally, "two-sided shape." Only one plane can separate animals with this kind of symmetry into two similar parts. These parts are not identical; however, they are mirror images of each other. Animals with bilateral symmetry have a definite right and left side. They have an upper, or **dorsal,** surface and a lower, or **ventral** surface. They also have a definite front, or **anterior,** end and a hind, or **posterior,** end. All the vertebrates and many invertebrates have this type of symmetry. See Figure 3-36c.

Many organisms with bilateral symmetry have a concentration of nerves and sense organs at the anterior end. Thus, as they move forward, they can sense and readily react to their environment. In general, animals for which a sense of orientation is important are bilaterally symmetrical.

FIGURE 3-36
Three types of symmetry.

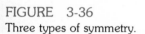

a. spherical

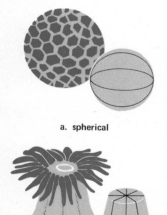

b. radial

c. bilateral

The flatworms

Of the three phyla of worms, the least complex are of the phylum **Platyhelminthes,** the flatworms. These organisms have three layers of cells; the ectoderm and endoderm, as in the coelenterates, and a middle layer called the **mesoderm** (Figure 3-37). The tissues of the organisms with two cell layers are formed from the ectoderm and endoderm. Organs and organ systems are found in animals containing all three cell layers. This is true in all the other animals to be considered, including humans.

The Planarian   The most common examples of the free-living (non-parasitic) flatworms, are the **planarians.** Planarians are aquatic, and are found under stones in fresh-water ponds and streams. Most planarians are between six and twelve millimeters in length, but certain tropical species reach 60 centimeters. Planarians are bilaterally symmetrical, blunt at the anterior end, and pointed at the posterior end (Figure 3-38). Two *eyespots* on the anterior end give the planarian its nickname, "the cross-eyed worm." The eyespots are **photosensitive;** that is, light striking them stimulates nerves in this area, and the animal, which usually avoids bright light, responds accordingly.

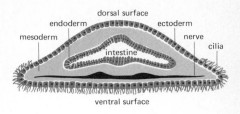

FIGURE 3-37
Cross-section of a flatworm showing the relationship of the three layers of cells.

FIGURE 3-38
The digestive and nervous systems of a planarian. The pigmented eyespots can distinguish light and dark.

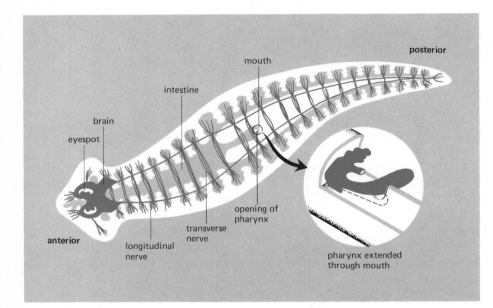

The **pharynx** is a tube located on the ventral surface of the planarian. When extended, this tube sucks up microscopic particles, including tiny organisms and organic debris. The planarian is a scavenger. When food is drawn into the digestive cavity, it enters any of the three main branches of the intestine and then passes into one of the side branches, where it is digested. Indigestible materials are eliminated through the pharynx. The planarian lacks a circulatory system; its cellular wastes are collected by tubules that branch throughout the body. This system has several tiny excretory pores that open on the surface of the worm's body.

Compared to the animals studied so far, the nervous system of the planarian is well developed. A mass of nerve tissue, the "brain," lies just beneath the eyespots. Two **longitudinal nerves** extend from the brain along the sides of a planarian's body, near the ventral surface. These nerve cords are connected by **transverse nerves,** giving the nervous system a ladderlike appearance. Many small nerves extend from the longitudinal nerves to the surface of the body. Its nervous system gives the planarian coordination of movement and permits it to respond to stimuli on all parts of its body.

Planarians may reproduce either asexually, by fission, or sexually. Each individual is **hermaphroditic;** it possesses both male and female reproductive organs. Cross-fertilization occurs, and the zygotes are shed in capsules. In two to three weeks, the eggs hatch, and tiny planarians emerge. Like sponges and coelenterates, planarians are capable of regenerating lost parts (Figure 3-39).

The Tapeworm—a Parasitic Flatworm   In the course of evolutionary history, certain flatworms have become specialized for a parasitic way of life. Among

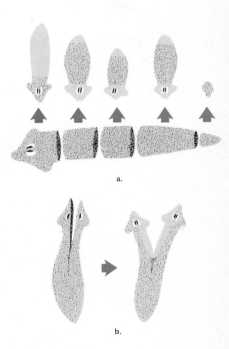

FIGURE 3-39
Regeneration by planarians. (a) All but the last part of each section will form a new head. (b) This animal, which has been cut along its plane of symmetry, has formed two heads. It will complete the separation and become two planarians.

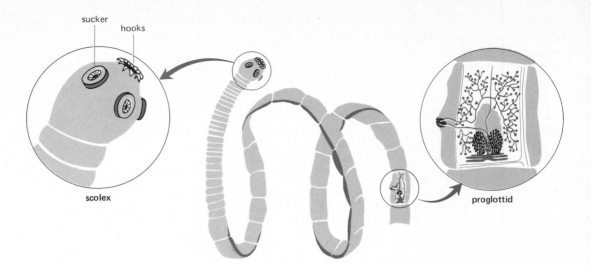

sucker
hooks
scolex
proglottid

FIGURE 3-40

The structure of a tapeworm. The tapeworm infects humans via insufficiently cooked meat that is infested with scolex-containing cysts.

FIGURE 3-41

Ascaris, a parasitic roundworm. Side views of a male and female are shown at the right. The dissected worm at the left is a female.

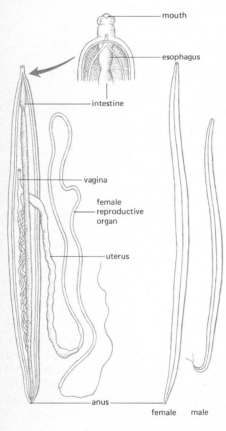

mouth

esophagus

intestine

vagina

female reproductive organ

uterus

anus

female   male

these is the tapeworm. An adult tapeworm has a flat, ribbonlike body (Figure 3-40). The knob-shaped "head," called a **scolex,** is equipped with suckers and in certain species with a ring of hooks. It lacks a mouth and has no digestive structures. Clinging to the inside of its host's intestine, a tapeworm absorbs the already digested food, in which it is bathed, through its body wall.

Below the slender neck, a number of sections, known as **proglottids,** extend to as great a length as thirty feet. New proglottids are formed at the anterior end; the oldest ones are at the posterior end. The proglottids are essentially masses of reproductive organs. Tapeworms are hermaphroditic, and eggs formed in a proglottid are fertilized there. When the eggs mature, the proglottids break off and are eliminated in the host's feces. Proglottids released in this way may be eaten by domesticated animals, such as pigs and cows. In the body of the new host, the eggs hatch into larvae that burrow into the muscles and form cysts.

Tapeworms enter the human body in the cyst stage if the insufficiently cooked flesh of an infected animal is eaten. Each cyst contains a fully developed tapeworm scolex. In the human intestine, the scolex is released from the cyst, attaches itself to the intestinal wall, and begins to grow. Since a tapeworm robs its host of nourishment, the victim may lose weight and vitality. In recent years, human infestations by tapeworms have been declining because of improved detection and treatment methods and because meat is inspected for tapeworm cysts.

The nematodes

Members of the phylum **Nematoda** are slender, smooth worms, tapered at both ends. They may be as short as one-fourth of a millimeter to over one meter. They occur in soil, fresh water, salt water, and as parasites in plants and animals. Among the parasitic nematodes are the hookworm, pinworm, and whipworm. Perhaps as many as one-third of the human population, mostly in warm regions of the world, is infected with parasitic nematodes. Harmless nematodes include those that live in vinegar and those living in droplets of water on moss. Certain species that live in soil are beneficial, while others cause serious crop damage.

Like flatworms, nematodes are bilaterally symmetrical and have three cell layers. They are, however, more complex than the flatworms. Their digestive system is a type with an opening at each end (Figure 3-41). This arrangement enables a nematode to take food in through an anterior mouth and to eliminate wastes through a posterior **anus.** Thus, digestion occurs in an orderly progression through the digestive tract—an evolutionary advance over the systems of the animals studied so far which take in food and eliminate wastes through the same orifice.

The segmented worms

The segmented worms, phylum **Annelida** are the most evolutionarily advanced of the worms. Most of the segmented worms live in salt water; some live in fresh water; and others, including the common earthworm, live in soil.

If you examine an earthworm, you will notice that its body consists of many segments (Figure 3-42). You will also see that the anterior end is darker and more pointed than the posterior end. There is no separate head, nor are there any visible sense organs. Like roundworms, earthworms have an anterior mouth, a posterior anus, and a digestive tract that runs the length of the body.

Four pairs of bristles, or **setae,** project from the undersurface and sides of each segment, except the first and the last. The setae assist the earthworm in moving and in clinging to the walls of its burrow. The earthworm moves by pushing the setae of its anterior segments into the soil, then shortening its body by using a series of **longitudinal muscles** that stretch from the anterior to the

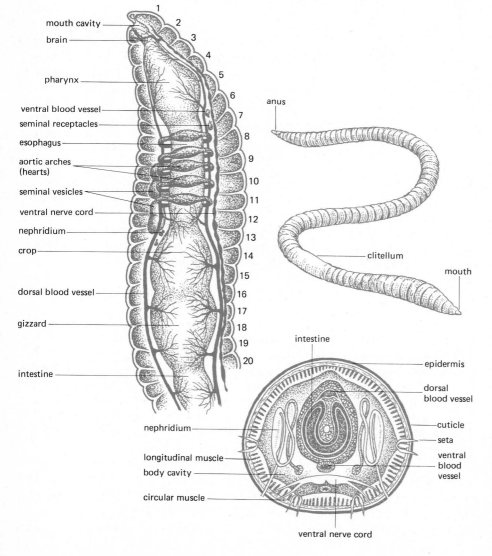

FIGURE 3-42
The external and internal structure of the earthworm. The anterior portion above is dissected to show the well-developed nervous and circulatory systems. Note that the segments are numbered beginning at the anterior end. One of the abdominal segments is shown in cross section.

posterior end. The worm then pushes the setae of its posterior segments into the soil, withdraws its anterior setae, and pushes forward through the soil by making itself longer. It does this by constricting the **circular muscles,** which surround its body at each segment.

The earthworm possesses several complex organ systems. Among these are the digestive, circulatory, excretory, nervous, muscular, and reproductive systems.

The Digestive System of the Earthworm   An earthworm has no jaws or teeth, but it uses its muscular **pharynx** to suck in soil containing organic particles which can be digested, as well as inorganic components. The soil then passes through a long **esophagus** into a round organ called the **crop,** which acts as a temporary storage place. From the crop, nutrients are forced into a muscular organ called the **gizzard.** The rhythmical contractions of the gizzard cause grains of sand to rub the nutrient particles together, thus grinding them into smaller pieces. From the gizzard nutrients move into the intestine, where enzymes break them down chemically. Then they are absorbed by the blood circulating through the intestine walls. By processing soil in the preceding manner, earthworms loosen it and convert inorganic compounds into forms that are easily utilized by plants. Thus, earthworms are important agents in improving soil consistency and fertility.

The Circulatory System of the Earthworm   As food is digested, the circulatory system picks up the nutrients for distribution throughout the organism. In simpler animals, digested food has to diffuse only a short distance in order to reach all the cells of the body. But in higher forms, the distances are greater. **Blood** is the specialized tissue of nutrient transport in these more complex animals.

The blood of the earthworm moves through a series of closed tubes, or **vessels.** It flows forward to the anterior end in a **dorsal blood vessel** and moves to the posterior end in a **ventral blood vessel.** Small vessels connect dorsal and ventral vessels throughout the animal, except in segments 7–11. There the five pairs of connecting vessels are large and muscular. By means of alternate contraction and relaxation, they keep the blood flowing. Not true hearts, they are called **aortic arches.**

Respiration and Excretion in the Earthworm   The earthworm absorbs oxygen and gives off carbon dioxide through its thin skin. This skin is protected by a **cuticle** secreted by the epidermis and kept moist by a slimy mucus also produced by epidermal cells. A moist surface is necessary for oxygen to be absorbed and carbon dioxide to be given off. If the worm is dried by the sun, it will die because the exchange of gases can no longer take place.

Nitrogen-containing waste materials from cell activities are removed to the outside of the body by little tubes, called **nephridia.** There are two nephridia in each segment except the first three and the last one. Each corresponds to a kidney tubule in humans which assists in waste removal also.

Earthworm Sensitivity   The nervous system coordinates the movements of the animal and sends impulses received from sense organs to certain parts of the body. There is a small nerve center in the third segment. From it run two nerves that form a connecting collar around the pharynx. These nerves join to become a long **ventral nerve cord.** There are also enlarged nerve centers, called **ganglia,** in each segment. The earthworm has neither eyes nor ears, but certain cells in the skin are sensitive to light and sound. From them the impulses are carried rapidly to the muscles of the earthworm to initiate movement.

Reproduction by the Earthworm   Although earthworms are **hermaphroditic,** forming both eggs and sperms, the eggs of one worm can be fertilized only by sperms from another worm. **Seminal vesicles** store a worm's own sperm and **seminal receptacles** store sperms from another worm. As they lie side by side, earthworms transfer sperms to each other's seminal receptacles. Here they are stored until eggs are laid. When the eggs mature they pass from the ovaries and are deposited in a slime ring secreted by the clitellum, a conspicuous swelling on the body. As this ring moves forward, sperms are released from the seminal receptacles, and fertilization occurs. The slime ring slips from the body and becomes the cocoon in which the young worms develop.

The mollusks

Members of the phylum **Mollusca** (which include such animals as oysters, clams, snails, and squids) live in fresh water, as well as in marine and terrestrial environments. Some are adapted to live buried in sand or mud where the oxygen content may be too low for a more active animal. In abundance of species, the mollusks are surpassed only by the phylum that includes the insects.

Mollusks and their products have been used for food, money, eating utensils, jewelry, buttons, dyes, tools, and weapons since earliest times, and they have been of value in still another way. Shells, or their imprints in clay, may remain as a permanent record of the presence of a mollusk. Where layers of shells have accumulated in various strata of the earth, paleontologists have been able to utilize them as a tool for dating as well as for reconstructing changes that have taken place in the earth's surface. Aggregations of certain shells or their impressions are also used to help determine the possibility of finding oil in various regions.

Common Characteristics of the Mollusks   Although there are various forms of adult mollusks, all of them pass through an early stage of development, a **larval stage,** in which they are essentially identical. This young form of mollusk is called a **trochophore** (Figure 3-43). The trochophore has a tuft of cilia at one end and a ciliated band around its equator. A trochophore stage is also found in the development of annelids. It is because of this that biologists consider the segmented worms and the mollusks to have evolved from a common ancestor.

The body of an adult mollusk consists of a head, foot, and visceral hump (Figure 3-44). The visceral hump contains the digestive organs, excretory organs, and the heart. It is covered by a **mantle,** a thin membrane that in some species secretes the calcium carbonate shell. The **mantle cavity** is formed by the curtain-like mantle that hangs down over the sides and rear of the animal. The **gills,** which are the respiratory organs in the aquatic mollusks, are located in this cavity.

Mollusks With Hinged Shells   Such mollusks as the clams, oysters, scallops, and mussels are called **bivalves** because their shells are composed of two halves, called **valves** (Figure 3-45). A hinge connecting the valves allows for opening and closing. When a bivalve is threatened, two powerful muscles hold the valves tightly shut. Each valve is composed of three layers secreted by the mantle. The smooth, glistening layer next to the mantle is the **pearly layer.** If a grain of sand or an encysted parasitic worm becomes lodged in the mantle, this layer builds up around the particle and forms a pearl. The hard middle layer of a valve consists of calcium carbonate, and the outer layer is composed of a hornlike organic substance. It protects the middle layer from being dissolved by the small amounts of acid that may be in the water. The hinge that connects the valves is also composed of this hornlike secretion.

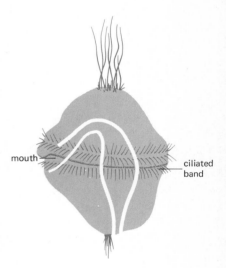

FIGURE   3-43
The trochophore larval form is found in the development of both mollusks and annelids. An outline of the digestive tract is shown in white.

FIGURE   3-44
Although no mollusk living today looks like this drawing, the major characteristics of the phylum are shown. The arrows indicate direction of water flow in the mantle cavity.

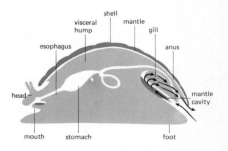

FIGURE   3-45
Two bivalve mollusks: a giant clam (left) and oyster (right).

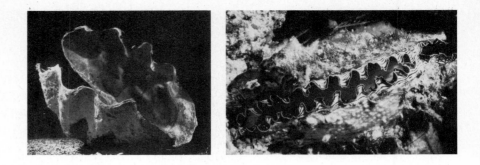

Mollusks With One Shell   Land and water snails, conches, and abalones have only one shell, and are called **Gastropods** or **univalves** (Figure 3-46). The familiar land snail has certain adaptations that enable it to live successfully out of water. Its mantle cavity serves as a modified lung. Oxygen diffuses through the thin, membranous lining of the cavity. Since this lining must remain moist for gas exchange to occur, snails are most active during the evening, night, and early morning hours.

The slug resembles a snail that has lost its shell. If you look closely at a slug, you will see the opening that leads to the mantle cavity, which is used in respiration. By controlling the size of this opening, the slug is able to conserve moisture in the mantle cavity. Like terrestrial snails, slugs are usually active at night, leaving trails of slime wherever they go.

FIGURE   3-46

Univalves are terrestrial, freshwater, and marine. Shown here are a snail (left) and a nudibranch (right).

"Head-Footed" Mollusks   The **Cephalopods** (literally, "head-footed") include the octopus, squid, cuttlefish, and chambered nautilus (Figure 3-47). The octopus has no shell, the nautilus has an external shell, and the squid and cuttlefish have internal shells. The giant squid is probably the largest of all invertebrates, sometimes reaching a length of 18 meters and a weight of 1800 kilograms.

The cephalopods have well-developed nervous systems and the eyes of the octopus and squid are remarkably similar to vertebrate eyes—perhpas nature's best example of convergent evolution.

FIGURE   3-47

The octopus (left), a mollusk with well-developed eyes, moves by pulling itself over rocks by its tentacles or by expelling a jet of water from its excurrent siphon. The squid (right) also moves by expelling a jet of water.

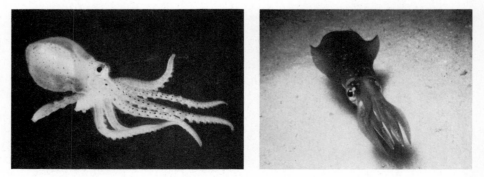

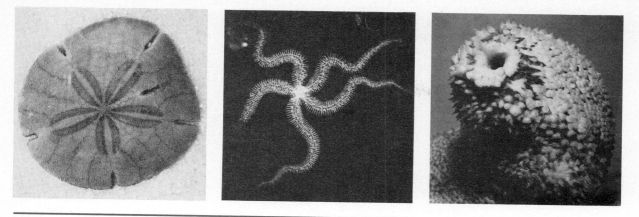

The echinoderms

The starfish, brittle star, sea urchin, and sand dollar are common members of the phylum **Echinodermata** (Figure 3-48). They have hard, radially symmetrical bodies covered with spines. The spines may be long, as in the sea urchin, or very short, as in the sand dollar. All the echinoderms are restricted to the marine environment, but they range from the shallowest of tidepools to the great ocean depths.

The characteristics of the echinoderms can be observed in one of the most familiar species, the starfish (Figure 3-49). Starfishes typically have five rays, or arms, which radiate from a central disk. In a groove on the lower side of each ray are two rows of **tube feet.** These are part of a **water-vascular system.** The tube feet are connected to canals that lead through each ray to a circular canal in the central disk. This **ring canal** has an opening to the surface, the **sieve plate,** on the dorsal side. When the starfish presses its tube feet against an object and forces water out of its canals through the sieve plate, the feet grip the object firmly by means of suction. Return of water to the canals releases the grip.

The starfish uses its water-vascular system to open the shells of clams and oysters, which are its principal foods. The body arches over the prey with the rays bent downward. The valves of the clam or oyster are gripped firmly by the tube feet, and at the same time a steady pull is exerted. The starfish then pushes its

FIGURE 3-48
Members of the phylum *Echinodermata.* From left to right: a sand dollar, brittle starfish and sea cucumber.

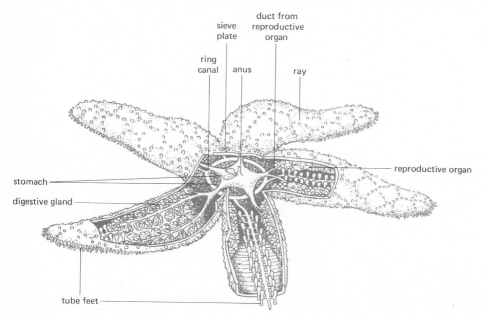

FIGURE 3-49
This dorsal view of a dissected starfish shows the location of its internal organs.

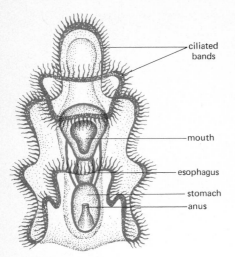

FIGURE 3-50

Free-swimming larva of an echinoderm. Called a bipinnaria, it is similar to the larva of some of the lower chordates. Compare it with the trochophore larval type, characteristic of annelids and mollusks, shown in Figure 3-43.

stomach out from a small opening in the center of the lower side. The stomach, turned inside out, squeezes through openings as small as a tenth of a millimeter between the mollusk's valves. Enzymes secreted by the stomach digest the tissues of the mollusk, and the absorbed nutrients pass from the stomach into digestive glands within the rays of the starfish, where further chemical breakdown occurs. After feeding, the starfish retracts its stomach and moves away, leaving behind the empty valves of its prey.

Starfish are either male or female. Small ducts lead from the reproductive organs to the outside. During the reproductive season, eggs and sperms are shed to the outside, and the fertilization which forms the zygote takes place in the water. The zygotes develop into free-swimming **bipinnaria larvae** (Figure 3-50). After swimming about for a period of time, the larvae settle to the bottom and gradually change into adults.

A phylogenetic tree

Before proceeding with the largest animal phylum, **Arthropoda,** and the phylum to which we belong, **Chordata,** we will consider some of the characteristics of organisms that have already been described; this will shed light on the course of evolution. Figure 3-51 is a diagrammatic representation of the way in which a taxonomist might fit the available bits of evidence together to explain the evolution

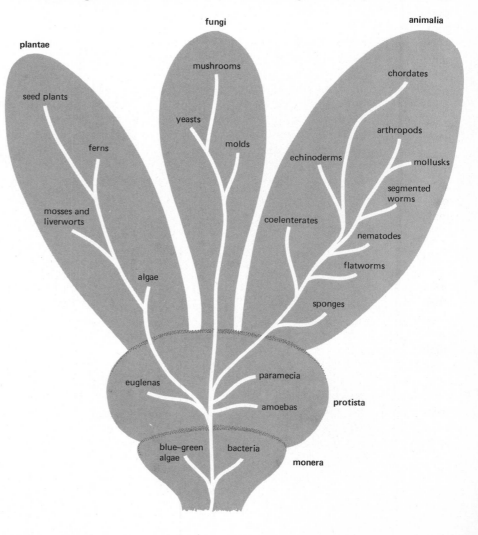

FIGURE 3-51

A phylogenetic tree based on five kingdoms. This figure should not be interpreted as indicating that a given group of organisms evolved from the group beneath it. Rather, two such groups probably had a common ancestor, with the lower group retaining a more primitive form.

78

of diversity in the living world. It is called a **phylogenetic tree.** There are many such schemes which are used to point out relationships based on incomplete fossil records and similarities of existing species (including both larval and adult forms). The phylogenetic tree represents a hypothesis of origins and relationships of various groups of organisms. It is based on the assumption that the more closely two groups resemble each other, the closer their relationships, and the more recent their common ancestry.

The trochophore larvae of the segmented worms and mollusks, for example, are similar. Thus, it has been hypothesized that they have a relatively common ancestor. Since the bipinnaria larvae of the echinoderms are quite different from trochophore larvae, echinoderms are believed to be more distantly related. However, similarities between bipinnaria larvae and the larvae of some of the lower chordates indicate a relatively close evolutionary relationship between the echinoderms and the chordates. As indicated in Figure 3-51, animal evolution is thought to have developed in two branches, with the segmented worms and mollusks developing along one line, and the echinoderms and chordates developing along the other.

In addition to the characteristics of larval forms, evolutionists also consider the structures of adult organisms. Recall from our discussion of the coelenterates that these simple organisms consist of two cell layers, the ectoderm and endoderm. Nutrients are merely absorbed by the cells of the endoderm. Flatworms have a somewhat more complex body structure. In addition to ectoderm and endoderm, they have a third cell layer, the mesoderm, in which their digestive organs are embedded.

In small organisms this form of body structure works quite well. But, in larger organisms, the digestive tract is suspended in a fluid-filled cavity, or **coelom,** which allows for looped intestines and aids in circulation of nutrients and waste removal. The coelom forms within the mesoderm of a complex organism and has a lining of specialized covering cells. This lining surrounds the internal organs like a pouch and covers the inner surface of the body wall. It forms a membrane-like structure, the **mesentery,** that suspends the intestine within the coelom.

Between the animals without a coelom and those which have one, are the animals possessing a **pseudocoel,** or "false coelom." This cavity is not surrounded by layers of muscle cells, and has no mesentery; the internal organs float freely within a pseudocoel. Among the animals possessing a pseudocoel are the nematodes.

The arthropods

From the standpoint of numbers, the arthropods are a most successful group of animals; over three quarters of the identified animals are arthropods, and of these, the vast majority belong to a single class, the insects. In addition to the insects, phylum Arthropoda includes a variety of other types of animals, including spiders, centipedes, millipedes, and crayfish.

To the casual observer, an air-borne butterfly has little in common with a crayfish lurking under a rock in a stream. But, careful study of these apparently different animals reveals that they have much in common. The structural characteristics that indicate that the butterfly is related to the crayfish also indicate its relationships to spiders, scorpions, and centipedes. All arthropods (which means "jointed feet") are similar in having the following characteristics:

- jointed appendages, which include legs and other body outgrowths
- a hard external skeleton, or **exoskeleton,** composed of a substance called **chitin**

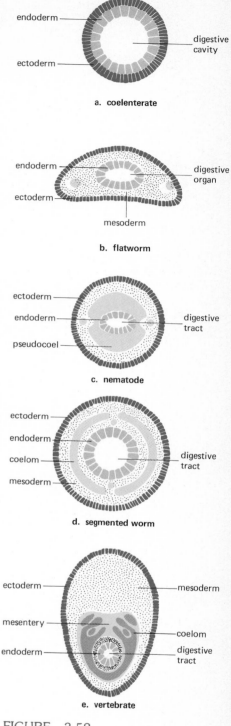

FIGURE 3-52
General body plans of organisms of five animal phyla.

- a segmented body; that is, a body with distinct divisions of the exoskeleton

The fact that the arthropod body is divided into segments indicates an evolutionary relationship to the segmented worms. But, the arthropods have been more successful than the segmented worms. The increased complexity of their sense organs gives them a greater degree of awareness of their environments. The grouped muscles and jointed legs permit more coordinated movements in search for food and escape from enemies.

The hard exoskeleton of the arthropods provides a protective advantage over the soft cuticle of the worm. It may seem strange to find the skeleton on the outside of an animal's body. But whether skeletons are internal or external, they serve the same function. They give the body form, protect delicate organs, and aid in motion by serving as attachments for muscles. Unlike an internal skeleton, however, an exoskeleton limits the size of an animal. A large exoskeleton with the powerful internal muscles required to move it would crush the animal under its own weight. Flying insects, for example, could never reach the size of birds, because their exoskeletons would be too heavy to be supported by wings. Growth of an organism with an exoskeleton can only take place by **molting,** whereby the old skeleton is shed and a new one is formed.

The arthropod phylum is commonly divided into five classes, according to the criteria outlined in Table 3-2. Members of each of these classes have all the fundamental characteristics of arthropods and, in addition, certain characteristics unique to the class to which they belong. For example, the **Crustacea** have two pairs of **antennae** (or "feelers") on the front of the body, two distinct body regions, five pairs of legs, and an exoskeleton which contains calcium carbonate in addition to chitin. Most have featherlike **gills** for respiration. Examples of the five more common classes of Arthropoda are illustrated in Figure 3-53.

General Characteristics of Insects   In a discussion of life's forms as brief as this must be, it would be impossible to consider each of the five classes of arthropods in detail. Therefore, we have singled out the insects for close examination.

TABLE 3-2   CLASSES OF ARTHROPODS

CLASS	BODY DIVISIONS	APPENDAGES	RESPIRATION	EXAMPLES
Crustacea	2-cephalothorax, abdomen	5 pairs of legs	gills	lobster, crab, water flea, sow bug, crayfish
Chilopoda	head and numerous body segments	1 pair of legs on each segment except first one behind head and last 2	tracheae	centipede
Diplopoda	head and numerous body segments	2 pairs of legs on each body segment	tracheae	millipede
Arachnida	2-cephalothorax, abdomen	4 pairs of legs	tracheae and air sacs	spider, mite, tick, scorpion
Insecta	3-head, thorax, abdomen	3 pairs of legs; usually 1 or 2 pairs of wings	tracheae	grasshopper, butterfly, bee, dragonfly, moth, beetle

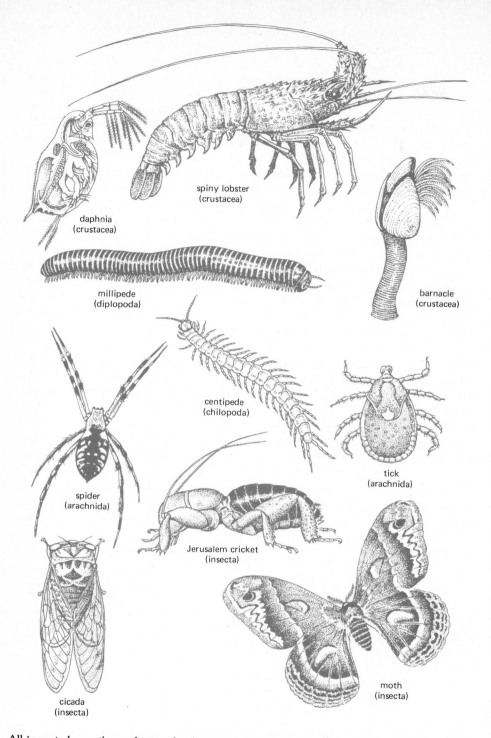

FIGURE 3-53
Various examples of the five classes of arthropods.

spiny lobster
(crustacea)

daphnia
(crustacea)

barnacle
(crustacea)

millipede
(diplopoda)

centipede
(chilopoda)

tick
(arachnida)

spider
(arachnida)

Jerusalem cricket
(insecta)

moth
(insecta)

cicada
(insecta)

All insects have three distinct body regions: a **head,** a **thorax,** and an **abdomen.** The head bears one pair of antennae and the mouthparts. The thorax bears three pairs of legs, and, if present, the wings. The abdomen has as many as eleven segments and never bears legs. The reproductive structures are usually in the eighth, ninth, and tenth segments. Respiration in insects occurs by means of branched air tubes called **tracheae.**

Why Are Insects So Successful?   It is estimated that insects have lived on the earth nearly 300 million years. During this time many **environmental changes**

FIGURE 3-54

This tree hopper resembles a leaf on a bush.

FIGURE 3-55

Although this fly looks like a honeybee, it does not sting.

FIGURE 3-56

Incomplete metamorphosis in the Harlequin cabbage bug. Notice that the newly-hatched young are very similar to the adult.

have occurred, and insects seem to have been the quickest of all the animals to make adaptive changes. They have managed to populate some of the harshest of the earth's environments. In the Antarctic, certain species have been collected where the winter temperatures reach $-65°C$; others live in hot springs where the temperature is $+50°C$. Some insects even thrive in crude oil pools around oil wells.

Another manifestation of the insects' potential for adaptability is their tendency to evolve forms which make them appear to be something they are not. Some species blend into their environments (Figure 3-54). Others may be harmless but have the appearance of species that can sting or otherwise defend themselves; they are avoided by would-be predators wary of the harmful species (Figure 3-55).

Also of survival value to the insects is their ability to utilize a wide variety of foods. This is made possible, in part, by specialization of their mouths. Some insects have chewing mouths with strong jaws to grind up leaves. Other have piercing and sucking mouths with which they can suck plant juices or, as do the mosquitoes, blood from animals. Still others have siphoning tubes for probing flowers to obtain nectar. Smallness is another virtue.

The success of the insects is also due to behavior more complex than that of lower invertebrates. Because they possess a relatively complex brain and well-developed sense organs, they are able to receive and interpret stimuli which aid them in escaping from enemies and finding food. Their jointed appendages permit them quick muscular responses.

With the development of complex sensory and muscular systems, insects have become able to perform coordinated patterns of behavior. Among other things they are capable of building various kinds of dwellings. The paper wasp builds nests of chewed wood pulp; the bee builds a comb of wax; the tarantula hawk, a large wasp, digs a burrow.

Their most interesting behavioral characteristic is the insects' tendency to act in social units. Colonies of ants and bees, for example, in which the workers are identical descendants of a single queen, function as highly cooperative units. As such they are able to defend themselves against powerful enemies (including humans) and maintain sophisticated life sytles. A species of African termites, for example, builds massive mounds of compacted earth in which they live and cultivate fungus as food. Thus, what is considered one of the most significant developments in human culture, agriculture, has also been mastered by insects.

Insect Metamorphosis   Most insects undergo several distinct stages during development from egg to adult. Such a series of stages in a life history is called **metamorphosis,** and there are two basic types of metamorphosis in insects. Grasshoppers, aphids, termites, and certain other insects undergo **incomplete metamorphosis.** This is a three-stage life history involving an egg, a nymph, and an adult. A **nymph** hatches from an egg in a form that resembles the adult except for size, absence of wings, and lack of development of reproductive organs. In most species, insects molt five times, each time becoming more like the adult (Figure 3-56).

Butterflies, moths, flies, and beetles are among the insects with **complete metamorphosis.** They pass through four stages in their development: egg, larva, pupa, and adult. The **larvae** that hatch from eggs are segmented and wormlike. Depending on the kind of insect, they are known as caterpillars, grubs, maggots, or wigglers. After a period of rapid growth, larva enters a **pupal** stage. This has been described as a resting phase because the changes occur within a case or cocoon, and cannot be seen externally. This is anything but a resting phase, however, because during it the tissues of the pupa are transformed into those of the adult (Figure 3-57).

FIGURE 3-57
Life cycle of the Monarch butterfly.

Respiration by the Grasshopper To illustrate some higher physiological processes common to insects, we will consider one common species, the grasshopper (Figure 3-58). As previously mentioned, an insect consists of three major body regions, a head, thorax, and an abdomen. It is the segmented abdomen that plays the major role in respiration (Figure 3-59). Each of the ten segments of the grasshopper's abdomen consists of two curved plates. The upper and lower plates are joined by a tough but flexible membrane that allows the segment to expand and contract in the process of respiration. The flexible membrane also joins each segment with the segments anterior and posterior to it, thereby allowing the segments to move.

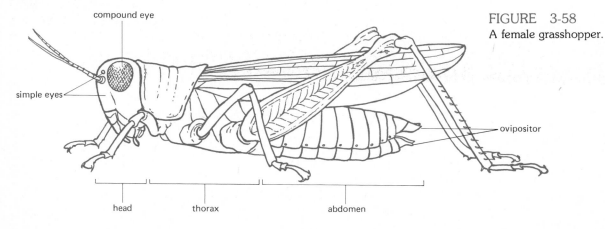

FIGURE 3-58
A female grasshopper.

FIGURE   3-59
Respiratory apparatus of a grasshopper.

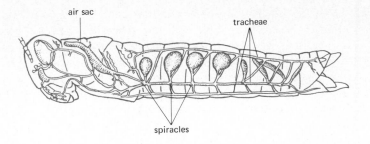

The first eight of the abdominal segments have pairs of tiny openings, the **spiracles,** which are also found on the second and third thoracic segments. The spiracles lead to the **tracheae,** or air tubes, which form a complex network inside the abdomen. Air is pumped in and out of the trachea by action of the wings and movement of the abdomen. Oxygen enters the tissues and carbon dioxide leaves the tissues through membranous **air sacs.**

Digestion, Excretion, and Circulation   With its chewing mouth, the grasshopper pinches off bits of grass, which are then sucked into the mouth. An **esophagus** carries the food to a **crop,** where it may be stored for a time (Figure 3-60). As in many animals, **salivary glands** secrete juices that enter the mouth. These

FIGURE   3-60
The internal structure of a female grasshopper.

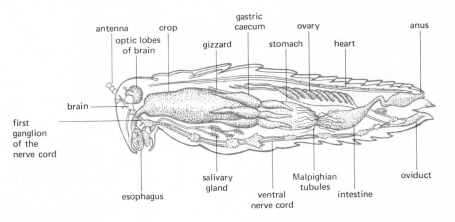

juices moisten the food to ease its passage to the crop. When a frightened grasshopper ''spits tobacco'' it is regurgitating food from its crop. Normally the food passes on to the **gizzard,** where it is shredded by plates of chitin-bearing teeth.

Partially digested food is screened through thin plates and passes into the large **stomach.** There are several double pouches on the outside of the stomach. These **gastric caeca** produce and pour enzymes into the stomach, where digestion is completed. Digested food is absorbed into the blood through the wall of the stomach. The material remaining in the stomach passes into the **intestine,** which terminates at the anus. Cellular wastes picked up by the blood are collected by a series of tubes, **Malpighian tubules,** which channel them back for another pass through the intestine.

The circulatory system of the grasshopper is an open system. Blood forced out the anterior end of the tubular heart passes through a larger blood vessel into the body cavity near the head. As the blood flows toward the posterior region, it gives up nutrients to, and takes on wastes from, the body organs. Eventually it returns to the heart.

Sensory Responses   Each side of the first abdominal segment of the grasshopper bears a membrane-covered cavity called the **tympanum,** which functions in hearing. Touch and smell are perceived by the many-jointed **antennae.** The grasshopper has two kinds of eyes (Figure 3-61). **Simple eyes** are located at the base of each antenna and in the groove between them. The large **compound eyes** project from a part of the front and sides of the head and are composed of hundreds of six-sided lenses. The shape, location, and number of lenses seem to adapt the insect for sight in several directions at one time, but the image formed is probably not very sharp. Most insects are thought to be nearsighted, yet some are able to distinguish colors.

The stimuli received through the sense organs are relayed to nerve centers called **ganglia,** which act as switches in directing the message to the proper structures controlling responses. The brain itself is composed of several fused ganglia, with nerves to the eyes, antennae, and other head organs, as well as the **ventral ganglia** of the thorax and abdomen. The **optic lobes** are the most prominent part of the brain. The image of an intruder approaching from any angle is enough to trigger a leaping push of the large posterior legs.

Reproduction   In insects the sexes are separate. Sperm cells are produced in the male's **testes,** and eggs are produced by the female's **ovaries.** The male deposits sperm cells in a storage pouch, the **seminal receptacle** of the female during mating. The extreme posterior segments of the female bear two pairs of hard, sharp-pointed projections, the **ovipositors.** With these the female digs holes in the ground to deposit her eggs.

FIGURE 3-61
The simple and compound eyes of a grasshopper. The compound eyes are made up of hundreds of six-sided lenses (inset). (Photo by Walter Dawn.)

The vertebrates

Phylum **Chordata** includes the most complex organisms of the earth, with our group, **Vertebrata,** being a major subphylum. Three unique characteristics distinguish chordates from all other animals.

1   The embryos of all chordates have a rod of stiff **connective tissue** running lengthwise along the dorsal side of the body. This is the **notochord.** The more primitive chordates retain the notochord throughout life (Figure 3-62). In other vertebrates, the notochord is present in the embryo, but is replaced early in life by the **vertebral column,** or backbone.

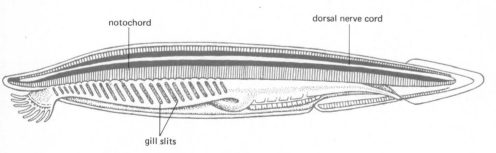

notochord   dorsal nerve cord

gill slits

FIGURE 3-62
The fishlike *Amphioxus* lives in tropical and temperate coastal waters. It retains its dorsal nerve cord, notochord, and gill slits throughout life. The notochord and gill slits disappear early in the development of the higher chordates.

2   All chordates have a tubular **spinal cord** that lies just above the notochord and extends from the brain. Together the brain and spinal chord compose the **central nervous system.**

3   At some time in life, all chordates have paired **gill slits,** which, for some species, transform to openings in the adult throat. These, like the notochord, disappear early in the development of higher vertebrates, including the reptiles, birds, and mammals. However, they develop into organs of respiration in the sharks and fishes.

Classes of Vertebrates   Fossil records indicate that many vertebrate forms have appeared and disappeared in successive geological periods.

Modern vertebrates are classified in seven classes (Figure 3-63). Listed in the order in which they are believed to have evolved (the order of increasing structural complexity) they are:

- **Cyclostomata,** the jawless fishes, including the lampreys and hagfishes.
- **Chondrichthyes,** the sharks, rays and skates.

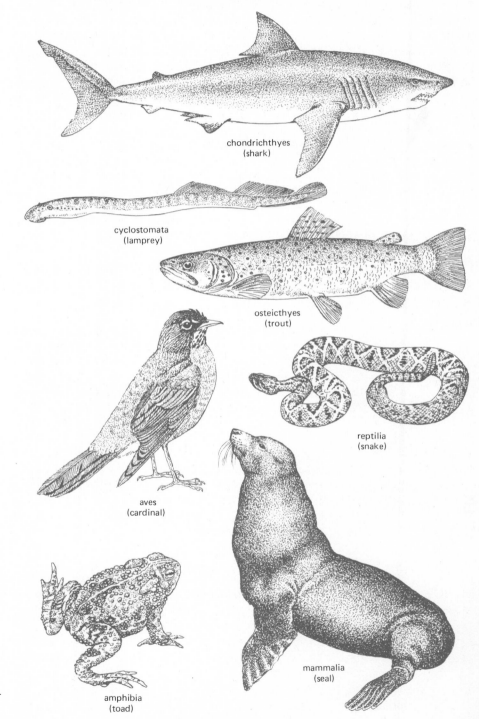

chondrichthyes
(shark)

cyclostomata
(lamprey)

osteicthyes
(trout)

reptilia
(snake)

aves
(cardinal)

mammalia
(seal)

amphibia
(toad)

FIGURE 3-63
Examples of the various classes of vertebrates.

- **Osteichthyes,** the bony fishes.
- **Amphibia,** the frogs, toads, and salamanders.
- **Reptilia,** the snakes, lizards, turtles, and crocodiles.
- **Aves,** the birds.
- **Mammalia,** the mammals.

During the evolutionary transition from life in water to life on land the vertebrates developed increasingly complex organ systems.

Nevertheless, the relatively primitive aquatic and marine species have retained predominance in their environments. Fish are abundant in nearly all waters of the earth. Amphibians and reptiles, while greatly reduced in numbers from earlier geological periods, still inhabit both aquatic and land environments. Birds are unchallenged in the air, and mammals are the most highly evolved land animals.

Characteristics of Vertebrates   Various structural characteristics distinguish vertebrates from other forms of animal life. They possess all of the chordate characteristics as well as the following features:

- An **endoskeleton** (internal framework) composed of the hard connective tissue, **bone,** and the somewhat flexible connective tissue, **cartilage** (Figure 3-64) A backbone consisting of cartilaginous and bony segments surrounds the vertebrate spinal cord.
- A body consisting of a head and trunk and, in many vertebrates, a neck and tail region.
- A closed circulatory system that includes a heart located in the anterior region of the body.
- Refined sense organs, eyes, ears, and nostrils on the head.

There are, of course, many aspects of vertebrate life which we have not touched upon, but these are covered elsewhere in the book, under a variety of topics.

FIGURE 3-64
The human endoskeleton.

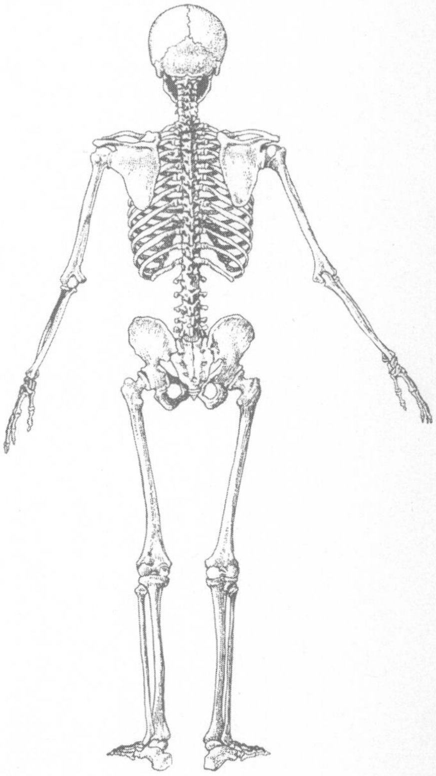

CHAPTER 3: SUMMARY

1 Biologists use agreed upon names and procedures for naming organisms to facilitate their study.

2 The most used scheme is classifying organisms is the five kingdom one: monera, protista, fungi, animals, plants.

3 The monera include the bacteria and blue green algae.

4 The protista include the protozoa and some algae.

5 The fungi include yeast and molds.

6 The plants include some algae and the mosses, liverworts, ferns, and seed plants.

7 The animals include the sponges, coelenterates, the worms, mollusks, echinoderms, arthropods and chordates.

8 Each of these kingdoms deals with its environment in its own structural way, but molecularly they are quite similar.

Suggested Readings

Bold, Harold C.   *The Plant Kingdom,* 4th Ed., Prentice-Hall, Inc., Englewood Cliffs, New Jersey, 1976 — Structure, physiology, and reproduction in the plant kingdom.

Christensen, Clyde M.   *Molds and Man,* 3rd Ed., University of Minnesota Press, Minneapolis, 1965 — A general account of fungi and their impact on humankind.

Farb, Peter and the Editors of *Life   The Insects,* Time-Life Books, New York, 1962 — A beautifully illustrated book on the natural history of insects.

Grzimek, H. C. Bernard   *Animal Life Encyclopedia,* Van Nostrand Reinhold Co., New York, 1974 — A 13-volume work for the advanced amateur covering all the animals from protozoa on up.

Hanson, Earl D.   *Animal Diversity,* 3rd Ed., Prentice-Hall, Inc., Englewood Cliffs, New Jersey, 1972 — Evolution and animal diversity.

Van Gelder, Richard G.   *Biology of Mammals,* Charles Scribner's Sons, New York, 1969 — How mammals are born and develop; how they find shelter, food, water, and how they defend themselves.

Zahl, Paul A.   "Where Would We Be Without Algae," *National Geographic,* March, 1974 — The importance of algae in the biosphere.

OBSERVATIONS ON THE THREE KINGDOMS OF NATURE
Carolus Linnaeus

Carolus Linnaeus' *Systema Naturae* (published in 1735) preceded the theory of evolution by more than 100 years; he devised his system of classification with the assumption that life forms were static. Nevertheless, the Linnaean system of classification, based on physical similarities between species, remains essentially the system that biologists still use.

Linnaeus believed himself to be divinely inspired, claiming that God had permitted him "a look into his secret council chamber." The following excerpt from *Systema Naturae* summarizes his view of the natural world.

1. If we observe Gods works, it becomes more than sufficiently evident to everybody, that each living thing is propagated from an egg and that every egg produces an offspring closely resembling the parent. Hence no new species are produced nowadays.

2. Individuals multiply by generation. Hence at present the number of individuals in each species is greater than it was at first.

3. If we count backwards this multiplication of individuals in each species, in the same way as we have multiplied forward (2), the series ends up in one single *parent*, whether that parent consists of *one single* hermaphrodite (as commonly in plants) or of a double, viz. a male and a female, (as in most animals).

4. As there are no new species (1); as like always gives birth to like (2); as one in each species was at the beginning of the progeny (3), it is necessary to attribute this progenitorial unity to some Omnipotent and Omniscient Being, namely *God*, whose work is called *Creation*. This is confirmed by the mechanism, the laws, principles, constitutions and sensations in every living individual.

5. Individuals thus procreated, lack in their prime and tender age absolutely all knowledge, and are forced to learn everything by means of their external senses. By *touch* they first of all learn the consistency of objects; by *taste* the fluid particles; by *smell* the volatile ones; by *hearing* the vibration of remote bodies; and finally by *sight* the shape of visible bodies, which last sense, more than any of the others, gives the animals greatest delight.

6. If we observe the universe, three objects are conspicuous: viz. α. the very remote *coelestial* bodies; β. the *elements* to be met anywhere; γ. the solid *natural bodies*.

7. On our earth, only two of the three mentioned above (6) are obvious; i.e. the *elements* constituting it; and the *natural* bodies constructed out of the elements, though in a way inexplicable except by creation and by the laws of procreation.

8. Natural objects (7) belong more to the field of the senses (5) than all the others (6) and are obvious to our senses anywhere. Thus I wonder why the Creator put man, who is thus provided with senses (5) and intellect, on the earth globe, where nothing met his senses but natural objects, constructed by means of such an admirable and amazing mechanism.
   Surely for no other reason than that the observer of the wonderful work might admire and praise its Maker.

9. All that is useful to man originates from these natural objects; hence the industry of mining or metallurgy; plant-industry or agriculture and horticulture; animal husbandry, hunting and fishing.
   In one word, it is the foundation of every industry of building, commerce, food supply, medicine etc. By them people are kept in a healthy state, protected against illness and recover from disease, so that their se-

From the English translation of C. Linnaeus, "Observations on the Three Kingdoms of Nature," in *Systema Natura* (1735), translated by M.S.J. Engel-Ledeboer and H. Engel, Nieuwkoop, Netherlands: de Graaf Publishers, 1964.

lection is highly necessary. Hence (8, 9) the necessity of natural science is self-evident.

10. The first step in wisdom is to know the things themselves; this notion consists in having a true idea of the objects; objects are distinguished and known by classifying them methodically and giving them appropriate names. Therefore, classification and name-giving will be the foundation of our science.

11. Those of our scientists, who cannot class the variations in the right species, the species in the natural genera, the genera in families, and yet constitute themselves doctors of this science, deceive others and themselves. For all those who really laid the foundation to natural science, have had to keep this in mind.

12. He may call himself a naturalist (a natural historian), who well distinguishes the parts of natural bodies by sight (5) and describes and names all these rightly in agreement with the threefold division. Such a man is a lithologist, a phytologist or a zoologist.

13. Natural science is that classification and that name-giving (10) of the natural bodies judiciously instituted by such a naturalist (12).

14. Natural bodies are divided into *three kingdoms of nature:* viz. the mineral, vegetable and animal kingdoms.

15. *Minerals* grow; *Plants* grow and live; *Animals* grow, live and have feeling. Thus the limits between these kingdoms are constituted.

16. In this science of describing and picturing many have laboured for a whole life-time; how much, however, has already been observed and how much there remains to be done, the curious on-looker will easily find out for himself.

17. I have shown here a general survey of the system of natural bodies so that the curious reader with the help of this as it were geographical table knows where to direct his journey in these vast kingdoms, for to add more descriptions, space, time and opportunity lacked.

18. A new method mainly based on my own authentic observations has been used in every single part, for I have well learnt that very few people are lightly to be trusted, as far as observations go.

19. If the Interested Reader should draw any profit from this, he should acknowledge that very famous Dutch Botanist Doctor *Joh. Fred. Gronovius,* as well as Mr. *Isaac Lawson,* the very learned Scotchman; as they were the ones who caused me to communicate these very brief tables and observations to the learned world.

20. If I find that this proves to be welcome to the illustrious and interested Reader, he may expect more, more special and more detailed (publications) from me soon, above all in botany.

Carolus Linnaeus
Doctor of Medicine

Given at Leyden, July 23, 1735.

THE LEAKEY LEGACY

Since the original discovery of hominid (humanlike) bones in the Olduvai Gorge, the Leakeys have periodically rewritten the evolutionary history of humankind.

One of the major preoccupations of physical anthropology is to prove that man is much older (and ostensibly wiser) than he readily admits. The race to push back our evolutionary clock is apparently still on in East Africa, where rival groups of scholarly manhunters, including the famous Louis S. B. Leakey family of Olduvai Gorge and independent bone-digger D. Carl Johanson, have each year come up with new and earlier dates to pinpoint the emergence of man. Now Mary Leakey, wife of the late Louis and a formidable anthropologist in her own right (though she generally remained in her husband's shadow until his death in 1972), has pulled the latest trump card with her recent announcement of the oldest and most firmly dated man discovered thus far—3.75 million years old, almost a million years older than any other man previously established by firm dating techniques.

Working at a volcanic ash site called Laeto-lil, about 25 miles south of Olduvai Gorge in Tanzania, East Africa, Leakey and her research team dug up the teeth and lower jaws of 11 hominid (manlike) individuals; which strongly resemble the 2.8-million-year-old, large-brained human skulls unearthed in 1972 by Richard Leakey (Mary's son) in Kenya. Both fossil finds are reportedly of the genus *Homo*, or true man, rather than early representations of near-man, the genus *Australopithecus*, previously thought to be the earliest ancestor to modern man.

In a recent press conference at the National Geographic Society (the major source of support for the recent excavations) in Washington, D.C., Mrs. Leakey also confirmed that the new fossils resemble the hominid bones found in late 1974 by D. Carl Johanson in north-central Ethiopia. Johanson had estimated the date of his find to be between three and four million years old—as old, if not older than the Leakey find. Johanson was unable to pinpoint a firm radiometric date to his fossils, however, and was thus forced to rely on comparisons with associated fauna. The Leakey fossils, dated by precise potassium-argon methods at 3.35 to 3.75 million years in age, is therefore the first reliable evidence that men roamed the African savannah more than three million years ago.

Mrs. Leakey located the teeth and mandibles in a heavily-eroded area of volcanic ash and rock which she and her husband had first explored as early as 1935. The fossils, located at different levels of the rockbed, which geologically span some 400,000 years, included several well-preserved adult jawbones and the mandible of a developing five-year-old. Dated by Professor Garniss H. Curtis of the University of California at Berkeley, the mandibles resemble those of modern man and are apparently smaller than the vigorous jaws of near-man, *Australopithecus*.

"This new evidence lends support to the view that *Australopithecus* is not in the direct line of human evolution and that the lineage of *Homo* extends much further back than believed a few years ago," Mrs. Leakey said during the press conference. What that means, according to some anthropologists, is that near man and true man evolved separately from two distinct ancestors, rather than from one common line.

CHAPTER

4

Introduction to Ecology

Objectives

1 Define the biosphere and ecosystems.

2 Understand the food chain and ecological pyramids.

3 See how populations grow and become ecologically balanced.

4 Explain how organisms cope with predators.

5 Understand the various physical influences upon organisms such as soil, water, light, air and temperature.

6 Be familiar with the various ecosystem cycles.

7 Understand how changes occur in ecosystems.

8 Observe what effect humans have had on the biosphere.

FIGURE 4-1a,b
Two ecosystems—a pond and a city.

THE BIOSPHERE—THE LAYER OF LIFE

The zone in which life on our planet is possible is the **biosphere.** Life exists upon and within soil, in bodies of water, and in the lower atmosphere. Relative to the diameter of the earth, life occupies but a thin film.

Several factors influence the distribution of life. Radiation from the sun, for example, is the biosphere's ultimate source of energy. Those organisms that utilize it directly, the green plants, must be in a position to receive it. Hence, they live only where photosynthesis is possible, at the surface of the earth or within the uppermost layers of bodies of water. Animals, however, have been found thousands of meters below the surface of the seas, thriving on the organic debris which falls from above.

Above the earth's surface organisms can survive only where there is sufficient oxygen, but the atmosphere does more than just supply this vital element in the useable form (O_2). High in the atmosphere a layer of molecular oxygen involving three atoms of oxygen rather than two, ozone (O_3), protects the biosphere from excess ultraviolet radiation from the sun. An illustration of the delicacy of the natural world is the fact that the ozone layer may be destroyed by a chemical reaction with the gaseous propellent of aerosol cans of hair sprays and underarm deodorants, and possibly even by supersonic aircraft.

Two ecosystems—ponds and cities

Although the entire biosphere may be considered one gigantic biological system, it is more convenient to conceive of it as an interrelated series of smaller units. Any environment in which living and nonliving things interact, and in which materials are recycled, is termed an **ecosystem.** Forests and coral reefs are natural examples. An aquarium and a human in a satellite orbiting the earth are also ecosystems. Mankind has altered many of the earth's natural ecosystems and created others.

In that both are centers for the transfer of energy and mass through living things, ponds and cities are ecosystems (Figure 4-1). Ecologists use the term **food chain** to describe such transfers, and all food chains begin with plants.

In pond ecosystems, plants include rooted pond weeds as well as several species of microscopic single celled plants, collectively termed phytoplankton (Figure 4-2). Phytoplankton live suspended in water.

Cities have few food-producing plants within their confines. Food must be brought in from outlying farms (hence, cities are not as self-contained as ponds). One of the most significant technological achievements of human beings has been the development of agricultural techniques that maximize the amount of food produced per amount of land area utilized. Unfortunately, agriculture has also resulted in some of our most disastrous ecological mistakes. For example, the Aswan Dam in Egypt, designed to irrigate agricultural land along the Nile, has held back mineral nutrients vital to populations of fish in the Mediterranean.

Food chains

In food chains, plants are the **primary producers.** At the next level are the plant eating animals, the **herbivores.** In pond ecosystems herbivores include the minute suspended animals of the zooplankton (Figure 4-3) as well as ducks and a few other larger animals. Cattle and sheep are among the city's absent herbivores.

Animals that eat other animals, or **carnivores,** are the next link in the food chain. In ponds the carnivores include large fish, water snakes, frogs, dragonflies, and certain birds. In cities there are very few pure carnivores, but dogs, cats, and humans eat meat as well as material produced by plants. Pet foods (often better labeled as to nutrients than human foods) are usually mixtures of plant and animal foods. Animals that eat both plants and other animals are termed **omnivores.** In ponds they include minnows, water beetles, turtles and carp.

Animals that feed predominantly upon dead, decaying plants and animals are called **scavengers.** In ponds this grouping includes crayfish and snails. In cities cockroaches, silverfish, and rats are scavengers, depending almost entirely on the waste products generated by humans for food (Figure 4-4).

The classifications herbivore, carnivore, omnivore, and scavenger are somewhat arbitrary. Often animals perform more than one function in any given ecosystem. Carp (in ponds) and dogs (in cities) can be both omnivores and scavengers, for example.

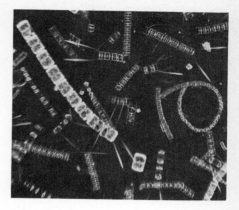
FIGURE 4-2
Microscopic phytoplankton.

FIGURE 4-3
Microscopic zooplankton.

FIGURE 4-4
Their wide dietary tolerance, speed and ability to slip into narrow crevices make cockroaches ideally suited to city life.

FIGURE 4-5

In ponds, food webs may be complex, with many direct and indirect food sources supplying consumers. All, however, can be traced ultimately to green plants, the primary food producers, which derive energy from the sun. This diagram shows several food relationships. (From *Pond Life* illustrated by Sally D. Kaicher and Tom Dolan, © 1967 by Western Publishing Company, Inc. Used by permission of the publisher.)

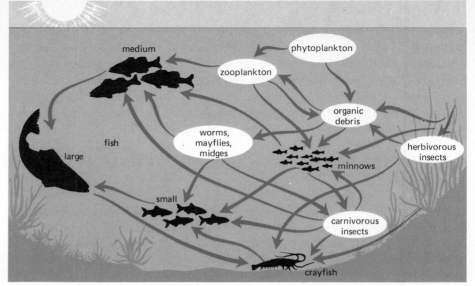

FIGURE 4-6
The practice of single crop agriculture has led to unstable ecosystems.

FIGURE 4-7
A pyramid based on the number of individuals in a pond.

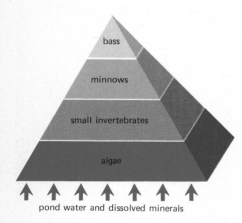

A final group of organisms which perform a crucial function in any ecosystem are the **decomposers.** Decomposition is the process by which organic material is broken down into smaller compounds, including carbon dioxide and water. Since plants require carbon dioxide and water to manufacture carbohydrates by phtosynthesis, the transfer of organic matter within ecosystems is a cyclical process. The same atoms are used over and over in the processes of life and we must therefore have atoms in our bodies that belonged to someone or something else.

Bacteria are the most numerous decomposers in any ecosystem. In ponds they live attached to suspended bits of organic matter in bottom mud, or they float freely. In urban ecosystems the process of decomposition occurs, to a large extent, in sewage disposal plants, where bacterial cultures are maintained for the purpose of breaking down organic waste materials in sewage into more simple chemicals. By keeping decomposition confined to small, isolated areas the spread of disease bacteria that may be carried in waste products is minimized.

COMPLEXITY AND STABILITY OF ECOSYSTEMS

Natural ecosystems are usually complex; they include many species at each level in the food chain. Therefore, natural ecosystems are more accurately conceived as **food webs** rather than food chains (Figure 4-5). The complexity of natural ecosystems gives them stability. If, for example, all of the sunfish in a pond were killed by a sudden outbreak of disease, other predatory fish, water snakes, frogs, and birds could partially fill the gap left in the web of food relationships.

It is distressing that many human activities have served to simplify ecosystems, thus destabilizing them. Large single-crop farms, for example, must be carefully tended to protect them from disease and insect infestations. Unfortunately the chemical agents used to protect crops sometimes kill harmless organisms. Thus, the practice of single-crop farming leads to instability, not only within farms themselves, but within surrounding natural ecosystems (Figure 4-6).

Ecological pyramids

To better understand flows of energy and matter, ecosystems are often diagrammed as pyramids, with primary producers at the base and the largest, or "top," carnivores at the peak (Figure 4-7). A **pyramid of numbers** represents the number of individuals at each level of a food chain. A count in a bluegrass field, for example, revealed the following averages per acre: 5,842,424 producers to 707,624 herbivorous insects to 354,904 ants, spiders, and predatory beetles to 3 birds and moles. For each top carnivore, therefore, there were over one million primary producers and hundreds of thousands of lower animals.

Food chains can also be diagrammed as **pyramids of mass,** on the basis of the total mass at each level. This gives a better picture of energy requirements. The energy utilized by each level of a food chain is more closely proportional to its mass than to numbers of organisms. Consider, for example, a food chain consisting of a willow tree, caterpillars feeding on the tree's leaves and birds feeding on the caterpillars. In this case a "pyramid" of numbers would be as shown in (Figure 4-8a); whereas a pyramid of mass would more accurately depict the passage of nutrients through the food chain (Figure 4-8b). Even a pyramid of mass, however, tells only a partial story.

Productivity of an ecosystem

An ecological pyramid indicates a condition within an ecosystem at a particular moment. But nature is not static. Thus, the rates of **productivity,** or turnover, at each level in a food chain reveal more than a pyramid of mass about the dynamics of an ecosystem. In the example of the willow tree, the rate at which leaves grow is more significant than the total mass of leaves on a particular day.

By measuring changes at the bottom of a food chain, it is possible to predict changes at the top. For instance, in pond ecosystems conditions in the springtime may or may not be favorable for rapid growth of phytoplankton. Since the rate and duration of the "spring bloom" of phytoplankton determine how much food is available to higher levels in the food chain, all populations are ultimately affected. Even populations of top carnivores, such as bass, fluctuate with the spring bloom (although the effects aren't observable until later in the year).

An ecologist monitors productivity by periodically counting and/or weighing organisms from a specific area over the course of one or more years. Such studies may provide valuable information. For instance, if a rancher knows when to expect changes in the productivities of the various kinds of grasses on which cattle graze, he can use natural pastures for maximum efficiency. All effective game laws are based on productivity studies. Game populations are carefully monitored, and hunting and fishing seasons are planned accordingly.

Obviously, it would not be practical to count all the clover plants in a pasture or all the bass in a lake. Instead, an ecologist selects several areas that are large enough to provide an accurate representative sample. He also observes physical conditions (like temperature or smog level), and looks for correlations between physical and biotic changes. Such studies help answer questions such as: To what extent are nutritional factors responsible for patterns of distribution of populations? Why do populations vary in size from year to year? How do sewage disposal plants affect the productivities of the waters into which they pump wastes?

POPULATION GROWTH

If organisms are placed in a new habitat that is favorable for their development, the population will increase at a rate depicted by an "S" curve (Figure 4-9). At first numbers will increase rapidly, but as time passes, the rate of population increase will gradually level off. If the ecosystem is balanced, the number of individuals will then fluctuate within a relatively small range.

The dynamics of population growth can be easily seen in nature's most simple form of life, bacteria. Because these one-celled organisms reproduce so

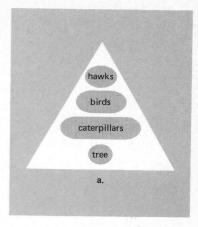

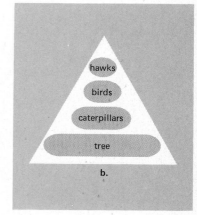

FIGURE 4-8
A pyramid based on numbers (a) and one based on mass (b). The latter reflects truer energy relationships.

FIGURE 4-9
The s-shaped curve (A) represents the growth of a population when a new habitat is opened to an organism. Once the organism is established in the ecosystem, the population density remains relatively stable, marked by only minor fluctuations (B). If a population is kept in an environment with limited resources, after a time the population is rapidly depleted (C).

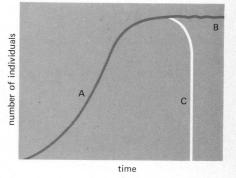

frequently, their populations expand rapidly to fill a new environment. It is not unusual for bacteria to reproduce every twenty minutes, by simply dividing in half. Thus, two bacteria may become four in twenty minutes, eight in forty minutes, sixteen in an hour, and thirty-two in eighty minutes. In three hours there could be 1024 bacteria and in five hours there could be 65,536. This assumes uninhibited reproduction within the culture, but such explosive growth cannot go on forever. Indeed, it is inevitably halted in a matter of hours by the limits of food supply and/or the accumulation of toxic waste compounds produced by the bacteria, themselves.

In a bacteria culture, a point is soon reached at which all of the nutrients are depleted, and the population slowly dies. However, in many naturally occurring bacterial communities, recycled nutrients keep populations at more or less stable sizes. In ponds, for example, plant and animal matter settling to the bottom nourish bacteria in the mud, while water circulation and diffusion carry away wastes.

Populations of higher organisms grow with the same dynamics as bacteria populations. For example, limited availability of vegetation for food will eventually stop the growth of population of deer (assuming it is protected from hunters). Since the natural predators (eaters) of deer have been wiped out in most areas of the country, either hunters or starvation ultimately limit population growth. Unfortunately, when deer are weakened by starvation they are usually susceptible to disease.

Density dependence and independence

Certain controls on population size vary in proportion to density. As a population becomes more dense these controls take a proportionately greater toll, and, for this reason, are called **density-dependent** factors. They include starvation, disease, and predation. Other factors, however, take a fairly constant toll no matter how dense the population. These are termed **density-independent** and include such changes in the physical environment as fluctuations in temperature or precipitation. For example, the first frosts of winter kill a fairly constant proportion of insect populations, no matter how dense or sparse they may be. There are no factors controlling population growth that are purely density-dependent or density-independent. However, biotic influences (such as starvation, disease, and predation) tend to be density-dependent, while climatic influences tend to be density-independent.

Overcrowding and undercrowding

Although there are disadvantages for organisms in overcrowded conditions, there are certain advantages to large population numbers. Most notable is protection from predators. In a large band of monkeys, for example, many sentinels can stand watch for predatory cats. A large flock of small birds presents a confusing picture to a hawk. In some cases a large number of organisms may be able to modify the environment favorably. On hot days, for instance, honeybees gather forces and fan the hive with their wings and keep it from melting.

Apart from protection from predators and mastery over unfavorable environmental forces there are other advantages to maintaining moderately high population densities. A low density reduces the reproductive potential of a species by making it more difficult for members to find mates. In highly social groups (such as colonies of termites, ants, and bees) efficiency in food-gathering is lost when the population is decreased. Thus, a balance between sufficiently high population density and the disadvantages of overcrowding determines optimum population size.

Ecological balance

Consider the events in a grassy meadow. Meadow mice characteristically eat seeds, and one might expect that they would reduce the next generation of grasses and other open-field plants. Under normal conditions, however, plants produce far more seeds than they need to maintain optimum population sizes. Owls prey upon mice, and were it not for their presence, mice might soon overrun the field. The mouse population might then reach a level injurious to the plants. However, by depleting the numbers of seeds, the mice would ultimately be doing damage to themselves. Fewer seeds would mean fewer plants to produce seeds in the next generation; hence, there would be less food for mice.

In a forest, a field, a lowland marsh, a rocky meadow or mountaintop — wherever life is found — there exists a close relationship between plant and animal and between prey and predator. Herbivores and carnivores play essential roles in maintaining ecological balance.

SYMBIOSIS

Within ecosystems certain interrelationships between differing species are more intimate than that of primary producer/herbivore or prey/predator. In **symbiosis** (meaning "living together") differing species live in direct association with each other. Biologists have classified three types of symbiosis: 1) parasitism; 2) mutualism; and 3) commensalism.

Parasitism

A **parasite** is an organism that saps nutrition from a living host. The parasite benefits from this association, while the host is weakened. Tapeworms, which are contracted via insufficiently cooked meat and cling with hooks in mammalian guts (Figure 4-10), and lampreys (eel-like fish that suck blood from other fish) are among the largest of parasites. Anyone who gardens is familiar with insects parasitic to plants, such as aphids. Other parasites include disease-causing bacteria, mildew on shower curtains, and the fungi that cause athlete's foot.

Parasites are not usually the direct cause of a host's death. (It is to a parasite's advantage that the host remain alive as a constant source of nutrition.) But, by weakening the organisms that they infest, parasites may pave the way for fatal secondary infections. Such is the case with the blood fluke, *Schistosoma japonicum*, which is common in China, Japan, and the Philippines. As the agent of the disease, schistosomiasis, this wormlike organism is responsible for more human deaths than cancer.

Most organisms harbor parasites (many parasites are themselves parasitized). In terms of numbers of individuals, there are more parasites in the biosphere than free-living organisms.

Mutualism

In **mutualism,** two different kinds of organisms live together to the advantage of each. In some cases they are so dependent on each other that neither can live alone. Termites, for example, can chew and swallow wood, but they cannot digest it. The cellulose of the wood is digested for them by microorganisms living in their digestive tracts. These microorganisms utilize some of the nutrition from the wood, but most of it is made available to their host. Thus, the termite is provided with a

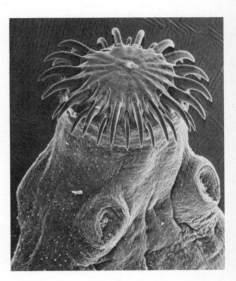

FIGURE 4-10
A tapeworm. A close-up of the head of a tapeworm is shown both in the diagram and the photograph.

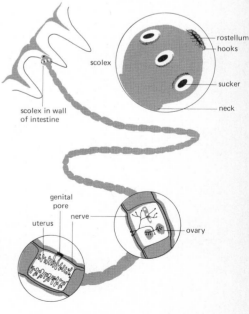

FIGURE 4-11

Lichens are composed of fungal and algal components which can survive separately in laboratory cultures but need each other to thrive in nature.

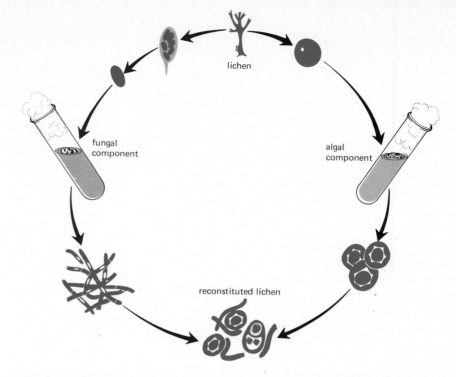

lichen

fungal component

algal component

reconstituted lichen

means of digestion, and the microorganisms receive room and board. The association of algae and fungus in lichens is another mutualistic relationship (Figure 4-11). The fungus, which is not photosynthetic, provides moisture and support for the algae, which, in turn, synthesizes food for the fungus and itself. A well-known example of mutualism is the relationship between flowers that produce nectar and insects that, in seeking nectar, pollinate the flowers.

Commensalism

In **commensalism,** one of the partners benefits, while the other is neither helped nor harmed. Such is the relationship between the remora and shark The remora is a small fish with an organ somewhat like a suction cup on its head. It attaches to the body of a shark and feeds on scraps of the shark's food. The remora thus benefits, although the shark does not. But the shark isn't harmed either. The term commensalism means, literally, ''common table.''

PASSIVE PROTECTION

The natural order of things is eat or be eaten, but evolution has provided animals with several means of protection from natural enemies. Claws, teeth, spines, stingers, pincers, and the ability to run fast are all used in defense. Camouflage also has survival value. By blending with their inanimate surroundings, animals hide from predators. A green katydid insect is nearly impossible to find among the green leaves of a lakeside tree, but if it should fall into the lake, it would be immediately vulnerable to birds, fishes, frogs, and other insects.

Animal camouflage

Animal camouflage is of several types. Sometimes an animal is colored or marked like its surroundings. This is called **protective coloration** (Figure 4-12). The orange background and black stripes of the tiger, for instance, blend with the

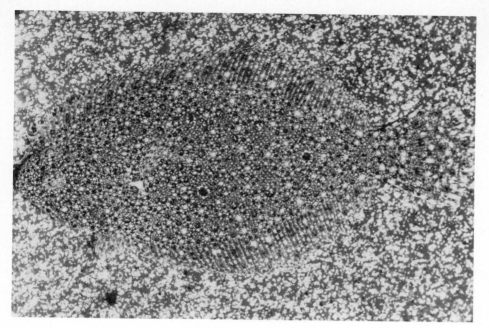

FIGURE 4-12
Protective coloration in the flounder.

grasses and shadows of its environment. Quail in a thicket go unnoticed until they become frightened and fly into the open. The common tree frog has irregular markings of brown and ashy gray that blend with the color of the bark of trees.

Many fishes have a form of protective coloration called countershading. Their backs are darkly colored while their bellies are of a lighter hue. When viewed from above they are undetectable against the dark background of the ocean abyss, while from below their light bellies blend against the sunlight filtering through the water.

Another kind of camouflage, **protective resemblance,** involves physical resemblance to an inanimate object in the environment. Certain butterflies resemble the dried leaves upon which they rest with their wings folded. The walking stick, a relative of the grasshopper, actually looks like a twig. **Mimicry** is yet another type of protective resemblance. In mimicry, however, the animal looks like another animal rather than a part of its environment. Several kinds of defenseless flies resemble stinging insects, and predators avoid them. Another example of mimicry is found in two butterflies, the viceroy and the monarch (Figure 4-13). The monarch has such an unpleasant taste that is is avoided by birds. The more tasty viceroy escapes because it looks so much like the monarch.

Animal camouflage is a striking example of the workings of natural selection. Today, we see the result of many years of survival by those animals best adapted to their surroundings. Through slight variations in form and color, certain individuals came to resemble their surroundings more than others. They had a better chance to survive and produce more of their kind. It has taken millions of years for the animals just mentioned to develop. In the process, countless ill-adapted life forms must have perished.

FIGURE 4-13
Mimicry. Note the resemblance between the Viceroy butterfly and the foul-tasting Monarch butterfly.

PHYSICAL INFLUENCES ON COMMUNITIES

To this point we have mainly considered **community interactions.** That is, interactions between the populations of species within a given area. But ecosystems consist of more than their living components, and to consider communities apart from their physical surroundings would be to consider but a fraction of ecology. The following sections deal with some of the nonliving components of ecosystems.

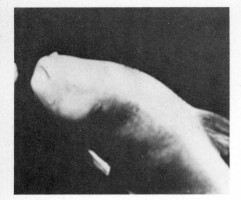

FIGURE 4-14
A blindfish.

Light

Light is needed by all green plants for photosynthesis. In large bodies of water, for example, plants are limited by the depth to which sunlight can penetrate. In the clearest of lakes green plants can survive only in the first 30 meters or so. In turbid waters, sufficient light for photosynthesis may reach but a meter beneath the surface.

Certain animals, however, are capable of living in total darkness. Blind fish, for example, with empty sockets where their evolutionary ancestors had eyes, survive in the total blackness of subterranean streams (Figure 4-14). Many bacteria survive without light; indeed they are killed by exposure to sunlight.

Light conditions vary from place to place. Deep valleys, forest floors, and the north sides of hills are locations where plants and animals with low light requirements thrive. Here there are snails, toads, and salamanders, as well as ferns and mosses. Open fields, southern slopes, deserts, and other exposed places offer habitats for plants, such as most vegetables, that need full sunlight. Some plants require intermediate exposure to the sun. Lettuce, for example, will survive a hot summer if it is planted in partial shade, but will wilt in an area that receives direct sunlight throughout the day.

Temperature

Many animals, including fish, amphibians, and reptiles do not maintain constant body temperatures. These are cold-blooded, or **poikilothermic** animals. Their temperatures fluctuate with the temperatures of their environments. Birds and mammals, however, maintain fairly constant temperatures, regardless of their surroundings. These are the warm-blooded, or **homeothermic** animals. Warm-blooded animals have extended thier habitats over wider ranges of temperatures than cold-blooded animals, and are capable of more intense activity in cold weather.

On a cold morning, a snake may crawl slowly out on a flat rock and lie in the

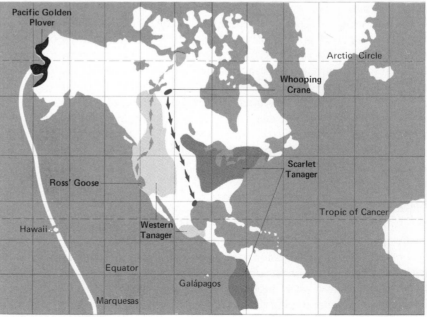

FIGURE 4-15
Migration areas of various birds.

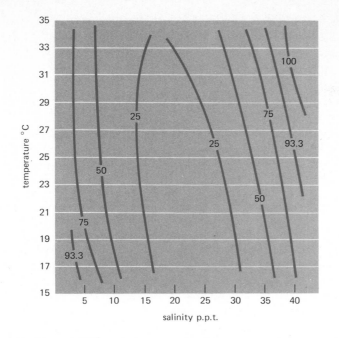

FIGURE 4-16

In addition to temperature, salinity (the salt content of water) also affects the survival of marine organisms. This graph illustrates the percentage of mortality under varying conditions of temperature and salinity (expressed as parts per thousand, p.p.t.) for the larval stage of a certain crustacean. Note that at an optimum salinity (about 19 p.p.t.) a wide range of temperatures is tolerable. (From *Involvement in Biology Today*. Copyright © 1972 by CRM Books. Reprinted by permission of CRM Books, a Division of Random House, Inc.)

sun. When its body temperature has increased, it can become more active. A meadow mouse living in the same area as the snake, however, can begin activity on a cold morning with no such "warm up."

In temperate regions of the earth (including most of North America) temperatures vary considerably. Here organisms must cope with daily fluctuations in temperature as well as the more extreme differences of summer and winter. Most trees and shrubs in temperate regions flourish through the warm weather of spring, summer, and fall. Then in the colder months, they enter a period of inactivity, or a **dormant phase.** The leaves of maples and other hardwoods fall, and sap flows to the roots. The leaves of the pine, spruce, and other evergreen trees remain throughout the winter, however, even though most activity in the plant has stopped. **Perennials** die to the ground and then grow again in the spring from dormant roots. **Annuals** start anew each spring from seed.

Some animals cope with temperature changes by moving. Many birds of the Far North migrate southward into the northern regions of the United States during winter months. Meanwhile, summer residents of these same areas migrate as far as the tropics of South America (Figure 4-15). Other animals burrow into soil or look for a protected cave, where they become inactive or sleep during winter. On the other hand, some desert animals remain in their burrows during the heat of the day, becoming active only as temperatures fall with sunset.

Aquatic and marine organisms may or may not have to cope with variations in temperature, depending on how deep they live. At great depths, temperatures change only slightly. But seasonal variations affect organisms living in shallow lakes or along shores (Figure 4-16).

The atmosphere

Air is the source of gaseous oxygen (O_2), which all organisms (with the exception of a few species of bacteria) must have in order to live. Land-dwelling organisms obtain O_2 directly from the atmosphere, while marine species survive on the O_2 that dissolves in seawater. In the oceans, oxygen content decreases with depth and is totally lacking in certain trenches (which may be several miles deep). In some areas at the bottom of Lake Erie (the shallowest of the Great Lakes) oxygen is also lacking. Much of Lake Erie's oxygen is tied up in the process of decomposition of organic pollutants.

Air movements have several influences on living things. Unusually heavy winds destroy plants and drive animals to shelter and persistent winds accelerate the rate of evaporation of water. Plants and animals whose habitats are open plains or exposed elevations must not only withstand buffeting winds, but must also survive the accompanying desiccation. Strong winds force trees to grow close to the ground and to form their branches on protected sides. Winds are an evolutionary pressure for reductions in leaf size. With a decrease in leaf size there is a decrease in the surface area from which water evporates. Winds also cause expansions of root systems to help hold the vegetation down.

Soil

Soil lies in a relatively thin layer on the surface of the bedrock of the earth. Erosion by running water, freezing and thawing, wind, and other forces of nature crumble rocks to form gravel, sand, and clay. These are the components of mineral soil, or subsoil. In most regions, the subsoil forms a layer several feet thick. This layer represents thousands of years of the slow disintegration of rock.

The organic portion of soil comes from the slow decay of dead plants and animals. The organic remains of land plants is **humus,** while aquatic plants form **peat** in lakes and bogs. Organic matter and mineral matter combine to form topsoil, or **loam** (Figure 4-17). Topsoil is the nutritional zone for plants. It forms slowly, at the rate of about one inch in five hundred years. Topsoil supports great numbers of bacteria, molds, and other microorganisms, as well as earthworms. Activities of soil organisms are essential to fertility. By breaking down dead plant matter into simpler chemicals they condition topsoil for new vegetative growth.

Soil varies from place to place, and the plant and animal life it supports varies accordingly. Some soils are compact because they are composed mostly of **clay,** (particles of weathered rock less than 0.002 mm in diameter) while others are loose because they contain mostly **sand** (particles greater than 0.05 mm in diameter). Particles of **silt** are intermediate in size between clay and sand. Loam is a mixture of clay, sand, and organic matter. Sandy soils support pine forests in Michigan, New Jersey, Georgia, and eastern Texas.

Well balanced loam supports beech and maple forests in Ohio and Indiana. Waterlogged soils of bogs and swamps provide ideal conditions for larch, white cedar, and cypress forests. The rocky, shallow soils of mountain slopes sustain forests of redwood, yellow pine, and spruce in the western states.

A "sour soil" is acidic, and a "sweet soil" is alkaline. The chemical states of acidity and alkalinity (described in Chapter 5) are such that as acidity decreases, alkalinity increases. Thus, soils can be ranked on the **pH scale.** On this scale 7 is neutral. Numbers from 7 to 1 indicate progressively stronger acidity, and numbers from 7 to 14 indicate increasing alkalinity.

Under cultivation, soil tends to become acidic, and to correct this condition, powdered lime may be dug into the surface. Certain plants, including beets, spinach, lettuce, cauliflower, onions, peas, alfalfa, and clover thrive best under slightly alkaline conditions (at a pH of about 7.5). On the other hand, rhododendrons, azaleas, and blueberries grow best in slightly acidic soil (at a pH of about 6.5); additions of acidic organic matter (such as pine needles or oak leaves) benefit them. In general, plants do best in soils that are close to neutrality.

A more difficult problem than improper pH is an excess of salts, which arises from irrigating with hard water. Approximately a quarter of the United States' twenty-nine million acres of irrigated land contains damagingly high concentrations of salts of sodium, calcium, and magnesium. Excesses of these elements reduce the rate at which plants absorb water and consequently retard growth.

FIGURE 4-17
A soil profile of sandy loam.

The character of soil is intimately linked to the manner in which it is farmed. Unless care is taken to maintain its content of organic matter, for example, soil will become highly compacted, and impenetrable to roots. Additions of composted animal manures and plant matter (such as dead leaves or lawn clippings) help to maintain soil's softness. Moreover, compost adds somewhat to soil's content of nutrients. For home gardeners, a program of fertilization with both organic refuse and concentrated forms of plant nutrients (such as blood meal, wood ashes, and powdered phosphate rock) will lead to good yields of crops.

Water

Water is the basic ingredient of life: it makes up 55 to 95 percent of the weight of organisms. The lower figure applies to some woody plants; the higher figure to jellyfish. The habitats of plants and animals vary from the depths of the oceans to sunparched deserts, but all organisms need water.

Plants and animals living in water are said to be **aquatic.** Aquatic plants and animals living in salt water are termed **marine.** Some species can inhabit both fresh and salt water, but most water-dwelling organisms are limited to one or the other.

Certain animals (including frogs and other amphibians) alternate between land and water, and various plants live at the interface. Ecologists classify plants that grow entirely or partially submerged as **hydrophytes.** Among them are pond lilies, cattails, bulrushes, and cranberries.

Plants that occupy neither extremely wet nor extremely dry areas are classified as **mesophytes.** The trees of the hardwood forests of the central and eastern states are mesophytes, as are most garden vegetables and flowers. In general, mesophytes have well-developed roots and broad leaves.

The driest environments—semideserts and true deserts—are occupied by **xerophytes.** Such species have widespread root systems and small leaves. Cacti are xerophytes whose leaves are reduced to spines, and whose thick stems are adapted for water storage (Figure 4-18).

FIGURE 4-18
The physiology of many desert plants is geared to take quick advantage of the infrequent rains. The seeds germinate and the plants grow rapidly when adequate water is available.

NATURAL CYCLES

Ecosystems are influenced by various cyclical phenomena—the cycle of day and night (diurnal cycle), for example, and the change of the seasons. Moreover, the materials of life flow in cycles. The decay of dead plants and animals produce nutrient compounds that are utilized by new generations, which, in turn, ultimately die. The influences of diurnal and climatic cycles are dealt with elsewhere in the book; the following sections consider cycles of materials.

The water cycle

Water oscillates between the earth and the atmosphere (Figure 4-19). Solar radiation heats the surface of the earth, causing water to be lost to the atmosphere by **evaporation.** As warm, vapor-laden air rises, it cools. As it cools, the water condenses into droplets, forming clouds. The droplets aggregate, become heavy and fall to earth in the form of rain or snow. In this way **precipitation** returns water to the earth. Some rainwater converges in streams and rivers. This is **runoff water** and it will reach a pond, lake, or the ocean, unless it evaporates first. A certain amount of precipitation soaks into the soil to become **ground water,** which may reach open bodies of water through springs and underground streams. Or, ground water may move upward through the soil during dry periods, returning to the atmosphere as vapor.

FIGURE 4-19
The water cycle.

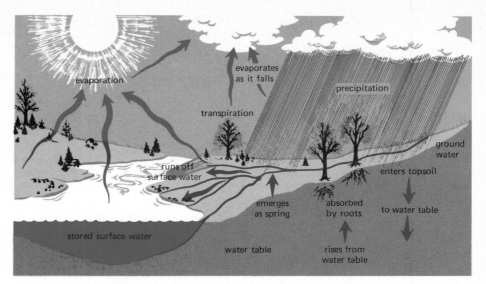

Another part of ground water moves downward through topsoil into subsoil, filling the spaces around rock particles. The upper level of subsoil that is saturated with water is termed the **water table** (Figure 4-19). The depth of the water table depends on the amount of precipitation, the consistency of the soil, the nature of the rock layers under the soil, and the proximity of large bodies of water. Where depressions occur, as in basins of lakes and ponds, the water table may be above the surface of the earth.

To a certain extent, both plants and animals are involved in the water cycle. Plants absorb water through their roots and release it through pores (stomata) in their leaves. Animals are involved to the extent that they drink water and give off a certain amount of it as vapor in exhalation or through pores in their skin. The amount of water that cycles through living things is small, however, compared to the amount that cycles through bodies of water, especially the oceans.

The carbon-oxygen cycle

As previously mentioned, **photosynthesis** is the conversion of carbon dioxide and water to organic substances (specifically carbohydrates) by plants. It requires the energy of the sun. **Respiration** is the conversion of organic substances to carbon dioxide and water by both plants and animals. It liberates energy that organisms utilize for life processes. Together photosynthesis and respiration comprise the **carbon-oxygen cycle** (Figure 4-20).

FIGURE 4-20

The carbon-oxygen cycle. In photosynthesis, green plants use carbon dioxide and release oxygen. Oxygen is used by animals and plants in respiration, and also in the burning of fuels. Carbon dioxide is released in both processes, as well as in the process of decay.

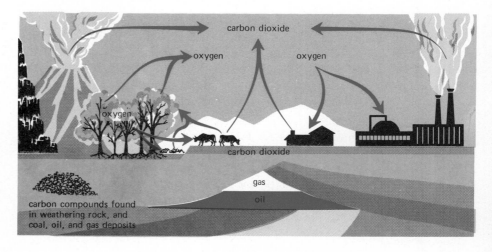

Oxygen is involved in this cycle in that it is generated by the process of photosynthesis and required in respiration. The carbon dioxide and water produced during respiration are available to plants for photosynthesis. The reverse chemical reactions of photosynthesis and respiration maintain a balance (equilibrium) between carbon dioxide, water, oxygen, and living substance in the biosphere.

The nitrogen cycle

Like the carbon-oxygen cycle, the **nitrogen cycle** is a circulation of vital chemical substances within ecosystems (Figure 4-21). It involves green plants, several kinds of bacteria, and may or may not involve animals. Plants play a key role in the nitrogen cycle in that they absorb simple nitrogen-containing compounds from soil or water and incorporate them into complex compounds of life. Nitrate (NO_3^-) and ammonia (NH_3) are taken in and utilized to eventually form proteins.* Animals become involved in the nitrogen cycle when they eat plants, and in this way absorb the proteins.

When an organism dies, decay by bacterial action begins. Nitrogen is released from decaying protein in combination with hydrogen as ammonia. This part of the

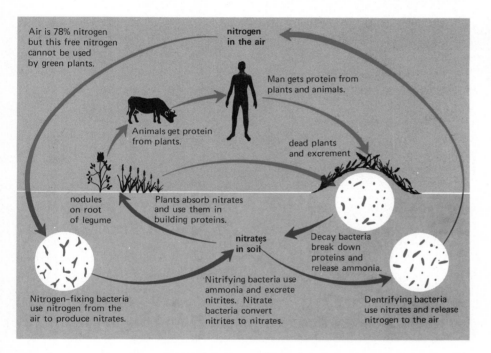

FIGURE 4-21

The nitrogen cycle. In addition to the steps shown, some free nitrogen in the air is changed to nitrates by lightning and carried to the soil by rain or snow.

nitrogen cycle is termed **ammonification.** Certain kinds of bacteria subsequently utilize ammonia and generate nitrite (NO_2^-), a form of nitrogen that cannot be utilized as a nutrient by plants. In fact, it is poisonous. However, further chemical reactions involving the addition of oxygen mediated by other kinds of bacteria, produce nitrate from nitrite, through a process termed **nitrification.**

In starting with nitrate and ending with nitrite, a complete sequence of the nitrogen cycle has been described; but, there are two significant offshoots.

In the form of a two-atom molecule, N_2, gaseous nitrogen composes 78 percent of our atmosphere. In spite of its abundance, however, plants cannot utilize N_2 as a nutrient. It is essentially biologically inert. But, certain bacteria can utilize atmospheric nitrogen. Among them is a group that lives in a mutualistic relation-

---

* NO_3^- is a negatively-charged chemical complex (*ion*). The nature of electric charge as related to biochemical activity is considered in the next chapter.

ship, within nodules on the roots of clover, alfalfa, peas, and other members of the **legume** family (Figure 4-22). Fortunately, legumes accumulate more than enough nitrates to meet their own requirements, and the excess builds up the nitrogen content of soil. Planting clover or alfalfa as part of a crop-rotation schedule, enriches soil from virtually unlimited supplies of atmospheric nitrogen.

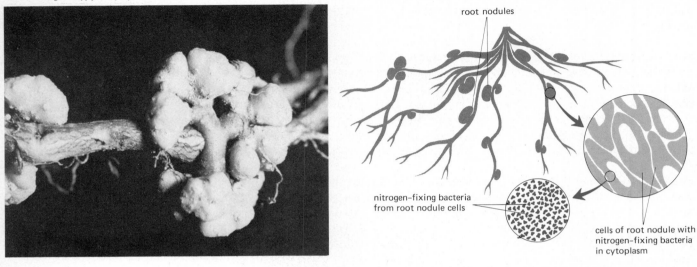

root nodules

nitrogen-fixing bacteria
from root nodule cells

cells of root nodule with
nitrogen-fixing bacteria
in cytoplasm

FIGURE 4-22
Nitrogen-fixing bacteria are found in the root nodules of legumes as shown in the photograph and in the diagram.

The other offshoot of the nitrogen cycle is unfavorable to agriculture. Certain bacteria generate gaseous nitrogen (N_2), by breaking down ammonia, nitrite, and nitrate in a process termed **denitrification.** In this way, nitrogen may be lost from the soil to the atmosphere. Fortunately, denitrification does not occur in well-drained, cultivated soil, since denitrifying bacteria are **anaerobic.** That is, they grow in an environment that has little or no oxygen. They thrive in soils that are waterlogged, or packed so tightly that air cannot penetrate.

SUCCESSION

Environmental conditions constantly change; and organisms must be able to cope with such inevitable changes as the cycle of the seasons. But, a harmoniously functioning ecosystem, which may have been established for thousands of years, can be destroyed in a matter of hours by catastrophic changes. Fire, heavy winds, volcanic activity, flood, drought, and careless use of insecticides, may wipe out the organisms of an ecosystem. But nature eventually repopulates devastated areas.

Succession is the gradual establishment of a stable biotic community. It is a process that may take hundreds of years, and, while it is occurring, there may be drastic changes in the kinds of organisms in an area. Eventually, however, a stable (or **climax**) community develops.

Forest climax communities

The natural climax community in many areas of northeastern United States is the "broad-leaved forest," in which such trees as beech and maple are the dominant species. Since colonial times, various areas of forest have been destroyed by fires, logging, and farming. Naturalists have watched and recorded the reestablishment of life in some of the undisturbed areas.

First, the seeds of grasses and other open-field plants that may be dormant in the soil or carried in by animals or winds find the environment hospitable. A meadow is produced by these pioneers, which may dominate the region for several years. Next, the seeds of elms, cottonwoods, and shrubs find their way into the meadow; a forest begins in which larger plants shade the shorter grasses and field plants. Thus, the environment of the once open field is changed into a low woods. Eventually the growth becomes too shady for the seedlings of the pioneer trees and shrubs, and the area again changes. The third stage is the arrival of seeds of oaks, ashes, and other trees whose seedlings grow well in shade. These trees grow among the elms and cottonwoods and gradually become the dominant vegetation.

Finally, a dense forest forms. As the earth becomes moist and fertile, beech and maple seedlings outdistance all other species in the competition for space. Eventually they crowd out most of the other trees. They are the climax species. When succession occurs on a ridge, the climax species are commonly ash and hickory. Grasses are the climax plants in the Great Plains of Kansas and Nebraska.

Pond climax communities

Ponds go through rapid sequences of succession, particularly if their supplies of water are slow or unsteady. The speeding up of succession in fresh waters due to an increase in nutrients caused by pollution is called **eutrophication**. Marked changes may occur within a few years, and change is most rapid at the interface between water and land. The cattails and water lilies around the edge of a pond hold soil around their roots, and as they die their decaying stalks produce a dense mat of humus. The pond grows smaller. Eventually, the plants at the edge may choke out their own living space by filling in the pond completely. Their populations then die out and are replaced by species more suited to dry land (Figure 4-23)

THE HABITAT, THE NICHE AND COMPETITION

The **habitat** of an organism is the part of the ecosystem that is its home. There are many habitats within the ecosystem of a lake, for example. Bullfrogs tend to stay in the shallow water at the shore, while bass live in deeper water. Yet, both contribute to the complex structure of the ecosystem. In fact, a bass may occasionally eat a bullfrog, since their habitats overlap somewhat. The term **niche** has a broader meaning than habitat. It includes what an organism does, as well as where it lives.

Habitats and niches of neighboring species may differ in subtle ways. Both fleas and ticks live on dogs. In fact, both of them suck blood from their hosts. But ticks have larger and more powerful piercing mouths than fleas; ticks can take blood through areas of tougher skin. Fleas must seek soft, vulnerable areas such as the groin. Since fleas can quickly jump to avoid the gnashing teeth of an irritated dog, they are well-suited to their taste for delicate skin. The more sluggish ticks, however, are adapted to a sedentary life on a part of the body which is less sensitive. Thus, slight differences may separate the habitats and niches of organisms.

FIGURE 4-23
Schematic diagram of succession in a pond (from top to bottom): pioneer, open-pond stage; submerged vegetation stage; cattail stage; sedge meadow stage, and climax forest stage.

Competition between species

Habitats and niches of differing species can overlap, in which case **competition** arises between them. In areas where mountain lions and wolves both live, for example, they compete for the animals on which they prey. If the numbers of prey species are sufficiently limited, and if either the wolves or lions are better hunters, then the less suited species will eventually die out or leave.

If an organism is to live in a certain habitat, it must be able to obtain what it needs for growth and reproduction. Anything that is essential to an organism and for which there is competition is called a **limiting factor.** Competition occurs between individuals of the same species (**intraspecific competition**) as well as between populations of different species (**interspecific competition**). Cattails, for example, compete with each other as well as other species of shore plants for space at the edge of a lake.

If there is sufficient overlap between the niches of co-occurring species, then interspecific competition will eventually drive the weaker of them to extinction. Intraspecific competition serves to maintain population sizes at optimum levels. Both types of competition are agents of natural selection.

THE LAND ENVIRONMENT

A journey from one of the earth's poles to the Equator would begin in a frozen zone nearly free of life and (if it were by land) end in an area of high temperatures and lush vegetation. As illustrated in Figure 4-24, an expedition up a mountain near the Equator would lead through similar climatic changes, ending at a peak that is frozen year round. Thus, climate can change rapidly or slowly with distance, depending on the terrain. In relatively flat areas, however, any given latitude is

FIGURE 4-24

Vertical climatic regions of the earth are similar to horizontal climatic regions. Life zones going up a high mountain can be compared to those found while traveling from the equator to either pole.

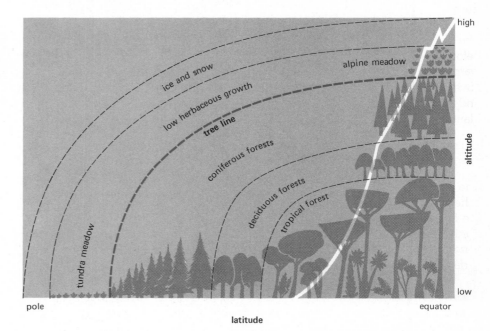

characterized by a **biome** (a climax plant formation and the animals it supports). The following sections describe some of the earth's major biomes.

The Poles

There is a basic difference between the earth's two polar regions. The ice of the North Pole is buoyed up by water, while at the South Pole there is a continent almost as large as North America. Most of the land of this continent (Antarctica), is never exposed to the light of day, however, for it is buried beneath a mile of ice. Winds up to 200 miles per hour rush across its lonely expanses of packed snow. There is no month in the year when the average temperature in the Antarctic rises above freezing.

Only three species of flowering plants live at the tip of the Antarctic Peninsula, and the lichens and mosses that are present have small populations. The animals of the Antarctic include penguins, a few visiting birds, mites, and several species of microscopic animals that can withstand incredibly low temperatures. The **tardigrade,** (Figure 4-25), after being dried-out, will survive freezing to within a few degrees of absolute zero (−273°C).* After water is added to them and they are rewarmed tardigrades become active, with no apparent ill-effects.

Of the five and one-half million square miles of Antarctica, only about three thousand are ever free of ice and snow. The nearest tree is about 700 miles to the north (at the tip of South America) of the farthest northward extension of the Antarctic peninsula. However, evidence excavated from beneath the Antarctic ice indicates the continent was not always the frozen desert it is today. Fossil leaves and coal deposits at depths of hundreds of feet suggest that a tropical climate once prevailed there. Thus, Antarctica once must have been at a more favorable angle to the sun for photosynthesis. "Continental drift" (a phenomenon described later in this chapter) may have been responsible for the climatic change. A shift in the earth's alignment to the sun may also have played a role.

In comparison to the Antarctic, the Arctic is surprisingly well populated. More than 100 species of flowering plants, many varieties of mosses, lichens, insects, birds, and mammals have been found there. Also within the Arctic live a million people, mostly the Eskimos of North America and the reindeer herdsmen of Europe and Asia.

Several factors contribute to the differences in the polar regions (Figure 4-26). The Arctic ice sheet is seldom more than 5 meters thick, and heat from the ocean below has a moderating effect on the temperature. The Arctic averages 30 degrees warmer than the Antarctic, and nine-tenths of the Arctic's lands lose their ice covering during the summer months. Temperatures may rise as high as 20.5°C and as the upper layers of earth thaw, water is available to the many flowering plants.

The tundra

The **tundra** is a large biome encircling the Arctic Ocean of the Northern Hemisphere. Since there is relatively little land at a corresponding latitude of the Southern Hemisphere, the area of southern tundra is very small compared to that of the north. The climate of the tundra is extremely cold and the ground is permanently frozen a few feet below the surface. During the continuous daylight of summer, the surface thaw produces saturated bogs, streams, and ponds.

Mosses and lichens form the prominent perennial vegetation, although some dwarf birches, alders, willows, and conifers make their homes on the tundra. Various rapidly growing annual plants produce brilliant flowers during the short summers. Most of the birds are summer migrants, but a few species are permanent residents. Herds of caribou visit the tundra to graze on the moss and lichens. Many of the inhabitants of the tundra, such as the Arctic hare, lemming, Arctic fox, and polar bear, have white coats that serve for protective coloration. The mosquitoes of the tundra, whose eggs are resistant to freezing, are infamous for their fierceness.

The coniferous forest

Coniferous forests lie below the **tree line** at the southern edge of the tundra in Europe, Asia, and North America (Figure 4-27). As with the tundra, there is no

* According to thermodynamic theory this is the temperature at which all atomic motion would cease and is the lowest temperature possible.

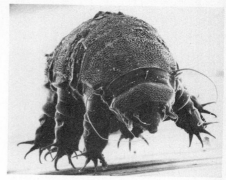

FIGURE 4-25
An active tardigrade is magnified in this scanning micrograph. Size, 200 micrometers.

FIGURE 4-26
Although the Southern Hemisphere is called the water hemisphere, its pole is covered by the large Antarctic continent. The Northern Hemisphere is called the land hemisphere, but its pole is located on a sheet of ice overlying the Arctic Ocean.

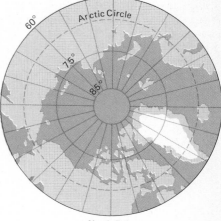

North Pole

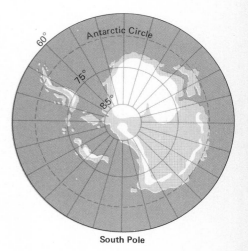

South Pole

large corresponding biome in the Southern Hemisphere, because there are no large land masses at corresponding latitudes. In coniferous forests the growing season is as long as six months although subzero temperatures persist for weeks on end in the winter.

A broad coniferous belt covers much of Canada, in which alders, birches, and junipers coexist in groves. Fires have destroyed certain areas of the coniferous forests, and where succession is in progress, pioneer grasses are followed by aspens and birches. These are eventually replaced by the spruces, pines and firs that form that climax community.

Magnificent stands of pine, spruce, and redwoods grow along the coastal ranges of Washington, Oregon, and California. Here, giant trees reach a height of two hundred feet or more. Rainfall may be as heavy as eighty inches per year along the northwest coast of the United States and when it is not raining it can be foggy.

Various animals inhabit the coniferous forests. Moose are plentiful in areas that have not been excessively hunted, and black bears roam the woods. Squirrels, chipmunks, rabbits, and mice are preyed upon by bobcats, foxes, wolverines, and wolves. Beavers and porcupines feed on vegetation. During the summer months, birds breed in the pines, but in the fall, they migrate south. The insects and other invertebrates of the coniferous biome lie dormant during the cold winter months.

The deciduous forest

In areas of the temperate latitudes, in which most of the United States lies, the growing season commonly lasts over half a year. Rainfall averages around forty inches per year, and where the soil is suitable, large deciduous forests grow. Eastern United States, England, central Europe, parts of China, and Siberia have large stands of deciduous trees. Although a similar zone occurs in South America, trees are stunted because of inadequate rainfall.

Local conditions of soil, drainage, and variations in climate throughout the temperate zone provide conditions necessary for different climax communities. In the United States, beech and maple forests are found in the north central regions, while oak and hickory forests are common in the western and southern regions. Although most of the native chestnut trees were destroyed by blight, oak and chestnut forests formerly covered much of the Appalachian Mountain chain. Other common deciduous trees found in the temperate zone are the elm (which is currently threatened by an epidemic of Dutch elm disease), sycamore, poplar, willow, and cottonwood.

Although the animal populations of deciduous regions are found in all the forests of this biome, deer are the most common herbivores. They peacefully coexist with foxes, raccoons, and squirrels. Wolves (whose populations are but a small fraction of what they were even a century ago) wander in packs, seeking prey. Woodpeckers, robins, and other tree-nesting birds are plentiful.

The deciduous forest undergoes magnificent seasonal changes. In late spring and summer, the trees are green, and shrubs produce beautiful blooms. During the fall, many areas are brilliant with multicolored leaves. In winter, bare branches contrast with the white snow.

The grasslands

Grasslands thrive in areas of midwestern North America, where annual rainfall is between ten and thirty inches. In the southern latitudes a similar biome exists in Argentina. The relatively low levels of precipitation in such areas is not enough to

FIGURE 4-27
A coniferous forest in North Carolina.

support large trees but is sufficient for many species of grass. Ancient grasslands were natural pastures for huge herds of wild grazing animals. Humans have taken over many grasslands for herds of domesticated cattle and sowed others with cultivated grains. Once robbed of their natural vegetation, grasslands are susceptible to erosion by wind. The "Dust Bowl" in America's midwest during the 1930s was a melancholy environmental symbol of the economic depression.

In the United States and Canada, isolated patches of grasslands still exist on the Great Plains and tall-grass prairies. The tropical grasslands of Africa support populations of giraffe, zebra, antelope, ostrich, and lion. A **savannah** (Figure 4-28) is a grassland with scattered trees. Oak grass savannah is the natural vegetation in much of western United States. South America also has large areas of savannah. In Australia, the grasslands and savannahs support cattle, sheep, kangaroos, and burrowing animals. Wild dogs are prominent predators.

In America, great herds of bison and antelope once grazed on the grasses of the plains. Burrowing mammals such as hares, prairie dogs, ground squirrels, and pocket gophers are still abundant. They are preyed upon by weasels, snakes, and hawks. Locusts and grasshoppers are important members of the insect population which wreck havoc on grain forms of vegetation.

FIGURE 4-28
There is fossil evidence that the earliest humans lived on an African savannah over one million years ago.

The desert biome

Deserts are characterized by low rainfall, but their temperatures can range from extremely hot to extremely cold. Death Valley, where the creosote bush is the climax vegetation, is a representative hot desert. In typical cold deserts, sagebrush is the dominant shrub. Cold deserts exist in certain areas of northwestern United States. In both hot and cold deserts, plants are adapted for living in areas where the rainfall may be only ten inches per year. Leaves are small, with thick, waxy outer layers that help retain water. Some desert plants, like cacti, have no typical leaves at all. Spines are leaf vestiges serving as protection from animals that would eat cacti for the water stored in fleshy stems.

Desert animals also have special adaptions to dry climates. The reptiles, some insects, and birds excrete nitrogenous wastes in the form of uric acid, which can be eliminated in an almost dry form, thus conserving valuable water. Mammals, however, lose water when they excrete nitrogenous waste in the form of urea dissolved in urine. Most desert mammals burrow during the day to prevent evaporation of water from body surfaces. Some rodents are able to live on the small amount of water in the seeds and fruits they eat. Other animals are capable of surviving on what is called **metabolic water.** They are able to make water through their own biochemical processes by combining hydrogen atoms extracted from food with oxygen.

Desert herbs, grasses, and flowering plants burst forth in growth and color in a surprisingly short time after a rain. Many of these plants are able to complete their cycle of growth, flowering, and seed production within a few weeks. Some of the larger perennial plants have extremely long tap roots to reach water sources deep beneath the surface.

FIGURE 4-29
A temperate rain forest on the northwest coast of Australia.

The rain forest

In areas of abundant water supply and long growing seasons, vegetation flourishes. These are **rain forest** environments (Figure 4-29). A lush temperate rain forest thrives on the northwest Pacific Coast of Washington, and tropical rain forests grow near the equator, including most of Central America, northern South America,

FIGURE 4-30
A fern jungle in Hawaii.

FIGURE 4-31
A map of the world showing the distribution of vegetation and use of the land.

central Africa, southern Asia, the East Indies, the South Pacific Islands, and northeastern Australia. In tropical rain forests the seasonal temperature variation is usually less than the temperature variation between day and night.

Plants produce a dense growth in a rain forest. Shorter trees grow beneath tall trees, but the canopy of leaves may be so dense that few plants can grow close to the ground. Even so, many smaller plants have become adapted to life in the tropical forest. Some have evolved long vines that make it possible for them to have roots in the moist ground and leaves high up toward the daylight. Other small, parasitic plants grow high among the trees and thereby receive sufficient light for photosynthesis. Most tropical rain forest plants have very large leaves to increase the area over which photosynthesis can take place. Conservation of water is not a problem for them. The critical problems in a rain forest are finding a place to grow and obtaining sufficient light for photosynthesis.

Epiphytes are plants that attach themselves to trees, sometimes one hundred feet or more above the ground. They have thick, porous roots adapted to catching and holding rainfall. Many epiphytes also have leaves arranged so that they catch water, insects, falling leaves, and other debris. As the organic material decomposes, essential nutrients are released for the epiphyte's use. Many species of orchids, mosses, ferns, and lichens, are epiphytes (Figure 4-30). Even though epiphytes do not take nourishment from the plants upon which they grow, they may cause minor injury by shading the leaves of the supporting plant or causing limbs to break from their weight.

Although animal life is plentiful in the rain forest, it may not be obvious to the casual observer. Much of it exists high in the trees. Except for an occasional birdcall or chatter from monkeys, the day in a rain forest is relatively quiet. But toward evening, everything comes to life. Ants, beetles, termites, and other insects come out in search of food, and are themselves eaten by other animals. The crickets and tree frogs begin singing, and tree-dwelling monkeys chatter excitedly before settling down for the night. The nocturnal carnivorous cats, such as the jaguar in South

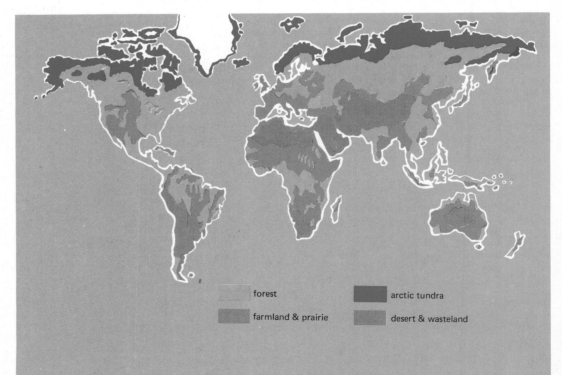

forest

arctic tundra

farmland & prairie

desert & wasteland

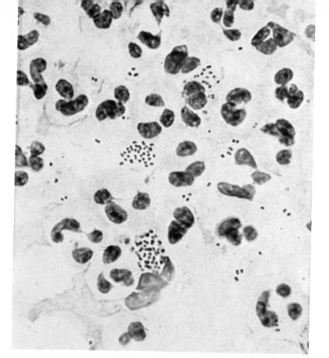

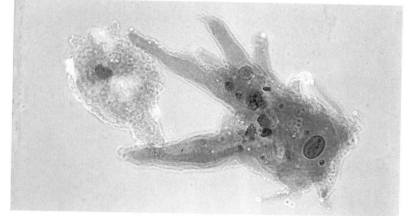

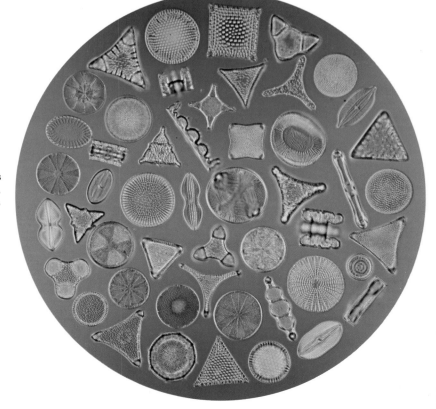

Gonococci bacteria *(Neisseria gonorrhea)* cause gonorrhea. Each bacterium is about 1 micrometer in diameter. The dots are the bacteria, and the darkly stained purple masses are cell nuclei. (Manfred Kage from Peter Arnold.)

Foraminifera are protozoans with shells about 600 micrometers in size; they live in the open ocean. (Manfred Kage from Peter Arnold.)

The amoeba *(Amoeba proteus)* is a freshwater protozoan with a flexible form. Its average linear dimension is about 500 micrometers. Several internal structures are clearly visible in this photograph of an extended and contracted amoeba. (Manfred Kage from Peter Arnold.)

Diatoms are unicellular algae with cell walls of silicon about 200 micrometers in size. They live in both fresh- and saltwater. (Manfred Kage from Peter Arnold.)

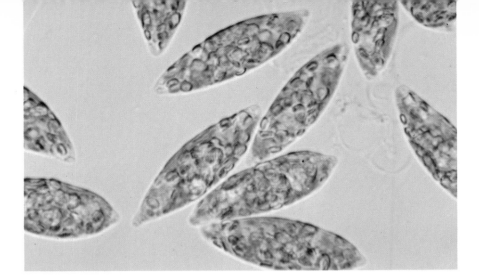

The *Euglena* is a flagellated protozoan, about 50 micrometers in size, that lives in freshwater. It is one of the few organisms capable of both photosynthesis and movement. The chloroplasts are the round structures in each cell. (Courtesy Carolina Biological Supply Company.)

A magnified view of a leaf infested with patches of rust fungi. (Manfred Kage from Peter Arnold.)

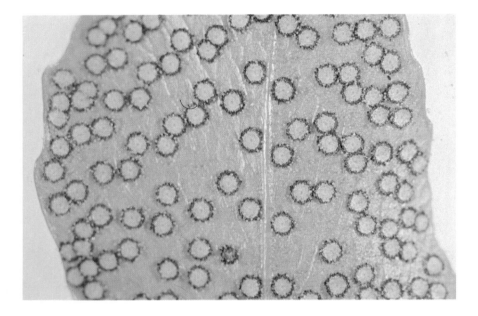

A mushroom—the most complex form of fungi—growing at the base of a tree. (Russell Dian.)

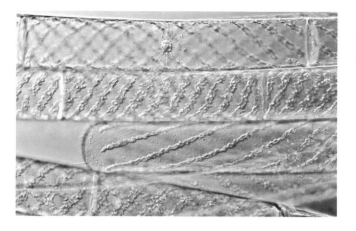

Filaments of spirogyra, a freshwater algae, about 150 micrometers wide. The cell nucleus is the round object in the center of one of the filaments. (Manfred Kage from Peter Arnold.)

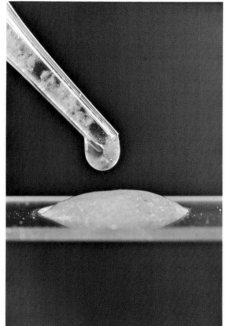

A pipette dispensing drops of water containing phytoplankton on a microscope slide. (Manfred Kage from Peter Arnold.)

Over the course of evolutionary history the spines of the cactus developed from leaves. (Manfred Kage from Peter Arnold.)

Millions of years ago ferns were the dominant form of land vegetation. (Russell Dian.)

The leaves of a maple tree, showing one of the many possible leaf structures. (Russell Dian.)

The pincushion lily, found in South Africa. (Richard Weiss.)

The petunia plant. There are many varieties both in color and size. (Russell Dian.)

Pollen-bearing anthers surround the pollen-receiving stigmas of this flower. (Manfred Kage from Peter Arnold.)

The insectivorous pitcher plant *(Sarracenia)* about to receive its prey. (Courtesy Carolina Biological Supply Company.)

The fruit (capsule) of the opium poppy contains a sticky white juice from which opium is produced. About 4000 capsules yield 1 pound of opium. (Courtesy Carolina Biological Supply Company.)

The pine is a gymnosperm ("naked seeded" plant) whose seeds develop beneath cone scales. (Courtesy Carolina Biological Supply Company.)

The starfish is an echinoderm commonly found in the intertidal zone. (Robert B. Evans from Peter Arnold.)

The spiny echionoderms in this photograph, sea urchins, are sharing a shallow marine environment with red sponges. (Robert B. Evans from Peter Arnold.)

The delicate tentacles of this sea anemone contain tiny harpoon-like stinging structures, nematocysts. (Robert B. Evans from Peter Arnold.)

The hermit crab has moved into the shell of a dead snail. (Marine Biological Laboratory.)

A spider, spinning its web. (Russell Dian.)

An assembly line of wasp pupas. (Hans Pfletschinger from Peter Arnold.)

A honeybee *(Apis)* covered with pollen in a flower. (Herbert Weihrich.)

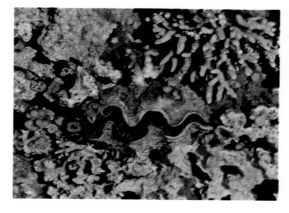

A giant clam's jagged "smile" may be several feet wide. Found in the Barrier Reef of Australia. (Richard Weiss.)

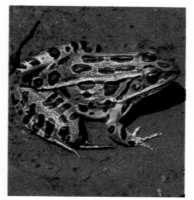

Rana pipiens, the common leopard frog. (E. R. Degginger.)

Marine iguanas *(Amblyrhynchus)* of the Galapagos Islands off the coast of Eucador. (Richard Weiss.)

Blue-footed boobies, which breed in the Gulf of California and the West Coast of Mexico. (Richard Weiss.)

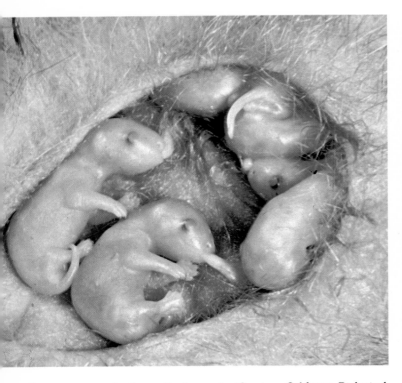

Opossum young in the mother's pouch. (Courtesy California Biological Supply Company.)

The bat is one of the few mammals capable of flying. (Courtesy Carolina Biological Supply Company.)

The common zebra *(Equus burchelli)* at a water hole in Ngoro Ngoro Crater, Tanzania. (E. C. Williams.)

America, the leopard in Africa, and the tiger in Asia, hunt for monkeys, deer, and other animals.

Many people confuse jungle growth with a rain forest, but there is a difference. A typical rain forest is climax vegetation. Jungle, however, is extremely dense ground growth that occurs along the edges of rivers or on land that was once cleared by man by some natural event like a flood or fire. If left alone, most jungle eventually becomes rain forest. A jungle is, therefore, a kind of immature rain forest. The worldwide distribution of these biomes is depicted in Figure 4-31.

THE MARINE ENVIRONMENT

Oceans cover more than two thirds of the earth's surface and support an abundance of life. Just as the continents are characterized by recurring ecosystems, so too is the sea (Figure 4-32).

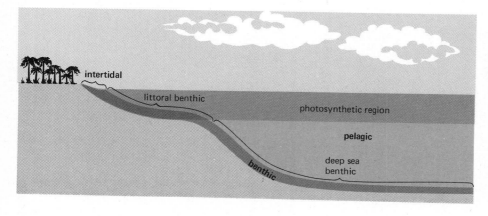

FIGURE 4-32
The ocean may be divided into several zones, each having characteristics that determine the kinds of organisms that are able to live in them.

The intertidal zone

The interface between the land and the sea is one of life's richest environments. The tides rise and fall in this zone, washing it with two tidal cycles a day. In areas of rocky coastlines, barnacles, anemones, mussels, and starfish cluster in tight communities. Crabs skitter across the rocks, and find shelter in cracks. When the tide is out, water caught in tide pools shimmers with brilliant animals and seaweeds. Overhead, seagulls hover in search of small animals and carrion to feed upon.

The sandy beach supports an entirely different kind of community. In addition to clams, small crabs and other crustaceans burrow into the sand. They are fed upon by long-billed, fast running shore birds. Bizarre microscopic animals cling to the sand grains.

The continental shelf

The sea floor slopes gently away from the intertidal zone, forming the **continental shelf.** This area of relatively shallow water may extend outwards for hundreds of miles, or it may abruptly end near shore, with the bottom plunging into trenches over 30,000 feet deep. As an average, however, the edge of the continental shelf is several miles from shore, and lies at a depth of about 600 feet. This is also the approximate depth to which sunlight penetrates, an important factor, since plants are capable of photosynthesis only above this level. Since no plant productivity occurs beneath the **photosynthetic region,** deep-dwelling organisms must depend on the dead organic matter that falls from above for nutrition.

The waters of the continental shelf support two different (but closely interacting) environments, the **littoral zone** and the **littoral benthic zone.** The life above the sea floor comprises the littoral community, while the bottom is the home of the littoral benthic organisms.

The littoral zone is one of the most productive areas of the earth. Phytoplankton form the base of the food chain here, and are fed upon by small crustaceans. These, in turn, are eaten by fish, which are eaten by larger fish. Fishing fleets ply the waters above the continental shelf for tuna, sardine, and anchovy.

In shallower areas, huge plants attached to the bottom dominate the littoral benthic zone. They sway gracefully in the shifting currents, and provide food and hiding places for a variety of fish. Beyond a depth of approximately 100 feet, these plants gradually become less prominent. The bottom, generally composed of fine mud or silt, is inhabited by clams, worms, and other burrowing animals. Scavenging fish search for the bodies of dead animals that have settled from above (Figure 4-33a).

FIGURE 4-33

At the left a sea robin appears to walk along the ocean floor with "fingers" feeling for food everywhere it goes. These "fingers" are really the first two or three spiny rays of the fish's pectoral fins. The rays are busy almost constantly, sifting the sand or hard-packed mud, turning over bits of gravel or small stones, seeking marine worms, tiny crustaceans, and some of the smaller mollusks. At the right, an underwater scene off the coast of southern California.

The ocean zone

Past the edge of the continental shelf, the sea floor slopes more rapidly until leveling off at an average depth of from one to two and one-half miles (for the great oceans). The waters of the open ocean comprise the **pelagic zone.** Beneath the level to which light penetrates, the pelagic is an environment of cold darkness and strange animals. Tiny fish, with gaping jaws, sparkle with patterns of **bioluminescence** (the light given off by chemical reactions occuring in animals) and giant squid search for prey.

The floor of the open ocean, or the **deep sea benthic zone,** is a relatively unknown area. But, photographs taken from deep diving vessels and samples of mud taken with coring devices have revealed that life does exist at the bottom of the sea (Figure 4-33c). Scavenging fish search for the remains of dead animals, and tiny worms, clams, and starfish live in the surface mud.

Bacteria living in the soft ooze on the bottom decompose plant and animal matter and, thereby, perform an important function for the cycling of nutrients in the sea. Just as on land, simple compounds are generated by decomposers, and reincorporated into living matter by plants. Deep water, rich in nutrients, is carried to the surface by upwelling currents, where phytoplankton and larger algae thrive in the photosynthetic zone. Upwelling is most intense at the western coasts of continents, which is why the offshore areas of Morocco, Southwest Africa, California, and Peru are the most productive waters of the earth. Nutrients brought into these areas stimulate the growth of plants, and the entire food chain subsequently flourishes.

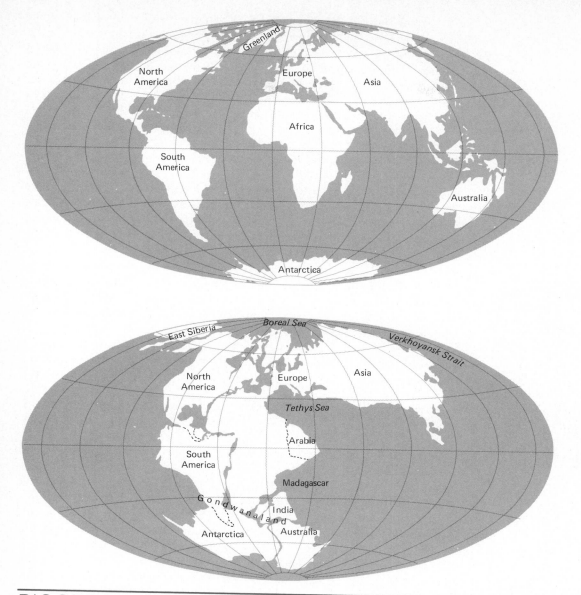

BIOGEOGRAPHY AND CONTINENTAL DRIFT

Life is easily disrupted by changes in the surface of the earth. A volcano can sweep an island clear of life, leaving it a sterile heap of lava. An earthquake can open a crack, which over many thousands of years can become a forest pond. The study of such geographical influences on distributions of species is called **biogeography.**

During the past few decades biogeographers and geologists have found evidence that changes in the earth on a greater scale than the mightiest earthquake have gradually altered life's environments. The continents are adrift on the surface of the earth, and are moving further apart (Figure 4-34).

Geologists have discovered that the earth's surface is divided into several vast "plates" that are growing at the centers of the oceans and plunging into deep submarine trenches. Although movement is only a matter of inches per decade, over millions of years the biogeography of the earth has profoundly changed.

Geologists believe that the land of the earth was once joined in a single "supercontinent." Gradually, however, land masses have drifted apart. Evidence for continental drift has come not only from geological studies, but also from fossils.

FIGURE 4-34
A single supercontinent, presumed to have existed some 150 million years ago, would have resembled that depicted in the map at bottom. A present-day map appears at top. In both maps the distortion of the continents is a result of the projection employed. (From "Continental Drift" by J. T. Wilson. Copyright © 1963 by Scientific American, Inc. All rights reserved.)

Ancient fossil plants and animals which have been unearthed on the eastern coast of South America are strikingly similar to those who found on the western coast of Africa. Biologists have concluded that the similarities could have arisen only if populations of these organisms were once in close contact. In the millions of years since the continents were joined, evolution has created marked differences between the organisms of these continents. Thus, the diversity of life on the face of the earth has increased as geological change has progressed.

HUMAN POPULATIONS

Humans aren't aloof from the forces that control the rest of life. However, more than any other species of animal, we have affected the processes that determine the sizes of our own populations and those of other organisms. To get a feeling for the extent of our influence, it is worth considering a short history of our peculiar evolution.

Studies of fossils of our ancestors have indicated that our humanlike ancestors first appeared on earth about two million years ago. Most likely, the scene of human evolution was Africa, since this is where the earliest humanlike fossils have been found. It is possible, of course, that similar fossils may eventually be found elsewhere in the world. Approximately 10 million years ago an extended drought destroyed many of Africa's forests, leaving a relatively barren savanna. It was on the African savanna that we were believed to have evolved from a line of scavenging and hunting apes.

Culture and population growth

Humans appear to have been an immediate evolutionary success. Populations grew rapidly and have continued to grow to the point where our presence threatens the world environment. The success of *Homo sapiens* can be summarized in a single word: **culture.** Culture is, in essence, a body of knowledge that is communicated from generation to generation. It is transmitted by symbols, not inheritance. No other species has developed a system of communication that approaches the complexity and power of ours. The ability to constantly develop and modify culture has given us increasing control of environments and has led to rapid growth in population sizes. In the early years of human evolution, culture probably consisted of knowledge necessary for survival. Methods of hunting and toolmaking may have been its essential aspects.

Anthropologists estimate that the total number of human beings of 10,000 B.C. was about 5 million (Figure 4-35). At this point, certain cultural developments

FIGURE 4-35
In this diagram, human population growth is plotted on logarithmic scales. Therefore, each division on the graph is 10 times different than the one next to it. Population growth is seen as occurring in three surges, as a result of the cultural, agricultural, and industrial-medical revolutions. (From "The Human Population" by E. S. Deevey, Jr. Copyright © 1960 by Scientific American, Inc. All rights reserved.)

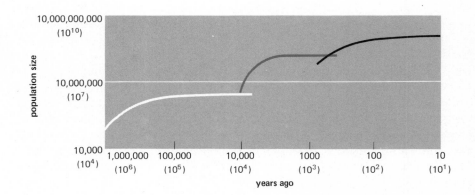

118

caused a surge in growth. Specifically, the development of agricultural techniques gave humans a more dependable source of food than hunting and gathering. Furthermore, agricultural societies could settle in communities, and protect themselves with permanent defenses against adverse environmental forces. A settled existence also permitted the development of better toolmaking techniques and other cultural advances that made life safer and easier. The average life expectancy gradually increased beyond the average of 25–30 years for early humans.

Following the development of agriculture, human populations grew at a relatively rapid rate. The number of individuals in 1650 A.D. has been estimated at approximately 500 million. Between 10,000 B.C. and 1650 A.D., therefore, the number of human beings increased approximately one-hundredfold. The rate of growth was by no means steady, however. At times, catastrophes, including famine, disease, and warfare caused temporary decreases in population sizes. The Black Death (bubonic plague), for example, resulted in a 25 percent decrease in the population of Europe between 1348 and 1350 A.D.

Between the sixteenth and eighteenth centuries, human populations grew more rapidly than ever before. This period was relatively peaceful, the plagues in Europe had ended, and the Industrial Revolution was underway. Death rates dropped considerably, and a period of rapid population growth continued into the nineteenth century.

The population explosion

At present we are in the midst of yet another period of accelerating population growth. At the time of World War II, greatly improved medical care was introduced on a large scale to the underdeveloped nations of the world. Modern drugs and information on health care were brought to countries in South America, Africa, and Asia by the technological societies in Western Europe and North America. Improved medical care has resulted in pronounced decreases in diseases such as malaria, cholera, and smallpox. The resulting decline in death rates has created an ironic problem, the so-called population explosion.

The population growth rates in the Third World nations are such that current methods of food production cannot keep up. Indeed, there is doubt as to whether any method of food production will be efficient enough to stop mass starvation. Even now, 10 to 20 million people are dying every year for lack of food.

Living in a nation that has more or less stabilized its population size, there is a temptation to blame the world's ecological problems on the rapidly reproducing, underdeveloped nations. But, the lavish lifestyles of our society make us at least as responsible for today's environmental dilemmas. Western society's insatiable demands for exotic foods, which are costly to grow and transport, are satisfied in spite of the hunger of the world's masses. Moreover, our fascination with travel and gadgetry is rapidly depleting the earth's store of fossil fuels and minerals. The developed, industrial nations of the world are primarily to blame for the pollution and depletion of natural resources.

Humans have drastically changed the environments of many organisms and consequently have influenced the process of natural selection. We are an evolutionary force that has driven hundreds of species to extinction. If we do not learn to live in balance with the other elements in our environments, we may well permanently alter the course of our lives. We must learn to exist more in harmony with the environment and more at peace with ourselves and the other living things of the earth.

CHAPTER 4: SUMMARY

1   Ecosystems are the basic units of analysis in ecology.

2   Food chains, along with biological cycles, determine energy flow in the biosphere.

3   Population growth helps determine the stability of a community and its ecological balance.

4   Organisms go to great lengths to survive predation, including symbiosis, protective coloration, and mimicry.

5   Light, temperature, the atmosphere, soil and water are physical factors affecting community development and survival.

6   Since environmental conditions constantly change, populations of organisms also change.

7   Different ecosystems offer different challenges to the organisms within them.

8   The growth of the human population had had profound effects upon ecosystems.

Suggested Readings

Hahn, James and Lynn Hahn   *Recycling: Reusing Our World's Solid Wastes,* Franklin Watts, Inc., New York, 1973 — The ecological and economic advantages of recycling.

Horwood, R. H.   *Inquiry into Environmental Pollution,* Macmillan Company, Toronto, Canada, 1973 — An information and activities approach to the pollution problem.

Newman, L. H.   *Man and Insects,* Doubleday and Company, Inc., Garden City, New York, 1967 — The significance of insects in the biosphere.

Odum, Eugene P.   *Ecology,* Holt, Rinehart and Winston, 1971 — The principles of ecology.

Page, Nancy M. and Richard E. Weaver   *Wild Plants in the City,* Quadrangle/The New York Times Book Co., New York, 1975 — A handbook about plants that grow in city environments.

Szulc, Tad   *The Energy Crisis,* Franklin Watts, Inc., New York, 1974 — Why we are in an energy crisis and why it will continue.

THE BIG HUNGER
George Wald

George Wald, who won the Nobel Prize in 1967 for physiology or medicine, calls himself "one of those scientists who does not see how to bring the human race much past the year 2000." In this excerpt from an article entitled "Arise, Ye Prisoners of Extinction," he sees the peoples of the "so-called free world" as pawns of giant enterprises. For the unskilled, unspecialized masses he predicts starvation.

The big hunger is now upon us, the great famines that scientists have been predicting for years past—hunger among the poor in the developed countries and starvation in Africa, South Asia, and South America.

The Green Revolution, so recently begun, has already collapsed. It depended on huge supplies of cheap oil and coal to prepare the artificial fertilizers and pesticides that alone made it work, and oil and coal are no longer cheap. The profits of the major oil companies—which also own most of the coal and are now developing nuclear power—doubled and tripled during the past year as the peoples of the Third World began to starve. It seems possible that 20,000,000 persons will die of famine during the next 12 months in India, Pakistan, and Bangladesh alone.

All of these problems are made more terrible by the population explosion. We have not yet quite taken in what that means. Even if all the developed nations reached the replacement level—an average of two children per producing pair—by the year 2000, and if all the nations of the Third World came to the same state by 2050 (both conditions highly unlikely), then the world population, which is now at about 3,700,000,-000, would rise by 2120 to about 13,000,-000,000.

Development, so-called, has meant mechanization. The work that used to be done by human and animal muscle is increasingly done by machines. That is true even in agriculture. It is another aspect of the Green Revolution. Farming is rapidly being replaced by "agribusiness."

In the U.S., the same huge corporations that make aircraft, control our oil and gas, and run our transportation also grow our food. Such agribusiness now controls 51% of our vegetable production, 85% of our citrus crops, 97% of our chicken-raising, and 100% of our sugar cane. That is happening all over the world. It means more food, but many fewer jobs, and only those who find work can eat and feed their families. Unemployment, that child of the Industrial Revolution, is rising throughout the world.

Moreover, a new phenomenon has developed that is much worse. With increasing mechanization, increasing numbers of persons have become not only unemployed, but superfluous. There is no use for them in the free-market economy. They are wanted neither as workers nor customers. They are not wanted at all. Their existence is a burden, an embarrassment. It would be a relief if they vanished—parents and children.

In his report to the International Bank for Reconstruction and Development (World Bank) in September, 1970, its president, Robert McNamara, former Ford executive and U.S. Secretary of Defense, spoke of such persons as "marginal men." He estimated that, in 1970, there were 500,000,000 of them—twice the population of the U.S.—and that by 1980 there would be 1,000,000,000; by 1990, 2,000,000,000. That would be half the world's population.

From G. Wald, "Arise, Ye Prisoners of Extinction," *Intellect*, April 1976.

WHAT IS THE "GREENHOUSE EFFECT"?
Isaac Asimov

Horticulturists have long used glass enclosures (greenhouses) to maintain the high temperatures necessary for plant growth. As described in the following article by the world's most prolific science writer, the atmosphere of a planet can act like a greenhouse. Changes in the atmosphere can ultimately affect the temperature at the surface of the planet. Some scientists believe that the input of carbon dioxide into the Earth's atmosphere through the burning of fossil fuels may result in enough of a temperature increase to cause a significant amount of melting of the polar ice caps.

When we say that some object is "transparent" because we can see through it, we don't necessarily mean that all kinds of light can pass through it. For instance, we can see through red glass, which is therefore transparent, but blue light won't go through it. Ordinary glass is transparent to all colors of light; it is, however, only slightly transparent to ultraviolet or infrared radiation.

Now imagine a glass house standing out in the sunlight. The visible light of the sun passes right through the glass and is absorbed by whatever is present inside the house. The objects in the house warm up as a result, just as do objects outside the house exposed to the direct light of the sun.

Objects warmed by sunlight give off that warmth again in the form of radiation. They are not at the temperature of the sun, however, so they don't give off energetic visible light. They give off, instead, the much less energetic infrared radiation. After a while, they give off as much energy in the form of infrared as they absorb in the form of sunlight, and their temperature remains constant (though they are warmer, of course, than they would be if the sun weren't shining on them).

Objects in the open have little trouble getting rid of their infrared radiation but the sun-warmed objects inside the glass house are in another situation altogether. Only small quantities of the infrared radiation they give off will go through the glass. Most is reflected, so that energy accumulates within. The temperature of the objects inside the house rises considerably higher than does the temperature of the objects outside. The temperature inside rises until enough infrared radiation can leak through the glass to set up an equilibrium.

Because of this, plants can be grown inside a glass house even though the temperature outside the house is cold enough to freeze them. The flourishing greenery inside such a glass house gives it the name of a "greenhouse." The additional warmth inside the greenhouse caused by the fact that glass is quite transparent to visible light and only slightly transparent to infrared is called the "greenhouse effect."

Our atmosphere consists almost entirely of oxygen, nitrogen and argon. These gases are quite transparent to both visible light and to the kind of infrared radiation the earth's surface gives off when it is warmed. The atmosphere also contains 0.03 percent of carbon dioxide, however, and this is transparent to visible light but not very transparent to infrared. The carbon dioxide of the atmosphere acts like the glass of the greenhouse.

Because carbon dioxide is present in such small quantities in our atmosphere, the effect is comparatively minor. Even so, the earth is a bit warmer than it would be if there were no carbon dioxide present at all. What's more, if the carbon dioxide content of the atmosphere were to double, the increased greenhouse effect would warm the earth a couple of additional degrees and that would be enough to bring about a gradual melting of the icecaps at the poles.

An example of an enormous greenhouse effect is to be found on Venus, where the very thick atmosphere seems to be mostly carbon dioxide. Astronomers expected Venus to be warmer than the earth since it is considerably closer to the sun. Not knowing the details of the composition of its atmosphere, they had not expected the additional warming of the greenhouse effect. They were quite surprised when they found that Venus' surface temperature was far above the boiling point of water and hundreds of degrees warmer than they had expected.

CHAPTER

5

Basic Chemistry

Objectives

1 Learn the structure of an atom and how atoms interact.

2 Understand radioactivity.

3 Understand the forces that hold molecules together.

4 Learn why water is an important chemical for life.

5 Learn to understand pH and acids and bases.

6 Define suspensions and colloids.

7 Distinguish among some major groups of biochemicals: carbohydrates, lipids, and proteins.

8 Understand enzyme function.

ATOMIC STRUCTURE

George Wald a Nobel laureate in Biology once suggested that life might be matter's way of knowing itself. But, if this is so, it has not yet succeeded. We consider ourselves the most self-aware species on earth (though, perhaps, we are overly self-confident in this assumption). Yet not one of us can claim to know truly the matter of which we are composed.

Atoms were once thought to be the ultimate, indivisible forms of matter—the smallest particles that could exist. Today, however, atoms are known to be composed of still smaller units. Physicists have shattered what was once thought to be the basic particle of matter, the atom, and have revealed a substructure. Three basic particles, **protons, neutrons,** and **electrons,** constitute all atoms.* The differences between oxygen, hydrogen, sulfur, iron, gold, uranium, and all other elements lie in the numbers and arrangements of these particles and depends upon no other properties that we know of.

The hydrogen atom has the simplest structure. It consists of a proton and an electron, and may contain neutrons. Each of the 92 naturally occuring elements have a characteristic number of protons and electrons, but atoms of the same element may have different numbers of neutrons. Neutrons and protons form the nucleus of the atom, about which electrons move. Most hydrogen atoms, for example, have no neutrons, but a very small proportion of hydrogen atoms contain either one or two neutrons. Atoms of the same element having different numbers of neutrons are called **isotopes.** Hydrogen, therefore, has three isotopes (Figure 5-1).

FIGURE 5-1

Representations of the three isotopes of the element hydrogen. While the numbers of protons and electrons are equal, the number of neutrons varies.

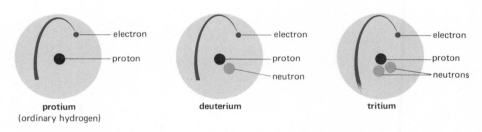

protium
(ordinary hydrogen)

deuterium

tritium

Protons and electrons have equal but opposite **electric charges,** while neutrons have no charge. Protons are said to have a positive (+) charge, and electrons possess a negative (−) charge. Neutrally charged atoms contain equal numbers of protons and electrons. Gold, for example, contains 79 of each. Since oppositely charged bodies attract one another, it is said that **electrostatic forces** bind the protons and electrons in atoms. However, since like charges repel, it is a mystery why atoms larger than hydrogen stay together when several protons may exist right next to each other. Physicists have suggested that there are "gluons" that hold all the particles together in the nucleus but we do not know how.

Another puzzling force inherent to matter, **gravitation,** results in attraction between bodies. The strength of gravitation is not determined by electrical charge, however, but by the masses of the bodies. A chair, for instance, isn't massive enough to attract itself to a table, but the earth and the sun are so massive that their mutual attraction binds them across a 93,000,000 mile distance.

In a relationship analogous to the revolution of the earth around the sun, the electrons of an atom move relative to the centrally located protons and neutrons but at a large distance away.

---

* The level of atomic structure described in this chapter is simplified. It should be kept in mind that the three "basic" particles described herein are themselves composed of many subparticles, whose physical interactions remain an elusive mystery.

The solar system analogy should not be taken too literally, however, because at the level of atoms things exist in a form which we cannot perceive. Physicists have discovered a totally new picture of matter in the realm of the atom. Because electrons have so little mass they have some characteristics which can be described as wave-like as well as other properties that are similar to particles. Thus, an electron is like a person riding on a wave.

Protons and neutrons are about equal in weight, while electrons are much lighter (a proton has approximately 2000 times the mass of an electron). The space occupied by the nucleus, however, is miniscule relative to the size of an entire atom. If a nucleus were on a scale the size of the period at the end of this sentence, a hydrogen atom that includes the electron would be as big as a house.* Thus, like the solar system, atoms seem to be mostly empty space.

Physicists have spoken of electrons as "clouds" hovering about the nuclei and having no precise location. But, no image from everyday life suffices to encompass all that has been discovered about electrons and the other constituents of atoms. The atomic world exists on a level to which our senses are blind, but since all things are composed of atoms, the ultimate explanation of ourselves must be at this level.

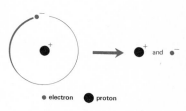

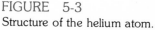

CHEMICAL ACTIVITY

Chemical reactions are changes involving the electrons of two or more atoms. Sometimes atoms merge to share electrons in aggregates called **molecules.** More simply, molecules are groups of two or more atoms held together. In other cases atoms gain or lose electrons to form electrically charged **ions** (Figure 5-2).

Some elements are reactive, while others are not. Even a small difference in atomic structure can mean a big difference in tendencies for atoms to be chemically active. Helium (Figure 5-3), for example, has two protons, two neutrons, and two electrons (making it the next larger element than hydrogen); yet helium and hydrogen have totally different chemical characteristics. Hydrogen is an extremely reactive element, which rarely exists as single uncharged atoms. But, atoms of helium almost always exist alone, losing their structural integrity only under the severest conditions—within the sun, for example, where protons and helium nuclei jostle in furious storms of free electrons.

FIGURE 5-2
In certain chemical surroundings, hydrogen (H) may ionize by losing its electron to become positively charged (H⁺). Some elements may gain electrons to become negatively charged.

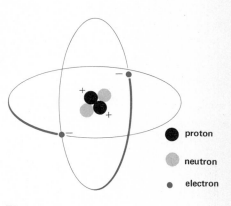

NUCLEAR REACTIONS

Chemical reactions do not usually involve nuclei. The mysterious forces that bind protons hold them aloof from normal chemical activity. However, in the sun and in man-made nuclear explosions nuclei of certain atoms react.

The reaction that produces the energy within the sun is essentially the same as that of a hydrogen bomb. Through a process termed **fusion,** hydrogen nuclei merge to form helium. Fusion in the sun involves a series of nuclear reactions, with the central one resulting from a collision between a proton and a proton bonded to a neutron. The reaction can be represented as follows:

$$^1_1H + ^2_1H \rightarrow ^3_2He + energy$$

The subscripts indicate *atomic numbers* (numbers of protons within nuclei), and the superscripts indicate *atomic weight* (total number of protons and neutrons). Thus, two isotopes of hydrogen react to form an isotope of helium that contains two pro-

FIGURE 5-3
Structure of the helium atom.

---

* This would represent an incredible expansion in actual size, since nuclei are about 0.0000000000001 (10^{-13}) centimeter in diameter; whole atoms are about 10^{-8} cm (or 1Å) in diameter.

tons and a neutron. In this process the total number of neutrons (1) and protons (2) does not change, only their distribution changes. In this rearrangement, energy is released in the form of heat, the heat of the sun.

Fusion occurs only at extremely high temperatures ($10^{7\circ}$C). In the sun it is a self-sustaining process, but on earth it must be touched off by a powerful explosion. Thus, fusion within a hydrogen bomb is initiated by another kind of nuclear reaction, one which involves the heavy atoms of uranium or plutonium which spontaneously break down into lower atomic weight atoms by the process of radioactive decay. In this **fission** process, the isotopes of uranium or plutonium break down to form lead, barium, and certain other elements. Simultaneously, neutrons are spewed from splitting nucleii at high speeds. Like bullets shot randomly, these neutrons are capable of doing damage to the nuclei of neighboring atoms including those of living organisms nearby. If a critical mass of the fissionable material is suddenly brought together a chain reaction of nuclear decomposition is initiated. A few kilograms explodes in a hail of neutrons and unstable nuclei. In a hydrogen bomb this critical mass is placed adjacent to a tank containing hydrogen; hence, the heat generated by fission touches off the process of fusion.

In nature a small proportion of nuclei are inherently unstable; they spontaneously decompose. But, such nuclei are isolated in seas of stable atoms, and do not initiate chain reactions. In the human body, for example, most carbon atoms are of the stable variety ($^{12}_{6}$C). Approximately one out of every trillion carbon atoms is of the unstable isotope, carbon-14 ($^{14}_{6}$C). An atom of carbon-14 may exist for thousands of years without breaking down, but eventually it will decompose.

"Nuclear Clocks"—Radioactive Dating   The rate at which carbon-14 breaks down is known, and paleontologists have put this knowledge to use to date fossils. Out of a given amount of carbon-14, only half will remain after 5,600 years; hence carbon-14 is said to have a **half life** of 5,600 years. While an organism lives, it incorporates carbon atoms (both carbon-12 and carbon-14) which first enter food chains as carbon dioxide from the atmosphere. In the atmosphere, a small concentration of carbon-14 is maintained as the result of an interaction of cosmic rays with nitrogen ($^{14}_{7}$N). After an organism dies, it no longer incorporates any carbon, and the concentration of carbon-14 gradually diminishes because it is radioactively decaying. The carbon-12 does not decay, however. In tissue that does not disintegrate (wood and bone) a "nuclear clock" begins to wind down at the moment of death. By measuring how much carbon-14 has decayed relative to how much carbon-12 is present, paleontologists can estimate the age of the material.

Carbon-14 dating is useful for calculating ages up to about 50,000 years. Older fossils are dated with a different technique, which involves analyzing the sedimentary rock that formed around them. Minerals that contain uranium (an isotope that decays to lead) or potassium (an isotope that decays to argon) provide nuclear clocks with half lives far greater than carbon-14. They are useful for dating fossils that are millions or even billions of years old.

Biological Effects of Radioactivity   Apart from dating fossils, biologists have found other uses for radioisotopes. Biochemical reactions can be analyzed, for example, by including a small amount of a radioisotope in the nutrients taken in by an organism and monitoring the radioactivity of the compounds consequently formed. Isotopes have been used with a certain amount of success in treating cancer. Biologists warn, however, of the powerfully disruptive potential of radioactivity. Exposure to highly concentrated sources of radiation (such as the waste from a nuclear power-generating plant) can result in a variety of ill-effects, ranging from loss of hair to death. Moreover, the offspring of an irradiated individual may suffer from debilitating genetic mutations. Thus, the scientific world is embroiled in a controversy over the current proliferation of nuclear energy balancing the probability of a nuclear accident versus our large demands for more energy.

ELECTRON DISTRIBUTIONS

The future may bring an increased amount of radioisotopes in the biosphere. But, presently nuclear reactions are relatively unimportant to the dynamics of living matter. Life, like other chemical activity, involves changes in electron distributions about nuclei (the vast majority of which are stable).

In an atom, electrons exist within distinct spatial zones, called **orbitals.** Some orbitals are spherical, with the nucleus at their centers, while others have more complex shapes (Figure 5-4). In general, electrons occupy the closest available orbitals to the nucleus. Energy is required to move an electron further from a

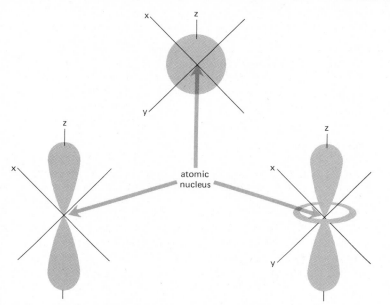

FIGURE 5-4
The shapes of three different electron orbitals. x, y, and z refer to the three dimensions in space. The darkened areas indicate the location of the electron clouds around the nucleus.

nucleus, and energy is given off when an electron falls from an outer to an inner orbital. For this reason, orbitals are also referred to as **energy levels.**[*]

Orbitals may contain either one or more electrons, and atoms in which all electrons are paired are usually more stable (that is, they are less likely to react chemically) than those in which one or more are unpaired. Hydrogen, for instance, has its single electron in the lowest energy level (the level nearest to the nucleus), while helium has two electrons in this orbital. Helium's lack of chemical reactivity is related to this paired electron configuration.

While paired electrons seem to give an atom stability, there is a pattern to the chemical characteristics of the 92 naturally occurring elements. Among them are six that have essentially no chemical reactivity. They are helium, neon, argon, krypton, xenon, and radon, which have 2, 10, 18, 36, 54 and 86 electrons, respectively. They are collectively termed the **inert** gases.

A discussion of the theories concerning the stability of inert gases is beyond the scope of this book. It is worth remembering, however, that recurring patterns in the distributions of orbitals about the nuclei of atoms leads to recurring patterns in their chemical natures. Inert gases possess similar electron configurations and constitute one of eight major groups of elements. Other elements tend to gain or lose electrons in order to achieve inert gas configurations and reach maximum stability. Atoms of calcium, for example, lose 2 of their 20 electrons to become

---

[*] Light is one form of energy involved in the shifts of electrons between orbitals. When it is absorbed an electron shifts to a higher energy level and light is emitted as the result of a fall to a lower level. Within any given element only certain electron shifts are possible, and each shift involves a specific color of light. Astronomers are able to make educated guesses of the chemical compositions of stars and other celestial bodies by breaking down the light from them into component colors.

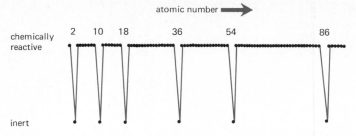

similar to argon. In so doing they become doubly charged positive ions (Ca^{++}). An atom of chlorine, which has 17 electrons, obtains the electronic configuration of argon by gaining an electron, becoming a singly charged negative ion (Cl^-). This pattern of chemical reactivity is reflected in Figure 5-5.

CHEMICAL COMPOUNDS

Over the past few centuries several general properties of chemical compounds have been discovered:

- Under certain conditions most elements will combine with one or more other elements. In other words, most elements exhibit the property of chemical activity. Only some of the inert gases are exceptions to this rule.
- Each element has a characteristic combining capacity for joining other elements. That is, a specific atom, such as hydrogen, has chemical properties that cause it to combine with certain other atoms, such as oxygen, but not with all other atoms.
- In forming compounds, elements combine in definite proportions. That is, the **empirical formula** for any given compound is always the same. Water (H_2O) always contains two hydrogen atoms and one oxygen atom, and the sugar glucose ($C_6H_{12}O_6$) always has 6 carbon, 12 hydrogen, and 6 oxygen atoms.
- Compounds exhibit unique properties that are unlike the characteristics of the elements composing them. Water, for example, is a liquid at room temperature, while hydrogen (H_2) and oxygen (O_2) are gases at room temperature. Similarly, sodium chloride (table salt) has none of the properties of either sodium (a soft metal) or chlorine (a poisonous gas).

Covalent and ionic compounds

Chemical bonds are linkages joining atoms in compounds. A bond is not an actual structure that connects atoms as commonly depicted in diagrams. It is rather

FIGURE 5-6

A representation of the formation of sodium and chloride ions.

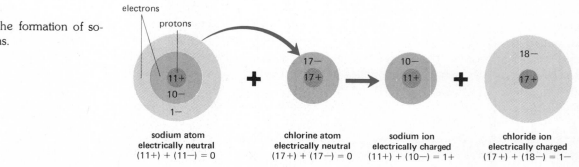

an electrostatic union. Compounds are of two basic types: **covalent** and **ionic**. In **covalent compounds** electrons are shared between two or more atoms. Some of the complex molecules of life (such as proteins and DNA) are covalent compounds including thousands of atoms. In **ionic compounds** electrons are lost by one component and taken on by another.

The electrons involved in both types of bonding are those that are furthest from the nuclei (or at the highest energy levels) of the component atoms. They are termed **valence** electrons. By forming either ionic or covalent compounds, atoms tend to achieve the electronic configurations of inert gases.

In a familiar ionic compound, table salt (NaCl), sodium atoms give up electrons to chlorine (Figure 5-6). In a grain of salt, sodium and chloride ions (Na^+ and Cl^-) are stacked in a rigid crystal, held together by electrostatic forces (Figure 5-7). A close look at a salt crystal reveals it to have a cubic structure, with perfect right angled edges; it contains millions of sodium and chloride ions. Smashing a few grains of salt with the underside of a teaspoon appears to produce a rough powder. But, if the powder is examined through a microscope, it will be found to have maintained a cubic structure. The grains of salt will have severed along planes separating layers of sodium and chloride ions. Thus, the electrostatic bonds linking the oppositely charged ions are easily broken.

The chemical bonds uniting atoms in covalent compounds are somewhat different than the forces within ionic compounds. Rather than involving transferals of electrons, covalent bonding involves electron-sharing. Gaseous hydrogen (H_2) is one of the simplest of covalent compounds (Figure 5-8). Within a hydrogen molecule the single electron from each of the component atoms is associated with both nuclei. Thus, through sharing electrons, each hydrogen atom achieves an electronic distribution similar to that of the inert gas, helium.

A similar phenomenon is involved in the bonding of oxygen atoms to form gaseous oxygen (O_2). However, in forming the covalent bond each oxygen atom contributes two electrons (Figure 5-9). Thus, by forming a "double bond," each of the oxygen atoms (which originally has eight electrons) obtains an electron configuration similar to that of the closest inert gas, neon (which has ten electrons).

The bonding within oxygen molecules is purely covalent and the bonding in sodium chloride is purely ionic. But, in most biochemical compounds the bonding is of both kinds. Such is the case with the most abundant compound of life, water.

THE NATURE OF WATER

Life probably originated in the ancient seas and has never severed its intimate ties to water, which composes well over half the mass of most living organisms. As the universal or, more precisely, earthly solvent of life, water is the medium in which biochemical processes occur. Moreover, it is the source of hydrogen and some of the oxygen required to form complex biochemical compounds. The unique chemistry of water deserves special attention.

Water molecules consist of two atoms of hydrogen bonded to a single atom of oxygen in a triangular arrangement (Figure 5-10). The nature of the bonding is essentially covalent. But, the bonding electrons aren't evenly distributed. The single electron from each of the hydrogen atoms is drawn toward the oxygen atom by the relatively powerful electrostatic pull of the oxygen nucleus (which contains 8 protons as compared to 1 proton in each of the hydrogen atoms). Furthermore, the nonbonding electrons of oxygen are held in orbitals opposite to the bonds with hydrogen. The overall effect is to give the oxygen atom a negative charge relative to the hydrogen atoms. For this reason water is said to be a **polar compound.**

The unequal distribution of electrons within molecules of water results in several chemical properties crucial to the functions of life. For instance, water is an excellent dissolver of ionic compounds. In blood, for example, positive ions (in-

FIGURE 5-7

A model of a crystal of sodium chloride. The sodium and chloride ions are held together in the crystal by ionic bonds.

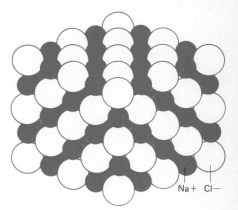

Na+ Cl−

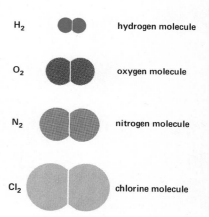

H_2 hydrogen molecule

O_2 oxygen molecule

N_2 nitrogen molecule

Cl_2 chlorine molecule

FIGURE 5-8

In these covalent compounds the bonding electrons are shared equally between the two atoms.

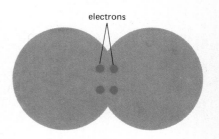

electrons

FIGURE 5-9

In a molecule of oxygen two pairs of electrons are shared, thus forming a double bond.

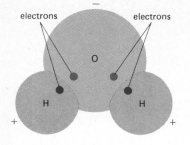

electrons   electrons

FIGURE 5-10

In a molecule of water the bonding electrons are held closer to the oxygen atom than to the hydrogen atoms. Thus, the oxygen atom has a relatively negative charge.

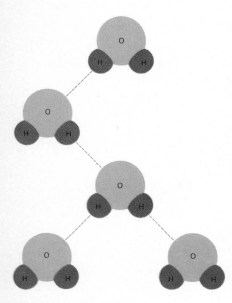

FIGURE 5-11

Hydrogen bonding can momentarily unite several water molecules in a "flickering cluster."

cluding Na^+, K^+, Ca^{++}, and Mg^{++}) are held in solution by the oxygen poles, and the hydrogen poles hold negative ions (including Cl^-, HCO_3^-, and SO_4^{--}) in solution. As will be discussed in later chapters, proper ionic balances are necessary for the functioning of nerves and muscles and the passage of materials in and out of cells.

Hydrogen bonding

Because of their polarity, water molecules not only interact with ions but also with each other. In a phenomenon termed **hydrogen bonding** neighboring molecules of water form loose bonds. A hydrogen atom of one molecule links with the oxygen atom of another. Relative to covalent bonds, hydrogen bonds are easily broken; hence in some substances they are short lived. They form and break in but a fraction of a second. At any given moment, however, several water molecules may be linked in a group by multiple hydrogen bonds (Figure 5-11). If it were possible to see what was happening at the molecular level in water, one would find a tumult of what have been called "flickering clusters" of water molecules.

When water freezes, hydrogen bonds no longer flicker but become static. Molecules assume an orderly array, and, similar to the way that ions of sodium and chloride form crystals of salt, molecules of water form crystals of ice. Water has a curious property in that when it freezes it becomes less dense. As the hydrogen bonds respond to the cold the molecules spread slightly apart (this is unlike the transformations of most liquids to solids, in which compaction results in greater density in the solid state). Thus, when a pond freezes in the winter, the ice floats rather than sinks. It's a lucky quirk of nature for pond organisms, which might otherwise be frozen or crushed by the ice.

Within organisms, hydrogen bonding not only involves water molecules, themselves, but is important in holding biochemical compounds in solution. Any compound that contains hydrogen, oxygen, and/or nitrogen is prone to forming hydrogen bonds. Such is the case with sugars, amino acids and DNA whose structures and functions in living systems are discussed later in this chapter.

Another important property of water is its high **specific heat,** defined as the amount of heat necessary to raise 1 gram of water 1°C. Water requires a great deal of heat input, relative to other liquids, to raise its temperature. Since most of an organism is water, this means the cells are protected somewhat against temperature changes that could severely effect life functions.

ACID/BASE EQUILIBRIA AND pH

Within water, a small proportion of hydrogen bonds result in dissociation—a hydrogen ion (which is simply a proton) is pulled away from one molecule and is momentarily held by another. It is a fleeting union, which can be represented as follows:

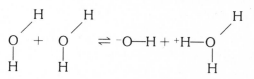

The double arrow indicates that this is a reversible reaction. The negatively charged **hydroxyl ion** (OH^-) may readily reacquire a proton from the positively charged **hydronium ion** (H_3O^+). Or, another water molecule may intervene by taking on a proton from the hydronium ion or giving up a proton to the hydroxyl ion. Thus, protons are in a constant state of flux in the jostle of the flickering clusters of water.

Reactions involving exchanges of protons (H^+) are termed **acid/base reactions. Acids** give up protons when dissolved in water whereas **bases** take them on. Water reacts as an aicd and a base, but certain compounds are either one or the other. Hydrochloric acid (HCl), for example, is a powerful acid. When added to water it quickly dissociates as follows:

$$HCl + H_2O \rightleftharpoons H_3O^+ + Cl^-$$

The HCl gave up a proton to the water molecule. This is a reversible reaction, but, as indicated by the top heavy arrow, most of the HCl comes apart rather than re-forms. Notice that there is no change in the total number of hydrogen, oxygen, and chlorine atoms. So much energy is liberated when a large proportion of HCl is added to water that the solution may momentarily boil.

Sodium hydroxide (NaOH) is as strong a base as HCl is an acid. It dissociates as follows:

$$NaOH \rightleftharpoons Na^+ + OH^-$$

At first glance this reaction may not appear to fit in with the definition of a base as a substance that takes on protons, but it must be kept in mind that hydroxyl ions have a strong affinity for protons. The donation of an OH^- to solution can also be defined as a property of a base.

Within the medium of water, the concentration of hydroxyl ions increases at the expense of the concentration of hydronium ions and vice versa. Pure water contains equal numbers of both, approximately one of each for every 600,000,000 molecules of water. The strongest of acid solutions has approximately one hydronium ion per 60 molecules of water and one hydroxyl ion per 6,000,000,000,-000,000 (60×10^{14}) molecules of water. The opposite is true for strongly basic solutions. As OH^- goes in one direction, H_3O^+ goes the opposite way.

Since the concentrations of H_3O^+ and OH^- are inversely related, only one number is needed to express both values (the concentration of one ion can be calculated if the concentration of the other is known). The convention used throughout the world for describing the acid/base nature of water solutions, the **pH scale,** was developed by a Swedish chemist, Jönenen Sørensen. Traditionally, scientists have referred to pH as an indication of proton (H^+) concentration. This is somewhat misleading since even in strongly acid solutions there are probably no protons that are not bonded to water molecules or to other chemicals. But, as a form of shorthand, it is a useful convention.

Sørensen noted that in going from strongly basic to strongly acidic concentrations, the proton concentration changed by a factor of 10^{14}. Since the range was so great, he decided to work with exponents of 10 rather than absolute concentrations. Thus, he arbitrarily established 7 as the pH of pure water, and the scale of from 1 to 14 as the range from strong acid to strong base (Figure 5-12). Therefore, a solution of pH = 5 has $10 \times$ less H_3O^+ than with pH = 4.

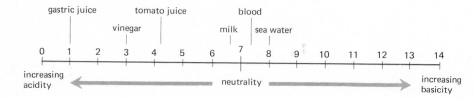

FIGURE 5-12
The pH scale.

The Carbonic Reactions   The pH of fluids within most living systems is slightly basic (around 7.2–7.4), and even small variations can have damaging effects. Within a certain range, however, organisms can handle fluctuations. If, for example, the pH of a person's blood falls slightly, that person will begin to pant. Consequently carbon dioxide is expelled from the blood via the lungs and the pH is raised. The following reaction is involved:

$$H^+ + HCO_3^- \rightleftharpoons H_2CO_3 \rightleftharpoons H_2O + CO_2$$

Within blood, carbon dioxide exists in combination with water as carbonic acid (H_2CO_3), which dissociates to form hydrogen and bicarbonate (HCO_3^-) ions.* Thus, carbon dioxide in solution in the blood lowers the pH by increasing the H^+ concentration. Panting results in the expulsion of carbon dioxide, the carbonic equation shifts to the right to correct for the loss. Hydrogen and bicarbonate ions combine to form the associated form of the acid; thus, the reaction shifts to a higher pH as H^+ has been removed.

People have learned how to take a more purposeful role in shifting the carbonic equilibrium toward neutrality by ingesting a salt of carbonic acid, sodium bicarbonate. More commonly known as baking soda or bicarbonate of soda, the formula of this popular remedy for acid indigestion is $NaHCO_3$. In stomach fluid, sodium bicarbonate dissociates to form sodium and bicarbonate ions. The latter, in turn, react with hydrogen ions to form carbonic acid. Since carbonic acid is a weak acid, carbon dioxide gas forms and bubbles out of the solution. Hence the familiar burp of relief.

SUSPENSIONS AND COLLOIDS

Substances composed of particles that are larger than ions or molecules are called **suspensions.** A suspension can be formed by stirring starch into water. The particles may remain dispersed through the water for a time, but they will eventually settle to the bottom, since the force of gravity is greater than the force that holds them in suspension. Thus, the size of the dispersed particles determines whether a substance will be dissolved in a solution or suspended. Generally, particles large enough to form suspensions can be seen with an optical microscope that magnifies about 1000 times.

Many substances are composed of particles that are larger than the small molecules that form solutions and smaller than the larger particles that settle out of suspensions. The particles of these substances may be very large molecules or groups of smaller molecules. They are dispersed in water in a special type of suspension known as a **colloid.** In describing colloids we speak of *dispersed particles* rather than solute (what is dissolved) and of *dispersing medium* rather than solvent (the dissolver). Particles in a colloidal suspension do not settle out. Living organisms embody many colloids along with suspensions and this combination forms the basis of cell **protoplasm.**

ORGANIC AND BIOCHEMICAL COMPOUNDS

Of the 92 naturally occurring elements 18 have been shown to be essential to human life although others may be important in very, very small amounts. Table 5-1 lists the approximate concentrations of elements in a 154-pound person. Each organism has unique requirements, but essentially the same elements are utilized by every organism on earth. The most abundant elements in living organisms are often also the most abundant elements on earth.

* Soft drinks are concentrated sources of carbonic acid.

TABLE 5-1	ELEMENTS ESSENTIAL TO HUMANS
ELEMENT	AMOUNT IN BODY (154-pound person)
oxygen	100.1 pounds
carbon	27.72
hydrogen	15.4
nitrogen	4.62
calcium	2.31
phosphorus	1.54
potassium	0.54
sulfur	0.35
sodium	0.23
chlorine	0.23
magnesium	0.077
iron	0.006
manganese	0.0045
iodine	0.00006
selenium cobalt copper zinc	minute traces

Within all organisms hydrogen and oxygen occur in the greatest concentrations (primarily as water); but it is the third most concentrated element, carbon, which is central to the structure of all organic compounds. Because of its unique chemical properties, carbon composes the backbone of most of the chemical compounds involved in life processes (Figure 5-13).* Carbon is extremely versatile in the manner in which it bonds. Its arrangement of valence electrons results in a

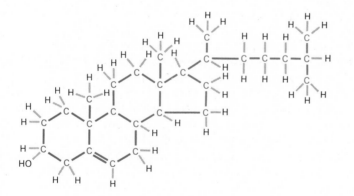

FIGURE 5-13
Cholesterol. The carbon "backbone" of the molecule is shown bonded to hydrogen and oxygen molecules. Note that there are always four bonds on each carbon atom because its valence is four.

tendency to form four covalent bonds, which may be combinations of single, double, or even triple bonds. Thus, a carbon atom may form four single bonds, or two double bonds, or two single and one double bond, or one single and a triple bond. Carbon combines with many other elements, but in biochemical systems hydrogen, oxygen, nitrogen, phosphorous, and sulfur (in decreasing order of abundance) are the most significant occurring elements found together. Carbon

* The chemical properties of carbon are closely approximated by another element, silicon, in the same chemical group. Since silicon and carbon have similar configurations of valence electrons, they react in similar ways. Writers of science fiction have hypothesized worlds in which silicon was the primary element of the compounds of life.

FIGURE 5-14
Structural formulas of three monosaccharides. All three molecules have the same number of carbon, hydrogen and oxygen atoms. Their properties are different because the arrangement of the atoms is different. Both chain (above) and ring forms are shown.

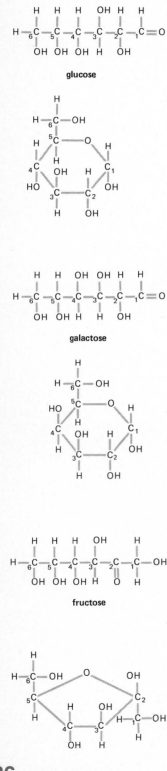

atoms also form bonds with each other, and their tendency to link in long chains results in some of the most significant biological properties of this vital element.

Originally "organic" chemistry was defined as the chemistry of those substances that were formed by living matter. It was assumed that the compounds of life couldn't be produced outside of organisms. In 1828, however, a chemist prepared a simple organic compound urea (which is excreted in human urine) by heating an "inorganic" salt (ammonium cyanate). Since the early nineteenth century, chemists have fabricated many of the compounds of life in laboratories including proteins and DNA. Moreover, chemists have composed variations on the themes of life; compounds have been created with structural similarities to those formed in living systems. Styrofoam, for instance, has carbon as the primary structural component, although no organism produces plastic cups. In the terminology of modern chemistry, styrofoam is **organic,** simply because it contains carbon.

In popular usage, however, *organic* still refers to compounds produced only by organisms, and certainly there is value in maintaining a distinction between compounds that are biologically active and those that bear only structural resemblances. Styrofoam has become an ecological eyesore simply because of its biological inactivity; bacteria can't even decompose it. To refer to compounds with biological activity, scientists use the term **biochemical.**

The following sections describe some of the most significant biochemical structures and reactions of life. This is by no means a complete structural survey of compounds essential to life. A description of the fourth major group, the nucleic acids, is left until Chapter 8, where they are considered in the context of their hereditary function. The biochemistry of energy production is described in Chapters 6 and 7, and various unique biochemicals are considered later in the book.

Carbohydrates

Carbohydrates are a major group of organic compounds composed exclusively of carbon, hydrogen and oxygen in the approximate ratio of 1:2:1, such as in glucose $C_6H_{12}O_6$. Biochemists have classified carbohydrates into three subgroups: *sugars, starches,* and *celluloses.*

Sugars   These are energy sources vital to all organisms. There are several types. Simple sugars, or **monosaccharides,** have five or six carbon atoms at the backbones of their structures. Three of the most significant monosaccharides in biological systems have the same empirical formula, $C_6H_{12}O_6$. One of these, **glucose** (which is produced in plants by photosynthesis) is utilized by both plants and animals as the primary cell fuel. It is also called dextrose, grape sugar, and blood sugar. Although **fructose** and **galactose** have the same empirical formula as glucose, their structures are different (Figure 5-14). Moreover, monosaccharides may occur either as chains or rings. In organisms they usually exist as rings.

Certain plants combine molecules of monosaccharides to form double sugars, or **disaccharides.** Two glucose molecules join to form **maltose,** or malt sugar, as illustrated in Figure 5-15. In a similar process, molecules of glucose combine with fructose to form **sucrose** (cane sugar or table sugar), produced by the sugar cane and sugar beet. A molecule of glucose joined to a molecule of galactose is **lactose,** or milk sugar.

The general formula for the formation of the preceding disaccharides is

$$C_6H_{12}O_6 + C_6H_{12}O_6 \rightleftharpoons C_{12}H_{22}O_{11} + H_2O$$

As is common with many biochemical reactions by which large molecules are formed from smaller constituents, a molecule of water is simultaneously formed and released. Hence the term **dehydration synthesis.** The reverse of dehydration,

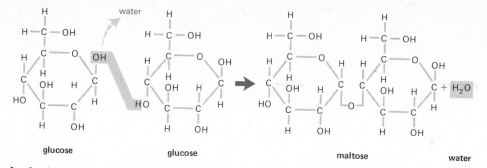

glucose   glucose   maltose   water

hydrolysis (by which simple molecules are generated from complex ones by the addition of water) is also a common biochemical process. Hydrolysis is the major reaction of digestion.

Starches   These are complex carbohydrates composed of monosaccharides linked in chains. The reaction that forms them is the same dehydration that generates disaccharides, repeated many times. This tendency to form chainlike molecules of the same or similar subunits (**polymerization**) is a characteristic of carbon chemistry which results not only in complex sugars (**polysaccharides**) but various other kinds of polymers. The chemical formula for a polysaccharide involving glucose may be abbreviated as $(C_6H_{10}O_5)_n$, in which the subscript n indicates a large number of glucose units—from dozens to several hundred.

Plant starches occur as the whites of potatoes and as the bulk of corn, rice, wheat, and other grains. Animal starch, or *glycogen*, is produced and stored in the liver and muscles. Starches are essentially inert and are not utilized directly by cells. When a need for energy arises, glycogen is converted to glucose, which enters the bloodstream.

Celluloses   Like starches, **celluloses** are polysaccharides, but their bonding is somewhat different. In celluloses, polysaccharide strands are bonded side by side, an arrangement that makes them fibrous. Cellulose is formed in the cell walls of plants, where it serves as support. It's commonly known in the forms of wood, paper, and cotton.

Lipids

Within this class of compounds are the *fats, oils, phospholipids* and *steroids*. Like carbohydrates, molecules of lipids contain primarily carbon, hydrogen, and oxygen. In lipids, however, the ratio of hydrogen atoms to oxygen atoms is much greater than two to one, and unlike carbohydrates, lipids are not soluble in water. They are soluble in such liquids as alcohol and ether. More energy is stored in a given weight of lipid than in the same amount of carbohydrate.

Fats   These occur as in butterfat in milk and form deposits beneath the skin when animals consume more food than needed for survival (as is common in our affluent society). Some fat molecules are composed of one molecule of glycerol and three fatty acids (Figure 5-16). **Glycerol** is a relatively small organic compound, based on a three-carbon chain. **Fatty acids** are polymers that commonly involve 10 to 20 $-\overset{\displaystyle H}{\underset{\displaystyle H}{C}}-$ units to which a **carboxyl group** $HO-\overset{}{\underset{\displaystyle \parallel}{C}}-$ is bonded at
                                                                                                            O

one end. Digestion disassembles these fat molecules.

Oils   These have essentially the same chemical structure as fats, except they have double bonds within the chains of carbon atoms of the fatty acids. The fatty

FIGURE 5-16

The dehydration synthesis of a molecule of fat. Three fatty acid molecules combine with a molecule of glycerol, and yield the fat molecule and three molecules of water.

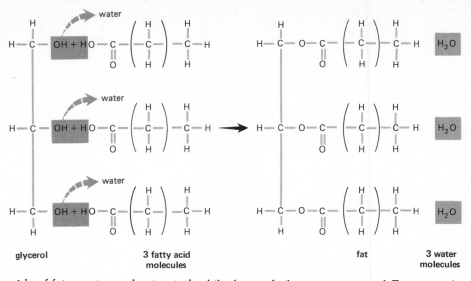

glycerol   3 fatty acid molecules   fat   3 water molecules

acids of fats are termed *saturated*, while those of oils are *unsaturated*. For example, C—C=C—C is unsaturated because if the double bond is broken, two more H atoms could be attached C—C—C—C making the carbons now saturated. This slight chemical difference is manifested as the difference between the relatively solid consistency of fats and the fluid consistency of oils.

In recent years diets rich in fatty acids (as in red meats and whole milk) have been cited as possibly causing heart disease. What has been shown to be important in the cause of heart disease is not actually the amount of these fats eaten, but the amounts that end up in the bloodstream. This amount varies from individual to individual. Nutritionists suggest replacing saturated animal fats with unsaturated vegetable oils (such as corn, peanut, and olive oils) whenever possible in cooking.

Phospholipids   These are similar to fats and oils in chemical structure except that a group of atoms including phosphorous occurs in place of one of the fatty acids. Phospholipids have a unique chemical property; the part of the molecule that contains phosphorous is soluble in water, while the portion with the fatty acids is water insoluble. Hence, when poured on water, phospholipids form a thin film, with the phosphorous groups beneath the surface and the fatty acids above. In organisms, phospholipids are important components of the membranes enclosing cells.

Steroids   The classification of *steroids* as lipids departs from the usual convention of grouping compounds according to structural similarities. Apart from the fact that carbon, hydrogen, and oxygen are the constituent atoms of steroids, they are quite different in structure from the other lipids. They do, however, have similar solubility characteristics. Like fats and oils, steroids are insoluble in water but soluble in ether and similar organic solvents.

All steroids are based on the structural format of four interlocking rings of carbon atoms (Figure 5-13). As vitamin D, hormones, and structural components of cell membranes, they perform a variety of functions. The diversity of steroid chemistry is all the more amazing when one considers the physical differences that result from seemingly minor differences in chemical structure. The difference between testosterone (a male sex hormone) and oestradiol (a female sex hormone) is but a single hydrogen atom (Figure 5-17). Surely there must be more to sex than one proton and one electron!

Proteins

Proteins are the most abundant organic compounds by weight in animals. As structural components, proteins form hair, skin, muscle, and other tissues. **Enzymes,** which regulate the chemical reactions of life, are always also proteins.

Protein molecules consist primarily of carbon, hydrogen, oxygen, and nitrogen. Sulphur, phosphorous, and iron (in hemoglobin, for example) are also included but in smaller amounts.

Amino Acids   Proteins are formed of long chains of molecules called **amino acids** which we have briefly discussed. You will remember that there are twenty-one different naturally occurring amino acids, all of which have the same basic structure. They all contain an amino group ($—NH_2$) and a carboxyl group (the same acid group included in fatty acids). Some amino acids are illustrated in Figure 5-18.

Amino acids link to form polymers through a dehydration synthesis (Figure 5-19). As with carbohydrates and lipids, the reverse reaction, hydrolysis, also occurs. The bond between amino acids involves the amino and carboxyl groups (the side groups aren't involved in the polymerization process). Biochemists term this a **peptide bond;** hence a polymer of amino acids is termed a **polypeptide.** Some protein molecules consist of a single polypeptide, while others involve two or more chains joined side by side as in hemoglobin. Most proteins aren't long, straight molecules; rather they are compressed like coiled springs and twisted back and forth upon themselves.

The numbers of amino acids in proteins range from three (glutathione) to over three thousand. Since proteins differ not only in kinds of amino acids present, but also in arrangement of them the possible number of different proteins is almost limitless. A single living cell may contain as many as five thousand different kinds of proteins.

The protein constituents of plants and animals vary from species to species, but within individuals of the same species proteins are quite similar. Nevertheless every individual has a certain uniqueness. While all human beings are alike in certain characteristics of protein makeup, for example, each of us has our own peculiarities.

To a certain extent, it appears that some proteins make us human, while others make us individuals. As will be related in Chapter 8, proteins are the products of the hereditary compound, DNA. Thus, proteins are the actors and DNA is the director of the molecular interplay of life.

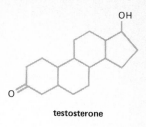

testosterone

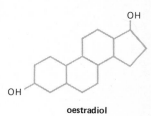

oestradiol

FIGURE 5-17
Testosterone and oestradiol. Only the carbon backbones of these steroid structures are shown.

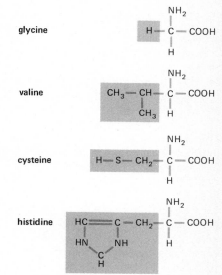

FIGURE 5-18
Some amino acids with side groups in boxes. Note the remaining identical structures.

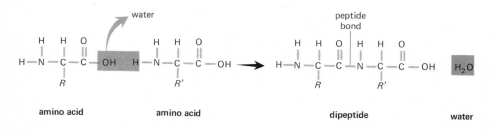

amino acid   amino acid   dipeptide   water

FIGURE 5-19
The dehydration synthesis of a dipeptide. As the water molecule is removed, the two amino acid residues become linked by a peptide bond. Long chains of peptides form polypeptides that compose protein molecules. R and R' indicate different side groups.

Enzymes   Biological systems manage to make the proper reactions occur in the right places at the right times with amazingly few errors. Obviously there must be complex mechanisms of control, and biochemists have discovered that the chemical agents of biological control are specialized proteins, called **enzymes.**

Enzymes are proteins that mediate chemical reactions. They facilitate the breaking and forming of chemical bonds by lowering the energy required to get a reaction going. If the temperature were raised, enough energy would be present to break the bonds apart, since an increase in temperature increases molecular motion and the motion would tear apart the bonds. But temperatures do not change much in cells. Something else is needed to provide this energy, and this is the role of enzymes. In a sense they "prime the pump" of biochemistry.

Because they mediate reactions but do not themselves change, enzymes are **catalysts.** Figure 5-20 illustrates a commonly used model of enzymatic activity. Enzymes are portrayed as fitting with **substrates** (the molecules that are to be

FIGURE 5-20
Model illustrating enzyme specificity.

active site

enzyme

substrate

enzyme + substrate ⟷ enzyme — substrate complex ⟷ enzyme + products

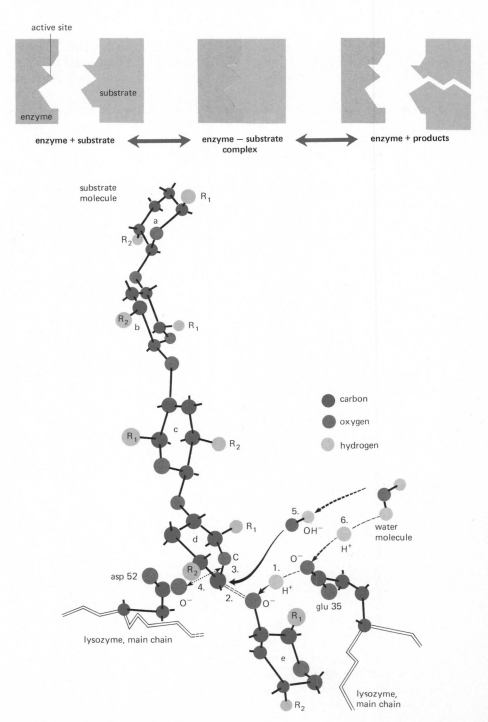

FIGURE 5-21
Enzyme action. The enzyme lysozyme is shown attacking a polysaccharide molecule (the substrate). Asp 52 and Glu 35 refer to amino acids in the 52nd and 35th position respectively of the enzyme. (Step 1) A hydrogen ion leaves the enzyme and binds to the oxygen of the polysaccharide, thereby breaking one of the bonds that hold the polysaccharide together (Step 2). A carbon atom is thereby left with a positive charge (Step 3), and to prevent it from reacting immediately, an oxygen in the enzyme binds to it (Step 4). An OH⁻ group from water now binds to the carbon of the polysaccharide (Step 5) and the remaining H⁺ from water replaces the one lost from the enzyme (Step 6). The enzyme has therefore ended up with the same structure as it started. Since bacteria have cell walls composed of polysaccharides, this illustration indicates how invading bacteria can be destroyed by enzymes of higher organisms. (From "The Three-dimensional Structure of an Enzyme Molecule" by D.C. Phillips. Copyright © 1966 by Scientific American, Inc. All rights reserved.)

substrate molecule

carbon
oxygen
hydrogen

asp 52

glu 35

lysozyme, main chain

lysozyme, main chain

water molecule

140

changed) like pieces of a jigsaw puzzle at a particular location called the active site. Forces exerted by an enzyme consequently cause the substrate to either break apart or form new chemical bonds. Once the reaction is complete the enzyme and the product(s) split, and the enzyme is free to interact with more substrates (Figure 5-21). This takes place in just a fraction of a second.

Enzymes are highly specific in their chemical activities. For example, an enzyme in human saliva (salivary amylase) acts on starches, breaking them down to sugar molecules. Thus, starchy foods begin to decompose in the mouth. Salivary amylase does not act on other foods, however. Proteins and fats are hydrolized later in the process of digestion, through the actions of other specialized enzymes.

CHAPTER 5: SUMMARY

1 Atoms are made up of protons, neutrons, and electrons that in turn are made up of many smaller particles.

2 Chemical and nuclear reactions involve the rearrangement of atoms and nuclear particles, respectively.

3 Radioactivity involves the emission of energy in some form from an atom.

4 Chemical reactions follow certain rules and resulting molecules are held together by chemical bonds.

5 Water is important to life because biochemical reactions take place in it and because water has a high specific heat.

6 pH is a measure of the acidity of a solution.

7 Acids and bases are important to cellular function.

8 Suspensions and colloids are part of every cell.

9 The major groups of biochemicals are: carbohydrates, lipids, vitamins, proteins, and nucleic acids.

10 Enzymes are always proteins (but may also contain vitamins), and speed up biological reactions.

Suggested Readings

Baker, J. J. and G. E. Allen   *Matter, Energy, and Life,* 3rd Ed., Addison-Wesley, Reading, Massachusetts, 1974—Reaction energetics, catalysis, reaction rates, equilibria, and thermodynamics.

Frieden, E.   "The Chemical Elements of Life," *Scientific American,* July 1972— The functions of the elements involved in life processes.

Grunwald, E. and R. H. Johnsen   *Atoms, Molecules, and Chemical Change,* 3rd Ed., Prentice-Hall, Englewood Cliffs, New Jersey, 1971—An introduction to basic chemistry.

Lehninger, Albert L.   *Biochemistry,* 2nd Ed., Worth Publishers, New York, 1975— One of the best written biochemical textbooks, but difficult.

Stroud, Robert M.   "A Family of Protein-cutting proteins," *Scientific American,* July 1974—A look at the function and evolution of a family of enzymes.

White, Emile H. *Chemical Background for the Biological Sciences,* 2nd Ed., Prentice-Hall, Inc., Englewood Cliffs, New Jersey, 1970—Life at the chemical level.

RUMOR AND CONFUSION FOLLOW OZONE THEORY REVISION

It came as a shock in 1974 when F. Sherwood Rowland revealed evidence that the fluorocarbon gases used as propellants for aerosol sprays were destroying the ozone layer in the upper atmosphere. Rowland's prediction was made on the basis of a computer model of the chemical activity in the atmosphere. Through manipulations of equations involving factors for movements of air masses and chemical equilibria, he concluded that ozone (which protects life from excesses of ultraviolet radiation) was being destroyed at a dangerous rate through reactions with fluorocarbons.

Recently scientists with different models for the physicochemical dynamics of the universe have concluded that there's not as much danger as Rowland had originally thought. At present the question is still up in the air, and it illustrates the difficulties involved in trying to translate natural phenomena into mathematical expressions.

Ruth Reck has been restructuring the earth's atmosphere again—a preoccupation not nearly as implausible as it sounds for a General Motors physicist. With commands typed neatly on keypunched computer cards, she wipes out massive amounts of ozone from the stratosphere or shifts the bluish gas around to different levels. Her latest set of models predicts some very interesting atmospheric behavior, and, moreover, promises to restructure another system—the current morass of confusion and uncertainty that was once an orderly theory of ozone destruction by fluorocarbons.

That theory and the supporting evidence, gathered by balloon, jet and infrared spectrometer, seemed so airtight last month that

rumor ran wild in Washington of an imminent ban on nonessential uses of fluorocarbons 11 and 12. Rumor is running rampant again this month that the theory was wrong, industry was right, the ozone layer is saved and fluorocarbons deserve an official reprieve. Or the inverse of all those statements, depending on who's talking. Committees impaneled by the National Academy of Sciences and a federal interagency task force called "IMOS" are meeting hastily to reconsider tentative recommendations. Scientists are shuttling in and out of the city to testify before the committees on new theories, models and evidence. And the founder of the original fluorocarbon-ozone destruction theory, F. Sherwood Rowland, nervously churns out yet more equations.

Rowland and co-worker Mario Molina proposed the now famous theory two years ago and, like the fluorocarbons themselves in the lower atmosphere, it had remained inert to serious attack until now. True to innovative form, Rowland, Molina and John E. Spencer of the University of California at Irvine made the first substantive revision in the theory themselves earlier this year.

Analyses, models and projections made since have touched off the latest paroxysm of speculation. Ruth Reck's study, reported in the May 7 *Science,* was begun and submitted before Rowland revised the theory, but, coming at this point, may shed considerable light on the murky situation as well as add force to the argument that all is *not* well with the ozone layer, theory revision notwithstanding.

Reck wanted to calculate the effects of ozone depletion on earth surface temperatures. Previous researchers had predicted that complete removal of ozone wouldn't really change the earth's surface temperature much (less than 1°C) but would completely eliminate the tropopause. The boundary between the troposphere and the stratosphere divides cold air below from warmer air above. This layering effect, in large part, controls weather and climate. Reck used a more sophisticated model of the atmosphere for her study. Plugging in ozone depletion rates from 10 percent to 100 percent, she, too, observed only a small surface temperature change up to 90 percent ozone depletion. Only at 100 percent was the tropopause eliminated.

This model, she told *Science News*, was "reassuring"; not even large ozone depletions —and certainly not the small ones estimated for current fluorocarbon release rates—would likely lead to drastic changes in surface temperature. But then Reck did a second, and ultimately disquieting, experiment. She lifted the "ozone profile" in her model. This profile, she explains, is the amount of ozone at each altitude. By arbitrarily changing the height of the maximum amount of ozone (the "bulge" in the ozone profile) but without removing any of the ozone, she saw a larger change in surface temperature than before.

Such a shift in the ozone profile, unfortunately, is predicted by both the old and new Rowland theories. And herein lies the importance of Reck's study for the current confusion. In Rowland's new theory, chlorine nitrate ($ClONO_2$) would be formed after reactive chlorine is kicked loose from fluorocarbons. It would then tie up the chlorine and prevent it from destroying as much ozone as originally predicted. Just how much prevention is afforded by $ClONO_2$, however, is a major point of contention now. Some industry-funded "modelers" calculate that with $ClONO_2$ in the picture, 90 percent less ozone would be destroyed. This practically exonerates the fluorocarbons. Others, however, like Rowland and Paul Crutzen of the National Center for Atmospheric Research in Boulder, Colo., calculate a 50-60 percent reduction in ozone depletion—less disastrous, but still a problem. The solution to this is still up in the air, so to speak.

The change in ozone profile, however, remains in the new model. $ClONO_2$ would not function effectively as a chlorine-atom-catcher at high altitudes (above 35 kilometers) due to the penetration of strong ultraviolet light. Thus, fluorocarbons reaching that height might break down a predicted 35 to 40 percent of the ozone above 35 km. Looking at a graph of the new ozone profile, were this upper stratospheric depletion to occur, Crutzen said, "This looks like the profile from another planet." And as Ruth Reck found, ozone profile changes theoretically could influence temperatures more strongly than even severe ozone depletion.

This new emphasis suddenly makes the fluorocarbons potential villains on the climatic scene to a greater extent, and villains on the skin cancer-crop damage scene to a lesser extent. And the uncertainty of the new theory and calculations now places a large part of the prediction problem in the hands of meteorologists —a group already saddled with such a complex, dynamic and poorly understood system that reliable prediction won't be forthcoming for a decade.

The co-chairman of the IMOS task force, Carol Pegler, told *Science News* that reevaluation is going on now and that the "troubling uncertainties raise new questions and research needs. But," she says, "as far as I am personally concerned right now, nothing in the new information removes the seriousness of the old."

Crutzen, one of the most highly regarded ozone researchers, told *Science News*, "I am sure that this industry will be phased out—if not for biological reasons, then climatological ones. There is no doubt that fluorocarbons will alter the ozone in some way; the longer we go on, the more it will be altered. Perhaps a two to three year phase-out is a reasonable compromise, I don't know. That's not my area. But I would be happiest if it were to go as soon as possible."

CHAPTER **6**

The Cell

Objectives

1 Distinguish between light and electron microscopes.

2 Learn about cellular processes.

3 Distinguish among the cellular organelles.

4 Define procaryotes and eucaryotes.

5 Determine the difference between plant and animal cells.

6 Learn how substances move across membranes.

7 Learn how cells divide.

8 See how cells are organized into organisms.

THE DEVELOPMENT OF CELL BIOLOGY

> I took a good clear piece of cork, and with a Penknife sharpen'd as keen as a Razor, I cut a piece of it off, and thereby left the surface of it exceeding smooth. Then examining it very diligently with a Microscope, me thought I could perceive it to appear a little porous . . .

So wrote Robert Hooke in 1665, at the start of his report, "Of the schematisme or texture of Cork and of some other such frothy bodies." Hooke went on to confirm his original observation and to describe cork as a microscopic "honeycomb" of adjoining "cells."

In the century and a half following Hooke's observation, microscopists discovered cells in other substances of life; but little was added to the knowledge of cellular functions. Cells were regarded simply as little boxes that contained the stuff of life. Today, however, it is known that even the most simple of cells is a complex, integrated piece of biological machinery.

Rapid strides in cell biology began early in the 19th century when a German botanist, Matthias Schleiden, observed various species of plants and concluded that all of them had a cellular structure. Soon thereafter a zoologist, Theodor Schwann, made a similar statement regarding animals. The work of Schleiden and Schwann led to the establishment of **cell theory,** which states that:

- The cell is the basic structural unit of all living things.
- Organisms grow and reproduce by cell division.
- A cell can be defined as protoplasm containing some internal structure all of which is surrounded by a membrane.

Microscopes

Robert Hooke made his observations of cork with a primitive forerunner of modern **light microscopes** (microscopes that utilize visible light). Over the years, sophisticated systems of optics have been developed which offer clear resolution up to a maximum 2000 times (2000 ×) actual size. But, because of the nature of light itself, this is as great a magnification as can ever be achieved with a light microscope. Light travels in waves, and, just as an ocean wave can pass essentially undisturbed over a small protruding rock, light will not be reflected by objects whose dimensions are smaller than the distance between peaks, or wavelength, of light (Figure 6-1). Thus, the resolution (the smallest object observable) of a light microscope is limited by the wavelengths of visible light which are about 5000 Å (½ micrometer).

A beam of electrons travels with one one-hundred-thousandth the wavelength of light. Therefore it will reflect back to the observer electrons of objects far smaller than the light microscope would do. Since we cannot see those reflected electrons directly, a phosphorescent screen (as on a television set) must be used to receive these electrons, which produce visible light where they hit the screen and produce the image.

In the 1930s physicists learned to control electron beams and use them instead of light to magnify objects. Biologists were given a tool, the **transmission electron microscope,** which could resolve the outlines of cellular contents at magnifications exceeding 200,000 times (Figure 6-2). More recently a new kind of electron microscope, the **scanning electron microscope** has been developed. It differs from the transmission electron microscope in that it is able to magnify the surfaces of cells. This ability, coupled with resolutions and magnifications

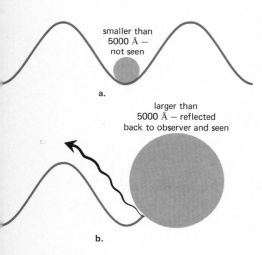

smaller than 5000 Å — not seen

a.

larger than 5000 Å — reflected back to observer and seen

b.

FIGURE 6-1

Light passes by objects whose dimensions are smaller than the wavelength (a), but it is reflected by objects with larger dimensions (b). The same is true of electron waves except the wavelengths are very much smaller.

This cross section of a one-year-old oak tree clearly shows its central conducting, vascular tissues. (Manfred Kage from Peter Arnold.)

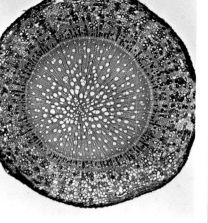

This bud of a spruce tree contains meristematic tissue, wherein cell divisions result in the growth of a stem. (Manfred Kage from Peter Arnold.)

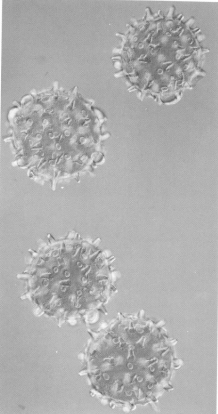

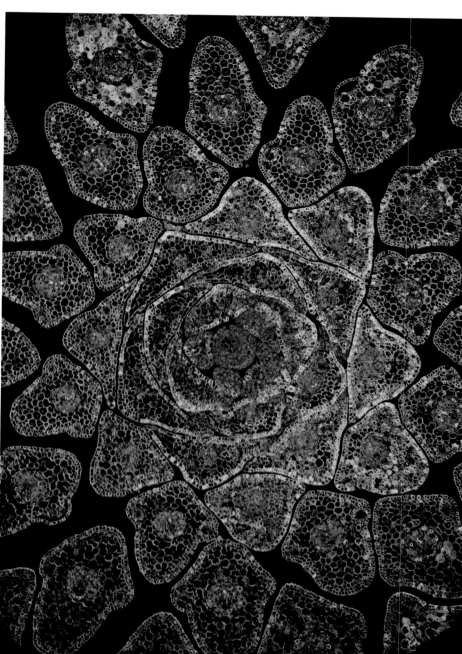

The protrusions from these hibiscus pollen grains help them to cling where they land. (Manfred Kage from Peter Arnold.)

The magnified leaf surface of the iris plant. (Manfred Kage from Peter Arnold.)

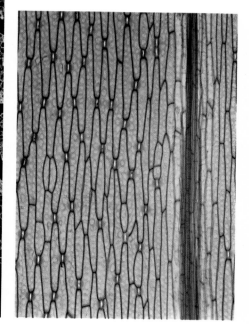

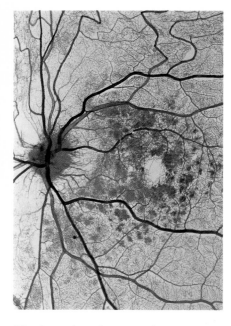

Blood vessels in the retina of an eye. (Manfred Kage from Peter Arnold.)

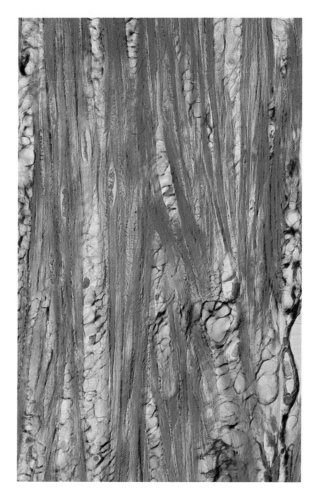

Long, fibrous cells of smooth muscle. (Manfred Kage from Peter Arnold.)

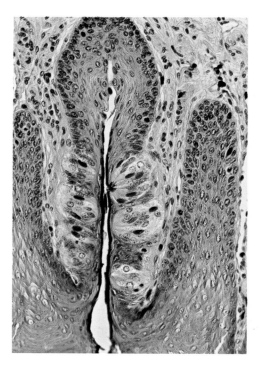

Cross section of a taste bud in the tongue. (Manfred Kage from Peter Arnold.)

This cross section of a spinal cord (far right) shows the complex interconnections of neurons. (Manfred Kage from Peter Arnold.)

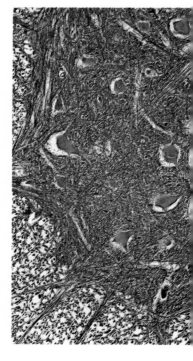

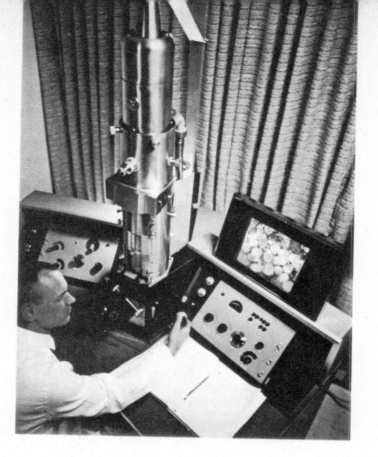

approaching that of the transmission electron microscope, has given us some fascinating views of the biological world such as the one in Figure 6-21.

Variations of these two types of electron microscopes have even allowed us to see images of individual atoms with magnifications of 20 million times (Figure 6-3). To work at the level revealed by the electron microscope, microscopists have developed techniques by which cells are embedded in plastic and sliced with diamond blades to thicknesses of a few millionths of an inch (500 Å). If the specimens are much thicker than this, the electron beam cannot penetrate and there will be no image on the screen. By means of microsurgery biologists have removed portions of cells such as nuclei for isolated study with microscopes. Other cell structures have been isolated by first crushing the cells to break them open, then centrifuging the solution of cellular components at high speeds (up to 100,000 rpm) to separate them according to density. The most dense cell particles, the nuclei, are spun down first, and the least dense cell particles, ribosomes, come down last.

CELLS—THE COMMON DENOMINATOR OF LIFE

The simplest organisms consist of but one cell. Bacteria, algae, protozoa, and yeasts are examples. Multicellular organisms may be composed of anywhere from just a few cells up to 10^{15} for ourselves.

The size of an organism is determined not by the size of its cells but by the number of them. Most of the cells of an elephant for example, are no larger than those that compose an ant. Large or small, simple or complex, the cell is the unit of structure and function of all living organisms, and all cells are the same in that they are units of living matter surrounded by a membrane. Because of special properties of this membrane, cells maintain a chemical composition different from

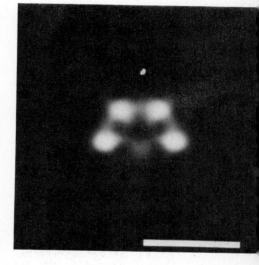

FIGURE 6-3
An electron microscope picture of a molecule showing mercury atoms (the light spheres) bonded to carbon atoms in a ring that contains a sulfur atom (the grey sphere at the bottom). The length of the white bar at the bottom of the picture corresponds to a distance of 15 Angstroms (0.0015 micrometers long).

THE CELL **147**

FIGURE 6-4a
The structure of a paramecium, a ciliate (200 micrometers long).

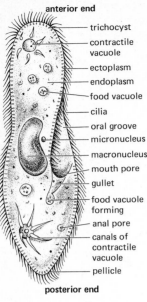

FIGURE 6-4b

Trypanosome, a flagellated protozoan (25 micrometers long). This protozoan is the cause of sleeping sickness.

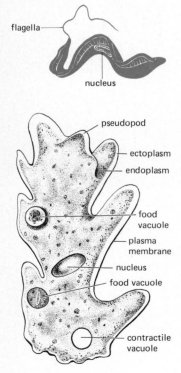

FIGURE 6-4c

The structure of an amoeba. This protist assumes many different shapes as it moves (size, 50 micrometers up to four millimeters).

the medium in which they live, such as blood in the case of animal cells, or sap in the case of plant cells.

To be sure, not all cells are the same in structure and function. Nerve cells, for instance, may be a meter long, whereas some bacteria are only a micrometer (10^{-6} meter) in diameter. Moreover, complex organisms contain many different types of cells. Nerve cells have long, branching projections, adapted for transmitting electrical impulses; red blood cells are platelike bodies, adapted for carrying oxygen. Although cells are of various forms, however, they have certain things in common.

Cellular processes

Among the cellular processes that occur throughout the living world—from bacteria to humans—are the following.

Absorption   Water and nutrients are taken into a cell from its environment.

Synthesis   The complex molecules of life are synthesized within cells from simpler nutrient molecules. Biosynthesis requires energy.

Respiration   Chemical energy is made available for biosynthesis and various other energy-requiring processes when certain nutrient molecules, especially glucose, are broken down by enzymes. While all cells respire, only certain cells are capable of trapping energy quite directly from the environment. Green plants are the primary energy-trappers of the biosphere. They convert light energy to biochemical energy by photosynthesis.

Secretion   Certain cells synthesize molecules that are discharged from the cell to perform functions elsewhere in the organism. The male hormone, testosterone, for example, is produced within the testicles but eventually reaches all parts of the body via the bloodstream. Deepening of voice and beard growth are among its many effects.

Excretion   Nonutilizable by-products of respiration and other wastes are eliminated by cells in various ways. Excretion, secretion and absorption are also called transport processes since they all involve the transportation of substances across cell membranes.

Movement   Different cells have different means of getting from one place to another. Locomotion may be by means of tiny oarlike cilia (Figure 6-4a), whiplike flagella (Figure 6-4b), or extensions of protoplasm (Figure 6-4c). Complex animals move by coordinated contractions and relaxations of many muscle cells.

Response   External stimuli such as heat, light, and gravity influence the activities of certain cells. The single-celled, photosynthetic *Euglena,* for example, seeks positions of optimum light intensity by swimming with lashing movements of its flagella.

Reproduction   Cells reproduce by division. This results in an increase in the number of cells within a multicellular organism or (for single-celled species) an increase in the number of organisms.

The nucleus

Cells contain several well-defined components of which the **nucleus** stands out as the most prominent. It is usually spherical or oval in shape, is centrally located and is surrounded by a membrane. Most cells have one nucleus, but some cells have two or more. Some small invertebrates like nematodes and rotifers have several nuclei per cell. Certain cells, however, lack a nucleus. Among these are bacteria. Red blood cells possess nuclei when they are developing in bone marrow but lose them after they mature and are released into the bloodstream.

The nuclear membrane is not a complete barrier. What appears to be a smooth covering under the light microscope is seen to be porous with the electron microscope (Figure 6-5). Certain substances pass freely through these pores, between the nucleus and the surrounding cellular substances.

The medium of nuclear activity is a protein-rich substance called **nucleoplasm.** Distributed in the nucleoplasm are numerous fine strands called **chromatin,** which are composed of DNA, RNA, and protein. When a cell is ready to divide, the chromatin changes shape to form the **chromosomes,** the cellular structures of definite shape that carry hereditary information. Chromosomes range in number from 8 for the fruit fly, 46 for humans, and up to 1600 for certain protozoans.

Also within the nucleus, is a spherical body called the **nucleolus** (Figures 6-6 and 6-7) in which certain cellular structures, ribosomes, are synthesized.

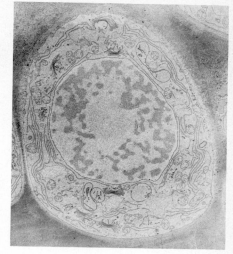

FIGURE 6-5
An electron micrograph of a cell in a maize root tip. Its structure is typical of the cells of many higher plants. The nucleus is in the center.

FIGURE 6-6
Model of an animal cell and organelles.

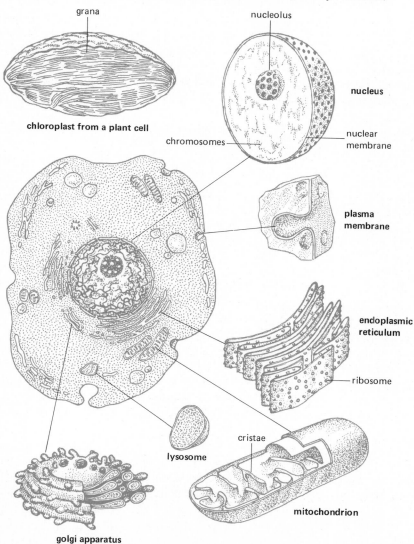

grana

chloroplast from a plant cell

nucleolus

nucleus

chromosomes

nuclear membrane

plasma membrane

endoplasmic reticulum

ribosome

lysosome

cristae

mitochondrion

golgi apparatus

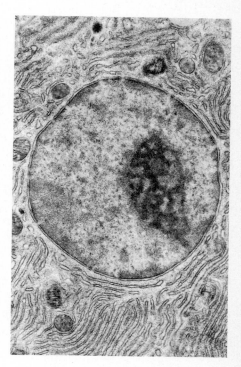

FIGURE 6-7
Electron micrograph of the nucleus. The nucleolus is the dark mass within the nucleus. Note the apparent pores in the nuclear membrane. The darker spheres in the cytoplasm are mitochondria.

Prokaryotes and eukaryotes

Before going further with a consideration of the similarities of structure and function of cells, it is worth considering one basic difference. Cells may be classified into two groups, the **prokaryotes** and **eukaryotes.** Prokaryotic cells lack a nucleus and possess only ribosomes. The hereditary material (DNA) is not segregated from the rest of the cellular contents by a nuclear membrane, although DNA may tend to gather in certain areas.

The prokaryotic cells of bacteria, blue-green algae, and certain other microorganisms are the smallest and simplest organisms, but each prokaryotic cell is equipped with the necessary biochemical machinery to maintain itself and divide. Although some prokaryotes attach themselves into groups of cells, such as chains of bacterial cells, they do not form cooperative units in which one cell depends upon others for its survival.

The eukaryotic cells, which include most of the cells of higher organisms, are larger and more complex than the other kind of cells. They have a nucleus surrounded by a membrane and contain many different types of organelles, (little organs), which perform specific functions. A single eukaryotic cell, like a prokaryotic cell, may exist independently of other cells as a complete organism. Other kinds of eukaryotes exist in cooperative units of from a few dozen cells (a strange group of animals called the Mesozoa), to form a colonial organism or a simple animal or plant, up to the higher animals and plants that are made up of many trillions of interdependent eukaryotic cells.

Prokaryotic and eukaryotic cells do have common features. Both, for instance, are limited at the outer boundary by cell membranes of similar thickness and gross structure, although the chemistry of the membrane differs for the two groups. Furthermore, both use DNA as the carrier of hereditary information. Much of what scientists have deduced about genetic control mechanisms comes from studies of prokaryotic cells. The fact that these findings also seem applicable to eukaryotic cells attests to the unity of life.

Eukaryotic cells are generally regarded as complex descendants of ancestral prokaryotes. It is unlikely that the present eukaryotes have descended from modern types of prokaryotes, but it does seem that both have descended from a common, now extinct prokaryotic ancestor.

The cytoplasm

The cell substances outside the nucleus are collectively termed the **cytoplasm.** Under a light microscope, cytoplasm appears as a semifluid colloid that fills the cell. It may shift in riverlike flows, a phenomenon called **streaming** which serves to mix up the cellular constituents and allow molecules to move from one place to another. The nucleus may be caught up in the streaming and may change shape as it moves, like a half-inflated rubber raft on a winding river.

Observation of cytoplasm under low magnification gives an erroneous impression of its composition. It is not a simple colloid; rather it contains many different kinds of organelles with specialized biochemical functions. Under magnification with the most powerful of modern electron microscopes it has been shown to contain a mesh of tiny filaments, **microfilaments** (50 A in diameter) that form a sort of cytoplasmic skeleton, giving rigidity to cells and aiding in movement.

The cell membrane

At the outer edge of a cell is a thin molecular layer (75Å), the **plasma membrane,** which isolates the cell from its environment. Like the nuclear membrane, the

FIGURE 6-8

A cell membrane model. The lipid molecules (small spheres with tails) are shown forming the membrane with larger protein molecules floating in the lipids.

membrane model

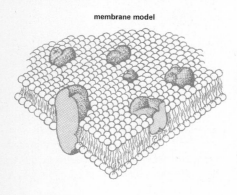

plasma membrane is not a total barrier; certain molecules pass through it. However, the plasma membrane is more selective; the pores are smaller than those of the nuclear membrane. The plasma membrane forms where the cytoplasm borders on another substance, such as the fluid outside the cell or on another cell.

High magnification by the electron microscope reveals that a plasma membrane is not as simple as it appears under the light microscope. The membrane consists of a sea of different types of lipid compounds in which protein molecules float (Figure 6-8). This is called the fluid mosaic model. Early representations of the plasma membrane portrayed it as a layer of lipids sandwiched by two layers of protein, but protein molecules are now known to extend all the way through the membrane. Movements of the proteins determine what goes into and out of the cells because as a protein moves, it leaves a pore where it was before.

Endoplasmic reticulum

At a magnification of about 30,000 times, the cytoplasm is seen to contain a complex system of double membranes lying parallel to one another. This network, the **endoplasmic reticulum,** serves as a system of canals through which materials are moved about in cells (Figure 6-9). The endoplasmic reticulum has been shown in some cases to link the nuclear membrane with the cellular membrane. Thus, materials can move from the nucleus to the outer surface of the cell.

Ribosomes

Attached to certain portions of the endoplasmic reticulum are organelles known as ribosomes. Ribosomes are spherical bodies, about 300 A in diameter. (Figures 6-6, 6-9). They are composed of about 30 different kinds of proteins and of what is called ribosomal RNA (a structural relative of DNA). There are millions of them per cell, and they are all identical. Ribosomes are protein "factories." Under direction by the nucleus they produce the proteins needed throughout the cell. Although ribosomes are among the smallest organelles, they are among the most vital cellular components. Ribosomes also exist free; not bound to endoplasmic reticulum.

Mitochondria

Mitochondria are rod-shaped organelles that vary tremendously in size. There is evidence that certain yeast cells contain one large mitochondrion; other types of cells are thought to contain thousands of small ones. They are the centers of cellular respiration, and they "package" energy in a form that can be utilized throughout the cell. A mitochondrion is bounded by two membranes, each characterized by the fluid mosaic model. The inner membrane folds inward at various places, forming partitions (Figure 6-10). These infoldings, called **cristae,** increase the surface of the membrane, thus increasing the area over which chemical reactions can occur. Small mitochondria are about 2 micrometers in size.

Mitochondria contain certain enzymes that split organic molecules (thereby releasing energy) and other enzymes that direct the reactions by which the unique energy-storing compound, ATP, is formed. ATP is released to the cytoplasm, where the energy is utilized in various cellular processes. This process is described in Chapter 7. Mitochondria have their own DNA and ribosomes, so they can divide independently of cell division and engage in protein synthesis as well.

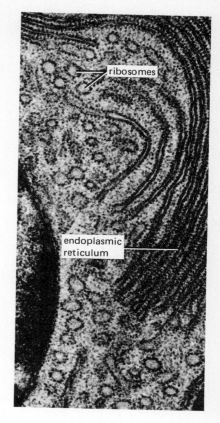

FIGURE 6-9

Electron micrograph of the endoplasmic reticulum and ribosomes. A small portion of the nucleus appears at the left.

FIGURE 6-10

Electron micrograph of a mitochondrion (size, about one micrometer), showing the outer and inner membranes and cristae.

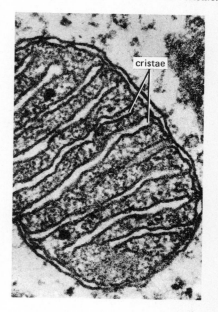

151

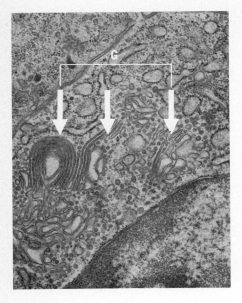

Electron micrograph of Golgi apparatus (G).

The Golgi complex

The **Golgi complex** was first observed in nerve cells by the Italian neurologist, Camillo Golgi in the late nineteenth century. For many years its fine structure and function was a mystery, but the electron microscope and sophisticated techniques of biochemical analysis have revealed its role in life.

Golgi complexes are small groups of parallel membranes in the cytoplasm near the cell nucleus (Figures 6-5, 6-6, and 6-11). They are most numerous in cells composing glands (although they also occur in other cells), where they store substances produced by other parts of the cell prior to secretion.

Substances to be secreted leave the endoplasmic reticulum and accumulate in tiny sacs of the Golgi complex. These sacs enlarge, separate from the Golgi complex, and move to the plasma membrane. The membrane of these sacs fuses with the plasma membrane, and the contents of the sacs are discharged to the exterior of the cell. This process has been observed in the production of plant cell walls and in the secretion of enzymes and other substances by animal cells.

Lysosomes

Lysosomes are spherical bodies a few micrometers in diameter surrounded by a single membrane. They occur in the cytoplasm of most animal cells (Figure 6-6). Lysosomes have been referred to as "suicide sacs," since release of the enzymes they contain would cause a cell to destroy itself by digesting its own proteins. Normally they function as destroyers of foreign particles and cellular components that have outlived their usefulness.

Lysosomes secrete enzymes similar to digestive enzymes into areas in the cell in which invading bacteria and other foreign protein-containing bodies are destroyed. Lysosomes also play a role in development. In a tadpole, for example, they cause a breakdown of the tissues of the tail during growth to an adult frog where the tail is not present.

When a cell dies, rupture of lysosomes results in rapid disintegration of cellular protein. Some biologists believe that lysosomes play a role in the aging process. Widespread rupturing of them by causes unknown may contribute to the disintegration of tissues with age.

Plastids

Among other microscopic bodies of the cytoplasm are various types of **plastids,** which function as chemical synthesizers and storage bodies. Plastids occur in greatest numbers in the cells of plants and in primitive single-celled organisms. The most common type is the **chloroplast** (Figures 6-6 and 6-12). Chloroplasts, like mitochondria, contain DNA and ribosomes; hence they can divide and synthesize proteins independent of the nucleus. Chloroplasts contain the green chlorophyll pigments that are vital to the process of photosynthesis; they impart the green color to plant foliage. Other plastids called **chromoplasts** produce blue, red, orange, or yellow pigments; they are the source of certain of the brilliant colors of flowers and fruits.

Chloroplasts vary both in size and shape but are about the size of mitochondria. In leaf cells they are commonly oval, spherical, or disk-shaped. In certain algae they are cuplike or ribbonlike. Within chloroplasts, chlorophyll is sandwiched

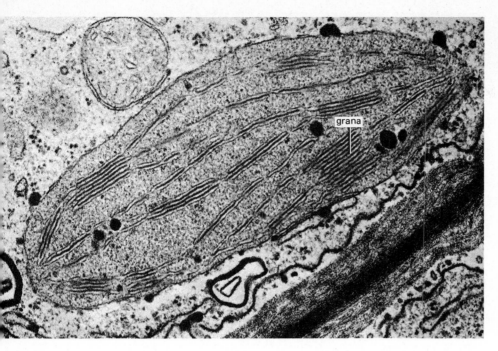

FIGURE 6-12
A chloroplast (size, about four micrometers). The grana are quite apparent in this electron micrograph.

between closely packed membraneous layers of proteins and lipids, in units called **grana** (Figure 6-12).

Leucoplasts are colorless plastids that serve as "storehouses" of nutrients in plant cells. They contain enzymes necessary to link glucose molecules together to form starch. Once formed, starch remains in leucoplasts until there is a need for glucose elsewhere in the cytoplasm. Leucoplasts occur in the cells of roots, stems, and in other storage areas of plants. High densities of them fill the cells of the white potato, for example.

Vacuoles

Plant cells commonly contain one or more fluid-filled cavities, or **vacuoles,** in the cytoplasm. In animal cells, vacuoles are mostly restricted to the simpler invertebrates. The bordering **vacuolar membrane,** formed by the cytoplasm, regulates the molecular traffic between the vacuole and the cytoplasm. A young cell sometimes contains several vacuoles that unite as the cell increases in size to form a single, large central vacuole. Pressure exerted by the fluids in the central vacuole may force the cytoplasm into a thin layer around the edge of the cell (Figure 6-13).

Water is the primary component of the fluid in vacuoles. Ions of mineral compounds, molecules of simple sugars, and other soluble substances are dissolved in this **cell sap.** Proteins and other organic materials are also included as colloids. Some vacuoles serve as storage centers for nutrient substances. Others serve to eliminate wastes and excess water. When the vacuoles become full they contract, forcing their contents through openings in their membrane.

Vacuoles of some plant cells contain water-soluble pigments. Among these are the **anthocyanins** which appear as shades of scarlet, blue, and violet. These pigments produce the varied colors of autumn leaves, the red of beets, and the petal coloration of tulips and certain other flowers. The formation of anthocyanins is determined by both internal and external factors. Accumulation of sugar within a cell for instance, stimulates their production. Environmental factors that trigger the formation of anthocyanins include low temperature and bright light.

FIGURE 6-13
A representation of a typical plant cell. In a whole cell, the central vacuole, which is filled with a clear *cell sap,* is completely surrounded by cytoplasm.

Centrioles and microtubules

Animal and lower plant cells contain a pair of organelles (Figure 11-4) called **centrioles,** which lie near the nucleus. Each centriole is a cylinder, about 0.2 micrometers in diameter, composed of nine parallel triplets of hollow, cylindrical **microtubules.** The two members of a centriole pair normally differ in length and are at right angles to each other. One is about 0.5 micrometers in length, the other usually about 0.2 micrometers long. The size and structure of centriole pairs are remarkably constant throughout all the various cells that contain them. Each member of the pair moves to opposite sides of the cell during cell division.

Apart from centrioles, microtubules are common in the cytoplasm of most eucaryotic cells. They are about 300 Å in diameter. In some animal cells they are oriented in regular patterns in the cytoplasm and appear to channel the flow of substances through the cytoplasm. In the cells of higher plants, microtubules are particularly abundant in the cytoplasm near the plasma membrane and are also found extending along streams of cytoplasm that move through the cell's interior and may serve to give rigidity to the cell structure. Microtubules are the building blocks of cilia and flagella which propel the motile forms of bacteria, protozoa and algae.

The cell wall

All cells are bound by a plasma membrane, but plant cells differ from those of animals in that they are encased by a rigid supporting structure, the **cell wall.** Where two plant cells lie against one another, each forms a portion of the wall separating them. Sandwiched in between is an intercellular layer, the **middle lamella** (Figure 6-14), which contains various substances, including the polysaccharide, **pectin.** Many fruits, including apples, contain large amounts of pectin. It is released during cooking and forms a jelly as it cools.

Adjacent cells form thin **primary walls** on both sides of the middle lamella. They are composed of **cellulose** and pectin. Cellulose is a tough, fibrous carbohydrate. In soft plant parts, such as flower petals and pulpy fruits, primary walls are only slightly developed. In the supportive and protective parts of plants, such as branches and nut shells, cellulose layers form thick **secondary walls,** which remain long after the cell is dead.

In a primary wall, cellulose fibers are arranged in a meshlike network. In a secondary wall they form parallel layers that are crisscrossed in sheets. This arrangement, somewhat like that of the thin layers composing a sheet of plywood, gives secondary walls strength. Spaces between the fibers contain pectin and **lignin,** a complex organic compound. Lignin, second only to cellulose in abundance in wood, adds hardness and rigidity to cell walls.

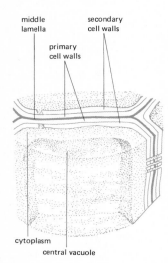

middle lamella

secondary cell walls

primary cell walls

cytoplasm

central vacuole

FIGURE 6-14

Structure of a plant cell wall. The cell walls are rigid, but they are also porous. This permits molecules of all sizes to pass through relatively easily.

MOVEMENT OF MOLECULES ACROSS MEMBRANES

Molecules and ions jostle against the plasma membrane. It is **permeable** to some, and they pass through freely. Others enter with difficulty, while some do not penetrate at all. Generally, small compounds move through the membrane most easily, but some relatively large molecules pass more readily than certain smaller ions. Apparently the charge of an ion (whether it is positive or negative and whether it is singly or doubly charged) has an effect. Also, the structure and composition of the membrane influence its permeability. In response to internal or external chemical changes, the nature of the molecular structure of the plasma membrane itself may change, with consequent changes in the sizes and shapes of the pores that permeate it.

Substances can be classified according to their rates of passage through a plasma membrane (Table 6-1). Because the plasma membrane lets certain things through more readily than others, it is said to be selectively permeable, or **semi-permeable.** The rate at which a given substance enters a cell may be quite different than the rate at which it leaves. Thereby, the plasma membrane selectively maintains high or low concentrations of certain substances within the cell.

Alcohol, ether and chloroform penetrate the plasma membrane very rapidly. These low-molecular-weight organic solvents dissolve into the lipid part of the membrane. It is this rapid penetration in brain cell membranes that is thought to account for the loss of sensation induced by anesthetics. By their action the lipid structure of the membrane is changed in some way so as to change the membrane permeability of brain cells in such a way as to alter the flow of ions which are crucial to nervous system function. Nerve function is discussed in Chapter 15.

Molecular diffusion in air

Visualize the fluid surrounding a cell as a mass of quivering molecules, bumping into one another like a crowd of people at a bargain counter. Molecular movements within the fluid are random, and concentrations of substances tend to be evenly dispersed. If a new substance is introduced at some point, random molecular movements tend to disperse it until it is evenly distributed.

Diffusion occurs in gases as well as in liquids. Consider, for example, a bottle of ammonia being opened in a room. Ammonia molecules are originally restricted to the bottle, but as soon as it is opened they start moving into the surrounding air. Soon the odor of ammonia fills the room, and as diffusion continues, the odor becomes stronger. More and more ammonia mingles with the molecules of gases in the air. If there is circulation in the room (which could be set up by opening and closing the door) the process occurs more quickly than if the atmosphere is still. Ultimately, when the ammonia molecules are distributed evenly among the other gases, diffusion (but not the random movements of molecules) ceases. A state of equilibrium has been reached. During diffusion two things happen. Ammonia molecules leave the bottle and enter the air, and air molecules enter the bottle. Both movements were from a region of greater concentration to a region of lesser concentration. This occurred not in a uniform flow but by random molecular movements. Diffusion continued until the concentration of air and ammonia was equal in all areas of the room. Once an equal distribution has been established, diffusion ceases, but the random movements of molecules never stops.

Diffusion through membranes

Figure 6-15 illustrates a simple laboratory apparatus that demonstrates diffusion through a permeable membrane (a membrane that does not selectively prevent passage of certain molecules). A piece of fine meshed cloth is tied over the bulb of a thistle tube, which is then filled with a sugar solution. When the bulb is submerged in a jar of water, sugar molecules diffuse outward, and water molecules diffuse inward. Diffusion continues until water and sugar are distributed equally on both sides of the cloth. At this point a state of equilibrium (balance) exists.

FIGURE 6-15

Diffusion through a permeable membrane. At the beginning of the experiment, a sugar solution in the thistle tube is separated from plain water in the jar by a permeable membrane. Water molecules (white) move through the membrane into the sugar solution as sugar molecules (blue) move through the membrane into the water in the jar.

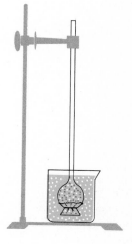

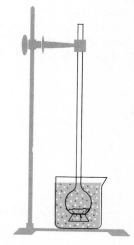

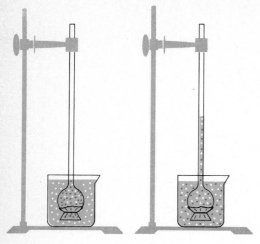

FIGURE 6-16

Diffusion through a selectively permeable membrane. At the beginning of the experiment, a sugar solution in the thistle tube is separated from plain water in the jar by a selectively permeable membrane. Over a period of several hours, water molecules move through the membrane rapidly into the sugar solution. Note the height of the liquid in the tube.

TABLE 6-1	THE CAPACITY OF VARIOUS SUBSTANCES TO PENETRATE THE PLASMA MEMBRANE	
RAPID PENETRATION	MEDIUM PENETRATION	LITTLE OR NO PENETRATION
gases	glucose	polysaccharides
oxygen	amino acids	starches
carbon dioxide	sucrose	cellulose
nitrogen	ions of	proteins
water	mineral salts	lipids (fats)
fat solvents		phospholipids
alcohol		
ether		
chloroform		

A variation of the preceding experiment is illustrated in Figure 6-16. The apparatus is the same except that a piece of sheep bladder is substituted for cloth. Sheep bladder is a **selectively permeable membrane** (large molecules cannot pass through); hence there is relatively little flow of sugar from the thistle tube. However, water passes through quite easily, with a net movement from the region of greater water concentration in the jar to the region of lower water concentration in the thistle tube where some of the water binds to sugar, effectively lowering water concentration here. Over a period of time the water level in the tube rises. But, as the level rises, the force of gravity comes into play. Gravity tends to pull the solution in the thistle tube downward so that pressure is exerted on the top surface of the membrane. An equilibrium is established when this presssure is equal to the **diffusion pressure** (the pressure generated by the tendency of molecules to flow from a region of greater concentration to a region of lesser concentration). At this point the water level in the tube ceases to rise.

This experiment reveals that forces can be generated at a semipermeable membrane and merits a second consideration in order to stress the concept of diffusion pressure. At the membrane, water molecules bombard from both above and below. However, water diluted with sugar contacts the upper side, while undiluted water contacts the underside. Therefore, a net diffusion pressure is exerted on the underside of the membrane, and more water molecules pass into the thistle tube than leave. An equilibrium is reached when pressure exerted by the column of water in the stem of the thistle tube prevents more water from entering from below. At the equilibrium point, water molecules still cross the membrane, but equal numbers pass through from above and below.

Cellular osmosis

The experiment just described involves **osmosis,** the diffusion of water through a selectively permeable membrane. Since plasma membranes are selectively permeable, living cells are tiny osmotic systems. As water diffuses into a cell, it builds up a pressure known as **turgor,** which is particularly important in maintaining firmness in plants. Since plant cells are encased in rigid cell walls of cellulose, they don't stretch or bulge as internal pressure increases. Instead the plasma membrane and cytoplasm are pressed firmly against the wall. An equilibrium is reached when turgor pressure equals diffusion pressure. In land plants the membranes separating the plants from water are in the roots (Figure 6-17). But, because water is trans-

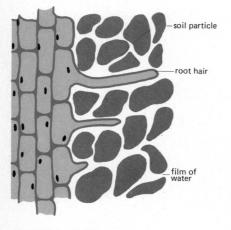

soil particle

root hair

film of water

FIGURE 6-17

Water enters roots via projections (root hairs) of the outermost cells.

ported upward through the stems, turgor pressure extends above the level of the ground.

Thin-walled plant tissues (in leaves, flower petals, and soft stems) maintain stiffness by turgor. As long as there is sufficient water this internal pressure is maintained. With insufficient water a plant wilts. With an excess of water it may have to eliminate some. Certain plants eliminate water through pores in the tips of their leaves (Figure 6-18) by a process called **guttation.**

Bear in mind that osmosis is purely a physical process, involving movement of water molecules by kinetic (motion) energy. No cell energy is involved. For this reason, osmosis and diffusion are also called **passive transport.**

Active transport

While certain substances move in and out of cells randomly, other molecules are actively absorbed or secreted. They pass through the plasma membrane against a diffusion pressure that would normally cause movements in the opposite direction. For example, root cells take in certain mineral ions (such as nitrates and phosphates) from ground water where the ions are in lower concentrations than already exist within the plant. Similarly, cells of certain algae living in the ocean absorb iodine even when the concentration of iodine within them is already millions of times higher than in the seawater surrounding them. To establish these relatively high concentrations requires energy, and this energy is generated by the chemical reactions of respiration, which are described in the next chapter.

An example of how **active transport** might function is shown in Figure 6-1 A carrier protein molecule in the membrane binds to the substance to be transported on the outside of the membrane. The binding process in some way moves the carrier-protein complex to the inside part of the membrane where the transported substance drops off, and the carrier protein returns to its original location. As long as there is energy available to power this process and a substance to be transported, active transport will continue.

Pinocytosis and Phagocytosis

So far, only penetration of the plasma membrane by small molecules and ions has been considered. Larger molecules, such as polypeptides and lipids cannot always pass through membranes; yet they are known to enter cells. Microscopic studies have revealed that large molecules are surrounded by pockets formed in the plasma membrane and are sealed in as the membrane closes behind them (Figure 6-20). This process is called **pinocytosis.** An amoeba can surround and take in food particles (including other single-celled species) up to several microns in diameter. Within the human body, amoebalike white blood cells engulf invading bacteria and destroy them with enzymes produced by lysosomes. These processes of taking in larger particles are called **phagocytosis.**

Hypotonic, hypertonic, and isotonic solutions

Because plant cells have rigid cell walls, the influx of water ceases when turgor pressure equals the diffusion pressure. Animal cells, however, lack a supporting structure. In a **hypotonic solution** (a solution that contains a lower concentration of soluble substances than cytoplasm) animal cells cannot stop the inward diffusion of water. For ocean-dwelling animals there is less of a problem, since concentration of salts in seawater are approximately the same as concentrations in cyto-

FIGURE 6-18
Certain plants eliminate excess water through pores on the edges of their leaves.

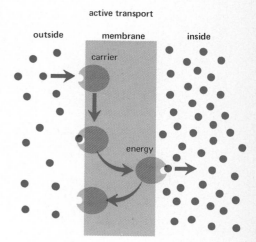

FIGURE 6-19
Active transport. Carrier proteins in the membrane move substances from the outside of the cell where the concentration is less to the inside of the cell where the concentration is greater.

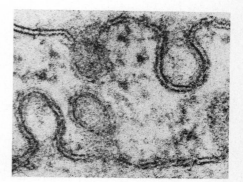

FIGURE 6-20
Electron micrograph showing the pouches that are involved in *pinocytosis.*

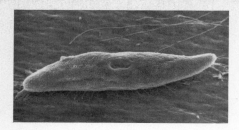

FIGURE 6-21

This one-celled organism *(Paramecium)* eliminates water through the membrane as rapidly as it diffuses into the cell. Otherwise, internal water pressure would burst the cell.

FIGURE 6-22

Red blood cells in three solutions.

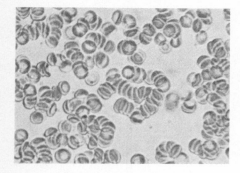

(a) Normal red blood cells in isotonic solution.

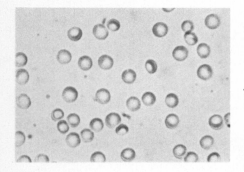

(b) Swollen red blood cells in hypotonic solution. Water pressure may burst the cells.

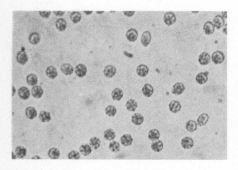

(c) Plasmolysis of red blood cells in hypertonic solution. Water loss has caused the cells to shrink.

plasm. Animals living in freshwater, however, must be able to eliminate excess water or they will die by **cytolysis** (bursting).

Certain unicellular organisms have water-eliminating organelles called **contractile vacuoles.** These tiny "pumps" accumulate water and then squeeze it out through pores in the plasma membrane (Figure 6-21). Fish and other aquatic animals with gills take in large amounts of water with the oxygen they absorb into the blood. In these animals, excess water is excreted from the body in urine. Kidneys, sweat glands, and lungs eliminate excess water from humans and other mammals.

In **hypertonic solutions** (solutions that contain more dissolved substances than cytoplasm) water diffuses from cells into solution. This results in loss of turgor pressure and shrinkage of the cell content, or **plasmolysis.** The threat of plasmolysis keeps shipwrecked men from drinking salt water, and gardeners know that too heavy an application of strong fertilizer will kill plants by preventing them from taking in the water they need.

An **isotonic solution** contains the same amounts of dissolved substances as a cell. The blood fluid, plasma, for instance, is isotonic to the red blood cells suspended in it (Figure 6-22). Isotonic salt solutions are used to bathe living tissues in biological experiments and surgical operations.

CELL DIVISION

Consider a tiny, tadpolelike sperm cell, its flagellum whipping furiously as it makes its way toward an egg within a duct of a human female's reproductive system. When an egg and sperm merge, **genes** from the parents come together to direct the formation of a primordial cell, the **zygote.** From that moment on the genetic constitution of the human-to-be is set for life. The cells of a human body are produced through a sequence of cell divisions beginning with a zygote. In the original division the zygote forms two cells, these become four, which then produce eight, and so on. By the time a human infant is born, it consists of many trillions of cells.

Division of reproductive and somatic cells

In all cells, genetic codes are carried within the structure of DNA molecules. But, there is a basic difference between cells of the reproductive system and **somatic** (of the body) cells. Eggs and sperm have only half the number of chromosomes as somatic cells. In humans, somatic cells contain 46 chromosomes.

Chromosomes occur in pairs, which is to say that for any given genetic trait there are paired genes. The genes of a given pair may or may not be alike; they may both contribute to the characteristics of the trait, or the effects of one gene of the pair may be masked by the other. The gene for brown eyes, for example, overrides the gene for blue eyes.

In divisions of somatic cells, **mitotic division,** (Figures 6-23 and 6-24) both of the cells generated by the splitting of the original cell in humans receive all 46 chromosomes. This seeming impossibility is possible because each of the 46 chromosomes duplicates between divisions. Thus, every cell composing a human body contains all the genes that code for the characteristics of that body, with the exception of reproductive cells. Eggs and sperm are produced by a unique type of cell division, **meiotic division,** by which only one chromosome of every pair is received by any given reproductive cell (Figure 6-25). Therefore, human reproductive cells contain only 23 chromosomes. When a sperm and an egg merge, they establish a unique combination of genes—a new molecular essence of individuality—all from 46 chromosomes.

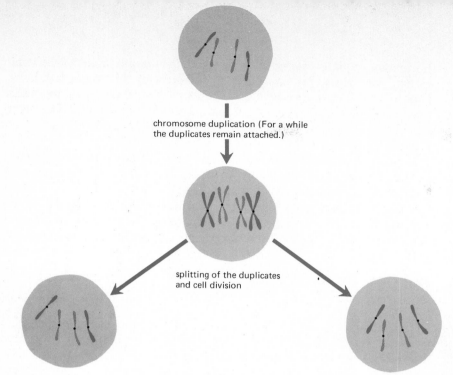

chromosome duplication (For a while the duplicates remain attached.)

splitting of the duplicates and cell division

FIGURE 6-23

Mitosis. During mitosis, chromosomes duplicate and each daughter cell receives the same amount of chromosomal material as the original parent. This figure illustrates mitotic division by a cell with two pairs of chromosomes (one member of each pair is blackened).

FIGURE 6-25

Meiosis involves two successive divisions. As in mitosis, chromosomes duplicate prior to the first meiotic division. However, unlike mitosis, each of the daughter cells receives an entire chromosome of each pair. Thus, the amount of genetic information is halved. The second meiotic division is the same as mitosis in that each daughter cell receives one of the duplicates of each chromosome.

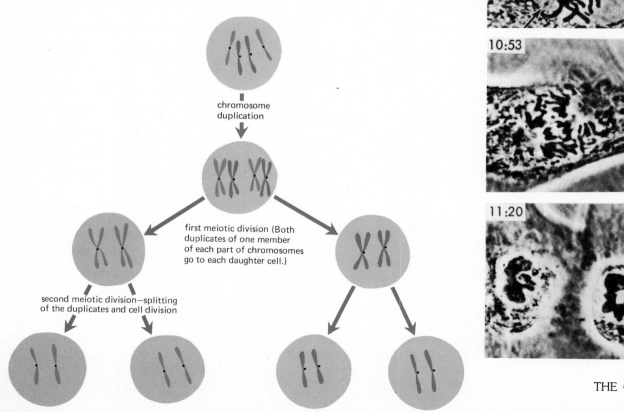

chromosome duplication

first meiotic division (Both duplicates of one member of each part of chromosomes go to each daughter cell.)

second meiotic division—splitting of the duplicates and cell division

FIGURE 6-24

The stages of mitosis in a living salamander cell in culture. Prophase: 9:17; metaphase: 9:34; anaphase: 10:53; telophase: 11:20. Time in hours and minutes.

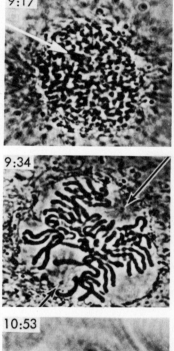

9:17

9:34

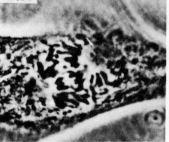

10:53

11:20

THE CELL **159**

The human mode of reproduction is, of course, not the only one. Certain single-celled green algae, for example, simply reproduce time after time by repeated mitotic divisions; then (when environmental conditions or an inner "program" call for it) two cells combine, chromosomal material is "scrambled," and a meiotic division produces two new individuals which begin new series of repeated mitotic divisions.

The sequence of divisions of growth and reproduction is unique for each species. But, for every species, mitosis maintains the number of chromosomes within cells, and meiosis halves the number. The precise way in which it is determined when and if a cell should undergo division is not yet understood but involves control of what is called the **cell cycle.** This is the movement in time of a cell from a resting state through DNA duplication and mitosis back to the resting state again.

LEVELS OF ORGANIZATION IN LIVING THINGS

The cell can be considered to be the first level of organization, in as much as every known species has a cellular makeup. The simplest of organisms consist of one cell. More complex plants and animals are composed of many specialized cells, working closely together in mutual dependence. Whereas cells of higher plants and animals have specialized functions in the organism as a whole, each individual cell must perform all of the functions necessary to maintain its own living condition. It must receive food, respire, synthesize proteins, and generally maintain all basic life activities. In this respect, it is like a unicellular organism.

When a cell is part of a higher organism, however, division of labor occurs. A muscle cell, for instance, can be a specialist in contraction because other cells supply it with nutrients and carry off its wastes. Such interdependence might be compared to a complex society, where a physician can devote his or her time to providing medical services because grocers, carpenters, machinists, and other specialists fulfill other human needs.

Cells are grouped to form tissues

A **tissue** is a group of structurally similar cells which act as a functional unit. This is the second level of organization of living things. In **colonial organisms** the tissue is the highest level of organization. **Gonium,** for example, is a type of algae which exists in cellular clumps of from 4-64 individuals (depending on the species). That some sort of coordination exists in **Gonium** is indicated by the organized fashion in which the flagella of all the cells beat in unison, thus enabling the colony to swim as a unit, even though each algal cell is an individual.

The human body consists of many kinds of tissues, including muscle, nerve, bone, and cartilage, each of which performs a unique function. Muscle tissue, for example, is composed of many adjoining contractile cells. Each muscle cell adds to the coordinated power of contraction.

Tissues are grouped to form organs

In higher plants and animals several kinds of tissue are combined in functional units called **organs.** This is the third level of biological organization.

A hand is an organ composed of skin, muscle, bone, tendons, ligaments, and nerves. The heart, stomach, liver, brain, and kidneys are internal organs. A tree trunk is an organ composed of bark, wood, pith, and other tissues, all working together. The trunk is an organ of support through which materials traverse between the roots and leaves.

Organs are grouped to form systems

Several organs may work together as a unit to perform a specialized activity. This is an **organ system,** the fourth level of biological organization. For example, several organs of digestion (including the mouth, stomach, and intestines) are involved in the digestive system that converts a meal to molecular forms that cells can use.

Systems are integrated in complex organisms

Organ systems function together in complex organisms. This is the fifth and highest level of our biological organization. Nutrient molecules released during digestion, for example, are transported from the digestive system throughout the body via the blood vessels of the circulatory system.

Beyond the level of the organism is the ecosystem and biosphere, the complex interrelationships of which were considered in Chapter 4. Environmental influences are felt at all levels of biological organization, however.

THE CELL IN ITS ENVIRONMENT

Living things maintain an intricate balance in the face of constantly changing conditions, both internal and external. Survival depends on making adjustments to these changes. This balance is **homeostasis.** Homeostasis occurs at all levels in the organization of living things. It involves, for example, behavioral responses to changes in the physical environment. A desert, for instance, is desolate under the broiling daytime sun, but it comes to life in the cool of the evening. Insects, birds, reptiles, and mammals leave their shelters as temperatures become more hospitable.

Within organisms continuous adjustments are made at the organ and system level to maintain stability (homeostasis). Organs function in close association with each other to maintain an optimum internal environment. For example, all organs depend on the heart to circulate blood, which bears oxygen and other vital molecules; and the rate of heart-beat varies with body needs. During periods of exertion, the heart speeds up to increase the supply of oxygen and nutrients to the tissues. Increased blood supply also speeds up the removal of cell wastes, which are produced in greater amounts during periods of heightened activity. The kidneys function as blood filters in removing excess mineral ions, water, and cell wastes from the blood. Thus, the kidneys and the heart work together in maintaining circulatory homeostasis.

Blood and other body fluids bathe an animal's tissues, maintaining a harmonious internal environment. In a similar way sea water provides a compatible environment for the single-celled plants of the suspended phytoplankton. These simple organisms have much wider ranges of tolerance for environmental variability than the highly specialized cells of muscle, bone, and nerve. In general, the more specialized a cell, the more sensitive are its homeostatic requirements.

While cells of specialized tissues normally couldn't survive the fluctuations in temperature and nutrient availability if isolated from the organism, biologists have been successful in growing many such cells in carefully controlled artificial environments, called **tissue cultures.** Cells in a tissue culture are grown in nutrient solutions that approximate the normal cellular environment within an organism. Oxygen is supplied, and waste products are removed. Only when the need for homeostatic adjustments is reduced to a minimum can the specialized cells of higher animals survive in a tissue culture.

CHAPTER 6: SUMMARY

1 The Cell Theory holds that the cell is the basic structural unit of life.

2 The electron microscope magnifies much more than the light microscope because the wavelength of electrons is less than that of light.

3 The important cellular processes are absorption, synthesis, respiration, secretion, excretion, movement, response, and reproduction.

4 A major division of cells is into prokaryotes and eukaryotes. The former lack a true nucleus with a membrane around it, while the latter have a true nucleus with a membrane around it.

5 The structural components of eukaryotes are the nucleus, nucleolus, cell membrane, cytoplasm, endoplasmic reticulum, ribosomes, mitochondria, the Golgi complex, lysosomes, plastids (in plants), cell wall (in plants) vacuoles, microfilaments, and microtubules.

6 Substances move across membranes by the processes of diffusion (passive transport), active transport, or pinocytosis.

7 Cell division occurs in a variety of ways including mitosis and meiosis.

8 Cells are grouped to form tissues, tissues are grouped to form organs, organs are grouped to form systems, and systems are grouped to form the organism.

Suggested Readings

Anderson, M. D.   *Through the Microscope,* Natural History Press, Garden City, New Jersey, 1965—The uses of the various kinds of microscopes.

Galston, Arthur W.   "The Membrane Barrier," *Natural History,* August-September, 1974—The cell membrane and the passage of materials through it.

Loewy, A. G. and P. Siekevitz   *Cell Structure and Function,* 2nd Ed., Holt, Rinehart and Winston, New York, 1970—General cell biology.

Mazia, Daniel   "The Cell Cycle," *Scientific American,* January, 1974—A fascinating account of cells and how they reproduce.

Novikoff, Alex B. and Eric Holtzman   *Cells and Organelles,* 2nd Ed., Holt, Rinehart and Winston, New York, 1976—A detailed introduction to cell biology with excellent illustrations.

Satir, Peter   "How Cilia Move," *Scientific American,* October, 1974—A detailed presentation on the hairlike appendages of cells.

WE ARE SHARED, RENTED, OCCUPIED
Lewis Thomas

As exquisitely described by Lewis Thomas, our life is intimately linked with what could be called suborganisms, which inhabit the cells of our bodies.

We are told that the trouble with modern man is that he has been trying to detach himself from nature. He sits in the topmost tiers of polymer, glass and steel, dangling his pulsing legs, surveying at a distance the writhing life of the planet. In this scenario, man comes on as a stupendous lethal force, and the earth is pictured as something delicate, like rising bubbles at the surface of a country pond or flights of fragile birds.

But it is illusion to think that there is anything fragile about the life of the earth; surely it is the toughest membrane imaginable in the universe—opaque to probability, impermeable to death. Man is the delicate part, transient and vulnerable as cilia. Yet it is not a new thing for man to invent an existence that he imagines to be above the rest of life; this has been his most consistent intellectual exertion down the millennia. As illusion, it has never worked out to his satisfaction in the past, any more than it does today. Man is embedded in nature.

The biologic science of recent years has been making this fact of life more urgent. The new, hard problem will be to cope with the dawning, intensifying realization of just how interlocked we are. The old notions most of us have clung to about our special lordship are being deeply undermined.

Item. A good case can be made for our nonexistence as entities. We are not made up, as we had always supposed, of successively enriched packets of our own parts. We are shared, rented, occupied. At the interior of our cells, driving them, providing the oxidative energy that sends us out for the improvement of each shining day, are the mitochondria, and in a strict sense they are not ours. They turn out to be little separate creatures, the colonial posterity of migrant procaryotes or unnucleated cells, probably primitive bacteria that swam into ancestral precursors of our eucaryotic or nucleated cells and stayed there. Ever since, they have maintained themselves and their ways, replicating in their own fashion, privately, with their own DNA and RNA quite different from ours. They are as much symbionts as the nitrogen-enriching rhizobial bacteria in the roots of beans. Without them, we would not move a muscle, drum a finger, think a thought.

Mitochondria are stable and responsible lodgers, and I choose to trust them. But what of the other little animals, similarly established in my cells, sorting and balancing me, clustering me together? My centrioles, basal bodies and probably a good many other, more obscure tiny beings at work inside my cells—each with its own special genome or chromosome formation—are as foreign and as essential as aphids in anthills. My cells are no longer the pure-line entities I was raised with; they are ecosystems more complex than Jamaica Bay, New York.

I like to think that they work in my interest, that they draw each breath for me. Perhaps it is they who walk through the local park in the early morning, sensing my senses, listening to my music, thinking my thoughts.

I am consoled somewhat by the thought that the green plants are in the same fix. They could not be plants, or green, without their chloroplasts, which run the photosynthetic enterprise and generate oxygen for the rest of us. As it turns out, chloroplasts are also separate creatures with their own genomes, speaking their own language.

We carry DNA in our nuclei that may have come, at one time or another, from the fusion of ancestral cells and the linking of ancestral organisms in symbiosis. Our genomes are catalogues of instructions from all kinds of sources in nature, filed for all kinds of contingencies. As for me, I am grateful for differentiation and speciation, but I cannot feel as separate an entity as I did a few years ago, before I was told these things; nor, I should think, can anyone else.

Item: The uniformity of the earth's life, more astonishing than its diversity, is accountable by the high probability that we derived, originally, from some single cell, fertilized in a bolt of lightning as the earth cooled. It is from the progeny of this parent cell that we take our looks; we still share genes, and the enzymes of grasses bear a family resemblance to those of whales.

The viruses, instead of being single-minded agents of disease and death, now begin to look more like mobile genes. Evolution is still an infinitely long and tedious biologic game, with only the winners staying at the table, but the rules are beginning to look more flexible. We live in a dancing matrix of viruses; they dart, rather like bees, from organism to organism, from plant to insect to mammal to me and back again, and into the sea, tugging along pieces of this genome, strings of genes from that genome, transplanting grafts of DNA, passing around heredity as though at a great party. They may be a mechanism for keeping new, mutant kinds of DNA in wide circulation among us. If this is true, the odd virus disease, on which we must focus so much of our attention in medicine, may be an accident.

Item: I have been trying to think of the earth as a kind of organism, but it is no go. I cannot think of it this way. It is too big, too complex, with too many working parts lacking visible connections. The other night, driving through a hilly, wooded part of southern New England, I wondered about this. If not like an organism, what is it like, what is it *most* like? Then, satisfactorily for that moment, it came to me: it is *most* like a single cell.

PEOPLE-PLANTS

In the summer of 1976, cell biologists at the Brookhaven National Laboratory achieved a bizarre "wedding" of cellular parts.

It sounds like a joke from the gaudy old days of science fiction, but it came true last week: a team of biologists at the Brookhaven National Laboratory announced they had fused living human cells with those of a plant, and they think the resulting crossbreed may grow.

The human part of this exotic experiment consisted of HeLa cells, themselves a scientific marvel. Henrietta Lacks, a 31-year-old Baltimore woman, died of cancer in 1951, but a smear of her tumor was successfully cultured in a test tube. The culture has provided the standard human reference cells for nearly all cancer research for 25 years.

The HeLa cells were crossed with cells from a hybrid tobacco plant. First the HeLa nuclei were tagged with a radioactive isotope so they could be identified. Then the thick outer walls of the tobacco cells were stripped away with a digestive enzyme. The human and plant cells were mixed in a solution that caused them to congeal. After new cell walls formed, radioactive analysis confirmed the existence of human nuclei within the hybrid. But there was no evidence of HeLa cytoplasm, the material inside the cell walls apart from the nucleus. Thus, says Brookhaven biologist Harold Smith, a hybrid will probably act more like a plant than an animal.

"There is no doubt that we can clone full-size plants," says Smith. Scientists worry, however, that the human material may be destroyed by plant enzymes. If the human nuclei survive, the behavior of genes can be observed in unique positions, and that could help unlock new parts of the human genetic code. What the hybrids will not do, the Brookhaven scientists promise, is grow hands and feet—or take over the world.

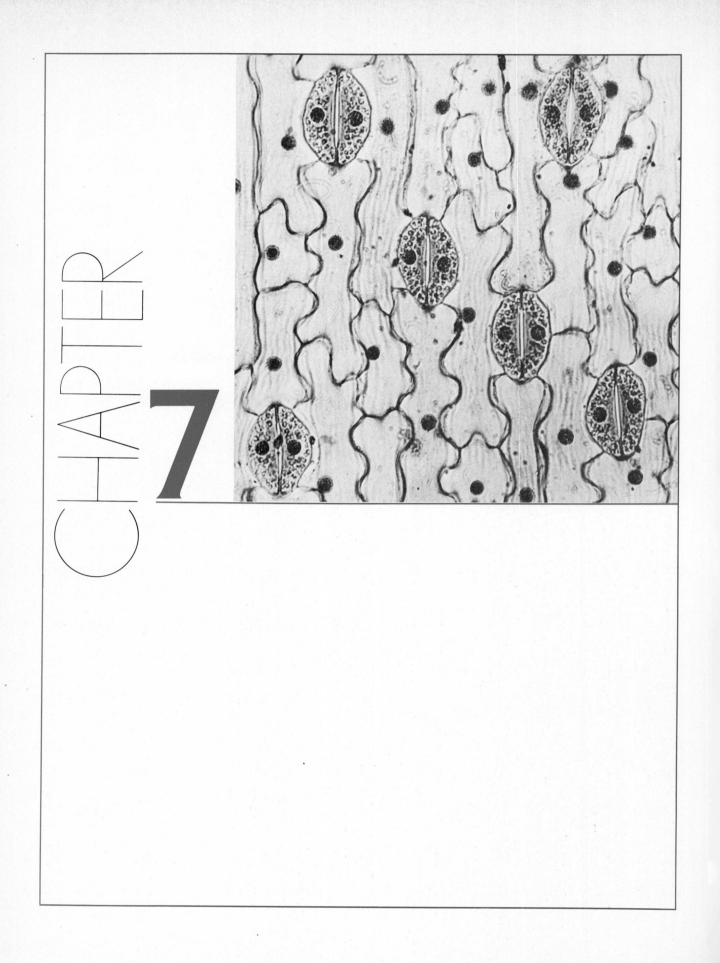

CHAPTER

7

Photosynthesis and Respiration– Life's Energy Flow

Objectives

1 Distinguish between heterotrophs and autotrophs.

2 Understand the nature of light and its relationship to the electromagnetic spectrum.

3 Learn what photosynthetic pigments do.

4 Distinguish between the light and dark reactions of photosynthesis.

5 Learn how ATP is synthesized by both plants and animals.

6 Understand the processes of respiration and fermentation.

7 Be able to explain gas exchange in the lungs and in plants.

CELLS—THE RECEIVERS AND USERS OF ENERGY

In chapter 4 we dealt with the nature of energy flow through ecosystems. This chapter considers the physical and biochemical level of this flow. Before turning to the nurturing of life by light, however, a brief reconsideration of the big picture is in order.

As the basic functional unit of life, the cell is the site of the reception and utilization of energy. By the process of **photosynthesis,** plant cells store light energy as chemical energy in organic compounds. By the process of **cellular respiration,** animal cells as well as plant cells break down organic compounds to release the energy necessary to support life. Each organism has unique biochemical subtleties, but the same basic chemistry of energy utilization recurs in every known organism with the exception of some bacteria.

Were it not for photosynthesis, life on earth might not have developed beyond the level of complexity of bacteria. There would be no forests or grasslands. Certainly there would be no animal populations, since animals need organic compounds produced by plants.

In classifying food requirements of organisms, green plants are designated as primary producers, or **autotrophs** ("self-feeders"). Organisms that do not photosynthesize are termed **heterotrophs** ("other-feeders"). Animals, non-photosynthesizing protists, and the fungi, are among these nutritionally dependent organisms. Although heterotrophs have remarkable biochemical abilities, none can capture energy from a physical source such as the sun.

LIGHT—LIFE'S ENERGY

The nature of light has long been a source of puzzlement to those concerned with the properties of the physical world. Although light is the only thing that is ever seen (we don't actually see objects; rather we see the light reflected from objects), we cannot see what light is. Our organs of visual sensation respond to light, but they do not detect its component parts.

Nevertheless, through various experiments, physicists have discovered some amazing properties of light. Early in this century scientists came to know light as both a particle-like and a wave-like phenomenon. It was found to travel through space at a constant speed (about 300 million meters per second) in units termed **photons.** It was also found to consist of alternately contracting and expanding electric and magnetic fields. When the electric component collapses it generates the magnetic component, and vice versa (Figure 7-1). An in-depth consideration of *electromagnetic radiation* is beyond the scope of this book. Regretably, the following characteristics of it must be accepted on faith (or be taken as a stimulus to look elsewhere for a more detailed description of this inscrutable universal phenomenon).

Since all electromagnetic radiation travels at the same speed, the frequency at which the fields contract and expand determines the wavelength of a given photon. The more rapid the frequency, the shorter the wavelength. Energy is directly related to frequency; the more rapidly the electromagnetic field oscillates, the more energy a photon carries. Furthermore, the product of the two, if frequency is measured in cycles per second and wavelength in meters, equals 3×10^8 (the speed of light in meters per second). Wavelength determines color.

Visible light constitutes only a small portion of the electromagnetic spectrum (Figure 7-2). But, it is by far the most significant portion to life processes. As the

FIGURE 7-1

Photons of light are composed of electric and magnetic components which are perpendicular to each other.

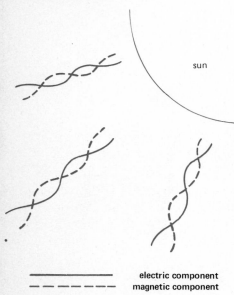

sun

——————— electric component
— — — — — magnetic component

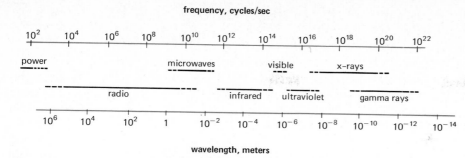

frequency, cycles/sec

wavelength, meters

FIGURE 7-2

The electromagnetic spectrum.

driving force of photosynthesis and the stimulus of the sense of sight, light exerts a profound twofold biological influence. It may seem puzzling that the same small slice of the electromagnetic spectrum would act in two such disparate ways. But, when one considers the nature of its effect on organic molecules, the reason for its varied biological activity is understandable. The wavelengths of visible light (between 10^{-7} and 10^{-6} meter) are such that the energy they impart to certain organic molecules causes electrons in these molecules to be elevated to higher or "excited" energy levels or orbits. Subsequently, in falling back to the original energy levels (equilibrium), the electrons release this excess energy. At the back of the eye, for example, nerve cells that are in close contact with light-sensitive pigments respond to the energy release by sending impulses to the visual center in the brain. In green plants the pigments of photosynthesis pass on the energy that they trap to other biochemical compounds that store energy.

As far as is now known, portions of the electromagnetic spectrum with longer wavelengths than microwaves have little, if any, biological impact. Radio waves, for instance, lack the energy to excite biochemical compounds. Infrared waves are essentially heat and can burn. Microwaves, such as might leak out of a microwave oven are thought to be injurious to tissues. Radiations with wavelengths shorter than visible light, however, can have devastating influences. While visible light merely excites molecules (leaving their basic structures intact), X rays and gamma rays may break them apart. Short wavelength, high energy electromagnetic radiation can burn living tissue, cause genetic mutations, and otherwise disrupt the processes of life.

Selective absorption

The rays of visible light that comprise sunlight appear as red, orange, yellow, green, blue, indigo, and violet in a rainbow or when passed through a glass prism (Figure 7-3). Red rays have the longest wavelength and the least energy of the visible radiations. Violet rays, at the opposite end of the spectrum, have the shortest wavelength and the highest energy (or frequency).

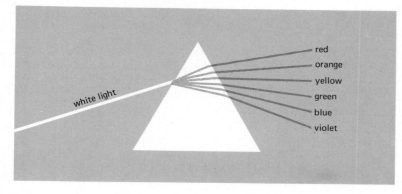

FIGURE 7-3

A spectrum of colors is formed when light passes through a glass prism. Light of all wavelengths is slowed down when it passes from the air into the glass of the prism. Violet light is refracted (deflected from the original course) the greatest amount and travels slower than red light which is refracted least.

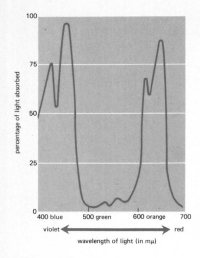

FIGURE 7-4

The light absorption spectrum of chlorophyll in alcohol. This curve is similar to the curve for light absorption by chlorophyll in leaves. Note that most of the green light is not absorbed; it is reflected away.

FIGURE 7-5

The amount of light at different wavelengths given off by three sources of light.

Plants selectively absorb particular portions of the light spectrum, and different plants may utilize different colors of light. Most land-dwelling plants absorb the greatest amount of energy in the form of violet and blue rays and a somewhat smaller amount of red and orange rays (Figure 7-4). Very little green and yellow light is absorbed; most of it is reflected or passes through the plant. This accounts for the green or greenish-yellow color of most plants.

Plants living beneath the surface of water receive much less light than land-dwelling plants. Water absorbs light energy of all wavelengths, but especially depletes wavelengths at the ends of the visible spectrum, namely, violet and red. Thus, submerged plants have evolved with photosynthesizing pigments that utilize more light from the middle of the spectrum (blue, green, and yellow rays) than land plants.

Growers of houseplants know that sunlight is not the only kind of light that will support photosynthesis. Artificial lighting also works, but care must be taken to give the plants a balanced spectrum. Incandescent bulbs supply light from the warm end of the spectrum (red, orange, and yellow light), but they are deficient in the cool wavelengths (blue and violet). Some kinds of fluorescent tubes, on the other hand, are good sources of cool but not warm light. Used together fluorescent and incandescent lights give a balanced light environment for photosynthesis (Figure 7-5).

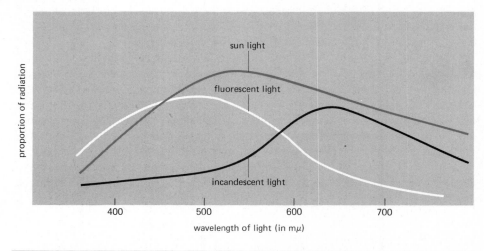

Light intensity and leaf arrangement

The rate of photosynthesis is directly proportional to light intensity, at least until the intensity reaches one-fourth to one-half that of full sunlight. Excessively bright light can even inhibit photosynthesis. Under low light intensities, however, plants orient themselves so that they receive as much as possible. Anyone who has ever watched a house plant grow toward a window knows this to be true. If the plant is turned so that its leaves face away from the window, it will slowly reorient itself toward the light over the course of several days. The physiology of this **phototropic response** is considered in Chapter 13.

Leaves are arranged on stems in configurations which maximize the amount of light received. Plants that have paired leaves generally grow so that alternating pairs of leaves are perpendicular to each other. Leaves growing singly spiral around the stem.

Although plants generally orient their leaves so as to receive as much light as possible, not all plants are equally efficient at utilizing light energy. Corn and sugarcane, for example, are highly efficient, while tobacco and beans are inefficient.

CHLOROPHYLL—THE AGENT OF PHOTOSYNTHESIS

To study the process in which energy-storing molecules are produced, almost any plant part can be chosen, as long as it contains the green photosynthetic compound, **chlorophyll.** A leaf cell, a cell from a green stem, or hairlike strands of algae from a pond could be considered. Any of the seaweeds could be used, since, although they may be brown or red, the green of their chlorophyll is only masked by other pigments.

In most plant cells photosynthesis takes place in organelles called **chloroplasts.** Within these subcellular bodies, chlorophyll molecules are sandwiched between layers of proteins and lipids in **grana** (Figure 7-6). Some primitive plants, such as the blue-green algae, have no organized chloroplasts, but they do have grana distributed throughout their cytoplasm.

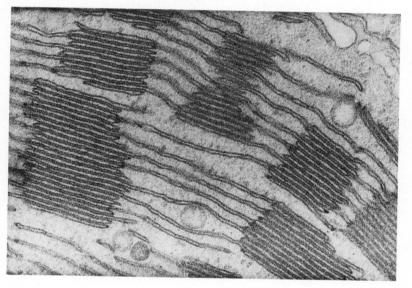

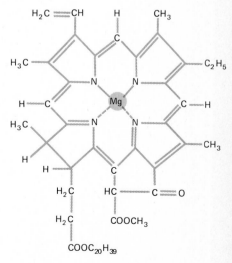

FIGURE 7-6
The intricate organization of pigment molecules in the grana of the chloroplast is probably responsible for the leaf's ability to trap light energy for photosynthesis. This electron micrograph shows the edge view of the grana, whose appearance has been compared to stacks of coins.

FIGURE 7-7
The structural formula of the chlorophyll *a* molecule.

Various combinations of four kinds of chlorophyll (slight variations on the same biochemical theme) may be found mixed in the grana of any given plant. The most common kind is chlorophyll-a, (found in all green plants) which is a bright blueish-green pigment composed of carbon, hydrogen, oxygen, nitrogen, and a single atom of magnesium at the center of the molecule (Figure 7-7). The chemical formula of chlorophyll-a is $C_{55}H_{72}O_5N_4Mg$.

Chlorophyll-b is a yellowish-green variant, with the chemical formula $C_{55}H_{70}O_6N_4Mg$. Chloroplasts in the cells of seed-producing plants usually contain both of these compounds, in a ratio of about three parts of chlorophyll-a to one part chlorophyll-b. The other two variants are found in plants that are more primitive than those that produce seeds. Sulfur bacteria (so-called because they utilize an unusually large amount of sulphur in their biochemical processes), photosynthesize with a unique form of chlorophyll. Most other bacteria do not contain chlorophyll.

Chloroplasts that have been removed from cells may be capable of photosynthesis, but chlorophyll alone is not. In addition to chlorophyll, chloroplasts contain enzymes that are essential as catalysts of photosynthetic reactions.

THE NATURE OF PHOTOSYNTHESIS

The net reaction of photosynthesis is summarized quite simply by the equation:

$$6CO_2 + 12H_2O + \text{light energy} \rightarrow C_6H_{12}O_6 + 6O_2 + 6H_2O$$

carbon   water   glucose   molecular water
dioxide   oxygen

However, although this equation accounts for the materials required and the products formed, it misses the subtleties of the biochemical flow from carbon dioxide and water to glucose and gaseous, molecular oxygen. Photosynthesis is a complex series of biochemical reactions, the nature of which was unknown until recent years. Within the past few decades biologists have come to see photosynthesis as a two-step process, one of which requires light and the other which does not. The simplified mechanical process can be seen in Figure 7-8.

FIGURE 7-8
Photosynthesis: an energy source, raw materials, products. The light, or *photo*, reactions occur on the left; the dark, or *synthesis*, reactions occur on the right as depicted in this simplified mechanical model.

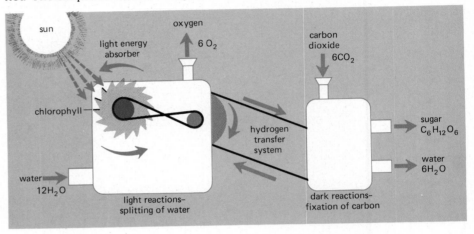

The light reactions

In the early 1940s the biochemist, Samuel Ruben and associates utilized a heavy isotope of oxygen, O^{18}, to find the source of the oxygen that is released from photosynthesizing plants. O^{18} is not a radioactive isotope, but it can be differentiated in the lab from the commonly occurring oxygen, O^{16}, with the use of the mass spectrograph (an instrument that can distinguish elements on the basis of their mass). Plants utilize O^{18} as readily as O^{16}.

In one phase of his experiments Ruben introduced carbon dioxide composed of O^{18}. He found no O^{18} released from plants, only O^{16}. In another experiment Ruben grew algae in water which was composed of O^{18}. This time when the gaseous oxygen from the plant was analyzed it was found to be the heavy isotope O^{18}. The conclusion was that water is the source of the oxygen released during photosynthesis.

In addition to contributing the oxygen atoms which form O_2, water also gives off electrons to chlorophyll-b by the following reaction:

$$H_2O \rightarrow 2H^+ + O + 2e^-$$

Water $\rightarrow$ Hydrogen + Oxygen + electrons
ions   Gas

These electrons, now part of chlorophyll-b, plus electrons that are part of the chlorophyll-a molecule are raised to higher energy levels by light falling on the plant. Though it is difficult to visualize changes at the infinitesimal level of molecules, the energizing of an electron can be likened to moving a book from a lower level in a bookcase to a higher level. If the book were to fall from the higher shelf, it would crash to the floor with a greater force than if it fell from the lower shelf. This is equivalent to saying that the book on the higher shelf has greater potential energy than it would have on the lower shelf.

The energy imparted by light remains in the chlorophyll molecules as long as they remain in an "excited" condition. Energy is released as the electrons drop back to the base energy level. The chlorophyll is no longer energized, but with more illumination the process may be repeated instantaneously. The energy released is channeled directly into the formation of adenosine triphosphate (ATP), as will be seen shortly.

The hydrogen ions and the electrons that have come from the splitting of water (and from other sources) are picked up by a hydrogen acceptor, nicotinamide adenine dinucleotide phosphate (NADP). NADP combines with two hydrogen atoms to form $NADPH_2$. This is a fleeting chemical union; the hydrogen atoms are soon drawn away from NADP into the cycle of dark reactions.

The processes referred to here are the **light reactions** of photosynthesis because light is required for the processes, they cannot occur in the dark. Figure 7-9 summarizes these reactions.

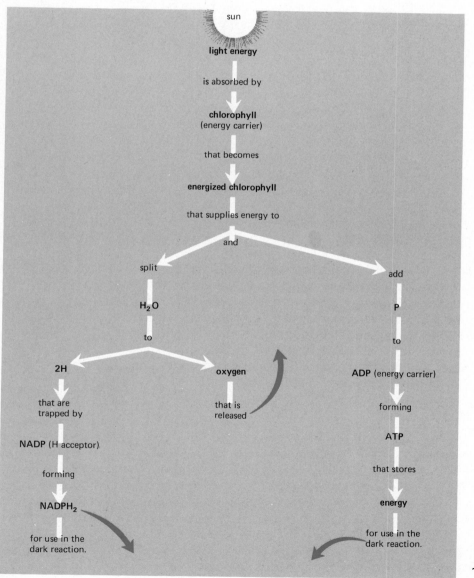

FIGURE 7-9
The light reactions of photosynthesis.

The dark reactions

A few years following Ruben's experiments, Melvin Calvin and his co-workers at the University of California conducted a series of experiments to monitor the reactions of carbon in photosynthesis. As his experimental organism Calvin chose the tiny green alga, *Chorella* which he grew in a solution of carbon dioxide with radioactive C^{14}. Calvin soon discovered that the reactions by which carbon was incorporated into glucose involved several intermediate compounds as he could follow the radioactive C^{14}. To determine the sequence of events, he devised a set-up involving a "lollipop" flask (Figure 7-10). Cultures were grown in the flask for various time intervals after introduction of C^{14} carbon dioxide. They were killed by release into boiling alcohol, which halts photosynthesis.

By analyzing products formed at succeeding time intervals, Calvin discovered that carbon atoms from six molecules of carbon dioxide were linked, in a process involving several stages, to form the framework of glucose (Figure 7-11) which contains 6 carbon atoms. Some of the oxygen atoms originally included in carbon dioxide were retained in the structure of glucose, while others reacted with hydrogen to form water. Because the reactions involving carbon dioxide do not require light, they are collectively termed the **dark reactions.** This designation, however, should not be interpreted as indicating that the dark reactions must occur in the absence of light. In fact, they generally occur concurrently with the light reactions.

The role of ATP

Involved in both the light and dark reactions of photosynthesis is the amazing compound adenosine triphosphate (ATP) which has been called the "universal energy currency of life." ATP has been found in every organism yet analyzed, and its function is not only essential in photosynthesis but in all cellular processes that involve transfers of energy. The structure of ATP can be quickly changed in order to release energy for the processes of life. Most of an organism's ATP is formed within the mitochondria during the process of cellular respiration. ATP that is in-

FIGURE 7-10
Calvin's "lollipop," used by Dr. Melvin Calvin to investigate photosynthesis. Radioactively labeled carbon dioxide was bubbled through suspensions of *Chorella* in the "lollipop" for varying amounts of time, after which they were released into a flask of boiling alcohol, which killed them. Analyses were made to determine what substances had been synthesized in the differing periods of exposure to the $C^{14}O_2$.

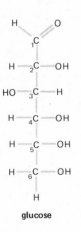

glucose

FIGURE 7-11
The structure of glucose.

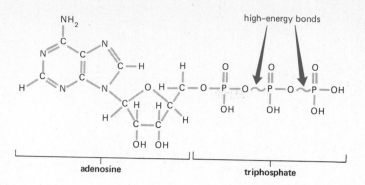

high-energy bonds

adenosine

triphosphate

FIGURE 7-12
A structural diagram of an ATP molecule.

volved in photosynthesis, however, is generated in the chloroplasts during the light reactions and utilized there in the dark reactions.

A molecule of ATP consists of a complex **adenosine** unit to which phosphate groups are linked (Figure 7-12). Three phosphates are included in each molecule; hence the name adenosine *tri*phosphate. ATP stores energy in the bonds of the phosphate groups in a form that is utilizable in biochemical processes. While the bonds of all other molecules also retain energy, it is not biologically available. Energy in other complex molecules must first be transformed to ATP before it can be utilized in such processes as muscle contraction, neural transmission, biosynthesis, and active transport. The biologically active bonds in ATP contain somewhat more energy than most bonds in organic molecules; hence, they are commonly termed "high-energy" bonds. We will follow the same convention in this text. But, keep in mind that it is the availability of the energy to biological processes, not the amount of energy, which is the important characteristic.

In the build-up of ATP from adenosine, the first phosphate group (the one closest to the sugar ribose) is firmly attached; the bond is not a biologically active, high-energy bond. This biochemical is adenosine *mono*phosphate (AMP). Addition of a second phosphate group forms adenosine *di*phosphate (ADP). The bond between the first and second phosphate groups is a high-energy bond. The third phosphate, which is linked to the second one with another high-energy bond, completes the phosphate "tail" of the ATP molecule. When the third phosphate group is removed from ATP, energy is released to power cellular activities (Figure 7-13). Energy for the formation of ATP may come from glucose, or it may come directly through a reaction with chlorophyll during the light reactions.

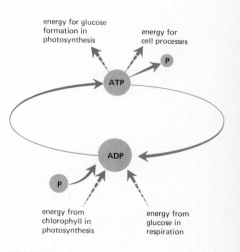

energy for glucose formation in photosynthesis

energy for cell processes

P

ATP

ADP

P

energy from chlorophyll in photosynthesis

energy from glucose in respiration

The dark reactions continued

The transfer of energy from chlorophyll to ATP during the light reactions "charges" the chloroplast for the reactions to follow. With biochemical energy available, this phase does not require light or any other form of external energy. The net result of the dark reactions is the assembly of glucose, which occurs in a series of cyclical steps (Figure 7-14):

Within a chloroplast carbon dioxide combines with **ribulose diphosphate** (RDP), a compound composed of a 5-carbon sugar molecule to which two phosphate groups are attached. The immediate product of this reaction is an unstable 6-carbon complex, which spontaneously splits into two identical molecules of **phosphoglyceric acid** (PGA), a relatively stable 3-carbon complex.

PGA reacts first with ATP and then with hydrogen atoms supplied by $NADPH_2$ (which was formed during the light reactions) to form **phosphoglyceraldehyde** (PGAL). Because PGAL is energy-rich relative to PGA, this is an energy-requiring reaction. The energy is supplied by ATP from the light reactions.

FIGURE 7-14
The dark reactions of photosynthesis.

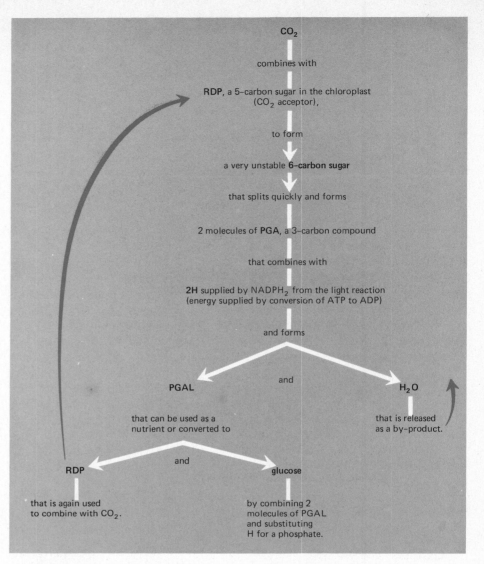

FIGURE 7-15

An abbreviated equation for the formation of one molecule of glucose from two molecules of PGAL. Notice that PGAL includes three carbon atoms and a phosphate group.

PGAL can be utilized directly in cellular processes, and for this reason, might be considered to be a principal product of photosynthesis. Plants nourished with PGAL can survive for extended periods in total darkness, without any other organic nutrients (mineral nutrients are, of course, required). Within the overlapping flows of life's chemistry, it is involved in several sequences of biochemical reactions. In the primary dark reaction two molecules of PGAL react in a series of steps to form one molecule of glucose. Two phosphate groups are simultaneously given off (Figure 7-15). A similar reaction results in the formation of fructose, which is sugar with the same empirical formula as glucose but a somewhat different chemical structure. PGAL is also involved in the reactions that form lipids.

A certain amount of PGAL reacts in several steps to form RDP; thus, since RDP is the acceptor of carbon dioxide with which the process begins, the cycle of dark reactions comes full swing.

CHEMOSYNTHESIS

A discussion of carbohydrate synthesis would not be complete without a mention of a small but unique group of organisms that do not rely on photosynthesis. By a process known as **chemosynthesis,** certain bacteria synthesize carbohydrates without utilizing light energy. The enzyme systems of these bacteria trap energy released during inorganic chemical reactions. Certain of these bacteria obtain energy by combining oxygen and hydrogen to form water. Others oxidize ammonia to nitrite and nitrate. Still others receive energy from reactions involving oxides of iron and sulfur. Energy from these reactions, rather than light, is used in synthesizing carbohydrates.

From the standpoint of cell structure and specialization, chemosynthetic bacteria are among the world's most primitive organisms. But, their unique chemical systems have helped them survive perhaps a billion years of evolutionary history, while many species of more complex organisms have gone to extinction. Thriving on what to other organisms is useless, the chemosynthetic bacteria have avoided the thick of the evolutionary battle.

RESPIRATION—ENERGY RELEASE

Only certain organisms are capable of trapping energy by photosynthesis or chemosynthesis, but all organisms require energy in the form of ATP. The process by which plants and animals break down "fuel" molecules and transfer energy to ATP is termed **respiration.** Essentially, the result of respiration is the reverse of photosynthesis:

$$C_6H_{12}O_6 + 6O_2 \rightarrow 6CO_2 + 6H_2O + \text{energy (ATP)}$$

glucose · oxygen · carbon dioxide · water

But this chemical equation is only a generalization. The reactions of respiration are unique for each species; indeed, they vary from cell to cell within an individual organism. Nevertheless, almost every known living thing respires.

The reactions of respiration are **oxidative reactions.** In biochemical processes oxidation involves either the addition of oxygen to a molecule or the removal of hydrogen. Both processes release energy. Complete oxidation of complex organic compounds yields the stable, low energy compounds, carbon dioxide and water, which cannot be utilized as energy sources by any organism.

Oxidative reactions with the same net result as those of respiration are involved when fossil fuels are burned for heat, light, and mechanical energy. Oxidation of gasoline to carbon dioxide and water, for instance, is the power-yielding reaction of the internal combustion engine. More appropriately known as internal combustion, the process within the automobile engine also generates carbon monoxide and various other poisonous compounds.

Although there are similarities between the breakdown of fuels during combustion and the utilization of nutrient molecules during respiration, there are two basic differences. First, the high temperature of fuel combustion would be deadly to any organism. Second, the process of combustion is chaotic while respiration is controlled. Fires burn indiscriminately, but respiration releases energy through a series of controlled steps. The intricate chemical mechanisms of living cells accomplish the break down of organic molecules at controlled temperatures and rates.

To most people the term "respiration" is synonymous with "breathing." Because inhalation brings oxygen into the body and exhalation expels carbon dioxide and water vapor, breathing certainly plays an important role in respiration. However, breathing is merely the beginning and the end of a complex biochemical process which takes place within cells.

Any organic molecules present in a cell can be utilized for respiration if the necessary enzymes are present. Sugars, fatty acids, proteins, nucleic acids, and various other biochemical compounds can be broken down to liberate the energy necessary for life processes. Complex molecules are storehouses of chemical-bond energy. As long as their bonds remain unbroken, they retain the energy that was required to form them. But, when they are broken down to atoms or simpler molecules, this energy is set free. For example, the land planarian under starvation conditions will reduce itself in size by successively digesting its reproductive and digestive systems and utilizing the energy in the molecular structures of these systems. It eats itself down to 1/300 of its original volume!

The stages of cellular respiration

Although various organic molecules can be utilized in respiration, glucose is the most common cell fuel. To make use of the energy stored in glucose, most organisms must have access to molecular oxygen. The break down of glucose during respiration occurs in two stages.

The Anaerobic Stage   The first stage of cellular respiration is termed **anaerobic,** because it does not require molecular oxygen. Also referred to as **glycolysis,** it occurs outside the mitochondria, in a series of biochemical reactions involving 12 enzymes. The result is a step-by-step decomposition of glucose (6 carbons) to a 3-carbon acid, **pyruvic acid** (Figure 7-16). Energy from two molecules of ATP is used in the initial phases of this series of reactions. However, in later phases, 16

FIGURE 7-16

A diagrammatic summary of the chemical changes that occur in cellular respiration.

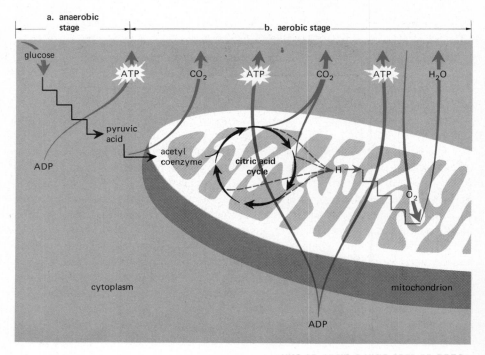

ATP molecules are generated from ADP, phosphate, and energy released in the splitting of glucose molecules. There is a net gain of 14 ATP molecules.

The Aerobic Stage   This stage of respiration requires molecular oxygen. In an enzyme-mediated sequence of reactions, pyruvic acid is broken down to carbon dioxide and water, and considerable energy is made available for storage as ATP.

At the onset of aerobic respiration, one hydrogen atom and one carbon dioxide molecule are removed from each molecule of pyruvic acid to form a 2-carbon compound, acetyl coenzyme-A. This enters a mitochondrion where it is joined to a 4-carbon acid, oxaloacetic acid, to form citric acid (6 carbons). This is the beginning of a cycle of reactions referred to as the **citric acid cycle,** or **Krebs cycle.** Early in the cycle, a carbon dioxide molecule splits off, leaving a 5-carbon acid. In a latter reaction, a second carbon dioxide molecule is removed, leaving a 4-carbon compound that evolves through a series of chemical changes to oxaloacetic acid, the compound with which the Krebs cycle begins.

Concurrent with the splitting of carbon dioxide molecules from the intermediates of the Krebs cycle, hydrogen atoms are also liberated in a highly excited state. The process releases so much energy that hydrogen atoms are split into their component electrons and protons. They eventually get back together again (though the chance of the exact same proton and electron recombining is remote) and merge with oxygen to form water. Prior to the formation of water, however, the electrons are passed through a series of proteins called **cytochromes** (Figure 7-17), which are collectively termed the **electron transport chain.** With each transfer, energy is released which produces a total of 24 ATP molecules.

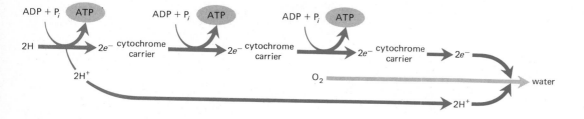

The transfer of electrons through the cytochrome series is the primary energy-yielding reaction of respiration. Electrons enter the cytochrome series at a high-energy level (so high that they are removed from association with any particular atomic nucleus) and are sequentially dropped to lower and lower levels as they are passed from cytochrome to cytochrome. Finally, the "unexcited" electrons recombine with protons and bond with oxygen to form water.

Cellular respiration, including both the anaerobic and aerobic states yields 38 molecules of ATP for each molecule of glucose. About 40 percent of the bond energy in a molecule of glucose is thereby transferred to ATP. Living organisms are a good deal more efficient than the internal combustion engine, which utilizes about 25 percent of the energy in gasoline.

FIGURE 7-17
The cytochrome system accounts for most of the ATP formation.

Other metabolic pathways

Not all organisms live in oxygen-rich environments. Certain species of bacteria, for example, thrive in the mud at the bottoms of stagnant ponds, where there is essentially no oxygen. Brewer's yeast lives in vats of malt which are carefully tended so that only a trace of oxygen seeps in. The fact that these organisms survive indicates that they are utilizing energy. But, the ways they do it (described below) are far less efficient than respiration.

Fermentation   As in the anaerobic stage of respiration, fermentation begins when glucose is broken down to pyruvic acid by glycolysis. It proceeds with further anaerobic reactions and can lead in two directions. In yeasts and certain other microorganisms, a carbon dioxide molecule is stripped from pyruvic acid, giving an

FIGURE   7-18
A diagrammatic summary of the chemical changes that occur in alcoholic fermentation.

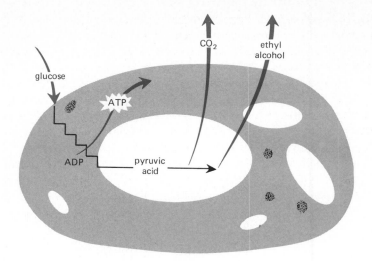

intermediate product that combines with hydrogen to form ethyl alcohol (drinking alcohol). Without molecular oxygen, this is as far as oxidation proceeds during **alcoholic fermentation** (Figure 7-18):

$$C_6H_{12}O_6 \rightarrow 2C_2H_5OH + 2CO_2 + energy$$

glucose　　　 ethyl　　　 carbon
　　　　　　 alcohol　　 dioxide

This is a much less efficient process than continuing on with the Krebs cycle. That there is still a good deal of energy left in alcohol is evidenced by its flammability. Fermentation stops at 12 percent alcohol which kills the yeasts.

Certain microorganisms produce fermentative end products other than ethyl alcohol because they have an additional battery of enzymes. Anyone familiar with the acid tang of wine turning to vinegar has experienced this form of fermentation. Winemakers add minute amounts of sulfur to prevent these vinegar making bacteria from surviving.

Anaerobic Respiration in Muscle   Within animals that normally respire aerobically, oxygen may be lacking in certain tissues at certain times (within muscles at times of strenuous activity, for example). Under such circumstances anaerobic respiration is temporarily necessary. Like fermentation, it begins with glycolysis, but unlike fermentation the end product is lactic acid, not ethyl alcohol. An energy release comparison of these processes can be found in Figure (7-19).

Respiration by Sulfur Bacteria   Certain bacteria respire anaerobically by a unique biochemical mechanism. Sulfur rather than oxygen is utilized as the final hydrogen acceptor. The product, hydrogen sulfide (H_2S) is infamous for its "rotten egg" odor. Sulfur bacteria thrive within the nearly anaerobic environment of an uncracked egg, where there is no niche for aerobic microorganisms.

Evolution and energy utilization

Because of the relative inefficiency of anaerobic respiration, it is not surprising that the evolutionary trend throughout the history of life has been toward aerobic respiration. Billions of years ago, before the evolution of photosynthesis, the atmosphere was without molecular oxygen. Primitive life was probably limited to fermentativelike biochemical systems. With the evolution of photosynthesis, and

FIGURE   7-19
A comparison of the end products and energy released in cellular respiration and alcoholic and lactic acid fermentation.

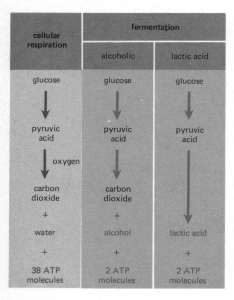

cellular respiration	fermentation	
	alcoholic	lactic acid
glucose	glucose	glucose
↓	↓	↓
pyruvic acid	pyruvic acid	pyruvic acid
↓ oxygen	↓	↓
carbon dioxide	carbon dioxide	
+	+	
water	alcohol	lactic acid
+	+	+
38 ATP molecules	2 ATP molecules	2 ATP molecules

the consequent liberation of oxygen to the atmosphere, oxidative biochemical processes became possible with an increased energy efficiency.

Today, aerobic respiration is the rule among all higher organisms; anaerobic biochemical pathways are reverted to only when unusual circumstances limit the availability of molecular oxygen. Nevertheless, environments still exist which provide all the requirements for life except oxygen. Within stagnant bodies of water, in compacted soil, inside eggs, and in various other places where contact with the atmosphere is lacking, primitive anaerobic organsisms still thrive. Since anaerobic niches have not disappeared, neither have anaerobic organisms.

THE PHYSIOLOGY OF HUMAN RESPIRATION

What happens to the energy released from glucose during respiration? Some of it is given off as heat, and in humans and other **homeothermic** (warm blooded) animals, this heat helps maintain a constant body temperature. The remaining energy, which is trapped in ATP, is utilized in biochemical processes. In animals these include biosynthesis of polysaccharides, fats, nucleic acids, and proteins. Additionally ATP is used in muscular contractions, active transport, cell division, and the transmission of nerve impulses.

All organisms require energy, and for plants and animals that respire aerobically, this means that oxygen must be taken in and carbon dioxide given off. In structurally simple organisms (such as bacteria, phytoplankton, jellyfish, and sponges) cells are in contact with the environment, and exchange of these gases between the cells and their surroundings occurs directly. But, the internal cells of more complex organisms are not in contact with the external environment. Higher organisms, therefore, must have special organ systems to take in oxygen and to eliminate carbon dioxide and water. In humans this function is served by the circulatory system and the lungs.

Breathing

Air may enter the body via the nose or the mouth. Except during times of intense physical exertion, when a large volume of air is required, the nose is the better route. The nostrils contain hairs that catch dust particles and other foreign matter that lodges on the moist mucous membranes in the nasal passages. All of this can be blown out of the nose. Along the length of the nasal passages air is warmed before it enters the lungs. All these advantages of breathing through the nose are lost in breathing through the mouth.

From the nasal cavity, air passes through the **pharynx** (throat) and enters the **trachea** (windpipe) (Figure 7-20). During swallowing, the end of the trachea is closed by a flap of cartilage, the **epiglottis.** The **larynx,** or Adam's apple, is the enlarged upper end of the trachea. Inside it are the paired vocal chords, whose vibrations produce the sounds of speech. Horseshoe-shaped rings of cartilage hold the trachea open for the free passage of air.

Cilia, which line the trachea, are in constant motion to carry particles taken in with air upward toward the mouth. Coughing, sneezing, or clearing one's throat eliminates foreign matter from the upper trachea. In smoggy, industrial areas, the cilia cannot keep up with the task. Consequently the lungs of a city person tend to grow black with debris over the years. White blood cells ingest a certain amount of foreign matter, but some remains permanently lodged in the inner tissues of the lungs.

At its lower end, the trachea divides into two branches, the **bronchi.** One bronchus extends to each lung and subdivides into countless small tubes called

FIGURE   7-20

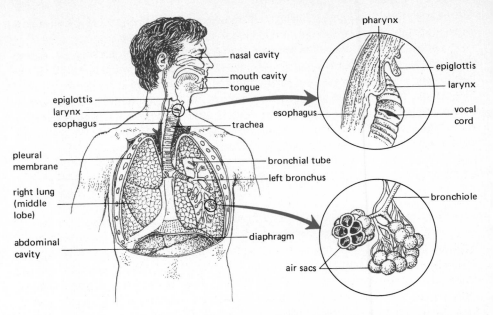

The organs concerned with breathing and external respiration in humans. In the lungs, the air passages increase in number, decrease in size, and end in tiny air sacs.

bronchioles, which end in clumps of microscopic **alveoli** (air sacs), like bunches of tiny grapes. The membranous walls of the alveoli are thin enough to allow oxygen to pass through them and flexible enough to expand with inhalation and contract during exhalation. The total area of the alveoli is about two thousand square feet—sufficient surface to take in enough oxygen for the needs of the billions of body cells having no direct access to air.

The spongy, porous tissue of the lungs consists mainly of bronchial tubes, blood vessels, and alveoli held together by connective tissue. The lungs are covered by a double **pleural membrane;** one part adheres tightly to the lungs and the other part adheres to the inside of the chest cavity. The membranes secrete mucus that acts as a lubricant, permitting the lungs to move freely in the chest during breathing.

Lungs themselves contain no muscle tissue; the movements of breathing involve muscles of the chest and upper abdomen. Inhalation occurs when muscle contractions increase the size of the chest cavity. The dome-shaped muscle beneath

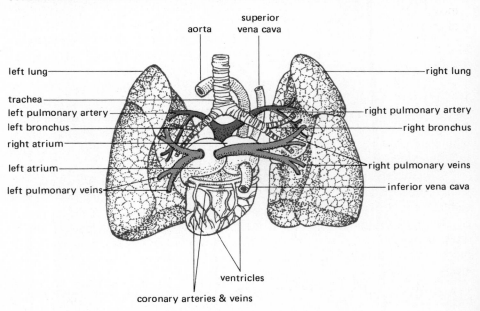

FIGURE   7-21

This posterior view of the lungs and heart shows the branches of the pulmonary arteries and the pulmonary veins.

182

the lungs, the **diaphragm,** flattens when it contracts, adding volume from below. Concurrently, the abdominal muscles relax, allowing compression of the digestive organs by the diaphragm. Muscles between the ribs play a role; when they contract they pull the ribs upward, into a position perpendicular to the spine. Throwing back one's shoulders increases the potential volume of the chest cavity, and a pumping action of the diaphragm fills the lungs to their greatest capacity. Try it; it's invigorating.

During exhalation, the muscles of the rib cage relax and the ribs lower. The diaphragm relaxes and resumes its original position as a dome at the top of the abdomen. Contractions of muscles of the lower abdomen push the diaphragm further up into the chest cavity, forcing air out in a rush.

Gas exchange in the lungs

Blood returning to the lungs from the body tissues arrives via the **pulmonary arteries** (Figure 7-21). Within the lungs blood is distributed into an extensive network of capillaries, surrounding the alveoli (Figure 7-22). Oxygen diffuses from the atmosphere into the bloodstream while carbon dioxide and water (the results of cellular respiration) diffuse in the opposite direction.

Within the red blood cells (Figure 7-23) is a complex, iron-containing protein called **hemoglobin,** which gives blood its color. It has a strong affinity for oxygen (Figure 7-24). Oxygen enters the red blood cells from the plasma (noncellular part of blood) in which they are suspended and becomes bonded to hemoglobin. At sea level each liter of blood leaving the lungs carries a fifth of a liter (200 ml) of oxygen to the interior tissues.

From the lungs blood circulates back through the heart, and it is distributed by arteries to capillaries throughout the body. When blood reaches tissues where the concentration of oxygen is low, hemoglobin releases a portion of the oxygen that it has brought in. Not all of the oxygen is given up, however. Blood returning to the lungs may contain from 80 to 140 ml of oxygen in each liter of blood.

At the same time oxygen is utilized in respiration, carbon dioxide is given off. The blood functions in the removal of carbon dioxide by carrying it to the lungs, from which it is exhaled. Carbon dioxide begins its trip out of the body by diffusing from the body cells into the capillaries. Some of it dissolves in the plasma, but most of it passes into the red blood cells, where it reacts with water in the presence of an enzyme to form carbonic acid:

$$CO_2 + H_2O \rightleftharpoons H_2CO_3$$

This weak acid rapidly ionizes to form hydrogen and bicarbonate ions:

$$H_2CO_3 \rightleftharpoons H^+ + HCO_3^-$$

When blood reaches the lungs, the reverse process occurs. Carbon dioxide and water are regenerated; they diffuse into the alveoli from the blood, thence to the atmosphere. Simultaneously oxygen is absorbed into the lungs, and the cycle of gas exchange is repeated.

Oxygen debt

During periods of strenuous activity, muscles utilize oxygen at a rate faster than the blood can supply it. At these times, muscle cells partially switch to anaerobic

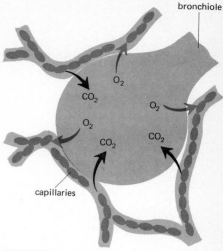

FIGURE 7-22

Gas exchange between an air sac (alveolus) and capillaries.

FIGURE 7-23

Human red blood cells photographed with a scanning electron microscope.

FIGURE 7-24

Structure of a heme group. Each hemoglobin molecule is a combination of protein and four heme groups. The oxygen binds to the Fe (iron) part of the molecule.

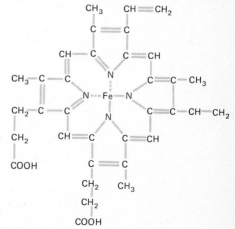

respiration where oxygen is not needed, with the consequent production of **lactic acid.** Athletes feel the build-up of lactic acid as muscle pain. In this state the body is said to have an **oxygen debt.** When activity slows, and a normal level of oxygen is reestablished, lactic acid is broken down to carbon dioxide and water by aerobic respiration. Runners sometimes speak of getting a "second wind," but as Frank Shorter, the winner of the marathon run in the 1972 Olympics, put it, "They call it second wind, but it's only your first all over again." In order to "pay" the oxygen debt, enough relaxation is needed to allow the build-up of oxygen in the blood. For a runner, this would mean slowing the pace.

In spite of the pain of oxygen debt, there is a certain advantage to it. The affinity of hemoglobin for oxygen decreases with increasing acidity. Therefore, in areas of the body in which lactic acid is being produced, hemoglobin releases a greater proportion of oxygen. For a time, the rate of aerobic respiration can rise parallel with the onset of anaerobic respiration. Ultimately, however, the detrimental effects of lactic acid outweigh the good. Muscles become fatigued and cramped.

The ill-effects of lactic acid aren't limited to physical impairment. Physiologists have found that injecting lactic acid into resting humans can induce feelings of anxiety. People who experience spontaneous attacks of anxiety also show high internal levels of lactic acid. Anxiety, therefore, may be rooted in biochemical malfunctioning. Interestingly, people deep in meditative states have lower than average levels of lactic acid in their bloodstreams. This suggests that a consciously induced meditative state may have anxiety-relieving effects at the biochemical level. Stomachs destroy milk's lactic acid so milk doesn't produce anxiety!

High altitude respiration

At high altitudes, where there are fewer molecules of oxygen per given volume of air than at sea level, resident populations have become physiologically adapted to obtaining sufficient oxygen. Indians in the Andes mountains of Peru are capable of physical activities that would quickly exhaust a healthy person accustomed only to living in the relatively dense atmosphere at sea level. Research has revealed that these people have greater lung capacities and more red blood cells (more oxygen available) than people who live at lower altitudes.

Additional research has shown that these physiological traits can be acquired through conditioning. Astronauts trained in atmospheres with low oxygen levels develop similar adaptations. Their bodies reach the point where they can work at a normal pace in an atmosphere with an oxygen content half that of the atmosphere at sea level.

MEASURING METABOLISM

Photosynthesis and respiration are the complementary energy-trapping and energy-releasing components of the complex biochemistry of life. But, organisms embody many other chemcial processes. The sum of all the chemical activities of an organism is its **metabolism.** The "constructive" phase of metabolism, **anabolism,** includes syntheses of carbohydrates, nucleic acids, lipids, and proteins. The "destructive" phase, **catabolism,** involves the oxidation of these compounds with consequent energy release. The rate of metabolism increases in proportion to the increase in activity of the body. This activity may be muscular or mental, but the mental changes are only a very small portion of the physical ones. Other factors governing the metabolic rate include exposure to cold and whether or not an individual is digesting food.

One way to find the metabolic rate is to measure the amount of heat given off by the body surface. This can be done with a **calorimeter.** The person to be tested enters a closed compartment that is equipped to accurately measure all the heat given off by the body. The subject may lie quietly or exercise vigorously, depending on the nature of the activity to be tested. The amount of heat given off is a direct indication of the rate of oxidation in the body tissues since heat is a by-product of cellular respiration. One study showed, for example, that an average of 200 calories is expended by a partner in sexual intercourse.

Even when the body seems completely inactive, as it does in sleep, energy utilization continues. Brain metabolism, for instance, does not decrease during sleep. But, with cessation of muscular activity, the rate of oxidation is greatly reduced.

The energy required to maintain life during complete rest is an organism's **basal metabolism.** It can be determined with a calorimeter, but another method, commonly used in hospitals, measures the amount of oxygen consumed in a given period. The test is preceded by a fifteen hour fast and at least one hour of inactivity. At the onset of the test, the subject's nose is plugged to prevent breathing from the atmosphere, and a mouthpiece connected to a tank of oxygen is fitted into the mouth. Thus, all oxygen inhaled during the test period comes from a measured tank (exhaled air is released separately) and the rate of oxygen utilization is used in the calculation of internal oxidation (respiration).

RESPIRATION BY GREEN PLANTS

Although the physical characteristics of plants and animals are quite different, the biochemical level of respiration is essentially the same for both kingdoms. As in animals, oxygen is the ultimate hydrogen acceptor in plants since it combines with hydrogen to form water, and carbon dioxide is released as the carbon skeleton of glucose breaks down. During daylight hours, the oxygen released from photosynthesis is more than enough to support a plant's respiration. At night, however, atmospheric oxygen is needed. The energy released in plants by respiration performs various functions. It powers streaming of protoplasm, for example, and is required for biosynthesis of cellulose, fats, proteins, and nucleic acids.

Transpiration

Rooted plants have no circulatory fluid analogous to blood. Instead, a one-way flow of water serves as the medium of transport. During its growing season, a rooted plant conducts water and minerals up through its roots and stems to the leaves. Some of the water is used in photosynthesis, but this accounts for only about one percent of the water that enters. The remainder escapes from the leaves as vapor (a single corn plant may lose over 50 gallons of water in the growing season) via pores called **stomata.** This flow is called **transpiration.** During transpiration, water first diffuses from leaf cells into the spongy intercellular spaces. It then passes to the atmosphere through the stomata (Figure 7-25). As water escapes to the atmosphere, more rises into the leaves to replace it. The rate of transpiration is regulated by the opening and closing of stomata (stoma, singular).

The opening and closing of stomata

In most plants (cacti are an exception) the stomata are usually open during the day and closed at night. They may also close at certain times during the day, especially on hot afternoons, when water is precious. A stoma consists of two facing **guard cells** (Figure 7-26). As the guard cells absorb water from adjacent tissue, their turgor increases. Pressure within guard cells causes the thin outer walls to

FIGURE 7-25
Stomata. The two bean-shaped guard cells that surround each opening contain chloroplasts.

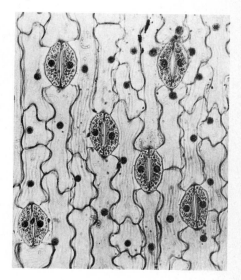

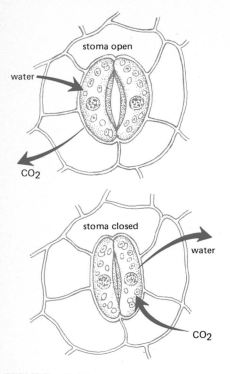

FIGURE 7-26
Changes in carbon dioxide in the guard cells affect water movement and turgor change that open and close the stoma.

bulge and pulls apart the thickened inner walls, forming a pore. Diffusion of water from the guard cells reduces turgor pressure, causing them to shrink. The inner walls then straighten and close the pore.

It would seem that turgor changes in guard cells and the opening and closing of stomata should be related directly to the water content of a leaf. If this were true, however, the stomata would be open at night when the water content is highest and closed during the day when a high rate of evaporation reduces it. Since stomata are, in fact, usually open during the day and closed at night, they work against a totally efficient water balance. The reason for this apparent paradox is that the carbon dioxide necessary for photosynthesis also enters a leaf through stomata. Because photosynthesis can occur only during the day, that is when the stomata must be open.

Investigations of the mechanism of the opening and closing of stomata indicate that the concentration of carbon dioxide in the guard cells affects their turgor changes. For some plants stomata close when the concentration of carbon dioxide in the air exceeds about 0.04 percent and open when the concentration falls below this amount. More recently it has been shown that during daylight, potassium ions are actively transported into the guard cells from the surrounding cells. Water follows into the guard cells, increases their turgor and the stomata open. As darkness approaches, potassium ions diffuse back out of the guard cells. Water then moves out of the guard cells, the cells lose turgor, and the stomata close.

PHOTOSYNTHESIS AND RESPIRATION — A FINAL CONSIDERATION

Since carbohydrates are built up during photosynthesis and broken down during respiration, the processes are essentially the reverse of each other. The requirements of one are the products of the other. This is a basic aspect of the biochemical balance maintained in the living world. The complementary relationships of photosynthesis and respiration are listed in Table 7-1. In comparing the two processes, notice that matter flows in a cycle. Atoms alternate between living and nonliving things. But, the flow of energy is not cyclical. Light energy is used in photosynthesis, whereas heat, which is the wasted form in which energy is released by life, is dissipated into the cosmos. Thus a constant source of energy, the sun, is needed to support life.

Plants photosynthesize at a rate much faster than they respire. In bright sunshine, energy fixation may occur ten times faster than energy dissipation. Although photosynthesis ceases at night, total energy fixation by plants exceeds energy loss. This places the green plants in the position of energy trappers for the dependent, heterotrophic organisms. The further an animal is from the plants at the base of its food web, the more dependent it is on other forms of life.

TABLE 7-1   COMPARISON OF PHOTOSYNTHESIS AND RESPIRATION

PHOTOSYNTHESIS	RESPIRATION
food accumulated	food broken down (oxidized)
energy from sun stored in glucose	energy of glucose released by oxidation
carbon dioxide taken in	carbon dioxide given off
oxygen given off	oxygen taken in
produces glucose from PGAL	produces CO_2 and H_2O
light necessary	light not necessary
occurs only in presence of chlorophyll or photosynthetic pigments	occurs in all living cells

CHAPTER 7: SUMMARY

1   Autotrophs are self-feeders compared to heterotrophs which are other feeders.

2   Photons are the particles of light that are absorbed by plants.

3   Only certain wavelengths of light are utilized by plants.

4   Chlorophyll is the major plant pigment that absorbs light.

5   Chloroplasts are the cell organelles where chlorophyll is located.

6   Photosynthesis has two major chemical reaction sequences — the dark reactions which synthesize sugars and the light reactions that generate energy.

7   ATP is the energy-rich molecule in the cell.

8   Respiration in cells is the extraction of energy from sugars and the release of CO_2 and H_2O.

9   Fermentation is the production of alcohol and CO_2 from sugars.

10 Breathing is a form of respiration involving gas exchange in the lungs and oxygen-carrying red blood cells.

11 Oxygen debt occurs when oxygen is utilized faster than it is obtained.

12 Body metabolism is measured by an instrument called a calorimeter.

13 Plants, as well as humans, engage in respiration.

14 Transpiration is the process whereby plants release excess water to the atmosphere.

Suggested Readings

Goldsby, Richard A.   *Cells and Energy,* 2nd Ed., Macmillan, New York, 1976 — A brief, easy-to-follow introduction to energy in biology.

Jensen, W. A.   *The Plant Cell,* 2nd Ed., Wadsworth, Belmont, California, 1970 — Photosynthesis and respiration by plants.

Lehninger, Albert L.   *Bioenergetics,* 2nd Ed., W. A. Benjamin Inc., Menlo Park, California, 1971 — The role of energy in biological processes clearly presented.

Levine, R. P.   "The Mechanism of Photosynthesis," *Scientific American,* December 1969 — What the title implies.

Weier, T. Elliot, C. R. Stocking and M. G. Barbour   *Botany,* 5th Ed., John Wiley and Sons, New York, 1974 — A comprehensive but clear introduction to plants.

ENERGY CAN BE GREEN
Roger Lewin

As described in this article, published by the British journal *New Scientist* in the summer of 1977, biologists are playing a part in relieving the energy crisis.

Three years ago it simply wasn't respectable to suggest that plants might make an important contribution to our stockpot of usable energy. But things have changed. For instance a senior official of the US Energy Research and Development Administration (ERDA), just recently rated biological solar energy conversion systems (which, roughly, is another way of saying plants) as one of the two most promising areas in future solar energy. And a recent series of international meetings reflects a growing awareness of the subject among scientists who at the very least are more than pleased to tap a new and for the moment fashionable source of research funds. Already, facts (and speculations) such as "the amount of carbon dioxide fixed into plants each year represents 10 times the energy from fossil fuels we currently consume annually," "an ocean farm producing kelp in an area slightly bigger than Texas but smaller than Alaska could supply the US with all its food and energy requirements," and "six per cent of US electricity consumption could be generated from readily available agricultural and urban wastes" are rapidly becoming part of common lore among energy-oriented biologists.

But bench top calculations can be dangerous: they can transform attractive pilot projects into readily destructable fantasies, a pastime through which everybody loses. No one would expose their reputation to severe risk were they to propose that by the year 2000 plant systems will be contributing by some small but significant extent to the energy budget and that by that date the prospects for important expansion will be very promising. And if there is a single element in plant biology that makes this certain

it is the fact that the products of biological solar energy conversion are *stored*: plants harvest electromagnetic energy from the Sun and manufacture chemicals, such as starch, cellulose, and all the other constituents that go to constructing a plant. Biologists admit that plants are particularly profligate in their use of the Sun's energy: the maximum energy conversion of incident radiation by a green plant is a mere five per cent, and it is usually much closer to one per cent. This figure looks even more discouraging when compared with the 12-15 per cent efficiency of photovoltaic systems (solar cells). But the product of photovoltaic systems is electricity which must be stored separately, a process which immediately knocks the overall efficiency to a figure below that of plants. In plants, energy conversion and storage go on in the same package. . . .

The most recent gathering on this topic, organised earlier this month with the enthusiastic support of the French Government in a mountain resort near Grenoble, focused on the variety of options offered by plant systems and tried to dissect out the important factors which limit their productivity. . . . One overall impression is that in the race to plug the energy gap that will be left as oil supplies are exhausted towards the end of the century, biology will not be offering one *single* technology as a replacement: there will be a matrix of technologies, each unit exploiting different aspects of plant production.

It is now almost a truism to point out that biological systems harvest resources that, though locally variable, are virtually unlimited: sunlight, carbon dioxide, and water. Plants trap the Sun's energy to build complex chemicals from simple ones, the basis of which is a carbon backbone. When we make use of the plant material, either as a source of heat or of chemicals, we sooner or later break it down again to carbon dioxide and water: biological energy conversion systems therefore represent a *renewable* energy source, and this is an important contrast with

This article first appeared in *New Scientist, London,* The Weekly Review of Science and Technology.

fossil fuels (coal, oil, and natural gas). The real point, then, is how well can plants, large and small, substitute in a world built around oil as a convenient liquid fuel and chemical resource? Ultimately the answer will depend on basic biology and inventive technology.

Each year the Earth's surface receives 3×10^{24} Joules of solar radiation of which $3 \cdot 1 \times 10^{21}$ J are trapped by plants and converted into chemicals (stored energy): $99 \cdot 9$ per cent of the radiation is therefore unused. Of the energy stored by plants each year only about $0 \cdot 5$ per cent finishes up on our plates as food while the rest cycles more or less rapidly through the world's biomass. There therefore appears to be scope for improvement. . . . Indeed, both US and European research programmes are searching for crops that will grow where currently nothing is grown systematically: for instance, ocean kelp farming is being investigated in the US, and the Irish are studying the possibility of growing willow trees in worked-out peat bogs. . . .

Improving photosynthesis

As well as attempting to develop new plants, either through more or less conventional plant breeding or by fancy genetic engineering, basic biologists are looking for ways by which to improve the efficiency of photosynthesis. Half the Sun's spectrum is useless to plants because their pigments absorb in only two main regions (400-500 nm and 600-700 nm). More light is wasted through reflection and dissipation in the leaf. . . . Theoretically there is great scope for enhancing photosynthetic efficiency as well as finding ways of encouraging plants to divert higher proportions of their material resources to economically important parts of the plant (away from stalks and into seeds, for instance). And the recent discovery that many important crops are apparently energetically extremely profligate offers another opportunity for boosting photosynthetic yield: as well as manufacturing chemicals in the light many plants indulge in what is termed photorespiration—they burn up in an apparently wasteful way at least half of their photosynthetic products. Wheat, spinach, tobacco, and hay are in this category (known as C3 plants), whereas maize, sugar cane, and sorghum are much more economical and do not photorespire (they are termed C4 plants). Currently there is a lot of head-scratching about ways to block at least partially the photorespiration, thus enhancing net photosynthetic productivity at a stroke.

In drawing up the energy budget in any kind of energy farming the cost of fertiliser rapidly establishes itself as a major input: the figure is often one-third of the total fossil fuel energy consumed. For this reason many people are interested in engineering nitrogen-fixing genes into plants. There are countless genetic and biochemical fences to be cleared before this attractive prospect is truly in sight, but once it is reached it will certainly impact significantly on energy budgets. One slick analysis of the energy budgets involved in farming hardwoods for fuel has recently been carried out by the MITRE Corporation in Virginia. On a six-year rotation using high technology management Bob Inman calculates the budget to be 15 units of energy produced for every unit put in. The balance is in fact somewhat less favourable than this figure suggests because the quality of the input energy is higher than that produced: you can't equate a barrel of oil directly with a pile of logs.

There are many problems with silviculture (wood growing), not least of which is making sure that the plantation is not overrun by unwelcome invaders. But Inman's analysis shows it to be close to commercial viability, a situation that is bound to improve as the cost of oil continues to rise. The most immediate impact of biomass on energy production, however, is likely to be use of wastes. Indeed, the sugar cane industry is already showing the way. Practically all cane factories generate their own electricity (and usually an excess) by burning the pulped fibre. Incidentally, sugar cane has featured in an Australian study on energy farming: the sucrose can be used as a chemical feedstock and the fibre as an energy source. Currently millions of tons of straw are burned in farm fields every year: in the UK the figure is $4 \cdot 5$ million, and in the US around 200 million. In most countries there is enough straw burned to provide the farms with most of their electricity. Questions of collection and transport are of course important here, as indeed they are in all aspects of waste utilisation.

Microbiologists are also interested in energy problems and there are a number of schemes involving, for instance, growing algae in sewage. Algal culture inevitably has problems of separation and high water content (both of which imply energy expensive processes), and a more profitable exploitation of these organisms may very well be in specialist areas. One such is a remarkable alga (*Dunaliella salina*) which grows in the Dead Sea and produces large amounts of glycerol as a defence against the high osmotic pressure. Under the right conditions this creature can be persuaded to manufacture glycerol as 85 per cent of its dry weight, a property that allows algal glycerol production at commercially very favourable prices. Glycerol could serve as an important chemical feedstock, and the Israel government is investing $0 \cdot 8m$ in developing the system.

Algae sometimes figure in projects for producing hydrogen (a potential fuel itself or for making fuels) from biological systems, as too do "synthetic" plant systems. This latter approach is currently at a very early stage of research, and yet it is conceptually very appealling. The idea is to construct systems based either on totally synthetic components (modelled on the biological structures) or on a mixture of biological and synthetic components. Either way, one great advantage of the system is that many of the natural inefficiencies of plant photosynthesis could be avoided. A research plaything it may be at the moment, but in the future it could be very important.

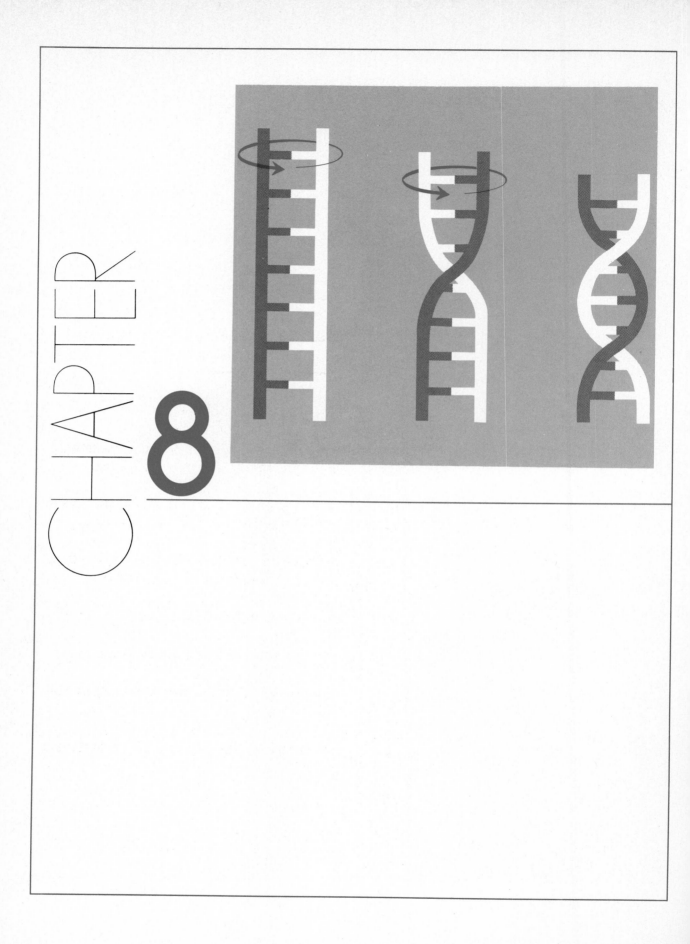

DNA and Proteins

Objectives

1 Learn what proteins do in a cell.

2 Describe the steps of protein synthesis.

3 Learn the structure of DNA.

4 Describe how DNA replicates.

5 Explain the primary, secondary, tertiary, and quaternary structures of protein.

6 Demonstrate how amino acid sequences in proteins reflect evolutionary changes.

7 Realize that mitochondria and chloroplasts, as well as the nucleus, contain DNA.

8 Be able to define a gene.

9 Understand the function of an operon.

10 Explain what mutations are and learn how they occur.

11 Understand the process of recombinant DNA.

THE IMPORTANCE OF PROTEINS

Proteins are the most diverse compounds of life. Some are the primary components of living tissues, and others direct the myriad of constant chemical changes within all active organisms. A given organism may contain tens of thousands of different proteins, with many copies of each.

The products of some biochemical processes do not vary from species to species (photosynthesis, for instance, always generates PGAL), but some proteins are unique for each species. Indeed, every individual produces certain proteins that are found nowhere else in the living world. While all proteins have the same basic structural format, there are significant differences even between individuals of the same species. The major problem in organ transplantation, for instance, is that the proteins of an implanted organ, which are only slightly different in chemical structure than those of the original organ, are destroyed by the natural chemical defenses of the recipient. Thus, the donated organ is rejected.

What proteins do

Proteins serve as structural components, enzymes, and hormones. The most common structural protein in animals is **collagen,** which occurs in skin, cartilage, and bone. Collagen forms a fibrous matrix which can withstand high forces. Less coarse structural proteins help to form cellular membranes and organelles.

Enzymes are protein molecules essential to all biochemical activities. Some enzymes perform their functions as catalysts of chemical reactions within the cells in which they were synthesized. Other enzymes are extracellular; these are secreted from the cells that produce them and catalyze reactions in body cavities. Digestive enzymes, for example, function within the mouth, stomach, and elsewhere along the digestive tract. Among them is **ptyalin,** which is secreted into the mouth where it initiates the break down of starch.

Hormones are secreted into the bloodstream by the endocrine glands. The activities that hormones regulate may be far removed from the cells that produce them. **Gonadotropins,** which are secreted by the pituitary gland (deep within the brain), are involved in the production of eggs and sperm in the reproductive organs.

Because proteins are so central to life processes, it isn't surprising that biologists have found that protein production changes with the development of an organism. As a plant or an animal matures, new kinds of proteins may be synthesized as those necessary in youth are phased out. Changes in protein production even come with behavioral changes. Brain researchers have found that while an animal is learning a new pattern of behavior, protein synthesis in certain parts of the brain increases. Research in progress is directed to determining the significance of these changes.

DNA AND PROTEIN SYNTHESIS

Protein synthesis is life's expression of individuality at the chemical level. Recent discoveries relating to the mechanisms of protein formation, therefore, have been major breakthroughs in the science of life. The protein factories of a cell are the ribosomes, which lie along the folds of the endoplasmic reticulum (Figure 8-1) in eukaryotes. While the ribosomes are the site of protein formation, the blueprint for the process comes from the nucleus.

For many years biologists have looked to the nucleus as the center of control of cellular activities. Nuclear components known as **chromosomes** have been known to somehow direct cellular chemistry. Studies before World War II had postulated that the chemical activities of chromosomes were directed by the protein, the nucleic acids, or both. Studies at the Rockefeller Institute during

FIGURE 8-1
Electron micrograph of a portion of a cell from the pancreas of a bat. The mitochondrion is the oblong structure, and it is surrounded by the endoplasmic reticulum. Small, spherical ribosomes lie along the endoplasmic reticulum.

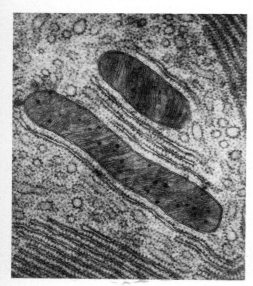

World War II provided evidence that nucleic acids alone directed the chemical activities of organisms. This was done by isolating the nucleic acid, DNA, from one type of bacteria and transferring it to a different type. Subsequently the recipient bacteria developed characteristics of the donor. This experiment is described in more detail later in the chapter.

The presence of DNA in the nucleus of cells can be demonstrated by feeding cells a radioactive constituent of DNA. A photographic emulsion is placed over the cells in the dark and the emulsion will be darkened where the radioactivity emanates from in the cells. After photographic development (just like film development), black dots will appear just over the nucleus indicating DNA's presence just in the nucleus (Figure 8-2).

Other experiments revealed the chemical composition of nucleic acids, but their physical structure—the arrangement of their atoms—stumped scientists for several years. In 1953, two young biologists working in the Cavendish Laboratory at Cambridge University made a breakthrough that has proved to be one of the most important scientific discoveries ever made. One member of the team was an

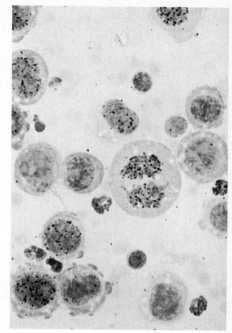

FIGURE 8-2
Radioautograph of cells from a mouse tumor. DNA, labeled with radioactive tritium, appears as black dots. Note that there is a concentration of dots in the nuclear material of the dividing cell (center) where DNA is concentrated.

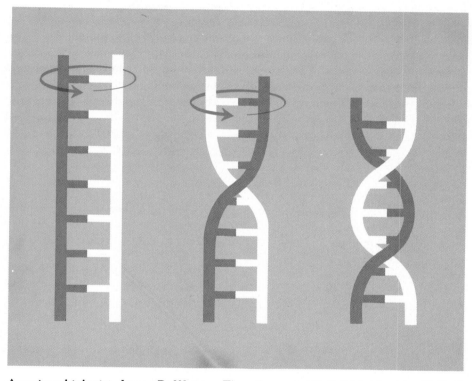

FIGURE 8-3
The formation of a double helix can be compared to the twisting of a flexible ladder.

American biologist, James D. Watson. The other was a British biophysicist, F. H. C. Crick. Together they proposed a model for the structure of DNA and its relationship to protein synthesis.

In the years following Watson and Crick's original findings, protein synthesis has been the subject of study of many investigators. Today, the study continues. The following sections summarize what is now generally accepted as the basic characteristics of protein synthesis. Given the complexity of the process and the inscrutable level of matter at which it occurs, however, it is quite possible that certain elements of this description will one day have to be revised.

The structure of DNA

Watson and Crick described the DNA molecule as a double alpha-helix—two strands of atoms wound around each other like a flexible "ladder" twisted into a spiral (Figure 8-3). Actually, the ladder is in two halves, each providing a side

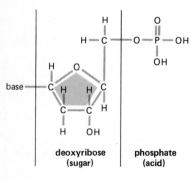

piece and half of a rung (Figure 8-4). Each side piece is composed of alternating units of phosphate and a five-carbon sugar, **deoxyribose** (Figure 8-5). A nitrogen-containing **base** is attached to each deoxyribose unit, forming part of each rung (Figure 8-6). The bases from opposing side pieces are joined by weak hydrogen bonds. A three-part unit (consisting of a phosphate, a sugar, and a base) make up a **nucleotide.** Thus, a DNA molecule is a double strand of nucleotides bonded by their bases. A single molecule of DNA may be composed of hundreds of thousands of nucleotides. The width of a DNA molecule is only 22 Å.

Notice in Figure 8-6 that there are four kinds of bases in DNA: **adenine, guanine, thymine,** and **cytosine;** they are indicated by the code letters *A, G, T,* and *C. The chemical natures of the bases are such that cytosine bonds only with guanine and adenine bonds only with thymine.* Adenine and guanine have similar chemical structures, and are classified as **purines,** while thymine and cytosine are designated **pyrimidines.**

The bonds between the bases in DNA are unlike the covalent bonds that join most of the atoms in organic molecules. As discussed in Chapter 5, covalent bonds involve the sharing of one or more pairs of electrons. Since hydrogen has only one electron, it can form only one covalent bond. However, when hydrogen forms a covalent bond, its electron is pulled to one side, leaving the positively charged proton of its nucleus exposed on the other side. This positive charge can interact with the electrons of either nitrogen or oxygen in a neighboring molecule to form a relatively weak **hydrogen bond.** In DNA there are two hydrogen bonds between adenine and thymine and three hydrogen bonds between guanine and cytosine (Figure 8-7)

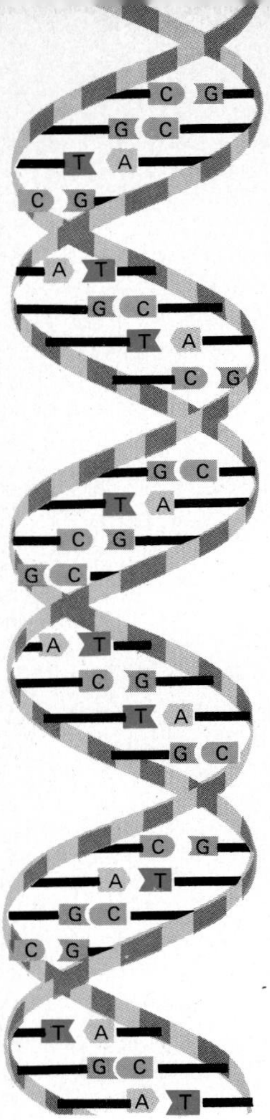

The replication of DNA

FIGURE 8-4
Model of double helix of DNA. A, T, G, C refer to the bases. Dark area on strands is deoxyribose. Light area is the phosphate group. Width about 22 Å (0.002 micrometers).

DNA possesses the remarkable ability to duplicate, or **replicate,** its own structure. During replication, certain enzymes separate the two strands of a DNA molecule, while other enzymes guide bases from the nucleus to the portions of the strands being duplicated (Figure 8-8), base A matching with T, and G with C. Thus, two double-stranded molecules of DNA are generated from the original double stranded molecule. Prior to a cell division, each DNA molecule in the nucleus duplicates itself, and each of the daughter cells receives one of the duplicates.

DNA's vocabulary in protein synthesis

FIGURE 8-5
A nucleotide of DNA, showing a phosphate bonded to deoxyribose. The base joined to deoxyribose may be either G, C, A, or T.

FIGURE 8-6
The four bases representing the four code letters in DNA.

Between cell divisions DNA directs the synthesis of the proteins required by the cell. The units of the "vocabulary" or "code" by which this is accomplished are

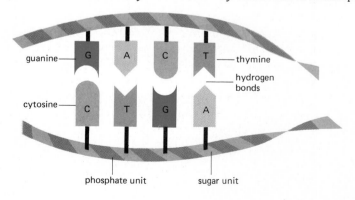

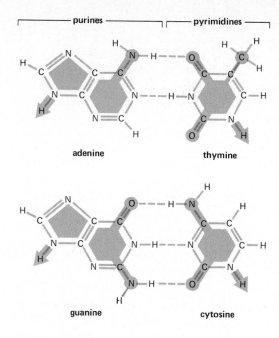

purines ───── pyrimidines

adenine   thymine

guanine   cytosine

sequences of three bases, known as triplets, along one of the strands in a DNA molecule. In all there are 64 different triplets that can be formed from the four nucleotide bases. For example, one triplet is composed of the sequence cytosine, guanine, and adenine; it is abbreviated as *CGA*. A given DNA triplet codes for a specific amino acid, and the order in which the triplets occur in DNA ultimately determines the order in which amino acids are joined to form protein. A single molecule of DNA codes for the structure of thousands of different proteins.

Transcription of the DNA code to RNA

DNA is contained in the nucleus; yet it controls a process that occurs outside the nuclear membrane, in the cytoplasm. If DNA controls the synthesis of proteins, how is the blueprint delivered from the nucleus to the ribosome factories? This function is performed by the mobile nucleic acid, messenger ribonucleic acid (mRNA).

In a manner similar to the way DNA replicates its own structure, DNA also directs the synthesis of a near duplicate of its structure messenger ribonucleic acid by a process called **transcription.** The structural format of mRNA is essentially the same as that of DNA, but the sugar in mRNA is **ribose,** which contains one more oxygen atom than **deoxyribose** (Figure 8-9). mRNA also differs from DNA in the composition of one of its bases. **Uracil** (*U*), a pyrimidine, is substituted for thymine (Figure 8-10) when mRNA is being synthesized. mRNA usually exists as a single strand, and, as a product of DNA, serves as a traveling agent in carrying the codes for protein synthesis to the ribosomes.

Consider an example in which three adjoining triplets in a strand of DNA are coded *CGT, TGC,* and *AAA.* Transcription would result in the synthesis of a strand of mRNA with the triplet sequence: GCA, ACG, and UUU (Figure 8-11). Notice that transcription is essentially the same as replication, except the uracil is substituted for thymine. Thus, the 64 DNA triplets give rise to the 64 RNA **codons.** Once formed, a strand of messenger-RNA detaches from its parent strand of DNA. It may be stored a short time in the nucleus; or it may immediately move through the nuclear membrane into the cytoplasm and attach to a ribosome.

FIGURE 8-7
Strands of the DNA molecule are held together by hydrogen bonds (dashed lines) between the purine-pyrimidine base pairs. Arrows indicate where each base is bonded to a DNA strand.

FIGURE 8-8
The process of DNA replication.

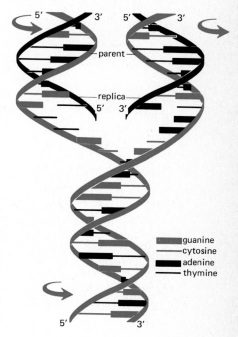

guanine
cytosine
adenine
thymine

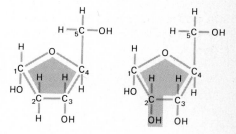

FIGURE 8-9
Deoxyribose (left) and ribose (right) compared.

FIGURE 8-10
Thymine and uracil compared. Thymine is found only in DNA; uracil only in RNA.

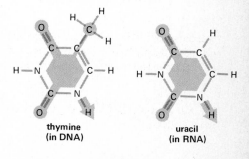

thymine
(in DNA)

uracil
(in RNA)

DNA AND PROTEINS **197**

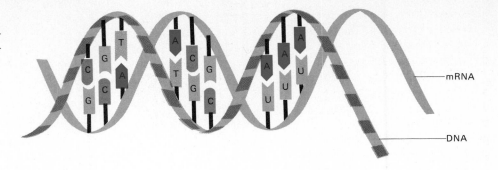

mRNA

DNA

FIGURE 8-11
Messenger RNA (mRNA) is being synthesized using the DNA as a template (pattern).

The assembly of polypeptides

There are 20 amino acids, each of which corresponds to one or more of the 64 RNA codons. Not all codons, however, have a corresponding amino acid. Some of them have been called punctuation marks, which are known to include UAA, UAG, and UGA. Punctuation marks function as starting or termination points in mRNA; they signal the beginning or the end of a chain of amino acids, like a capital letter signals the beginning of a sentence, and a period signals the end. Some amino acids are coded by only one codon, whereas others are coded by as many as six different codons. However, for any given codon there is only one corresponding amino acid (Table 8-1). Codons refer to the bases of mRNA, not DNA.

TABLE 8-1   MESSENGER RNA CODONS FOR THE AMINO ACIDS

AMINO ACID	CODONS
alanine	GCA, GCG, GCC, GCU
arginine	CGA, CGG, CGC, CGU, AGA, AGG
asparagine	AAC, AAU
aspartic acid	GAC, GAU
cysteine	UGC, UGU
glutamic acid	GAA, GAG
glutamine	CAA, CAG
glycine	GGC, GGU, GGA, GGG
histidine	CAC, CAU
isoleucine	AUC, AUU, AUA
leucine	CUC, CUU, CUA, CUG, UUA, UUG
lysine	AAA, AAG
methionine	AUG
phenylalanine	UUU, UUC
proline	CCA, CCG, CCC, CCU
serine	UCA, UCG, UCC, UCU, AGU, AGC
threonine	ACA, ACG, ACC, ACU
tryptophan	UGG
tyrosine	UAC, UAU
valine	GUA, GUG, GUC, GUU

When a strand of messenger-RNA reaches a ribosome, it acts as a pattern for the assembly of amino acids. This process is known as **translation.** Several **polypeptides** (a polypeptide is a chain of amino acids; a single protein molecule may be composed of two or more bonded polypeptides) can be translated simultaneously along a strand of mRNA. All ribosomes in a particular cell seem to function identically; there are not specific ribosomes for each kind of protein synthesized by a cell. Translation begins when a ribosome attaches at a position on an mRNA molecule where a chain of amino acids will be started. These positions are indicated by certain sequences of codons, the punctuation marks.

In addition to messenger-RNA, two other forms of RNA play roles in the translation process. Ribosomes themselves contain **ribosomal-RNA** (rRNA) which functions in the attachment of ribosomes to mRNA. The third type of RNA, **transfer-RNA** (tRNA) acts as a carrier in that it brings amino acids from the cytoplasm to the messenger-RNA. Strands of tRNA are relatively short, usually consisting of only 70 to 80 nucleotides.

A strand of tRNA is highly convoluted (Figure 8-12). The three nucleotides at the bottom of the central lobe comprise an **anticodon,** which is the mate of a codon along a strand of mRNA. For instance, the anticodon, AAG, is the comple-

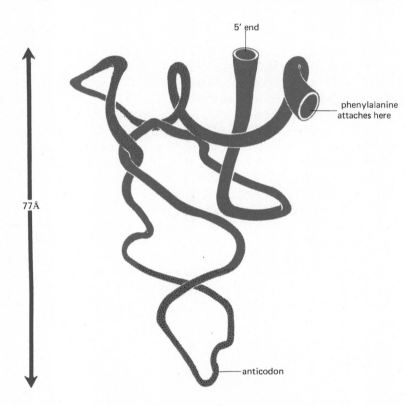

5′ end

phenylalanine
attaches here

77Å

anticodon

FIGURE 8-12
A three-dimensional model of yeast transfer RNA (tRNA) for phenylalanine showing where phenylalanine binds and where the tRNA binds to mRNA at the anticodon site. The 5′ end is the phosphate end of the molecule. Other parts hold the tRNA to the ribosome.

ment of UUC in mRNA. At the opposite end of the tRNA molecule is the binding site for the amino acid corresponding to the mRNA codon that matches the anticodon site. In the above example, the amino acid bound to the tRNA would be phenylalanine. Therefore, the anticodon determines which of the 20 amino acids are picked up. An activating enzyme in the cytoplasm recognizes the correct amino acid and bonds it to the tRNA molecule. According to theory, there is a specific enzyme for each of the 20 amino acids. Energy from ATP (the universal "energy currency" of life, described in the preceding chapter) is required for this process.

During formation of a polypeptide chain, a ribosome moves along the mRNA, from one end to the other, and "reads" each codon in succession. On each codon, the proper anticodon of tRNA binds, and on the ribosome, places the amino acid carried by the tRNA in proper position to be added to the growing protein chain. After delivering an amino acid to the ribosome, the tRNA unit detaches and moves back into the cytoplasm. Other amino acids are brought into position, until, after several seconds, the polypeptide is completed. Stages in the assembly of a polypeptide are diagrammed in Figure (8-13).

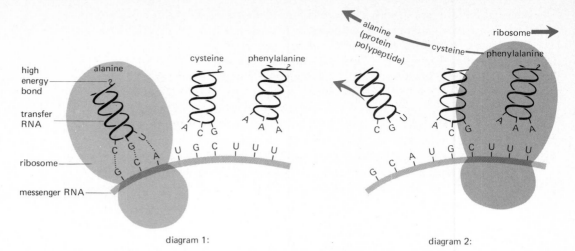

diagram 1:   diagram 2:

FIGURE 8-13

Diagram 1: Transfer RNA units with attached amino acids are moving to correct positions on the messenger RNA template. Diagram 2: Amino acids are being bonded by a ribosome to form a growing protein chain.

Once completed the polypeptide may be released from the ribosome and function on its own, or it may be linked with other polypeptides to form a more complex protein. An average protein contains about 3000 amino acids. Given this figure, one can easily compute the size of the gene involved. Assume a protein of 3000 amino acids. This requires 3000 codons on the DNA. A codon is about 10 Å long, so the length of DNA required is 3000×10 Å or 30,000 Å (3.0 microns). Thus this particular gene is 30,000 Å long. It has been estimated that the DNA length in humans is 3 meters (3×10^{10} Å). If our genes were each 30,000 Å long, then our DNA *in theory* would contain 3×10^{10} Å/30,000 Å or 1,000,000 genes.

Electron micrographs have revealed that several ribosomes, each contributing to the synthesis of the same protein molecule, may move along a single mRNA molecule at the same time (Figure 8-14). There is also evidence that a strand of mRNA may function as a pattern for protein synthesis more than once. Part of a mRNA molecule is thought to indicate how many times the translation process should be repeated. However, after use, the mRNA's are destroyed by enzymes so as not to interfere with the synthesis of other proteins needed by the cell. tRNA units probably function more than once, shuttling back and forth in the cytoplasm to pick up amino acids.

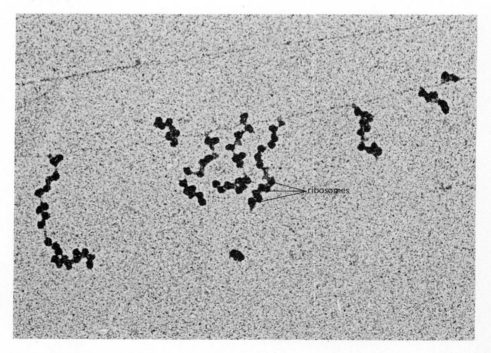

FIGURE 8-14

This photograph shows a DNA molecule in *E. coli* synthesizing mRNA which is shown attached to ribosomes. Several ribosomes are translating the same mRNA at different points in time, but all ribosomes are moving along the mRNA.

200

PROTEIN STRUCTURE

At this point you may be visualizing protein molecules as long, chainlike aggregates of amino acids. This, however, is only part of the picture. Certain protein molecules are linked side by side, with bonding between adjoining chains of amino acids. Others are twisted and folded, and the manner in which they are contorted affects their biochemical properties. Proteins can be described as having four levels of structure:

- Primary Structure   The combination of amino acids in a sequence predetermined by DNA is a protein's primary structure (Figure 8-15).

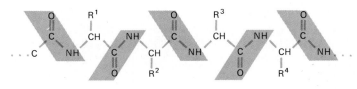

FIGURE 8-15
Primary structure of a protein. As described in Chapter 5, there are about 20 naturally occurring amino acid R groups. Peptide bonds in blue.

- Secondary Structure   Just as nucleic acids are twisted in alpha-helixes, so too are proteins. However, while DNA is a double helix (formed of two strands of nucleotides), protein molecules are single helixes (Figure 8-16).

- Tertiary Structure   A protein's spiral chain of amino acids may be compressed and folded into a globular form (Figure 8-17). The stability of the tertiary structure has been ascribed to electrostatic attractions between amino acids along the chain, as well as interactions of the protein with cell fluids.

- Quaternary Structure   Protein molecules often combine with one another and express their chemical activity as a complex. For example, some enzymes are formed of two or more protein molecules which alone would be chemically inactive (Figure 8-18). The molecules of quaternary structure are not always identical. Indeed, some complexes include ribonucleic acids with protein units. In the function of ribosomes, for example, ribosomal-RNA and enzymatic proteins act in close association.

Not all proteins have the same degree of structural complexity. In general, those that are components of fixed tissues (such as collagen) tend to be fibrous, while those that are biologically active (enzymes) tend to be globular.

PROTEINS AS "SIGNPOSTS" OF EVOLUTION

At the chemical level, proteins are the expressions of evolutionary change. Biochemical evolution begins with changes in the structure of DNA, and these changes are reflected in the proteins that are subsequently generated. A change in the sequence of the nucleotide bases in a DNA molecule results in a change in the sequence of amino acids in the corresponding protein. A structural protein may

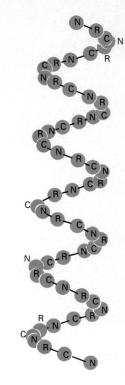

FIGURE 8-16
Representation of a polypeptide chain as an α-helical configuration. The C-N is the peptide bond, and R is an amino acid.

FIGURE 8-17
The complicated folding of a globular protein stabilized by electrostatic bonds.

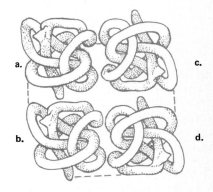

FIGURE 8-18
A tetramer of protein units showing the quaternary structure of a complex globular protein.

American Museum of Natural History

be strengthened, but, more likely, it will be weakened. The rate at which an enzyme catalyzes a reaction may be changed. Ultimately chemical changes are reflected in changes of physical or behavioral characteristics.

Biologists now believe that all phases of an organism's existence, from skin color to mating behavior, are affected by the evolutionary heritage stored in its nucleic acids. Proteins are signposts of evolution in that the closer two organisms are related, the more similar are the proteins in them which perform similar functions. Molecular biologists are capable of determining the amino acid sequences in proteins, through chemical procedures that remove amino acids, one by one, for analysis. For example, the amino acid sequences in the protein, **cytochrome-C** (which is part of the electron transport chain) have been worked out for a variety of organisms. Although the amino acids in this protein vary from species to species, it performs a similar function in the respiratory processes of all organisms in which it has been found. Cytochrome-C is composed of 104 amino acids, and in this entire sequence there is only a difference of one amino acid between a human and a rhesus monkey. The cytochrome-C in dogs, wheat, and yeast differ from that in humans by 11, 43, and 51 amino acids, respectively. There is no difference at all between human and chimpanzee cytochrome-C (Figure 8-19).

FIGURE   8-19
Human and chimpanzee cytochrome C are the same. This indicates a very close evolutionary relationship.

EXTRANUCLEAR DNA AND CELLULAR EVOLUTION

It has been known for several years that not all of a cell's DNA is contained in its nucleus. Some of it is associated with organelles, including mitochrondria and chloroplasts (Figure 8-20). The genetic material they contain makes it possible for them to replicate themselves and synthesize some of their own proteins, independent of the nucleus. In short, while they cannot survive on their own, they have some of the properties of independent cells. These properties have led contemporary biologists—most notably Lynn Margulis of Boston University—to theorize that certain organelles have evolved from what were once simple, free-living organisms. At some long-past stage in evolutionary history, certain primitive cells may have internalized others, with a symbiotic relationship developing between them. While the theory is still speculative, most cell biologists agree there is strong evidence to support it.

As long ago as the nineteenth century, when chloroplasts were first microscopically examined, biologists noticed a resemblance to blue-green algae (Figure 8-21). Comparisons of biochemical systems have added evidence suggesting

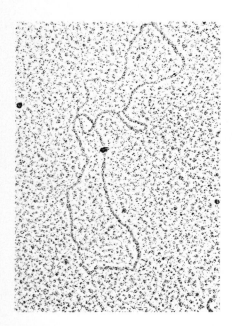

FIGURE   8-20
Electron micrographs of circular DNA molecules extracted from mitochondria by immersing them in a hypotonic solution which caused them to burst open. Width of the DNA strand is about 22 Å.

FIGURE   8-21
A cell from a fern showing the internal chloroplasts. Some blue-green algae cells are similar in appearance to these chloroplasts.

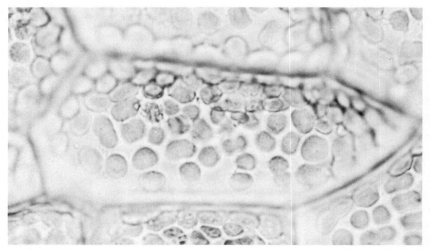

Carolina Biological Supply Company

that chloroplasts are evolutionary descendents of organisms which were also ancestral to blue-green algae. Other studies indicate there was a bacteria-like organism that evolved into a mitochondrion. These studies are of basic importance to evolutionary theory, since they indicate how an early step toward complex organisms could have occurred.

The cells which compose complex organisms are, themselves, complex. They contain systems of membranes, as well as several kinds of organelles. The most primitive forms of life—bacteria and blue-green algae—lack this complexity in their cytoplasm, and are referred to as **prokaryotic.** The cells of more complex organisms are termed **eukaryotic.** The theories of Lynn Margulis shed significant evidence on how eukaryotic cells could have evolved from prokaryotic precursors.

GENES AND HEREDITY

In controlling protein synthesis, DNA determines the nature of the enzymes, structural proteins, and other protein substances synthesized by an organism. These materials, in turn, are involved in the expression of genetic traits—the inherited physical and behavioral characteristics of an organism.

In the remainder of this chapter and in subsequent chapters, the term *gene* will come up again and again. It is a term worth a lot of consideration, because it is at once one of the most significant and ill-defined words of the biological vocabulary. Actually, there are two quite different definitions now applied to a gene:

The Historical Definition   Since the late nineteenth century, when biologists came to realize that units of heredity were passed from parents to offspring via reproductive cells, the gene has been thought of as a portion of hereditary material which determines the nature of a given characteristic. The father of the science of genetics, Gregor Mendel (whose work is described in the next chapter) considered such characteristics as seed color and stem length of pea plants. He used the term *factor* when referring to units of hereditary material.

By the early twentieth century, it was generally agreed among biologists that hereditary factors were carried by chromosomes. The term *gene* was coined in 1911 and referred to the portion of a chromosome which determined a given characteristic. In subsequent years, genes have been described as controlling (or, at least, influencing) such diverse characteristics as the biochemical activities of digestive enzymes, the structures of internal organs, characteristics of mating behavior, and human intelligence.

The Molecular Definition   Since the discovery of the structure of DNA and the nature of protein synthesis, biologists have sought to describe genes in molecular terms. Presently the mainstream of this school of thought defines the gene as a segment of DNA which codes for a complete polypeptide chain. This definition has been particularly useful in considerations of enzymes, whose functions are closely related to the sequences of the amino acids which compose them (and, hence, to the sequences of nucleotides in the DNA which coded for their structures).

Environmental versus genetic influence

It is not at all clear how genes exert their influences on certain hereditary characteristics. The complex mating dances of some species of insects, for example, are thought to be genetically rooted, but the biochemical transition from gene to dance is unknown. Obviously, this is a significant gap in knowledge.

Certainly, genes have influences on all levels of existence. A human genetic deficiency that results in the absence of a single enzyme involved in the metabolism of lipids, for example, causes the horrible physical and mental symptoms of Tay-Sachs disease. Idiocy, blindness, convulsions, or paralysis may be suffered by children born with this deficiency. However, genes are not the only influences on the characteristics of human existence. Intelligence, for example, is known to be influenced by the experiences of early childhood. Children who are frequently talked to and encouraged to explore their environments, for example, tend to learn more easily than those who are ignored.

Thus, while it is known that polypeptide synthesis affects every facet of human existence, the extent to which heredity determines certain characteristics is unknown. Moreover, even when genetic control of a given characteristic has been established, there is often more than one polypeptide involved. Even a trait so seemingly elementary as eye color is affected by several polypeptides (that is, genes in the molecular sense).

GENETIC EXPRESSION

Chromosomes, the gene complex in eukaryotes, consist of DNA molecules entwined about a core of protein. The nature of the bonding of the DNA to the protein core seems to partially determine which segments of the strand are genetically active. It is believed that the portions that function as genes protrude from the surface. The action of genes in controlling protein synthesis via mRNA is reasonably well understood. But, the mechanisms involved in the expression of genetic traits, possibly by enzyme action, remain a mystery for most traits.

Genetic traits — Everywhere at once, but expressed only somewhere

Every cell in an organism's body contains its complete genetic code. A human's genetic makeup, for example, is determined at the moment a sperm from the father combines with an egg within the mother. Thereafter, every mitotic division replicates this original genetic endowment. If you have blue eyes, every cell in your body contains genes for blue eyes. But only in the cells of the irises is this trait expressed. The cells of an iris also contain genes that affect the color and texture of hair, but these genes act only within the cells of hair follicles.

A given gene apparently requires a specific type of cellular environment in order to express itself. Evidently physicochemical properties within or around a given tissue call for the expression of certain genes. Hormones and enzymes have been cited as arousers of genetic activity in some cases. At present theories on the mechanism(s) of genetic expression are somewhat speculative, but certain cases are fairly well understood. The following section considers the activity of the genes which control utilization of a certain sugar in a species of bacteria that inhabits the human intestine.

Lactose metabolism by *E. coli*

One aspect of gene control has been illustrated beautifully in the human intestinal bacterium *E. coli*. A group of adjacent genes called an **operon** (Figure 8-22) have been shown to be turned on and off by the chemical environment of the bac-

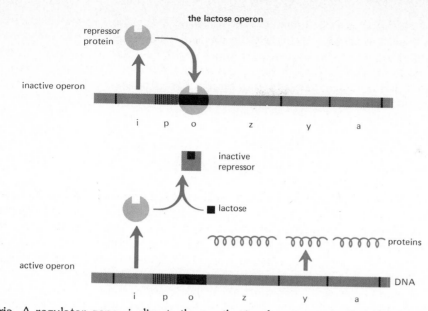

the lactose operon

repressor
protein

inactive operon

i   p   o   z   y   a

inactive
repressor

lactose

proteins

active operon

DNA

i   p   o   z   y   a

FIGURE 8-22

Top: The regulator gene *i* makes a repressor protein which binds to the operator gene o and prevents the "z," "y" and "a" genes from functioning because the RNA polymerase cannot move beyond the promoter gene *p. Bottom:* An inducer molecule binds to the repressor and changes its shape so that the repressor cannot now bind to the operator. The polymerase is free to move from the *p* gene and read the "z," "y" and "a" genes. The "z" gene synthesizes β-galactosidase, "y" the permease, and "a" an acetylase enzyme. Lactose is the inducer molecule.

teria. A regulator gene, *i*, directs the synthesis of a protein, called the repressor, which binds to the operator gene (*O*). This binding prevents the RNA polymerase enzyme (the enzyme responsible for synthesizing mRNA) from leaving its position on the promotor gene (*p*) and transcribing the z, y, and a genes. The z gene (along with the y and a genes) when turned on produces an enzyme B-galactosidase which breaks down the sugar lactose into simpler sugars the bacterium can utilize for food.

In the absence of lactose, there is no need for the bacterium to produce B-galactosidase so the operon is turned off. However, when lactose becomes available to the bacterium, it binds to the repressor protein and prevents the repressor from binding to the *O* gene. Now, the RNA polymerase is free to transcribe the z, y, and a genes since the repressor protein is out of its way. The B-galactosidase synthesized then acts upon the lactose molecules.

This total process is called **enzyme induction.** Eventually, the free lactose in the cell will be used up, B-galactosidase will then attack the lactose bound to the repressor protein, and the repressor will be free to bind again to the operator gene and turn the system off.

It is interesting to note that in this lactose operon the p and o genes do not make any proteins—they are control genes. It is thought that a good portion of the genes in eukaryotic cells may be just such control genes producing no protein product. Operons could control the expression of human traits also.

External environmental influences on genetic expression

The expression of a gene can be altered by conditions external to the organism, a phenomenon made obvious by the Himalayan rabbit (Figure 8-23). Normally, this breed is white, except for its nose, ears, tail, and feet, which are black. Apparently the temperature of a given portion of the body determines its pigmentation; cool portions are black and warm portions are white. Experiments have supported this conclusion. For example, if hair is shaved from an ear, and the ear is then wrapped with gauze to raise its temperature, the hair which grows in will be white. Thus, coloration of the Himalayan rabbit is influenced by temperature. However, no change in the genetic material is indicated. When the bandages are removed, coloration eventually returns to normal as the new hair replaces the old.

FIGURE 8-23

A genetic trait altered by environmental conditions. (a) The normal color pattern of the hair of the Himalayan rabbit. (b) After the hair was removed from a portion of the rabbit's back, an ice pack was kept in position while the hair grew in. (c) New hair in region kept at lower-than-normal temperature is black.

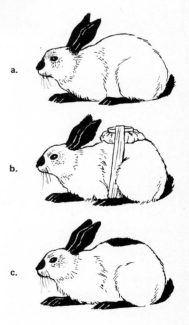

a.

b.

c.

A similar temperature influence has been found in the fruit fly, *Drosophila*. A gene for curly wings will produce this trait if the flies are raised at a temperature of 25°C. However, when the flies are raised at a temperature of 16°C, the wings are straight. The gene for curly wings hasn't been destroyed, but it has somehow been rendered ineffectual by the low temperature. The next generation will develop curly wings at 25°C also. Only gene expression is affected here.

Transformation in pneumococcus— proof of the influence of DNA

Lest the forces of the environment be overestimated, consider the phenomenon of genetic transformation from one species to another. The following study began half a century ago, when DNA was yet to be discovered. In 1928, Frederick Griffith, a British bacteriologist, was working with two types of *Pneumococcus* bacteria. The cells of the two strains were similar—tiny spheres, usually joined in pairs. However, there was a sharp distinction between them. The one that formed a slimy external capsule caused pneumonia, and the one that formed no capsule was harmless (Figure 8-24). The capsulated strain was resistant to white blood cells, which engulfed the naked variety and destroyed them.

In one phase of his investigation, Griffith inoculated mice with a mixture of dead capsulated bacteria (killed by heating) and living noncapsulated bacteria. Much to his surprise, the mice soon died of pneumonia, and, when he examined their blood, he found *Pneumococcus* with capsules. None of the capsulated bacteria he had injected into the mice had been alive. Yet, here were living capsulated bacteria. Apparently, the noncapsulated strain had been transformed into a capsule-producing strain within the mice. Griffith never found out exactly what had happened.

Early in the 1940's, Oswald T. Avery and his associates began to search for an explanation of Griffith's results. They prepared an extract of heat-killed capsulated bacteria and mixed it with living noncapsulated bacteria. When this mixture was cultured in a sterile medium, it was found that most of the colonies were still of the noncapsulated type. But, in a few of the cultures, the investigators found flourishing colonies of capsulated bacteria. These newly transformed cells were then isolated and cultured (Figure 8-25). When injected into mice, they caused pneumonia. Some factor in the extract from dead, capsulated bacteria had entered the noncapsulated individuals and had produced a change that was passed on to offspring. Avery and his associates were able to isolate and purify this transforming substance. It was DNA.

The discovery of the phenomenon of **transformation** in *Pneumococcus* was of significance for two reasons. First, it was evidence that DNA is the material of genetic inheritance. Furthermore, it demonstrated that DNA could be transferred from one organism to another and continue to express its genetic qualities.

MUTATIONS—CHANGES IN THE GENETIC CODE

Normally, DNA replicates with remarkable accuracy. Millions of duplicate strands of DNA can be generated from an original molecule without an error. Variations may occur in offspring as new gene combinations are established in sexual reproduction, but individual genes usually remain unchanged.

From time to time, however, an organism may appear with a genetic trait totally unlike anything seen in other members of its species. A new pattern of coloration, perhaps, or an altered digestive enzyme are possible physical mani-

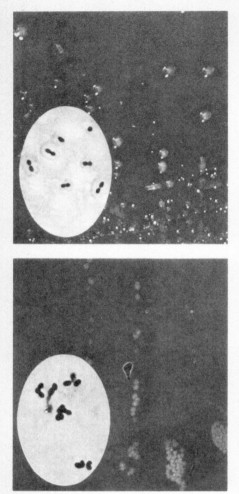

FIGURE 8-24

Pneumococcus (now called *Streptococcus*). Cells of the capsulated, infectious strain (above), and cells of the noncapsulated, noninfectious strain (below). Size of each bacterium is about two micrometers.

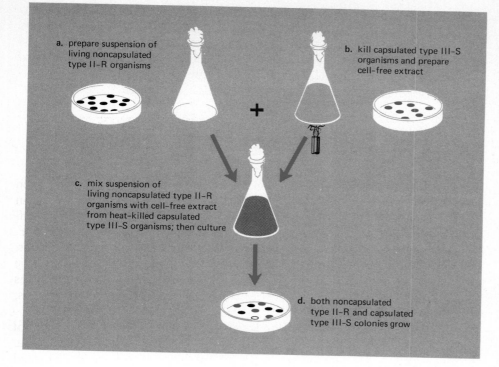

a. prepare suspension of living noncapsulated type II–R organisms

b. kill capsulated type III–S organisms and prepare cell-free extract

c. mix suspension of living noncapsulated type II–R organisms with cell-free extract from heat-killed capsulated type III–S organisms; then culture

d. both noncapsulated type II–R and capsulated type III–S colonies grow

FIGURE 8-25

The discovery of the transforming substance—Avery and associates' investigation. Noncapsulated Type II-R and capsulated Type III-S pneumococci were again cultured, and the capsulated Type III-S organisms were heat-killed (a) and (b). Suspension of living noncapsulated Type II-R cells was mixed with extract from heat-killed capsulated Type III-S cells (c). When this was cultured on sterile medium, some capsulated Type III-S colonies appeared (d). It was later determined that DNA from the heat-killed capsulated Type III-S organisms had entered some of the noncapsulated organisms and transformed them into capsulated cells.

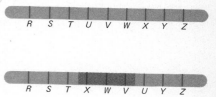

FIGURE 8-26

The upper chromosome has the normal gene sequence; the lower chromosome has undergone an inversion—the UVWX part has broken out, turned around, and fused back in. Letters stand for specific genes. (Illustration is reprinted from *Biological Science*, Second Edition, by William T. Keeton, illustrated by Paula Di Santo Bensadoun, with the permission of W.W. Norton and Company, Inc. Copyright © 1972, 1967 by W.W. Norton and Company, Inc.

festations of changes at the biochemical level. A change in just one base sequence of a DNA molecule may alter or destroy the trait associated with the gene. This is a **gene mutation,** also referred to as a **point mutation.**

Less common but potentially more devasting are **chromosome mutations,** which involve breaks in, or rearrangements of, chromosomes (Figure 8-26). When chromosomes are undergoing movements and changes in a cell (as during cell division), abnormalities may occur. A portion of a chromosome may be broken off and lost, a segment may be duplicated so that its genes are represented twice, or a chromosomal segment may be inverted. The most severe of all mutations is the loss of all or a major part of a chromosome. If this occurs in the process by which reproductive cells are formed, the offspring may not survive.

Inheritance of mutations

Most mutations are not passed on to offspring. If a mutation occurs anywhere but within the reproductive cells, the tissue that descends by mitosis from the original altered cell will receive a mutated DNA; but the mutation will cease to exist when the parent organism dies (Figure 8-27). Such mutations are **somatic** (of the body). However, if mutation occurs in reproductive cells (sperm and eggs in humans) they are transmitted to offspring. These are **germ mutations,** which may be passed from generation to generation.

The effects and frequency of gene mutations

Minor mutations may occur with little or no visible effect. In fact, this is the case with most mutations. However, some of them result in drastic changes. Organisms have delicately balanced systems that are upset by most genetic changes, and mutations are often harmful. Nevertheless, certain changes are for the better, and the rare mutations that act for the good of a species may lead to new evolutionary directions.

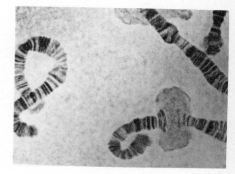

FIGURE 8-27

The giant chromosomes in the nucleus of a cell from the salivary glands of the midge, a small delicate fly. Mutations here have no effect on offspring. Only mutations of the chromosomes in sex cells are passed from generation to generation.

Some genes are more stable than others. Unstable genes mutate as often as once in a thousand or so cell divisions, whereas stable ones mutate only once in millions of cell divisions. Rates of mutation are difficult to calculate, because most mutations have no readily apparent effects on an organism. Most observable changes are the gradual cumulative effects of a series of gene mutations.

Induced mutations

Among the most potent **mutagens** (agents of mutation) is high-energy radiation. Penetrating, high-frequency radiation from unstable isotopes (such as given off by thermonuclear weapons or faulty X-ray machines) can break or otherwise alter a strand of DNA. Radiations from outer space would wreck havoc with DNA were it not for the ozone in the upper atmosphere which protects us. Mutations have been induced experimentally by exposing organisms to X rays, gamma rays, beta particles (high-energy electrons), and ultraviolet light. Certain chemicals are also mutagens; among them are formaldehyde, nitrous acid, peroxide, and mustard gas. Damaged DNA leads to faulty protein synthesis and damage to life.

Radiation-caused Mutations   The first proof that radiation causes mutations came from research conducted by Hermann J. Muller in 1927. As the subject for his experiments, Muller chose the fruit fly, *Drosophila*, a convenient organism because of its prolific reproduction and the ease with which it can be raised in the laboratory. Some of the mutations which Muller found in the descendents of irradiated *Drosophila* are illustrated in Figure (8-28).

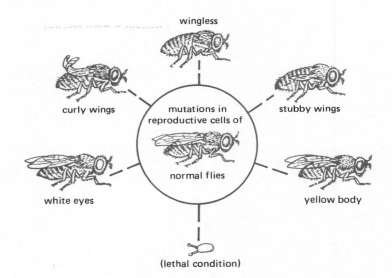

FIGURE 8-28
Some of the changes caused by mutations in the reproductive cells of wild-type *Drosophila*.

With radioactivity and other mutagens, it usually is not possible to predict exactly what kinds of mutations will result; they usually affect genes randomly. However, certain recurring mutations have been observed. Descendents of the survivors of the atomic bomb attacks on Japan in 1945, for example, suffer from higher incidences of thyroid cancer and leukaemia than their non-exposed peers. Perhaps awareness that cancerous diseases could be the world-wide heritage of nuclear war will serve as a deterrent to such outrages in the future.

Recombinant DNA Mutations   A completely different form of an induced mutation comes from **recombinant DNA** experiments. Here, DNA from one kind of organism is treated with enzymes to separate out specific sequences of genes. These genes are then added to another kind of organism and this added DNA recombines (with the help of certain enzymes) with DNA already present in the recipient organism. Consequently, the recipient can be considered a mutant in that it contains a gene compliment not found in the wild. For example, the gene for insulin synthesis has been transferred in this fashion from a mammal to a bacterium. The bacterium has no biological need to synthesize insulin and therefore did not contain the gene to do it until the recombinant experiment.

As you can imagine, experiments of this kind could be potentially dangerous if those viral genes that have been shown to cause cancer in animals are transferred to bacteria that inhabit us. If these bacteria were to escape from the laboratory and infect us, they could cause a cancer epidemic. Many cities and states are adopting legislation to set standards on laboratory safety and the kind of experiments permitted.

THE GENE REVISITED— GENETIC RESEARCH TODAY

Much of the research in genetics today is conducted with organisms even more primitive than *Drosophila*. Molds and bacteria are studied because of their short life cycles—many generations may develop in a matter of hours. Furthermore, because of their structural simplicity, their metabolic processes can be readily studied to gain insights into the nature of gene action in cellular biochemistry.

The red mold, *Neurospora crassa*, has been the subject of extensive genetic studies in recent decades. This common mold, a relative of the yeasts, thrives on bread and certain other foods. As it grows, it forms a mass of white threadlike projections, **mycelia,** through which it receives nourishment. Within a short time, stalks arise, and the tip of each stalk produces a capsule in which eight salmon-pink spores materialize. Each spore can initiate a new colony.

Although *Neurospora* is a primitive organism from the standpoint of overall evolutionary development, it is highly developed in its biochemical capabilities. In fact, it is capable of utilizing much simpler sources of nutrition than more highly evolved organisms, including all animals. *Neurospora* will survive on a medium containing only salts, sugars, and biotin (a vitamin). From these substances alone, it can synthesize amino acids, vitamins, carbohydrates, lipids, nucleic acids, and the other essential components of protoplasm. All of these biosynthetic processes require specific enzymes, which, as proteins, are generated under the direction of DNA.

In the 1940s biologists were just becoming aware of the role of DNA in heredity, but they had been experimentally inducing mutations for several years. George Beadle and Edward Tatum of Stanford University devised a method by which mutations induced in *Neurospora* were utilized to explore the nature of genetic influences. They used X rays to cause mutations affecting the mold's ability to synthesize vital compounds. In every case where the mutant had lost the ability to synthesize a particular molecule, they found that the genetic trait was inherited in a manner that indicated the action of only one gene. This finding led them to formulate the "one gene-one enzyme" hypothesis, which held that the production of each enzyme is controlled by one and only one gene. Since enzymes are proteins, this was, in effect, a one gene-one protein hypothesis (Figure 8-29).

Beadle and Tatum's work preceded by a decade the discovery of Watson and Crick which established the structure of DNA and the manner in which it codes for the assembly of polypeptides. Since enzymatic proteins may consist of two or more polypeptides, it has been necessary to modify Beadle and Tatum's hypothesis to one gene-one polypeptide. Though this form of the hypothesis is in accord with what is now known about most aspects of protein synthesis we have already seen an exception in the functioning of the lactose operon where the O gene only functions to hold the repressor protein.

FIGURE 8-29
The technique used by Beadle and Tatum to detect and isolate a nutritional mutant in *Neurospora* and to identify the amino acid it could not synthesize. The bottom tube on the left is an arginine mutant; its gene for arginine synthesis was damaged by X-rays. The bottom tube on the right has arginine in it and the mutant will grow.

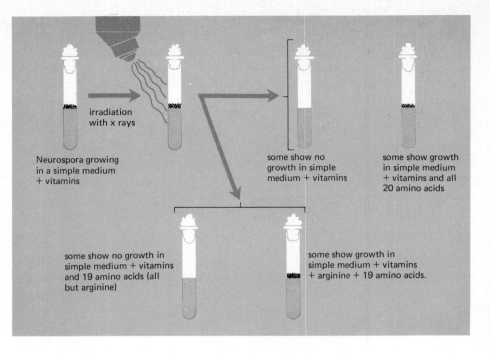

irradiation with x rays

Neurospora growing in a simple medium + vitamins

some show no growth in simple medium + vitamins

some show growth in simple medium + vitamins and all 20 amino acids

some show no growth in simple medium + vitamins and 19 amino acids (all but arginine)

some show growth in simple medium + vitamins + arginine + 19 amino acids.

CHAPTER 8: SUMMARY

1   Proteins serve as enzymes, hormones, and as the building blocks of cells.

2   DNA is composed of four bases (adenine, guanine, thymine, and cytosine), phosphates, and sugars.

3   In DNA, adenine always pairs with thymine, and guanine always pairs with cytosine.

4   From DNA, messenger RNA is synthesized, and from messenger RNA acting on a ribosome, protein is synthesized.

5   Proteins have a primary, secondary, tertiary, and quaternary structure.

6   Changes in the amino acid sequences of proteins reflect evolutionary changes in organisms.

7   Mitochondria and chloroplasts contain their own DNA.

8   A gene is a sequence of DNA.

9   Mutations are changes in the genetic code of an organism.

10 Operons are gene clusters involved in regulating metabolic processes.

11 Transformation experiments have shown that DNA can be transferred from one bacterium to another.

12 Recombinant DNA experiments involve the insertion of DNA from one organism into the DNA of another organism.

Suggested Readings

Barry, J. M.   *Molecular Biology: An Introduction to Clinical Genetics,* Prentice-Hall, Englewood Cliffs, New Jersey, 1973—Protein synthesis.

Crick, F. H. C.   "The Genetic Code," *Scientific American,* October, 1962—How nucleotide bases function as triplets, by one of the co-discoverers of the structure of DNA.

Maniatis, Tom and Mark Ptashne   "A DNA operator-repressor system," *Scientific American,* January, 1976—A look at how protein synthesis is controlled.

"Peanut Butter Protein," *Saturday Review,* 1973—A comparative study of the costs of various sources of protein.

Watson, J. D.   *The Molecular Biology of the Gene,* 3rd Edition, Benjamin, New York, 1976—The function of the gene in protein synthesis, by F. H. C. Crick's partner in the discovery of the structure of DNA.

Watson, J. D.   *The Double Helix,* Atheneum, New York, 1968—A personal history of the discovery of the structure of DNA.

Excerpts from THE DOUBLE HELIX
James Watson

James Watson was a freshly minted Ph.D. in his early twenties, and Francis Crick was a late-blooming graduate student in his thirties when they set out to discover the structure of DNA in 1951. They were an unlikely pair to match wits with Linus Pauling of Cal Tech, who only months earlier had correctly described the helical structure of protein and was known to be in pursuit of the secret of DNA.

But boundless exuberance was with Watson and Crick, and at Kings College of the University of Cambridge they were in the company of some first-rate scientific minds. Maurice Wilkins and Max Perutz were there, and they ultimately shared a Nobel Prize with Watson and Crick in 1962. Also at Cambridge was Peter Pauling, Linus's son, who was studying biochemistry and served as an agent of communication between the gentlemanly competitors.

The drama also involved Rosy Franklin, whose X-ray crystalographic photographs of DNA were essential to determining its structure. The eminent Sir Lawrence Bragg was Crick's thesis advisor, who came close on at least one occasion to cutting short his volatile student's career.

The following are excerpts from James Watson's entertaining account of what happened on the way to DNA.

From my first day in the lab I knew I would not leave Cambridge for a long time. Departing would be idiocy, for I had immediately discovered the fun of talking to Francis Crick. Finding someone in Max's lab who knew that DNA was more important than proteins was real luck. Moreover, it was a great relief for me not to spend full time learning X-ray analysis of proteins. Our lunch conversations quickly centered on how genes were put together. Within a few days after my arrival, we knew what to do: imitate Linus Pauling and beat him at his own game.

Pauling's success with the polypeptide chain had naturally suggested to Francis that the same tricks might also work for DNA. But as long as no one nearby thought DNA was at the heart of everything, the potential personal difficulties with the King's lab kept him from moving into action with DNA. Moreover, even though hemoglobin was not the center of the universe, Francis' previous two years at the Cavendish certainly had not been dull. More than enough protein problems kept popping up that required someone with a bent toward theory. But now, with me around the lab always wanting to talk about genes, Francis no longer kept his thoughts about DNA in a back recess of his brain. Even so, he had no intention of abandoning his interest in the other laboratory problems. No one should mind if, by spending only a few hours a week thinking about DNA, he helped me solve a smashingly important problem. . . .

Most unexpectedly, Francis' interest in DNA temporarily fell to almost zero less than a week later. The cause was his decision to accuse a colleague of ignoring his ideas. The accusation was leveled at none other than his Professor. It happened less than a month after my arrival, on a Saturday morning. The previous day Max Perutz had given Francis a new manuscript by Sir Lawrence and himself, dealing with the shape of the hemoglobin molecule. As he rapidly read its contents Francis became furious, for he noticed that part of the argument depended upon a theoretical idea he had propounded some nine months earlier. What was worse, Francis remembered having enthusiastically proclaimed it to everyone in the lab. Yet his contribution had not been acknowledged. Almost at once, after dashing in to tell Max and John Kendrew about the outrage, he hurried to Bragg's office for an explanation, if not an

From J. D. Watson, *The Double Helix,* New York: Atheneum Publishers, 1968.

apology. But by then Bragg was at home, and Francis had to wait until the following morning. Unfortunately, this delay did not make the confrontation any more successful.

Sir Lawrence flatly denied prior knowledge of Francis' efforts and was thoroughly insulted by the implication that he had underhandedly used another scientist's ideas. On the other hand, Francis found it impossible to believe that Bragg could have been so dense as to have missed his oft-repeated idea, and he as much as told Bragg this. Further conversation became impossible, and in less than ten minutes Francis was out of the Professor's office. . . .

Upon my return to Cambridge in mid-January, I sought out Peter [Pauling] to learn what was in his recent letters from home. Except for one brief reference to DNA, all the news was family gossip. The one pertinent item, however, was not reassuring. A manuscript on DNA had been written, a copy of which would soon be sent to Peter. . . .

Two copies, in fact, were dispatched to Cambridge—one to Sir Lawrence, the other to Peter. Bragg's response upon receiving it was to put it aside. Not knowing that Peter would also get a copy, he hesitated to take the manuscript down to Max's office. There Francis would see it and set off on another wild-goose chase. Under the present timetable there were only eight months more of Francis' laugh to bear. That is, if his thesis was finished on schedule. Then for a year, if not more, with Crick in exile in Brooklyn, peace and serenity would prevail.

While Sir Lawrence was pondering whether to chance taking Crick's mind off his thesis, Francis and I were poring over the copy that Peter brought in after lunch. Peter's face betrayed something important as he entered the door, and my stomach sank in apprehension at learning that all was lost. Seeing that neither Francis nor I could bear any further suspense, he quickly told us that the model was a three-chain helix with the sugar-phosphate backbone in the center. This sounded so suspiciously like our aborted effort of last year that immediately I wondered whether we might already have had the credit and glory of a great discovery if Bragg

had not held us back. Giving Francis no chance to ask for the manuscript, I pulled it out of Peter's outside coat pocket and began reading. By spending less than a minute with the summary and the introduction, I was soon at the figures showing the locations of the essential atoms.

At once I felt something was not right. I could not pinpoint the mistake, however, until I looked at the illustrations for several minutes. Then I realized that the phosphate groups in Linus' model were not ionized, but that each group contained a bound hydrogen atom and so had no net charge. Pauling's nucleic acid in a sense was not an acid at all. Moreover, the uncharged phosphate groups were not incidental features. The hydrogens were part of the hydrogen bonds that held together the three intertwined chains. Without the hydrogen atoms, the chains would immediately fly apart and the structure vanish.

Everything I knew about nucleic-acid chemistry indicated that phosphate groups never contained bound hydrogen atoms. No one had ever questioned that DNA was a moderately strong acid. Thus, under physiological conditions, there would always be positively charged ions like sodium or magnesium lying nearby to neutralize the negatively charged phosphate groups. All our speculations about whether divalent ions held the chains together would have made no sense if there were hydrogen atoms firmly bound to the phosphates. Yet somehow Linus, unquestionably the world's most astute chemist, had come to the opposite conclusion.

When Francis was amazed equally by Pauling's unorthodox chemistry, I began to breathe slower. By then I knew we were still in the game. Neither of us, however, had the slightest clue to the steps that had led Linus to his blunder. If a student had made a similar mistake, he would be thought unfit to benefit from Cal Tech's chemistry faculty. Thus, we could not but initially worry whether Linus' model followed from a revolutionary re-evaluation of the acid-base properties of very large molecules. The tone of the manuscript, however, argued against any such advance in chemical theory. No reason existed to keep secret a first-rate theoretical breakthrough. Rather, if that had occurred Linus would have written two papers, the first describing his new theory, the second showing how it was used to solve the DNA structure.

The blooper was too unbelievable to keep secret for more than a few minutes. I dashed over to Roy Markham's lab to spurt out the news and to receive further reassurance that Linus' chemistry was screwy. Markham predictably expressed pleasure that a giant had forgotten elementary college chemistry. He then could not refrain from revealing how one of Cambridge's great men had on occasion also forgotten his chemistry. Next I hopped over to the organic chemists', where again I heard the soothing words that DNA was an acid.

By teatime I was back in the Cavendish, where Francis was explaining to John and Max that no further time must be lost on this side of the Atlantic. When his mistake became known, Linus would not stop until he had captured the right structure. Now our immediate hope was that his chemical colleagues would be more than ever awed by his intellect and not probe the details of his model. But since the manuscript had already been dispatched to the *Proceedings of the National Academy,* by mid-March at the latest Linus' paper would be spread around the world. Then it would be only a matter of days before the error would be discovered. We had anywhere up to six weeks before Linus again was in full-time pursuit of DNA. . . .

When I got to our still empty office . . . I quickly cleared away the papers from my desk top so that I would have a large, flat surface on which to form pairs of bases held together by hydrogen bonds. . . . Suddenly I became aware that an adenine-thymine pair held together by two hydrogen bonds was identical in shape to a guanine-cytosine pair held together by at least two hydrogen bonds. All the hydrogen bonds seemed to form naturally; no fudging was required to make the two types of base pairs identical in shape. Quickly I called Jerry over to ask him whether this time he had any objection to my new base pairs.

When he said no, my morale skyrocketed, for I suspected that we now had the answer to the riddle of why the number of purine residues exactly equaled the number of pyrimidine residues. Two irregular sequences of bases could be regularly packed in the center of a helix if a purine always hydrogen-bonded to a pyrimidine. Furthermore, the hydrogen-bonding requirement meant that adenine would always pair with thymine, while guanine could pair only with cytosine. Chargaff's rules then suddenly stood out as a consequence of a double-helical structure for DNA. Even more exciting, this type of double helix suggested a replication scheme much more satisfactory than my briefly considered like-with-like pairing. Always pairing adenine with thymine and guanine with cytosine meant that the base sequences of the two intertwined chains were complementary to each other. Given the base sequence of one chain, that of its partner was automatically determined. Conceptually, it was thus very easy to visualize how a single chain could be the template for the synthesis of a chain with the complementary sequence.

Upon his arrival Francis did not get more than halfway through the door before I let loose that the answer to everything was in our hands. Though as a matter of principle he maintained skepticism for a few moments, the similarly shaped A-T and G-C pairs had their expected impact. His quickly pushing the bases together in a number of different ways did not reveal any other way to satisfy Chargaff's rules. A few minutes later he spotted the fact that the two glycosidic bonds (joining base and sugar) of each base pair were systematically related by a diad axis perpendicular to the helical axis. Thus, both pairs could be flipflopped over and still have their glycosidic bonds facing in the same direction. This had the important consequence that a given chain could contain both purines and pyrimidines. At the same time, it strongly suggested that the backbones of the two chains must run in opposite directions.

The question then became whether the A-T and G-C base pairs would easily fit the backbone configuration devised during the previous two weeks. At first glance this looked like a good bet, since I had left free in the center a large vacant area for the bases. However, we both knew that we would not be home until a complete model was built in which all the stereochemical contacts were satisfactory. There was also the obvious fact that the implications of its existence were far too important to risk crying wolf. Thus I felt slightly queasy when at lunch Francis winged into the Eagle pub to tell everyone within hearing distance that we had found the secret of life. . . .

Linus arrived in Cambridge on Friday night. On his way to Brussels for the Solvay meeting, he stopped off both to see Peter and to look at the model. Unthinkingly Peter arranged for him to stay at Pop's. Soon we found that he would have preferred a hotel. The presence of foreign girls at breakfast did not compensate for the lack of hot water in his room. Saturday morning Peter brought him into the office, where, after greeting Jerry with Cal Tech news, he set about examining the model. Though he still wanted to see the quantitative measurements of the King's lab, we supported our argument by showing him a copy of Rosy's original B photograph. All the right cards were in our hands and so, gracefully, he gave his opinion that we had the answer.

CHAPTER

9

Principles of Genetics

Objectives

1 Learn of Mendel's contribution to genetics, the science of genes.

2 Understand Mendel's hypotheses and how he arrived at them.

3 Learn Punnett square analysis.

4 Understand the principle of dominance.

5 Learn about sex chromosomes.

6 Describe the relationship between genetics and evolution.

7 Describe how animal and plant breeding is carried out.

8 Learn what happens to the genes in egg and sperm production.

FIGURE 9-1
Gregor Mendel in his experimental garden. Mendel was a contemporary of Charles Darwin, but Darwin was not aware of Mendel's work.

INFLUENCES ON INDIVIDUALITY

Since before you were born, two kinds of influences have been interacting to determine your individuality. The first of these is heredity, the transmission of genetic material from parents to offspring. In humans inherited traits include the biochemistry of metabolism, eye color, and, to a certain extent, state of mind. But, genes alone do not create an individual. Forces from the environment also contribute to an organism's nature. The basic format of the brain, for example, is the same throughout the human species. Yet, the nature of one's thoughts and emotions are unique. Past experiences as well as genetic endowment contribute to an individual's psychic response to a given situation. However, where the influences of heredity on human mental activity end and the forces of environment begin remains a mystery.

Nevertheless, scientific awareness of the nature of heredity and environmetal influences on organisms has grown by leaps and bounds in the past century. We have come a long way since an Austrian monk, Gregor Mendel, published his book on the nature of inheritance in 1865.

MENDEL AND THE MATHEMATICS OF INHERITANCE

Gregor Mendel (Figure 9-1) was not the first to be concerned with heredity. Hippocrates in the fifth century B.C. and Aristotle about a century later had thought about inheritance. However, Mendel was the first to perceive the mathematical relationship by which genes are passed from parents to offspring. His paper, representing years of breeding experiments with peas, was published by the Natural History Society of Brno, Czechoslovakia. It was ignored by Mendel's contemporaries, including Darwin, to whom Mendel sent a copy.

Mendel had been dead for sixteen years when scientists discovered his work and began to make use of his findings. The conclusions that Mendel drew from his experiments stand today as the basis of the science of genetics. His work is especially remarkable since his concept of inheritance was developed without a knowledge of chromosomes, let alone of DNA.

Mendel's methods

During his years as a secondary school teacher in Brno, Mendel tended an experimental garden at the monastery where he lived. He observed several kinds of plants, but his most fruitful experiments were conducted with peas.* Mendel chose pea plants because he had observed that they were predictably variable in several contrasting characteristics. Some plants were short and bushy, while others were tall and climbing. Some produced yellow seeds, and others produced green seeds; the seed coats of some were smooth, and others were wrinkled. Altogether Mendel identified seven pairs of contrasting traits (Figure 9-2).

Flowers are the reproductive organs for peas as well as other seed plants (Figure 9-3). Every flower of a pea plant has **stamens** (which produce sperm-bearing pollen grains) and a **pistil** (which contains egg cells at its base). Because of the close proximity of the stamens to the pistil, pea plants usually **self-pollinate** (pollen is transferred from stamens to pistil within the same flower).

* Mendel also worked with bees, until an outcry from neighbors stopped the breeding experiments that had accidentally led to a particularly vicious strain.

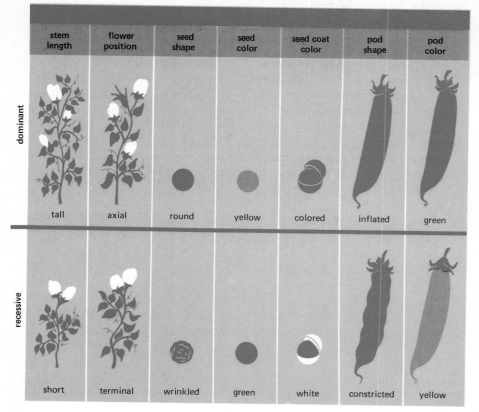

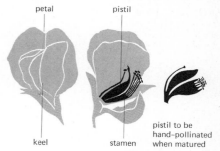

FIGURE 9-2
Mendel's seven pairs of contrasting traits in garden peas.

FIGURE 9-3
Mendel made an ideal choice for hand-pollination when he selected the garden pea. The keel of a pea blossom surrounds the stamens and pistil. If the keel is spread and the stamens are removed before natural pollination occurs, the flower may be hand-pollinated when the pistil is mature.

Cross-pollination, involving flowers on two different plants, rarely occurs for peas. However, Mendel found that cross-pollination could be assured by removing the stamens from immature flowers on one plant, and, when the pistils matured, transferring pollen to them from another plant. The hand-pollinated flower was then carefully protected from any pollen grains that might be transferred to it by wind or insects.

Mendel's conclusions were based on data accumulated from the study of thousands of plants. He kept accurate records of all the crosses he made. In recording crosses, Mendel designated the **parent plants** used in the first cross as P_1, and he referred to the generation resulting from the first cross as the **first filial,** or F_1. Crossing plants of the F_1 generation produced a **second filial** (F_2), and so forth.

Mendel began by allowing several generations of peas to self-pollinate, and he found that the offspring of a parent plant with a long heritage of a given characteristic always inherited that trait. Seeds from interbred tall plants produced more tall plants, and yellow seeds developed into plants that, in turn, yielded yellow seeds. Mendel's next step was to see what would happen if he crossed two plants with contrasting traits. Selecting tall and short ones for the P_1 generation, he made hundreds of crosses by transferring pollen from the stamens of tall plants to the pistils of short ones. When the seeds matured on the short plants, he sowed them, wondering if the F_1 plants would be short like one parent, tall like the other, or of an intermediate height. He discovered that all the plants were tall, like the plant from which he had taken the pollen.

Mendel then considered if it made any difference which plant he used for pollen and which he used to produce seed. To find out he reversed the process of pollination, using a short plant for pollen and a tall one for seed-production. He found that the results were the same as before—all the offspring were tall.

In other experiments, Mendel mated plants with a heritage of yellow seeds to plants arising from green seeds; all of the offspring produced yellow seeds. In still more experiments he crossed a round-seeded variety with a wrinkled-seeded variety, finding that all of the seeds of the offspring were round. He performed crosses with all seven pairs of contrasting traits, and, in every case, one of the traits seemed to be lost in the next generation. What would happen if he permitted these offspring to self-pollinate?

Mendel was surprised by the results. By allowing the tall plants of the F_1 generation to self-pollinate, he produced an F_2 generation of approximately 75 percent tall plants and 25 percent short ones. The reappearance of short plants revealed to Mendel that the hereditary factor for shortness had not disappeared. The F_1 plants had possessed it without showing it.

Mendel's first hypothesis

Mendel reasoned that something within peas carried indestructable hereditary messages. He called these unknown influences *factors;* today they are termed *genes*. Mendel's experiments with paired characteristics led to the formulation of Mendel's first hypothesis, the **concept of unit characters,** which holds that hereditary characteristics are controlled by factors and that these factors occur in pairs.

The second hypothesis

Mendel reasoned further that the tall plants of his F_1 generation were not like the pure tall plants. The F_1 plants carried a masked factor for shortness that could reappear in succeeding generations. According to Mendel's second hypothesis, now known as the **principle of dominance and recessiveness,** one gene in a given pair may mask the expression of the other gene. Mendel designated as **dominant** the characteristic, such as tallness, that always appeared in the off-spring of a cross between parents with contrasting traits. He termed **recessive** the characteristic, such as shortness, that did not appear in the F_1 generation but appeared in the F_2 generation.

The third hypothesis

According to Mendel's third hypothesis, now called the **law of segregation,** a pair of factors (genes) is segregated, or separated, during the formation of sex cells. That is, a **gamete** (an egg or sperm) contains only one factor of a pair, the other having gone to another gamete. Furthermore, the composition of one factor is not altered by the presence of another factor of a pair. The factor for shortness, for example, is not altered by the factor for tallness. As previously described, two tall plants (each with a masked recessive gene for shortness) can generate a genetically "pure" short plant (Figure 9-4).

Genetics abbreviated

In Mendel's first crosses, one parent was pure tall, having both genes for tallness. The other was pure short, having both genes for shortness. The members of the F_1 generation were all tall but were **hybrids** (crosses between two different genetic

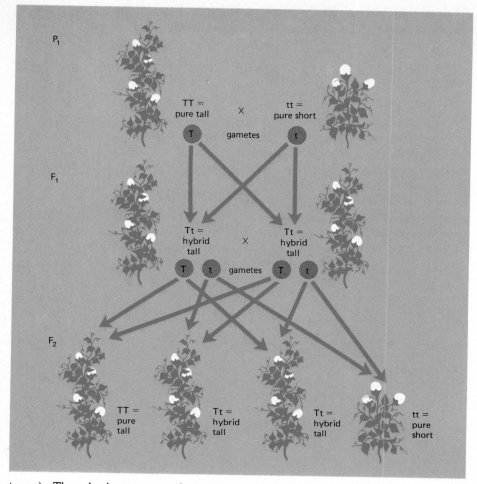

FIGURE 9-4
Mendel's law of segregation. While the F₁ generation consists only of tall plants, the recessive gene for shortness is again expressed in the F₂ generation.

types). They had one gene for tallness and one for shortness. They grew tall because the gene for tallness was dominant over the one for shortness.

If we let the letter *T* stand for tall, a pure tall plant would be designated *TT*, indicating that both of its genes for height were for tallness. The capital *T* indicates that tallness is dominant over the contrasting characteristic, shortness, which is designated by a small letter, *t*. An individual which is *Tt* is tall, and one which is *tt* is short (Figure 9-4).

The genes of any organism can be designated by paired symbols. These symbols indicate the **genotype** of the organism. The term, **phenotype,** is applied to the effects of genes (such as size, color, and metabolic uniqueness). For hybrid tall peas, the genotype is *Tt* and the phenotype is tallness.

If the paired genes for a particular trait are identical, the organism is said to be **homozygous** for that trait (*tt* or *TT*). An organism having contrasting genes is termed **heterozygous** (*Tt*). Any of two or more contrasting genes for a given trait is termed an **allele** (*t* is the allele for shortness, and *T* is the allele for tallness).

In peas, as well as other sexually reproducing organisms, chromosomes are present in pairs in the **somatic cells** (all cells of the body, except gametes). Gametes have only one member of each pair of chromosomes (and, hence, only one member of each pair of genes). Thus, the eggs at the base of a flower's pistil and the sperm of the pollen grains each have a single gene for height. When eggs or sperm are formed by meiosis in a pure tall pea plant, each gamete receives only *T* alleles. Similarly, all gametes of short plants embody only *t* alleles.

During fertilization, when a sperm and egg combine, chromosomes become paired. The pairing is maintained through the cell divisions by which an organism develops from the fertilized cell, the **zygote.** Life for all sexually reproducing organisms, including humans, begins in the same way.

DIAGRAMMING MENDEL'S CROSSES

For determining the probable outcomes of experiments in crossbreeding, geneticists commonly use a grid system called a **Punnett square** (Table 9-1). Possible combinations of genes can be determined by filling in the squares of the grid, each square representing a single gene. Mendel's work with the genes for seed color is summarized in the grids of the table.

A cross between a homozygous yellow-seeded plant (*AA*) and a homozygous green-seeded plant (*aa*) is diagrammed in grid A. It indicates there is only one possible outcome for this cross, namely, hybrid plants with the genotype *Aa* and the phenotype of yellow seeds because *A* is dominant over *a*.

TABLE 9-1   PUNNETT SQUARES FOR SEED COLORS

A.   Results of Crossing AA and aa

Female → (Genes) Male ↓	a	a
A	Aa	Aa
A	Aa	Aa

B.   Results of Crossing Aa and Aa

Female → (Genes) Male ↓	A	a
A	AA	Aa
a	Aa	aa

C.   Results of Crossing Aa and aa

Female → (Genes) Male ↓	a	a
A	Aa	Aa
a	aa	aa

If the heterozygous *Aa* plants are crossed, three genotypic combinations *AA, Aa,* and *aa* are possible, as indicated in grid B. This grid also shows the real power of the Punnett square method of diagramming genetic crosses. It indicates not only the kinds of possible genotypes but also the probable ratios in which they will occur. Since fertilization is a random process, the ratios of genotypes which will actually result from a cross may vary somewhat from the predicted ratio. This is particularly true when only a few plants are considered. However, the larger the number of offspring, the closer will the predicted ratio approximate that which is actually observed (Mendel may have fudged his data to fit the predicted ratios).

Just as it is impossible to predict whether heads or tails will come up with the flip of a coin, it is impossible to predict the results of any given **monohybrid cross** (a cross involving one pair of genes). However, if a coin is to be flipped a thousand times, it is safe to predict that it will land heads-up close to five hundred times.

Grid B shows how mating heterozygous yellow-seeded parents results in offspring that are approximately one-fourth pure dominant (AA), one-half hybrid (Aa), and one-fourth recessive (aa). This can be abbreviated as 1:2:1. Phenotypically, the ratio is 3:1, that is, three yellow-seeded plants to every green-seeded plant because yellow is dominant.

The cross between heterozygous, yellow-seeded plants (Aa) and homozygous, green-seeded plants (aa) is illustrated in grid C. Phenotypically, the ratio is now 1:1, that is, one yellow-seeded plant for every green-seeded plant.

Figure 9-5 depicts the results of crossing peas homozygous for particular traits (P_1 generation) to give heterozygous traits (F_1 generation), which are then crossed

FIGURE 9-5
Results of Mendel's monohybrid crosses. Axial peas flower along the entire stem, while terminal ones flower at the tips.

P_1 cross		F_1 generation	F_2 generation	actual ratio	predicted ratio
round X wrinkled		round	5,474 round 1,850 wrinkled	2.96:1	3:1
yellow X green		yellow	6,022 yellow 2,001 green	3.01:1	3:1
colored X white		colored	705 colored 224 white	3.15:1	3:1
inflated X constricted		inflated	882 inflated 229 constricted	2.95:1	3:1
green X yellow		green	428 green 152 yellow	2.82:1	3:1
axial X terminal		axial	651 axial 207 terminal	3.14:1	3:1
long stem X short stem		long stem	787 long 277 short	2.84:1	3:1

with each other to produce the F_2 generation. The F_2 generation results correspond to those of grid B in Table 9-1.

Incomplete dominance

Mendel's contributions to the science of genetics have not, of course, been the last word on the subject. Since his time, various deviations from his simplistic portrayal of inheritance have appeared. For example, genes are not always completely domi-

FIGURE 9-6
Incomplete dominance in four-o'clocks.

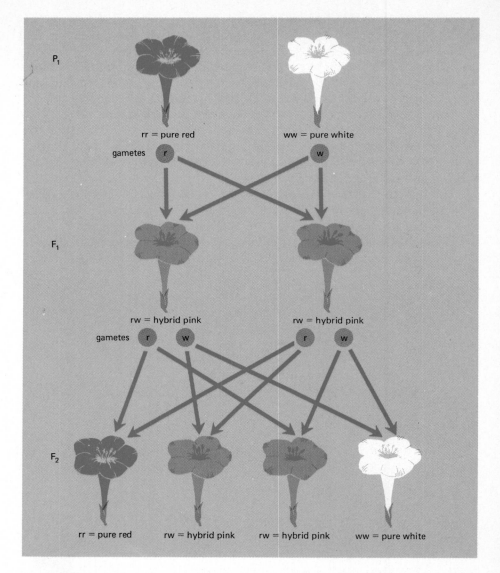

nant or recessive. In some characteristics, both alleles can be expressed in the same individual. Such **incomplete dominance** is found in crosses of different-colored four o'clock flowers (Figure 9-6). When pure red four o'clocks (*rr*) are crossed with pure white ones (*ww*), all of the first filial generation are pink (*rw*). Neither red nor white is completely dominant, so both colors are expressed as pink in the heterozygous F_1 offspring. When two pink flowers are crossed, the offspring include one-fourth red, one-half pink, and one-fourth white individuals.

Inheritance of coloration by Shorthorn cattle also illustrates incomplete dominance. A homozygous red bull mated with a homozygous white cow produces a blend of red and white called roan. When two roan animals are mated, the expected ratio of the offspring is one-fourth red, one-half roan, and one-fourth white —the 1:2:1 ratio of a heterozygous, monohybrid cross (Figure 9-7).

SEX-LINKED TRAITS

Early in the development of the science of genetics, biologists began to use the fruit fly, *Drosophila*, as an experimental subject. Several characteristics of *Drosophila* make it an ideal object of genetic research. It is easily raised in jars with foods as simple as mashed bananas, for example, and its life cycle is short, varying from

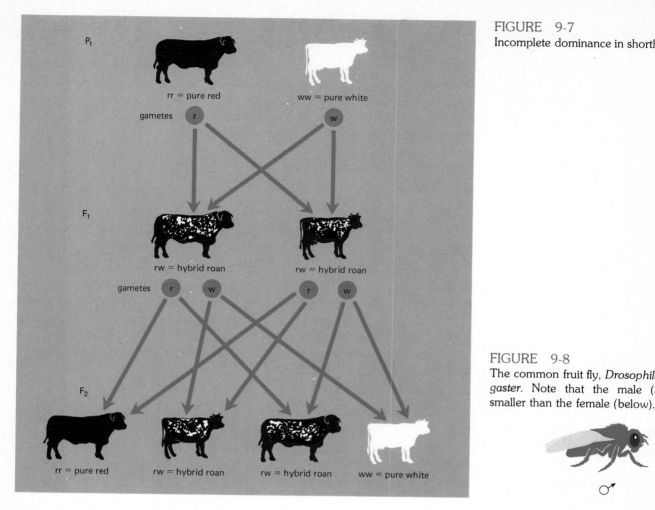

FIGURE 9-7
Incomplete dominance in shorthorn cattle.

P₁ rr = pure red   ww = pure white

gametes

F₁ rw = hybrid roan   rw = hybrid roan

gametes

F₂ rr = pure red   rw = hybrid roan   rw = hybrid roan   ww = pure white

FIGURE 9-8
The common fruit fly, *Drosophila melano-gaster.* Note that the male (above) is smaller than the female (below).

ten to fifteen days, depending on the temperature. Thus, many generations can be observed in a short time. Furthermore, the sexes of *Drosophila* are easily distinguished. The male, which is usually smaller than the female, has characteristic combinations of dark bristles on its first pair of legs, and a black-tipped, blunt posterior (Figure 9-8).

Eye color, body color, and wing structure are but a few of the thousands of genetic variations in *Drosophila* with which geneticists are familiar. In 1910 Dr. Thomas Hunt Morgan, in examining a large number of *Drosophila,* found one that had white eyes instead of the normal red ones. The fly was a male, and it was mated with a red-eyed female. All of the F₁ generation had red eyes. By applying Mendel's principle of dominance, Morgan concluded that red eyes are dominant over white eyes. The investigation was continued by mating flies of the F₁ generation. About three-fourths of the F₂ flies had red eyes, and one-fourth had white eyes. This conformed with the results Mendel had obtained with peas. But, Morgan observed something unexpected. All of the white-eyed flies were males. This could not have been the result of chance. There was a definite association of eye color and sex.

The discovery of sex chromosomes

Morgan's next step was to examine microscopically the chromosomes of *Drosophila,* in search of a difference between those of the male and those of the female. He found that the somatic cells of females contained four matched pairs of chro-

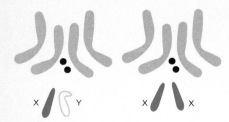

FIGURE 9-9

Drosophila has four pairs of chromosomes. The sex chromosomes are represented in color. The male has one *X* chromosome and one hook-shaped *Y* chromosome (left), whereas the female has two *X* chromosomes.

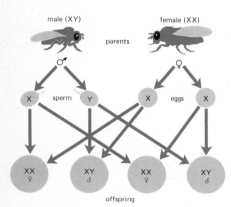

FIGURE 9-10

The inheritance of sex in *Drosophila,* indicating that males and females should occur in equal numbers.

mosomes. Examinations of male chromosomes revealed that seven of them were like those of the female, but one was different. It did not match its mate. Instead of being rod-shaped, it was bent like a hook. The rod-shaped chromosome was designated the *X* **chromosome** and the hook-shaped one was called the *Y* **chromosome** (Figure 9-9). Morgan found that all males possessed the same unmatched pair of chromosomes, regardless of eye color.

Morgan's investigation into the phenomenon of white-eyed males led to the discovery of sex-determining chromosomes. He also discovered that traits unrelated to sex could be determined by genes on the sex chromosomes. Since Morgan's time many such traits (including color blindness in humans) have been discovered. They are termed **sex-linked traits.**

Sex chromosomes in other species

Following the initial work with *Drosophila,* studies were made of various other sexually reproducing organisms, and similar sex-chromosome differences were found in plants as well as animals. Females were found to possess two identical chromosomes (*XX*), while males were found to have an oddly shaped one (*XY*).

Although *X* and *Y* are used as designations for some species that possess sex chromosomes, this should not, of course, be interpreted as indicating that the sex chromosomes are the same for all species. Indeed, they are unique for each species. *Drosophila's X* chromosome, for example, is far different from the human *X* chromosome. Furthermore, among the many organisms that reproduce sexually there are some that vary from the *XX, XY* theme. The males of certain species, such as butterflies, have no *Y* chromosome; rather they have a single *X* chromosome.

The separation of pairs of chromosomes during the formation of eggs and sperm, and the recombination of chromosomes during fertilization is diagrammed in Figure 9-10. During meiosis, pairs of *X* chromosomes are split to yield eggs containing single *X* chromosomes. During sperm formation, pairs of *XY* chromosomes are split. Half of the sperm receive an *X* chromosome, and half of them receive a *Y* chromosome. The sex of the offspring, therefore, is determined by genetic constitution of the fertilizing sperm. A male results only if a Y is present.

In human beings the sex chromosomes are one pair of a total of twenty-three pairs of chromosomes. The remaining twenty-two pairs are termed **autosomes.**

THE GENETICS OF EVOLUTION

To this point we have been primarily concerned with the genetics of individuals. We have considered how the genetic traits of sexually reproducing organisms are established at the moment of the union of two gametes. Once established the genetic constitution of an individual is essentially set for life. Genes may mutate within its body, but, since the rate of mutation is only about once in every million or so gene duplications, there are only infinitesimal changes in its genetic characteristics. Moreover, these changes do not have evolutionary implications unless they occur in the formation of gametes and are passed to offspring.

The genetics of evolution involve many successive generations, during which selective pressures cause shifts in a population's gene pool. Some of the natural forces of evolutionary change were discussed in Chapter 2. The following sections consider what may be the most potent evolutionary force in the modern world, human design.

SELECTIVE BREEDING

In certain respects the genes of all members of a given species are alike. An individual normally inherits certain **species characteristics,** which make it like all other members of its species. An ability to walk erectly, grasping fingers, and a highly sophisticated nervous system, for example, are among the species characteristics of human beings.

Apart from the traits that make an individual like others of its species are certain **individual characteristics,** traits that set it apart. As an intelligent species, we recognized this long before we began to keep our history, let alone record scientific observations. Yet, our primitive ancestors used their knowledge to direct the hereditary development of domesticated plants and animals. By choosing individuals with desired characteristics ancient breeders developed high-yielding, hardy varieties.

The history of Western civilization's selection of national leaders offers an ironic parallel to the history of domestication of other organisms. At the end of the first millenium, B.C., our ancestors were dominated by an interbreeding class of nobility. It was a practice that (by concentrating recessive genes) ultimately led in a few cases to moronic monarchs and hemophilliac czars. At times, it is not clear that we have selected our leaders from the population gene pool as carefully as we have selected our crops.

Plant and animal breeding — from an art to a science

For centuries human-kind has directed the evolution of plants and animals as sources of food. The various cole crops (cabbage, broccoli, brussel sprouts, and cauliflower), for example, were developed from a single species resembling modern kale. Five thousand years ago wheat was grown as a cereal crop by the early Egyptians. And since before recorded history special breeds of flowers, fruit trees, fowl, sheep, goats, and cattle were developed.

During the past century the science of genetics has served to refine the techniques of plant and animal breeding. Historically breeding originated as an art; its techniques came about by chance and intuition. With the development of genetics as a science, however, breeding has become more efficient. Mathematical precision has supplemented intuition as a guide to programs of selective breeding. Nevertheless, there is still a large element of chance in breeding, because it involves the manipulation of not one or two, but tens of thousands of genes. While the laws of heredity always apply, the results are never totally predictable.

The objectives of plant breeding

A plant breeder may have one or more purposes in producing new strains. Such characteristics as tasty fruit, vigorous growth, and resistance to disease are of primary concern. In recent years resistance to diseases and insect pests has become of particular importance to breeders. Insecticides and other poisons used in agriculture have reached concentrations in the environment which are harmful to many desirable species of wild and domesticated animals. It has reached the point where we are even poisoning ourselves. Agriculturalists are looking for a partial solution to the problem in plant breeds that are inherently resistant to insects.

Mass selection—the solution to blight

Imagine that the cornfields of Iowa have been struck by a mysterious blight, for which there is no cure. In walking through the acres of decimated plants, however, an agent of the state agricultural extension finds a few plants which are unharmed by the disease. Kernels from them are kept and planted the following season. The blight strikes again. But, a high proportion of the corn grown from the seeds of the resistant plants survives. A disease-resistant genetic trait has been passed to them. Again, seeds from the surviving plants are grown the following season. Each year a greater proportion of plants withstand the blight. The resistant trait has become more common in the offspring as they have been selected and bred. **Mass selection** (the choosing of seed from healthy parents) has propagated a hardy strain.

Hybridization

Years ago, farmers saved the best ears from their corn crops as seed for the next year. By mass selection, they tried to produce more corn like the best plants of the previous season. But, in corn each kernel is fertilized separately, and it is possible for pollen to come from several different plants. With no control over pollination, farmers had no idea about the identity of the parents. The seeds on a single ear often produced many different varieties of corn, some good and some poor. Some of the kernels might have resulted from self-pollination, while others might have been developed after fertilization by pollen from fields miles away. It was not unusual to find ears with a mixture of yellow, white, and red kernels. Before controlled breeding, a yield of 20 to 40 bushels per acre was the best that could be expected.

Today, hybrid varieties are produced by controlled cross-pollination. When two varieties of corn are cross-pollinated, the offspring are usually more vigorous than either parent. The desirable dominant characteristics from both parents mask the undesirable recessive traits. The hybrids produced today yield from 100 to 180 bushels of corn per acre.

Hybrid corn is commonly bred by a "double-cross," in which four "pure-line" (inbred) varieties are mixed in two crosses. Each parent is selected because of its yield, vigor, resistance to disease, or some other desirable trait. Plants resulting from the double-cross get characteristics of all four.

Figure 9-11 illustrates how hybrid corn is produced by the double-cross method. Four inbred plants, designated as *a, b, c,* and *d,* serve as the foundation. These varieties are the results of self-pollination, which is accomplished by covering the developing ears with a bag until the silks are ready to receive pollen (Figure 9-12). Then, pollen collected from the tassels of the same plant is dusted onto the silks. The plant breeder carefully avoids contamination by exposing the silks only during self-pollination. In the second step of the double-cross, plant *a* is mated with plant *b,* producing a single-cross hybrid, *ab.* To assure that contamination by self-pollination does not occur, the tassels of the plants which are to receive pollen are removed before they become mature. A similar first cross is made between plants *c* and *d,* resulting in the hybrid, *cd.* The following season the *ab* and *cd* varieties are crossed yielding the double hybrid, *abcd.* Through hybridization, plants with large root systems, sturdy stalks, broad leaves, and large ears have been developed.

One might imagine that in succeeding years farmers could use seed from the offspring of the original hybrid, but that would be unwise. Genes combine differently and the corn develops characteristics (often undesirable ones) that are unlike those of the hybrid parents. Thus, in order to maintain high standards, hybrid seeds

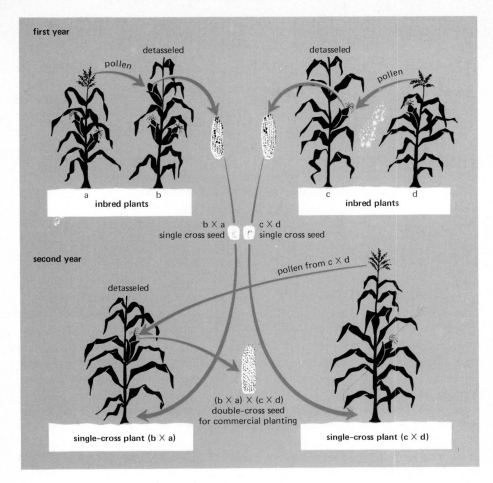

first year

pollen

detasseled detasseled

pollen

a b
inbred plants

c d
inbred plants

b X a
single cross seed

c X d
single cross seed

second year

detasseled

pollen from c X d

(b X a) X (c X d)
double-cross seed
for commercial planting

single-cross plant (b X a)

single-cross plant (c X d)

FIGURE 9-11

Crossing inbred corn plants, then crossing the resulting single crosses to produce double-cross hybrid seed. The four plants —a, b, c and d—represent the products of four different inbred lines. Strain a is crossed with strain b; strains c and d are crossed in a similar way. The F_1 generations of these two lines are then crossed to produce the hybrid corn seed that is used by farmers today.

FIGURE 9-12

Controlled pollination of corn in a research plot. To produce the necessary inbred lines required for hybridizing, care must be taken to insure that natural pollination does not occur. This is done by placing plastic bags over the silks.

must be planted every year. An example of the problems overcome with the breeding of wheat is illustrated in Figure 9-13.

The law of independent assortment

Crosses involving two or more traits are more complicated than monohybrid crosses, in which only one pair of contrasting traits is involved. The same prin-

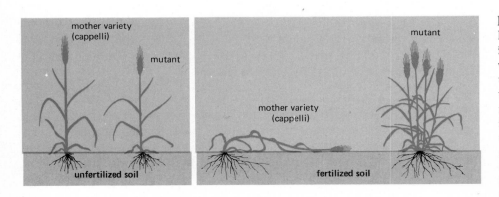

mother variety
(cappelli)

mutant

unfertilized soil

mother variety
(cappelli)

mutant

fertilized soil

FIGURE 9-13

Durum wheat was improved in Italy by induced mutation. The standard Cappelli variety was satisfactory when grown in unfertilized soil, but in fertilized soil it tended to lodge, or bend, during wind and rain. As a result of mutations induced by irradiating seeds, new varieties were developed that were shorter and had stronger stalks and so did not lodge. (From "Induced Mutation in Plants" by Björn Sigurbjornsson. Copyright © 1971 by Scientific American, Inc. All rights reserved.)

PRINCIPLES OF GENETICS **229**

FIGURE 9-14

A cross involving two traits. A pea plant that produces round *(R)* green *(y)* seeds is crossed with a pea plant that produces wrinkled *(r)*, yellow *(Y)* seeds.

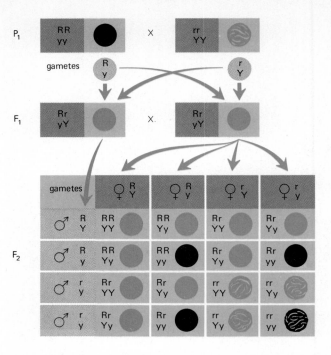

ciples apply, but the possible gene combinations are increased. When two pairs of traits are involved, the individuals possessing mixed genes for both characteristics are called **dihybrids.**

Consider two of the characteristics Mendel examined in peas. If a pea plant pure for round, green seeds (two traits) is crossed with one pure for wrinkled, yellow seeds, all the seeds produced are round and yellow. The recessive traits of green color and wrinkled seed coat are masked by the two dominant traits. In Figure 9-14, *R* represents a gene for round seed coat, *r* for wrinkled, *Y* for yellow color, and *y* for green (*Y* and *y* are not to be confused with the sex chromosomes). The F₁ dihybrid seeds all have the genotype *RrYy*, a gene for round seed coat (*R*) having come from one parent and a gene for wrinkled seed coat (*r*) having come from the other. In like manner, one parent supplied a gene for yellow color (*Y*), while the other supplied a gene for green (*y*).

When two plants grown from the dihybrid seeds are crossed, the situation becomes more complicated. Each dihybrid with the genotype *RrYy* can produce four kinds of eggs (♀) or sperm (♂). During meiosis, the pairs *R* and *r* as well as *Y* and *y* must separate and go into different cells. *R* may pair with *Y* to form *RY*, or *R* may pair with *y*, resulting in *Ry*. Similarly, *r* may pair with *Y* to form *rY* or with *y* to form *ry*. The nature of the offspring in such a cross depends on which eggs and sperm happen to unite during fertilization.

The possible offspring that can result from such a cross and the ratio of their occurrence is diagrammed in Figure 9-14. Note that all of the F₁ generation are alike, dihybrid round and yellow (*RrYy*). In the F₂ generation, however, four different phenotypes have been produced, as follows:

- 9/16 of the seeds are round and yellow (both dominant traits).
- 3/16 of the seeds are round and green (one dominant and one recessive trait).
- 3/16 of the seeds are wrinkled and yellow (the other dominant and the other recessive trait).
- 1/16 of the seeds are wrinkled and green (both recessive traits).

FIGURE 9-15
Secretariat and jockey Ron Turcotte.

This dihybrid cross illustrates the **law of independent assortment:** the separation of gene pairs on a given pair of chromosomes and the distribution of the genes to gametes during meiosis are entirely independent of the distribution of other gene pairs on other pairs of chromosomes. This law applies only when genes are on different chromosome pairs, since chromosomes, not genes, assort independently.

FIGURE 9-16
Two breeds of cattle: an example of a milk breed—a Holstein-Friesian cow (top); and an example of a prime beef breed—a Hereford steer.

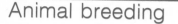

Animal breeding

The results of years of selective breeding can be seen in the graceful lines of a race horse (Figure 9-15) or in the modern breeds of poultry. The Leghorn, for example, has been bred for an ability to lay large numbers of eggs. Its flesh, however, is less palatable than that of other breeds. The Cochin and Cornish produce desirable meat but are poor egg producers.

The modern turkey, with its massive breast of white meat, is a far cry from the slender bird the Pilgrims found in the New England forest. As well as breeding their birds for meat production, turkey ranchers manipulate their birds' lifestyle toward the same end. Modern turkeys spend their caged, sedentary lives eating controlled diets and building up large, but little used muscles. Although modern methods of poultry breeding might appear to be unpleasant, there is no doubt that the techniques have resulted in amazing rates of production.

Using similar selective breeding methods, domestic cattle have been developed along two lines: one for beef, and the other for milk production (Figure 9-16). In livestock breeding, records of outstanding individuals are kept in pedigree and registration papers. Purebred animals can be registered at the headquarters of their respective breeds. Papers must include the names and registration numbers of both "sire" (male) and "dam" (female), as well as facts about their ancestries. Utilizing these records, it is possible to cross different strains of the same breed without introducing undesirable characteristics or losing good qualities. Even with careful breeding however, unwanted traits, such as bad temperament or susceptibility to disease, may develop.

Hybrid animals

Hybrids are crosses between two organisms of markedly contrasting characteristics. The mule is an animal that has resulted from the crossing of two different species. This hardy, useful beast is produced by crossing a female horse with a male donkey. A mule's size is inherited from the horse, and from the donkey the mule inherits long ears, surefootedness, great endurance, and the ability to live on coarse food. However, with all its hybrid vigor, the mule is **sterile** (unable to reproduce), and its reputation for stubbornness is well-deserved.

Several kinds of hybrid cattle have been produced by crossing domestic beef breeds with Brahman cattle. The Brahmans, which are native to India, have a heritage of tolerance to harsh climates. They can endure the humidity of the Gulf states and the dry summer heat of the Southwest much better than domestic breeds of beef cattle. One of the most successful Brahman crosses was conducted at the King Ranch in southern Texas. The original cross was a Brahman bull and a Shorthorn cow. Over a period of thirty years the Brahman-Shorthorn cross was perfected and a new breed of beef cattle, Santa Gertrudis, was established. The breed is now stabilized and carries three-eighths Brahman and five-eighths Shorthorn ancestry. Santa Gertrudis cattle possess the Brahman's resistance to hot climate and the Shorthorn's beef qualities.

Domestic cattle have also been crossed with the bison, or North American buffalo, to produce "beefalos." In this cross the bison bull is usually mated with the domestic cow. The hybrid offspring have the bison's stamina and endurance to climatic extremes as well as the quality flesh of the domestic breed.

It has been said that the success (in terms of survival) of the human race is due to the interbreeding of the many kinds of people over the last few thousand years. We may owe our existence to hybrid vigor.

CHAPTER 9: SUMMARY

1   Mendel was the first person to study genetics methodically.

2   Mendel was able to formulate three hypotheses to explain his experimental results. These hypotheses are still useful today.

3   A Punnett square is a system used to determine all possible gene combinations in mating.

4   A dominant gene takes precedence over a recessive one but genes are not always completely just dominant or recessive.

5   The sex chromosome carries the genes for sex determination as well as other genes called sex-linked genes.

6   Meiosis is the process occurring in the formation of eggs and sperms.

7   Animal and plant breeders select for desirable traits controlled by genes.

8   Breeders often cross organisms of differing characteristics—the process of hybridization.

Suggested Readings

Asimov, Isaac *The Genetic Code*, New American Library, New York, 1962 — How the arrangement of bases in DNA determines heredity.

Beadle, George and Muriel *The Language of Life: An Introduction to the Science of Genetics*, Doubleday & Co., Inc., Garden City, New York, 1966 — Some of the basic discoveries in genetics.

Head, J. J., Ed. *Readings in Genetics and Evolution*, Oxford University Press, New York, 1973 — A series of articles by leading authorities on genetics and evolution.

Jacob, François *The Logic of Life*, Pantheon Books, New York, 1973 — An easy to read history of heredity.

Kuspira, John and G. W. Walker *Genetics: Questions and Problems*, McGraw-Hill, New York, 1973 — A set of questions and problems in genetics, ranging from nucleus structure through theories of speciation.

Lerner, I. Michael *Heredity, Evolution and Society*, 2nd Ed., W. H. Freeman and Co., San Francisco, 1976 — Genetics and its implications.

CAUTION: GENE TRANSPLANTS

From Cambridge, Massachusetts, to San Diego, California, the furor rages over whether genes from one organism should be transferred to another species.

To its critics, it is an outrageous—and possibly lethal—case of meddling with the delicate balance of life. To its proponents it is a scientific tool, not unlike a well-tried chemical procedure, whose potential benefits far outweigh possible risks. At issue is the biological technology known as recombinant DNA research, which allows molecular biologists to transplant genes from one species of living organism to another.

Developed five years ago, this research proceeded quietly in the lab, largely unremarked by anyone except scientists who debated the risks and called a halt to certain experiments. But lately, recombinant DNA has erupted into a fiery public issue. Last June, the city council of Cambridge, Mass., learned that Harvard University was designing a new laboratory for recombinant experiments—and promptly voted a moratorium on all potentially dangerous DNA research. That ban was lifted last month, but only after much acrimony and long deliberations over guidelines for such experiments. Similar town-gown debates are under way in other cities, ranging from Princeton to San Diego.

Licensing? Within the last month, broad-ranging bills to clamp strict controls on "hazardous biological research" have been introduced in the New York and California state legislatures. This week, a Federal committee is expected to issue a preliminary report that will advocate an unprecedented step—national legislation to regulate the research. And a House of Representatives subcommittee will also start hearings on a bill to establish Federal licensing for all U.S. studies involving recombinant DNA.

Why such anxiety? At bottom, it is an understandable fear of the unknown. Recombinant DNA researchers are deliberately creating new types of life—producing organisms with genetic make-ups that don't exist in nature. One fear is cancer, which could conceivably be spread by bacteria into which cancer viruses have been transplanted. Another is that organisms containing genes from extremely diverse species, such as bacteria and mammals, might make humans susceptible to new diseases.

Looking further into the future, opponents of the research fear that efforts to create new microbes for the good of man could backfire. Two possible useful products of the research are made-to-order bugs for cleaning up pollutants, and new bacteria specifically engineered to convert nitrogen from the atmosphere into a biological form that plants can use. But if they manage to produce these organisms, critics ask, can the scientists guarantee that their creations will not play havoc with the environment? Pollution-gobbling bugs might go on uncontrolled binges, eating every chemical in sight; nitrogen-fixing bacteria could possibly devastate soil ecology. Possibly worst of all is the fact that, once created, the new bugs cannot be destroyed. "In the broadest sense we are here, through the creation of wholly new gene combinations, intervening profoundly in the evolutionary process," argues Robert Sinsheimer of the California Institute of Technology, himself an eminent molecular biologist.

The recombinant experts reply that these fears are more related to science fiction. They concede a very faint chance that unpleasant organisms could be produced, but assert that past experiments with dangerous disease-producing organisms have been well contained by simple laboratory safeguards. Furthermore, they say, no actual hazard of recombinant research has yet emerged. "What we're talking about," says Stanley Cohen of Stanford University Medical Center, "are hypothetical dangers that have never happened."

Critics: These arguments raged last week at an extraordinary meeting at the National Academy of Sciences in Washington. The demands of critics ranged from calls for an immediate halt to all recombinant DNA research to proposals that the experiments be carried out only in a few ultra-secure national facilities such as Fort Detrick, the Maryland center used until 1972 by the U.S. Army to develop chemical and biological warfare weapons. While researchers accepted the need for controls, including Federal regulations, they rejected the extreme limitations suggested, and the meeting ended without any agreement.

Techniques: Recombinant research, in principle, is not difficult. It is a form of sophisticated surgery on strands of DNA, the double-helix structure that is the basis of all life. The genes that impart all hereditary characteristics are nothing more than short lengths of DNA (deoxyribonucleic acid). Most of the genes in living cells are strung together in long chains known as chromosomes. But a few also combine in small loops of DNA that are called plasmids. The prime targets of the gene transplanters are plasmids.

Researchers first remove plasmids from bacteria, using chemicals that cause the bacteria to split apart. The scientists then apply enzymes, which open the closed loops of the plasmids and allow insertion of small stretches of DNA that represent one or more alien genes. The newly created, or recombined, plasmids then close up, and, in yet another chemical procedure, are inserted back into bacteria of the type from which the original plasmids were removed. Once inside, the transplanted foreign genes operate exactly as if they had been there all along, carrying out their hereditary tasks and reproducing themselves in every new generation of the bacteria. Molecular biologists aren't restricted to transplanting bacterial genes. They have found that bacteria will accept—and permit growth of—foreign genes from frogs and plants.

Infection: Most recombinant experiments are performed on the bacteria *Escherichia coli,* and this is a source of both reassurance and concern. On the one hand, molecular biologists

are comfortable with this bug; they understand its ways and feel confident that they can control transplanted forms of it if experiments go awry. On the other, *E. coli* live in the human gut. If a dangerous version of the bacteria did somehow get out of the laboratory, it would be more likely than most bugs to infect people.

For this reason, a group of molecular biologists headed by Paul Berg of Stanford University Medical Center proposed, in July 1974, a voluntary deferral of experiments that might create risky hybrids. Their objective was to allow scientists to estimate potential hazards of the research and devise ways of minimizing the risks. Last June, the National Institutes of Health issued a set of guidelines, to be applied to all recombinant research funded by NIH.

The guidelines ban completely some lines of investigation, such as those designed to transplant cancer viruses or genes that confer resistance to drugs. For all other recombinant experiments, the guidelines specify the physical and biological precautions that researchers must use. The biological safeguards generally require the use of recently developed *E. coli* bacteria that are genetically weakened, so that they cannot survive long outside the laboratory. Levels of safety range from ordinary microbiological laboratory practice (classed ''P-1'') to super-secure laboratories (''P-4''), with sealed cabinets, glove boxes and air locks, of the type once used in germ-warfare research. Last week, the NIH announced its plans to build such a laboratory at Fort Detrick. Scientists are due to start recombinant experiments at another P-4 facility, in a mobile trailer at NIH headquarters in Bethesda, Md., this spring.

Researchers generally welcomed the guidelines. Since they were issued, experts have started a number of experiments at the second most stringent level of safety, known as P-3. Their critics, who include scientists and legislators, have not been so easily satisfied. They cite two major objections. First, they argue, the transplanters don't know enough about the organisms that they study to guarantee that the precautions measure up to the potential hazards. Further, the critics say, the public, which stands to suffer from any recombinant accidents, had no say in setting the guidelines. Cambridge Mayor Alfred Vellucci, who strenuously opposed last month's decision to rescind his hometown ban on recombinant research, put the case succinctly: "The big question is, does a city have a right to control genetic engineering within its city limits?"

Adding to the concern of the critics are growing signs that industrial companies—which are not limited by guidelines at present—are eying DNA research as a potentially fruitful area of enterprise. Ananda Chakrabarty of General Electric, for example, has used a forerunner of the recombinant technique to create a "super-bug" for cleaning up oil spills. He combined plasmids from four different strains of *Pseudomonas putida* bacteria, each of which scavenges a different component of oil, into a single bacteria. Eli Lilly and other drug companies are studying recombinant technology as a possible means of mass-producing insulin and other rare medications.

Exchange: Actually, transplantation of genes from one species to another may not be quite so contrived as it sounds. Some recent research suggests that the process occasionally happens in nature. Scientists have long recognized that different species of bacteria can exchange genes. However, two San Diego researchers have found evidence of a far more unusual exchange. Certain bacteria, apparently, harbor genes similar to human genes that regulate the production of a substance known as chorionic gonadotropin. This suggests an unexpected transfer between humans and bacteria, and one implication is that such hybrids may not be as dangerous as has been feared.

Even so, is the risk presented by recombinant DNA worth the effort? Molecular biologists think so because the technique promises major strides in practical and fundamental science. Stanley Falkow of the University of Washington, for example, has recently used recombinant techniques to remove from *E. coli* part of the gene that produces diarrhea in pigs and cattle. He hopes that the advance will lead to a specific cure for this disease, which can kill young animals. Other scientists are concentrating on acquiring knowledge of the makeup of genes in animals and man—knowledge that, according to Paul Berg, "will make the diagnosis, prevention and cure of disease more rational and effective."

The recombinant DNA issue is fundamentally one of unknown risk versus unknown benefit. Neither risk nor benefit can be specified for many years. But there has already been one major result: scientists have agreed to accept a degree of public control of research that would have seemed unthinkable five years ago.

CHAPTER 10

Human Genetics

POPULATION GENETICS

The same laws of inheritance that apply to fruit flies and peas also apply to human beings, but the dynamics of heredity are not as obvious in humans as in less complex organisms. There are several reasons for this. One is the length of human life. A geneticist can observe several generations of *Drosophila* in a few months, or many generations of bacteria in a few days. But we do well to see five generations in a lifetime.

The quantity of human reproduction presents another problem. In many animals and plants, a single mating may produce hundreds of offspring. In comparison with such prolific reproduction, the average human family represents a very small sampling of genetic possibilities. Yet another complicating factor in studies of human inheritance is that most people come from mixed ancestry. Although many other animals limit their breeding to isolated populations, few human populations are genetically "pure." There are some notable exceptions (including some isolated African and South American peoples), but in general our breeding habits are more cosmopolitan than any other species.

Mendel established his ideas of heredity after observing thousands of selectively bred peas. Controlled breeding of human strains is, of course, morally intolerable. Nevertheless, researchers of human genetics have developed various sophisticated techniques for analyzing the genetic natures of individuals and populations. The benefits of such studies have included the ability to detect physically and mentally crippling genetic diseases prior to the birth of afflicted fetuses. With the benefit of genetic counseling, parents can now choose between a legal abortion or a life of service to a debilitated offspring. It is a hotly debated freedom, which is opposed by those concerned with the right of unborn humans.

Polls and population genetics

In recent years, opinion polls have become increasingly accurate indicators of political trends. Based on relatively small samples of the national population, polls can predict the winners of elections or indicate public opinion about office holders.

Poll-taking is a highly refined science, the accuracy of which depends on the selection of a representative group of people. The national population includes individuals of varied educational, professional, social, racial, cultural, and economic backgrounds. A sampling, therefore, must include people from all these groups in proportions representative of the entire population.

Techniques similar to those of poll-takers are used by population geneticists to determine the frequencies of genes. However, the parallel between opinion polls and genetic surveys must not be taken too far. What is a politically accurate sampling of a given population may not give an accurate genetic cross section. When sampling a population for its genetic characteristics, particular care must be taken to assure not only a representative sexual, racial, and national heritage cross-section but also a proper measurement of the trait.

One easily tested genetic trait is the ability to roll one's tongue (Figure 10-1). This trait is determined by a single pair of genes; and the allele for rolling (**R**) is dominant, while the allele for nonrolling (**r**) is recessive. Thus, a roller may be homozygous (**RR**) or heterozygous (**Rr**). A nonroller must be homozygous, recessive (**rr**).

Assume that a sample of 600 people in a certain interbreeding community were asked to try to roll their tongues, and of these, 384 could do it and 216 could not. Dividing the number of rollers by the size of the sample population (384/600) yields a proportion of 0.64. Similarly, the proportion of nonrollers is 216/600 =

FIGURE 10-1
Tongue-rolling: a genetic trait.

0.36. Having determined these proportions in the sample population, it is a simple matter to estimate the probable number of rollers and nonrollers in the total population. If the population of the community were 120,000, for example, one would expect that $0.64 \times 120,000 = 76,800$ of them would be rollers, and $0.36 \times 120,000 = 43,200$ would be nonrollers.

Furthermore, once the proportions of the phenotypes of a given genetic trait have been determined, it is possible to take the calculations a step further to determine the proportions of the alleles involved. If we let **p** represent the proportion of the **R** allele and **q** represent the proportion of the **r** allele, then **p** + **q** = 1 (This is merely a mathematical way of saying that when there are only two alleles of a given gene, the sum of the proportions in which they occur must be one). Furthermore, since the genotype of an individual must be homozygous, recessive (**rr**) in order for the phenotype of nonrolling to be manifested, we know that $\mathbf{q}^2 =$ the frequency of nonrollers. (This is the mathematical form of saying that the chance of two gametes with the same allele getting together is the proportion of that allele times itself, $\mathbf{q} \times \mathbf{q} = \mathbf{q}^2$). Going back to the data for the hypothetical community considered in the previous paragraph, $\mathbf{q}^2 = 0.36$. Therefore, **q** = 0.6. Utilizing this value in the expression **p** + **q** = 1 and solving for **p** yields the equation **p** = 1 − 0.6 = 0.4.

Thus, given a representative survey of phenotypes, it is possible to calculate the proportions in which the corresponding alleles occur in the population. In this case the proportion of the allele for nonrolling (0.6) is greater than the proportion of the allele for rolling (0.4), but the proportion of the phenotype of rolling (0.64) is greater than the proportion of the phenotype for nonrolling (0.36). This is so because most of the people who are rollers are heterozygous for the trait. It is impossible to tell if a person who can roll their tongue is homozygous (**RR**) or heterozygous (**Rr**). However, it is possible to calculate the proportions in which these genotypes occur in the population. Since the proportion of the allele for rolling is now known (**p** = 0.4), the proportion of those who are homozygous for the trait is this value squared ($\mathbf{p}^2 = 0.16$). Now, since there are only three possible genotypes for the tongue-rolling trait, and since the proportions of two of them in the population are already known, it is a simple matter to calculate the third value.

The following summarizes what has been calculated:

- p = proportion of the dominant (R) allele in the population = 0.4
- q = proportion of the recessive (r) allele in the population = 0.6
- p^2 = proportion of the population homozygous for the tongue-rolling trait (RR) = 0.16
- q^2 = proportion of the population which is homozygous nonrolling (rr) = 0.36
- pq = proportion of the population which is heterozygous rolling (Rr) = $0.24 \times 2 = 0.48$

A Punnett square indicates that after mating, two of the four possibilities are Rr (in addition to RR and rr). Therefore, pq should be doubled (2×0.24) to give the correct proportion (0.48) for pq. There are few genetic characteristics that are so simply determined, however. A step higher in complexity is illustrated by the genetic basis of human blood type.

INHERITANCE OF BLOOD TYPE

In 1900, while working in a medical laboratory in Vienna, Dr. Karl Landsteiner discovered that mixing blood from different people sometimes resulted in clumping, or **agglutination,** of red blood cells. In further investigations he found that the red blood cells of individuals whose blood agglutinated when combined, differed in the protein compositions of their cellular surfaces.

On the surfaces of some red blood cells are proteins known as **agglutinogens.** There are two major types of agglutinogens, designated as type A and type B (although others have been found). The red blood cells of some people have one of the agglutinogens, other people have both of them, and still others have neither agglutinogen.

Three alleles, which can be designated as A, B, and O, are involved in the inheritance of agglutinogens. Allele A codes for the presence of one agglutinogen, B codes for the presence of the other, and O codes for neither agglutinogen. Since more than two alleles are involved in the determination of blood type, the trait is said to be controlled by multiple alleles. The genotypes for the four basic human blood phenotypes are:

- AA or AO produce type A red cells (with type A agglutinogen)
- BB or BO produce type B red cells (with type B agglutinogen)
- OO produces type O red cells (with neither agglutinogen)
- AB produces type AB red cells (with both agglutinogens)

Different populations of people have different proportions of blood types. Such variations occur both among racial groups and in populations in different geographic regions (Table 10-1).

Blood types are commonly used in court cases involving questions of paternity. For example, if a man who has type O blood is charged with being the father of a child who is type AB, he will be acquitted, since the child must have received either an A or B allele from its father. If the child were type O there would be a possibility that the man was its father; however, this would not be absolute evidence. There are millions of men who possess the O allele. Thus, blood type can be accepted as absolute evidence for acquittal but not for conviction.

Blood type is better known for its involvement in hospital cases than legal cases. In giving transfusions of blood, care must be taken to make sure the blood types are compatible. Mixing noncompatible blood types results in an adverse antigen/antibody reaction, which is described in Chapter 13.

COMPLEX PATTERNS OF INHERITANCE

Beyond the level of complexity of multiple alleles are genetic traits that are determined by more than a single pair of genes. Eye color and skin color are among the human traits which are influenced by two or more gene pairs.

TABLE 10-1   BLOOD TYPE PERCENTAGES IN VARIOUS REGIONS OF THE WORLD

BLOOD TYPE	A	B	AB	O
U.S.A. — white	41.0%	10.0%	4.0%	45.0%
U.S.A. — black	26.0%	21.0%	3.7%	49.3%
Swedish	46.7%	10.3%	5.1%	37.9%
Japanese	38.4%	21.8%	8.6%	31.2%
Hawaiian	60.8%	2.2%	0.5%	36.5%
Chinese	25.0%	35.0%	10.0%	30.0%
Australian aborigine	44.7%	2.1%	0.0%	53.1%
North American Indian	7.7%	1.0%	0.0%	91.3%

For eye color there appears to be one major gene pair, whose expression is influenced by several modifier genes. For the major gene, the dominant allele, *B*, results in brown eyes, and the recessive allele, *b*, codes for blue eyes. Brown-eyed people have cells in the front layer of their irises which contain a pigment called melanin. Blue-eyed people lack melanin; the blue color is an effect of the interaction of light with black pigment on the back of the iris.

It is obvious, however, that differences of eye color are not simply a matter of blue or brown. They cover a wide spectrum, and patterns within the colors are even more variable. Eye color results from an interplay of effects of genes apart from the basic blue/brown pair. Perhaps the effects of these genes are merely by-products of their influences on other characteristics of the individual.

Skin color is also affected by more than one pair of genes. The genetics of skin coloration are not precisely known. Estimates of the numbers of pairs involved range from two to eight. In any case, variations in skin color (which include seemingly infinite shades of brown, red, and yellow) are the result of the effects of more than one pair of genes. As with eye color, the major pigment involved is melanin.

SEX-LINKED GENES

As discussed in the last chapter, certain genes that have no apparent effect on sexuality are carried on the sex chromosomes. (Figure 10-2 shows the sex chromosomes for human males and females as well as the other chromosomes.) The best known of these sex-linked traits in humans is **red-green color blindness.** While color blindness rarely occurs for females, about 8 percent of the male population has it. People with this condition cannot distinguish red and green; both colors appear as shades of gray. The fact that color blindness occurs more frequently in males than in females indicates that it is sex-linked. Geneticists have found that the gene associated with it is carried on the *X* chromosome. The allele for normal color vision is dominant. Using *C* for normal vision and *c* for red-green color blindness, the possible gene combinations are as follows:

- $X^C X^c$, a normal female, heterozygous for color vision
- $X^C X^C$, a normal female, homozygous for color vision
- $X^c X^c$, a colorblind female, homozygous for color blindness
- $X^C Y$, a normal male with a single gene for color vision
- $X^c Y$, a colorblind male with a single gene for color blindness

Figure 10-3 illustrates how it is possible for a mother who is a "carrier" of color blindness to have a son who is color blind, even though the father has normal vision. The male is either $X^C Y$ or $X^c Y$.

Hemophilia, or "bleeder's disease," is another sex-linked characteristic which is inherited mostly by males. This is a condition in which a blood protein

FIGURE 10-2

Chromosomes of the human female (top) and of the human male (bottom). The numbers are those used by geneticists for reference purposes. The sex chromosomes are *XX* for the female and *XY* for the male.

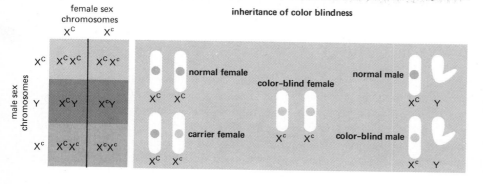

female sex chromosomes

X^C X^c

inheritance of color blindness

	X^C	X^c
X^C	$X^C X^C$	$X^C X^c$
Y	$X^C Y$	$X^c Y$
X^c	$X^C X^c$	$X^c X^c$

male sex chromosomes

normal female X^C X^C

carrier female X^C X^c

color-blind female X^c X^c

normal male X^C Y

color-blind male X^c Y

FIGURE 10-3

The gene for color blindness is indicated in color, and the chromosome that carries it is designated as X^c. Notice that two such chromosomes are necessary for a color-blind female, while only one is necessary in case of a colorblind male.

necessary for clotting is not produced because of the lack of the gene necessary for its formation. Victims of hemophilia may bleed to death from wounds that would quickly heal in a normal person. Hemophilia can appear in a female only when her father has the disease and her mother is a carrier.

Hemophilia spread throughout the families of European nobility via the descendants of the prolific queen, Victoria (Figure 10-4). In the royal family of Russia it played a role in the downfall of the czars. The son of Nicholas II, Alexis, who was in line to become czar, was a hemophiliac. Due to bleeding from a cut suffered in a fall, Alexis was near death when a "holy man," Grigori Rasputin, who was reputed to have miraculous powers of healing, was called in. Rasputin was actually successful in arresting the disease and this gained him influence in the royal court. But, in the meantime, the debauched lifestyle which he brought to court helped weaken it to the point that it fell to revolutionary attack.

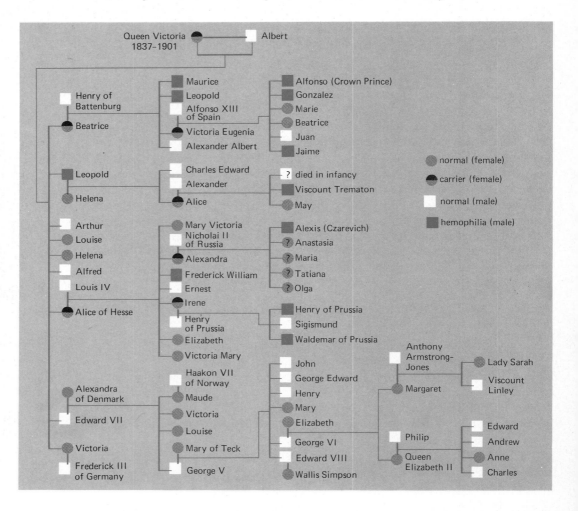

FIGURE 10-4

The inheritance of the sex-linked trait hemophilia by descendants of Queen Victoria. The mutant gene is not still present in the Windsor family, the royal family of Great Britain.

SEX-LIMITED GENES

The fact that some genes are sex-linked should not lead to the assumption that all genes for characteristics commonly associated with sex are sex-linked. They are not. Many genes that control sexual traits are located on the **autosomes** (chromosomes other than the sex chromosomes). Genes that control the development of the penis and vagina, for example, are carried by autosomes in people of both sexes. Many of the secondary sex characteristics, including the distribution of body hair and size of breasts are also determined by genes carried on autosomes. These are **sex-limited** genes, which apparently come into effect only in the presence of the proper sex hormone (testosterone in males and estrogen or progesterone in females). The expression of the allele for baldness, for example, is brought out by the presence of testosterone. Are bald men sexier?

Various other sex-limited characteristics occur throughout the living world. Certain male birds, for example, exhibit flamboyant feathers, whereas the females of the same species do not (Figure 10-5). The roosters of most breeds of chickens develop large combs and wattles (skin hanging down from chin) and long tail feathers; hens of the same breed develop less showy plummage. In human beings a similar phenomenon determines the growth of beard. The son of a man with relatively little facial hair might inherit a coarse beard from his mother, even though the trait is not manifested by her. As with other sex-limited traits, a beard develops only when the proper sex hormones course through the bloodstream. Thus, the first traces of facial hair do not appear until puberty, when the testicles begin releasing testosterone.

HEREDITY AND DISEASE

While some disease are directly caused by genetic abnormalities (such as hemophilia), other aspects of human health are influenced more subtly by genes. Tuberculosis, for instance, is a bacterial disease and cannot be inherited. However, body processes important in resistance to tuberculosis are inherited.

On the other hand, disease resulting from abnormal structures or functions of the body organs are likely to have more direct hereditary roots. Case-histories of human families indicate that various diseases and disorders are associated with genes. Among them are respiratory allergies, asthma, bronchitis, near-sightedness, far-sightedness, night blindness, and sugar diabetes.

FIGURE 10-5
Bright plumage and a large comb on top of the head are sex-limited traits expressed in the male (top) but not the female.

Sugar diabetes

Also known as *diabetes mellitus,* diabetes is a disease in which the pancreas fails to secrete a sufficient amount of the sugar-metabolizing hormone, insulin. Diabetes has been found to run in families, and it is thought to be caused primarily by a recessive gene. However, the disease is not equally serious in all people, and there is evidence that more than one pair of genes is involved in its inheritance. Furthermore, regulation of the diet and body weight may prevent or arrest the condition even when the genes are present.

Tay-Sachs disease

This affliction, which results in brain deterioration, is a genetic disorder found most frequently in populations of Eastern European Jewish descent. It is caused by an enzyme deficiency linked to a recessive gene. Therefore, both parents must carry the trait before it can be manifested in their offspring. The absence of the enzyme causes the accumulation of lipids in the brain which ultimately affect mental functions.

Sickle-cell anemia

This is one of several known hereditary blood disorders. It results from a recessive allele in a single pair of genes, and is most prevalent in Black populations. One

FIGURE 10-6

A scanning electron microscope photograph of sickle-cell anemia showing the red blood cells.

survey indicated that 8.5 percent of North American blacks carry the trait and 0.3 percent to 1.3 percent have the disease. Sickle-cell anemia is most common among natives of central and western Africa, where as high as 20 percent of Black populations carry the trait.

Sickle-cell anemia was first observed in 1910, when the blood of anemic patients was found to contain abnormal "sickle-shaped" red blood cells (Figure 10-6). However, it was not until 1949 that the red cells were found to contain abnormal hemoglobin. In 1957 the exact biochemical nature of the abnormality was pinned down.

The protein chains of normal hemoglobin are composed of 574 precisely arranged amino acids. In sufferers of sickle-cell anemia, only one of these amino acids is replaced by another. This slight variation alters the structural properties of hemoglobin, causing the red blood cells to buckle. Because of their distorted shape, the red cells jam in the capillaries and break thereby preventing proper circulation of the oxygen they carry.

Sickle-cell anemia usually appears during the latter months of the first year of life. It shortens life considerably by causing increased susceptibility to infections and cellular damage in various organs due to lack of oxygen. The allele associated with the disease is almost totally recessive; although individuals who are heterozygous for the trait sometimes show mild symptoms of the disease.

It would seem that natural selection would operate against the maintenance of any gene so obviously harmful, and that such a gene would be held at a very low frequency in any population. This seems to be true among Blacks in the United States, but it isn't true for many populations in Africa. The British geneticist, A. C. Allison, solved the paradox of sickle-cell anemia when he discovered that individuals heterozygous for the trait have a much higher than normal resistance to malaria. Thus, in parts of the world where malaria is unchecked, natural selection operates in favor of the heterozygous effects of the allele. The severity of the anemia suffered by people who are homozygous for the trait, however, operates against the allele becoming as common as the one for normal red blood cells. The balance between two opposing selection pressures determines the frequency of the sickle-cell allele in the population.

The allele for sickle-cell anemia is a dramatic example of a gene that has more than one effect. Such a gene is said to be **pleiotropic.** Pleiotropy is, in fact, the rule rather than the exception. Most genes probably have more than one effect on the organisms which carry them, although all but one of the influences may be too subtle to detect.

Phenylketonuria

FIGURE 10-7

The genetic disease phenylketoneuria results from the lack of a gene necessary to produce an enzyme that converts phenylalanine to tyrosine.

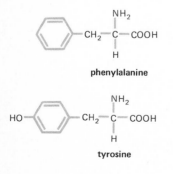

Termed PKU, for short, this disorder results from the lack of a gene necessary to produce an enzyme that converts the amino acid, phenylalanine, to a similar amino acid, tyrosine (Figure 10-7). As a result of this chemical block, phenylalanine is converted to a compound, phenylpyruvic acid, which is thought to be toxic to tissues of the brain. Severe mental retardation develops unless therapy begins soon after birth.

Studies of families in which phenylketoneuria has been transmitted indicate that it is caused by a recessive allele. However, while individuals who are heterozygous for the trait are apparently normal, they are unable to convert phenylalanine to tyrosine as readily as those who lack the recessive allele entirely.

Clinical tests can detect the presence of phenylpyruvic acid in the urine of an infant, and if the disease is detected early enough (preferably in the first few weeks of infancy), treatment can be given to prevent its accumulation. The child can be

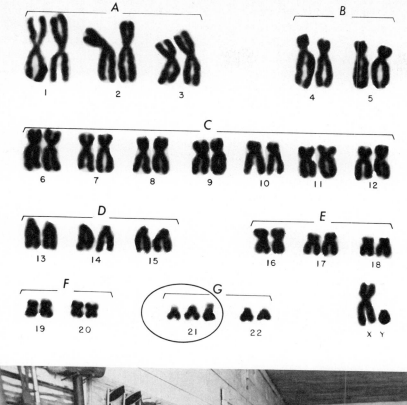

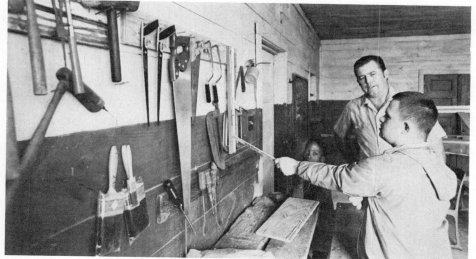

FIGURE 10-9
A boy afflicted with Down's syndrome.

kept on a diet that restricts intake of foods containing phenylalanine. This regimen does not cure the genetic defect, however. It only prevent the disease's tragic effects.

Down's syndrome

Another disease affecting mental development, **Down's syndrome** (formerly called Mongolism) results when an infant develops with one autosome too many (Figure 10-8). It has been traced to a malfunction in the process by which chromosomes are segregated during meiosis. Down's syndrome is characterized by mental retardation and abnormal physical characteristics, including an enlarged tongue, underdeveloped hands and weak musculature (Figure 10-9).

In the United States, about one in 600 babies is born with Down's syndrome. The number is lower for mothers under 35 (about one in 1,000), but increases to one in 60 babies in mothers over 45 years of age. The striking increase in the incidence of Down's syndrome with increasing age of the mother has been linked to the dynamics of egg production. When a female is born, she already possesses all the eggs she will ever have, although they are not yet in a mature state. As she grows older, her eggs begin to deteriorate.

Abnormal numbers of sex chromosomes

Physical and mental disorders have also been traced to irregular numbers of sex chromosomes. **Monosomy-X** (also known as Turner's syndrome), for instance, is the absence of one of the X sex chromosomes. With neither a second X chromosome or a Y chromosome, an abnormally short person develops, who has the external features of a female (Figure 10-10). Internally, however, the reproductive system is incomplete; she doesn't menstruate and is sterile. Monosomy-X occurs about once in 4,000 births.

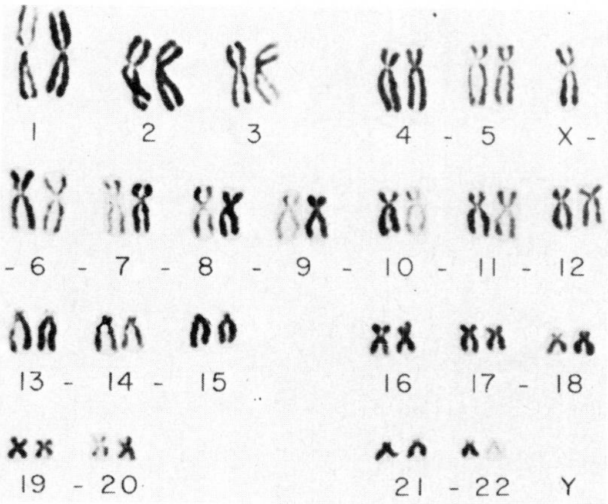

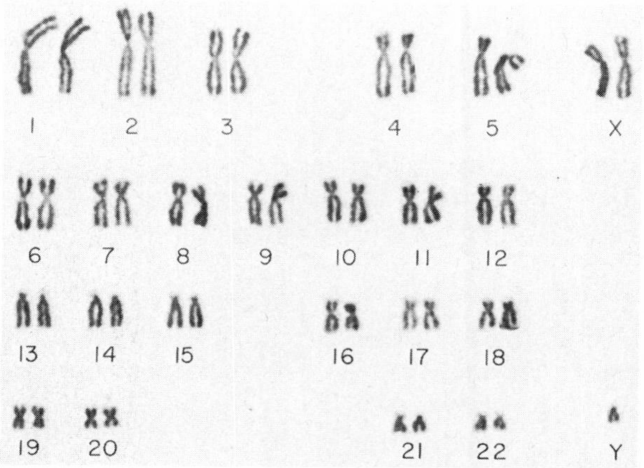

FIGURE 10-10
Chromosomes in Turner's syndrome (XO) above left.

FIGURE 10-11
Chromosomes in Klinefelter's syndrome (XXY) above right.

Trisomy-X results in females with three X chromosomes. Some have traits usually associated with males. Many are fertile and appear normal but are mentally retarded. Some females with trisomy-X develop symptoms similar to those of Monosomy-X. Trisomy X has an incidence of 1 in 1000 births.

XXY-syndrome (Klinefelter's syndrome) results when a normal egg, bearing an X-chromosome, is fertilized by an XY sperm (Figure 10-11). A person with this affliction is a male, usually long-legged, who is sterile and underdeveloped sexually. In some cases, he is mentally retarded. Occurrence is 1 per 500 births.

Far less debilitating than the three preceding conditions is **XYY-syndrome,** which results when a normal egg is fertilized by a YY sperm. It occurs about once in 2500 births and results in an abnormally tall male. Because the XYY-syndrome has been found in prison populations more frequently than in the general population, men who have it were reputed to be particularly prone to criminal acts. But, it has been argued that the high proportion of arrests is due more to their abnormal height (presumably making them easier to identify by victims and easier to spot by police) than to any genetic predisposition to crime.

Some of these gene and chromosome abnormalities can cause mental defects, but the people so afflicted are not called "mentally ill." They are said to suffer from a genetic disorder. It may well be that some of the people today that we call mentally ill will, as our biological knowledge increases, also be taken out of this rather vague category and be classified as having a gene disorder.

APPROACHES TOWARD TREATMENT OF GENETIC DISEASES

As humankind's understanding of genetic diseases has grown, various approaches toward treatment have been developed. Medical researchers have devised techniques by which missing hormones are artificially administered (as in diabetes) and accumulations of toxic substances are avoided (as in PKU). Perhaps in the future researchers will find ways to provide enzymes for which genetic instructions are missing. Some optimists speak of a coming age when it will be possible to provide enzymes for which genetic instructions are missing. Some even speak of a coming age when it will be possible to provide people with genes they are lacking or replace ones that are defective. Geneticists are quick to point out, however, that such procedures will not be available in the near future (if ever). We now have the technology of making directed changes in the structure of DNA of complex organisms with recombinant DNA procedures. At present the best we can do is treat the symptoms of certain genetic diseases. For others there is little that can be done yet.

Amniocentesis

It is critical that genetic abnormalities be detected early, preferably before birth. This is possible with a surgical technique called **amniocentesis.** This is a procedure in which a syringe is inserted through a pregnant woman's abdomen, and a small portion of the amniotic fluid surrounding the fetus is withdrawn (Figure 10-12). The amniotic fluid contains cells that have been sloughed off from the fetus. These are microscopically examined for chromosome abnormalities. Depending on the outcome, the parents may decide to continue pregnancy or end it with an abortion.

Amniocentesis should not presently be considered a routine test for all abnormalities. The genetic disorders for which it examines are rare, and the procedure involves a certain amount of risk to the mother and fetus. Thus, it has been applied only in special cases, such as if a mother has already had a child with a genetic affliction, if a woman or her mate are known carriers of a dangerous recessive gene, or if the prospective mother is of advanced age. Amniocentesis is advisable in the latter case, because the incidence of Down's syndrome increases so drastically with the age of the mother.

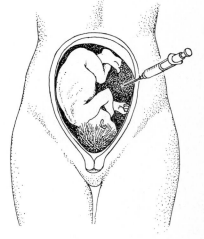

FIGURE 10-12
The procedure of amniocentesis.

Genetic counseling

As for all diseases, the best treatment for genetic disease is prevention. In the case of gene abnormalities, prevention means identifying the presence of the aberrant gene in the parents and avoiding the conception of a new diseased human being. Identification of genetic problems and giving advice based on that identification is called **genetic counseling.** Because many genetic diseases are due to recessive alleles, both prospective parents must be carriers for the future child to be affected. Early recognition of carrier individuals is often difficult. No doubts are left when a previous child has already been born diseased, or when the prospective father or mother had a diseased parent. In other cases a detailed study of the family tree of each spouse and tests performed on selected cells may reveal them to be carriers.

As you know from Punnett-square analysis, matings of two recessive heterozygous individuals result, on the average, in one dominant homozygous conception, two heterozygous conceptions, and one recessive homozygous conception. Thus, about one out of four children of heterozygous carriers of recessive, disease-causing genes would manifest the disease. Of course, chance has no memory. This same probability applies to every new pregnancy, regardless of what has happened with previous ones. Thus, for two carrier parents there is always a 25 percent chance of having an affected baby with every pregnancy. In most cases the prospective parents can begin a pregnancy, then rely on the results of amniocentesis for the decision of whether or not to go on with it.

NATURE OR NURTURE?—ANSWERS THROUGH STUDIES OF TWINS

What are relative influences of genes ("nature") and environments ("nurture") on human existence? This is a basic question of human genetics—a question that is far from being satisfactorily answered. Because of problems involved in studies of human genetics, mentioned at the beginning of this chapter, it is doubtful that it will ever be totally answered in our lifetime. Nevertheless, geneticists have found various means to gain at least partial insights. A quirk of reproduction, the phenomenon of twin births, has made some of the most fruitful studies possible.

Twins are of two types. **Fraternal twins,** the more common type, are of different genetic constitution. Often they are of opposite sexes. Fraternal twins develop from separate eggs that are fertilized by different sperm (Figure 10-13a). They are no more genetically similar than other brothers or sisters. However, fraternal twins are born together and usually grow up in the same household. Thus, their early environmental experiences are similar. Yet, they may be quite different in physical and mental nature.

Identical twins on the other hand, are almost totally alike. Having started life as the same cell, they have the same genetic make-up (Figure 10-13b). Identical twins result when the conglomerate of cells that has started development as a single embryo divides (why this happens is unknown) and continues development as two embryos of the same genetic constitution. Identical twins are of the same sex and are essentially mirror images of each other. Moreover, they are generally similar in temperament, mental and physical abilities, likes and dislikes, and many other personality traits. Nevertheless, many characteristics of their personalities can be quite different; one may be easy going, for example, while the other is aggressive.

One group of twin studies has shown that if one individual of a pair of identical twins has the mental disorder, schizophrenia, there is about a 50 percent chance that the other individual will also be schizophrenic. Compare this to an incidence of about one percent in the population at large. (Schizophrenia is a form of psychosis for which withdrawal, apathy, hallucinations, and delusions are symptomatic). The fact that there is such a high percentage of correlation indicates that genetic influences may predispose a person to schizophrenia. But where schizophrenia did not co-occur, it is obvious that the environment must play a role too.

It may be argued that since identical twins are often raised identically, as one becomes schizophrenic the other does too because they were exposed to the same environmental conditions. Studies have shown, however, that identical twins raised apart (by different foster parents) become schizophrenic much more frequently than fraternal twins raised under the same conditions.

Many geneticists believe that there is a gene for schizophrenia, perhaps even for manic-depressives, but what activates it is not known. Perhaps the gene is turned on in a manner similar to that for enzyme induction.

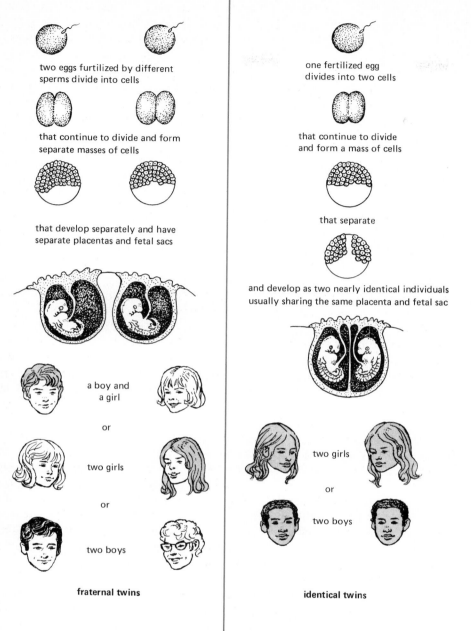

FIGURE 10-13

Fraternal twins (left), developing from different eggs and sperms, are no more alike than other brothers and sisters. Identical twins (right) develop from a single egg and sperm, and have the same genetic makeup.

two eggs furtilized by different sperms divide into cells

that continue to divide and form separate masses of cells

that develop separately and have separate placentas and fetal sacs

a boy and a girl

or

two girls

or

two boys

fraternal twins

one fertilized egg divides into two cells

that continue to divide and form a mass of cells

that separate

and develop as two nearly identical individuals usually sharing the same placenta and fetal sac

two girls

or

two boys

identical twins

INHERITANCE AND INTELLIGENCE

To what extent is intelligence determined by genes? Since there is yet to be an adequate definition of *intelligence,* this question is somewhat ill-founded. Furthermore, as should be obvious from the preceding section, when it comes to matters of the brain it is difficult to differentiate between the effects of environment and heredity. Nevertheless, attempts have been made to devise testing procedures to assess mental abilities and to determine to what extent these abilities are determined by genes.

The most widely accepted method of measuring intelligence is a test that involves quick memorization, mathematical calculation, word recognition, and other fundamental (to our culture) thought processes. The score is said to indicate the tested individual's "mental age," which is divided by age in years and multiplied by 100 to yield the **intelligence quotient,** or I.Q. If, for example, a child of 5 is assessed as having a mental age of 7, the I.Q. would be $7/5 \times 100 = 140$.

There are many families in which I.Q.'s are markedly above or below average. But, is this a genetic phenomenon, or is it the product of the home, the school, and other cultural influences? Studies of twins have shed a certain amount of light on this question. Since the I.Q.'s of identical twins are more nearly the same than the I.Q.'s of fraternal twins, it would seem that genetic influences are involved in intelligence. Whatever these influences are, however, they are not simple. Many genes seem to be involved in determining innate intellectual ability. Furthermore, there is evidence that in families where mental achievement is rewarded by encouragement and praise, innate abilities are developed to a greater extent than in families that are indifferent to intellectual achievement. Qualitative evidence for this has come from observations of human families, and quantitative evidence has come from experiments involving rats and other animals. Rats raised in "impoverished" environments (bare cages) have been found to be slower learners than rats of the same brood which were raised in "rich" environments (which included objects to explore and rewards of food for performing simple tasks).

Apart from genetic and environmental influences there are other factors which affect intelligence. During a difficult birth, for example, an infant may suffer from lack of oxygen, which might irreversibly damage brain cells at this delicate stage of development. Childhood malnutrition (particularly the deficiency of protein that is characteristic of poverty) can also lead to retardation of mental development. Thus, one might expect that children of impoverished families, whose diets are lacking in protein, would score lower on I.Q. tests than children of better-off families.

There is no greater area of dispute in human genetics than the question of the relative intelligence of the human races. Some racial groups in America score somewhat below whites on standard intelligence tests (the difference is about 15 I.Q. points). This has been interpreted by certain educational psychologists as indicating that they are somehow intellectually inferior. But I.Q. tests are not culture-free; that is, to score well on them one has to be part of the culture that values these tests. The contribution of genetics to I.Q. scores is not clear.

It has been argued that I.Q. testing is biased in favor of those who are steeped in the heritage of Western civilization, since it favors skills that have come to be valued in our society (for example, an ability to do rapid paper work). It can be argued that intelligence should not be defined in terms of adaptability to a society that is alienating many of its members. A fundamental question is what does an I.Q. measure? It is presumptuous to believe that one number could sum up so complex a concept as intelligence. There is, for example, no one number in physics that sums up the properties of something so relatively simple as an atom! Moreover, it is quite possible that in future ages, other mental characteristics which are possessed to a lesser extent by whites than other races will be of more social value than high I.Q. If society is subject to change (which it has certainly proved to be in the past) then ought not criteria of intelligence also be changeable? Perhaps, one day, what is colloquially known as "soul" will be integrated into measures of intelligence.

CHAPTER 10: SUMMARY

1 Population genetics is concerned with determining the frequency of a trait in a population.

2 The major blood types are inherited and because of incompatibilities among them, care must be used in proper blood matching in transfusions.

3 Most traits are determined by more than one gene pair.

4 The best known of the sex-linked traits in humans is color blindness, carried on a female sex chromosome but expressed mostly in males.

5 Not all genes related to sexual characteristics are on the sex chromosomes, however.

6 A variety of diseases and malformations have been shown to be related to genetic defects.

7 Genetic counseling involves the identification of human genetic problems and giving advice to prospective parents.

8 Both nature (genes) and nurture (environment) contribute to animal and plant development and maintenance.

Suggested Readings

Benzer, Seymour   "Genetic Dissection of Behavior," *Scientific American*, December, 1973 — A detailed analysis of behavior genetics.

Chapnick, Phillip   "Creeping Up on Eugenics," *The Sciences*, May, 1973.

Chasin, Gerald   "Doomed to Inequality," *The Nation*, July 2, 1973 — A consideration of research which attempts to determine the relative intelligence of the races.

Dobzhansky, T.   "The Present Evolution of Man," *Scientific American*, September, 1960 — How our control of our environments may be influencing our evolution.

McKusick, Victor A.   *Human Genetics*, 2nd Ed., Prentice-Hall, Englewood Cliffs, New Jersey, 1969 — An introduction to the complexities of human genetics.

Müller, H.   "Radiation and Human Mutation," *Scientific American*, November, 1955 — The discoverer of the genetic effects of radiation considers the influence of mutations on human evolution.

AN EPITAPH FOR SIR CYRIL?

Scientists aren't immune to the temptation to be dishonest which we all experience. But, since all research is subject to review, science is a poor field for those with a tendency to stretch the truth.

The following reading deals with the work of one of the pioneer researchers in the field of inheritance of human intelligence, Sir Cyril Burt. Apparently, Sir Cyril faked some of the data he reported. The fact that the dishonesty is now coming to light is a testament to the ultimate honesty of the scientific method. But, the fact that the discovery was so long in coming (Burt's erroneous reports date back as far as 1909) indicates that it is difficult to detect untruths in a field as complex as human genetics. It also makes one wonder if human eugenics could ever be achieved without bias.

When he went to his grave five years ago at the age of 88, Britain's Sir Cyril Burt left behind him an almost undisputed reputation as the most prestigious, powerful and influential psychologist since William James. Now Sir Cyril looms as the central figure in a scientific scandal that might have been lifted right from a C.P. Snow novel.

Burt was a scholar of enormous charm and erudition, and his classic studies of the IQ's of identical twins provided the main building block for the view that it is heredity, rather than environment, which is the most important determining factor in human intelligence. Now he stands accused of having faked much if not all of his research, of solemnly reporting tests and other studies that were never done and of signing fictitious names to a host of papers, all of them filled with praise for himself and his work. To scientists, the effect is rather as if someone had just discovered that Charles Darwin never made his voyage aboard the Beagle at all, but had confected the "Origin of Species" while smoking opium with Thomas De Quincey.

The result is a fire storm of controversy between the "nurturists" (those psychologists who think that environment is the most important factor in human intelligence) and the "naturists" (those who incline to the heredity determinism espoused by Burt, by Berkeley's Arthur R. Jensen and by Richard Herrnstein of Harvard). "Burt's work was fraud linked to policy from the word go," says Princeton psychologist Leon Kamin, who was one of the first to attack the accuracy of Sir Cyril's statistics. Herrnstein disagrees sharply. "All of Burt's conclusions," he says, "have since been replicated by independent researchers."

Kamin first began to scrutinize Burt's work about four years ago. He quickly found at least twenty glaring anomalies. One of the most important of these was the fact that Burt's statistical correlation between the IQ scores of identical twins in three separate studies remained exactly the same to the third decimal point—this despite the fact that the number of twins in each of the three studies was different. Kamin realized that this kind of statistical correlation was virtually impossible. He went on checking and found similar errors, some of them in work of Burt's dating back as far as 1909. So armed, Kamin took to the lectern and began a series of slashing attacks on the "naturists" in general and on Burt in particular.

Old: At about the same time, Jensen also had found himself puzzled by some of Burt's statistics, and reported his findings in a scientific paper—the first time Burt had ever been attacked in print by one of his protégés. But Jensen was inclined to take a more charitable view of the great man's failings. He attributed them to Burt's age, his crankiness and his admittedly failing memory. "It is almost as if Burt regarded the actual data as merely an incidental backdrop for the illustration of the theoretical issues," said Jensen. Like Herrnstein, Jensen thinks there are several other independent studies that provide ample support for the validity of Burt's original work.

A further blow to Burt's reputation came this autumn, when The London Sunday Times concluded that two of the major coauthors of many Burt papers, the Misses Margaret Howard and J. Conway, simply never existed. The paper charged that Burt invented them—and Burt's housekeeper, Grete Archer, has admitted that she was aware that Sir Cyril regularly used pseudonyms in his later years. Still further investigative work revealed that the two pseudonyms had appeared regularly as by-lines on reviews of books and on articles on Burt's work in the British Journal of Statistical Psychology—a publication edited by Burt himself for more than fifteen years. Invariably, these articles were rich with praise for Burt and his work and packed with scathing attacks on his enemies. They stopped appearing after Burt stepped out as editor in 1963.

Doubt: So it has gone, and each successive revelation has led to further questions. If the Misses Conway and Howard were pseudonyms, the critics ask, then who conducted the tests on which their articles were based? Professors Anne and Alan Clarke, a husband-and-wife team at Hull University, worked with Burt in the 1940s and '50s. They now admit that there were suspicions about Burt's integrity even then. "People had grave doubts long ago," Anne Clarke told Newsweek's Malcolm MacPherson last week, "but Burt was a fearsome figure. He was an autocrat of the old school, wrapped in a most charming style. Our own doubts did not become outright skepticism until quite recently."

The mystery of how much—if any—of Burt's work was fraudulent and why will probably never be solved. Six chestsful of his papers were burned right after his death, and many of his original records are thought to have been among them.

This will leave the environmentalists and the hereditarians to battle things out as best they can until still more research either supports or disposes of the validity of Burt's earlier work. In the meantime, those British and American psychologists who were there at the time are recalling Burt's performance at a symposium in London in 1960, where he dazzled all his hearers with the elegant style and delivery of his address. It was then, after Burt had finished, that the late L.S. Penrose, an eminent geneticist who had known Sir Cyril for 30 years, added a postscript of his own. Said Penrose: "I don't believe a word the old rogue says, but, by God, I admire the way he says it." What now remains to be seen is whether Penrose's epigram may eventually become Sir Cyril Burt's epitaph.

CHAPTER 11

Cellular and Organismic Reproduction

Objectives

1 Be aware of how cell size varies.

2 Be able to follow spermatogenesis and oogenesis.

3 Describe the various ways animal and plant cells reproduce.

4 Understand the development of a flowering plant.

5 Describe how seeds disperse and survive in the environment.

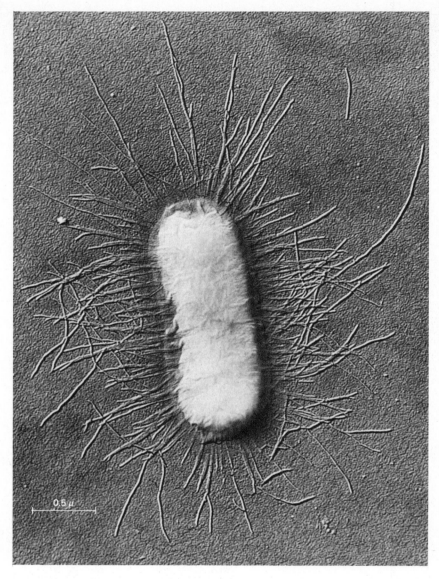

FIGURE 11-1
Electron microscope photograph of a bacterium with many pili filaments which it uses to attach itself to surfaces.

0.5 μ

LIMITS ON CELL SIZE

The smallest nonparasitic cells are bacteria (Figure 11-1), which may be as small as a micrometer in diameter, while the largest cell, the ostrich egg yolk, is 10 centimeters in linear dimension (Figure 11-2). Any active cell must be large enough to contain all the biochemical "machinery" necessary to sustain life, but not so large that it smothers its own existence.

A continuous exchange of matter passes through a cellular membrane. Nutrients are absorbed, and wastes are excreted. Therefore, in order for a cell to function properly, a critical ratio of the volume of its contents to the surface area of its membrane must be maintained. But, as a cell grows its volume increases as the cube of its diameter, while its surface area increases more slowly, as the square of its diameter (Figure 11-3). Thus, a point is soon reached at which a growing cell cannot take in nutrients or excrete wastes fast enough across its surface. The dilemma is solved by cell division. Division of a parent cell produces two cells with more favorable surface/volume ratios for exchange with the environment.

FIGURE 11-2
Ostrich eggs.

FIGURE 11-3
An illustration of scaling in animals. The elephant-sized animal on the left has been drawn to be the same size as the horse-sized animal on the right. However, the size of the bones of the legs indicates the animal on the left must be much bigger than the one on the right.

DIVISION OF SOMATIC CELLS

A cell that is about to divide is termed a **mother cell,** and the division results in two **daughter cells,** which are generally about equal in size. Division of **somatic cells** (all cells of the body except reproductive cells) occurs in two stages. The first stage is the duplication of nuclear materials, with each daughter nuclei receiving genes that are identical with those of the mother cell. Genetic duplication, **mitosis,** maintains an organism's hereditary code intact. During mitosis, the cytoplasm of the mother cell divides into two approximately equal parts, and a membrane grows between them. In plants a cell wall of cellulose also develops.

Mitosis

By convention mitosis is described as occurring in five phases: interphase, prophase, metaphase, anaphase, and telophase (Figure 11-4). Although these mitotic stages aren't characterized by abrupt changes, certain significant events distinguish each one.

- Interphase   Interphase is the stage between nuclear divisions, during which the nucleus is active in cellular processes. The cell is in a

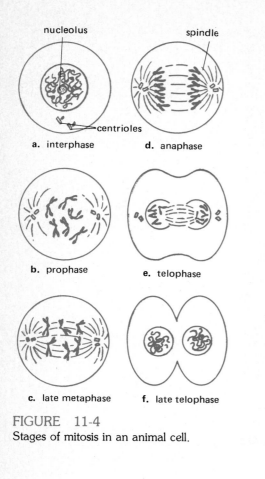

FIGURE 11-4

Stages of mitosis in an animal cell.

FIGURE 11-5

A late-prophase chromosome consisting of two identical chromatids united by a centromere.

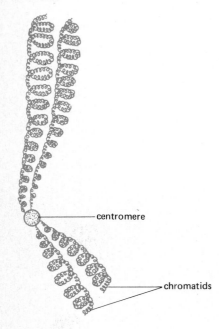

period of growth and maintenance, and the chromosomes are elongated and diffused in the nucleus as a network of fine threads. As prophase approaches, however, the chromosomal DNA duplicates, and the chromosomes become double strands.

- Prophase   In most animal cells a small, dense area of cytoplasm exists just outside the nuclear membrane. It contains two small granules, the **centrioles.** Examination with the electron microscope has revealed centrioles to consist of clusters of tiny tubules. Soon after the chromosomes become double, the centrioles separate, and migrate to opposite sides of the nucleus; threadlike fibrils begin to radiate from the centrioles in all directions. Concurrently, the chromosomes shorten and thicken. At this point they are double strands, **chromatids,** attached at a single point, the **centromere** (Figure 11-5). Late in prophase the nuclear membrane dissolves and the chromosomes, still consisting of paired chromatids, move toward the center of the cell.

- Metaphase   As prophase evolves into metaphase, fibrils grow between the centrioles. They bow outward in the center, forming a three-dimensional network shaped like a football; this is the **spindle** (Figure 11-4c). As the spindle forms, each double-stranded chromosome attaches at its centromere to a fibril. For a moment the chromosomes hang suspended in the middle of the cell as if hung on microscopic clotheslines. Then, the centromeres break. The former chromatids are now separate, single-stranded chromosomes.

- Anaphase   Immediately after separation, a process occurs that is similar to muscle contraction. The **chromosomes** begin to migrate to opposite centrioles. The mechanism involved isn't completely understood, but the chromosomes appear to be pulled by contractions of the spindle fibrils. Anaphase ends with the arrival of one duplicate of each chromosome at each centriole.

- Telophase   During this phase the chromosomes are enclosed by membranes to form two daughter nuclei, and the cytoplasm divides to enclose them in separate cells (Figure 11-4e). Soon after reaching the centrioles, the chromosomes lengthen and gradually take on the diffuse nature of interphase nuclei. Concurrently, the fibrils of the spindle disappear. With the formation of plasma membranes around the daughter cells, the cycle of mitosis comes full swing. The daughter cells enter a phase of growth and eventually reach a size at which they too must divide.

The division of cellular structures outside the nucleus isn't as exact as the division of chromosomes. Distribution of the endoplasmic reticulum, ribosomes, mitochondria, plastids, and other cytoplasmic structures in daughter cells is not exactly equal. Since the daughter cells contain all the necessary genetic information, however, they can soon assemble missing or deficient cytoplasmic structures.

The timing of mitosis

The time required for a mitotic division is different for different kinds of cells. The process may occur in less than a half hour (some chick cells divide in 23 minutes), or it may require much longer. Some cells never divide; this is the case with most of the nerve cells in the human body. Generally mitosis is most frequent in tissues that are least specialized. The nearly identical cells composing the microscopic sphere that develops into a human embryo, for example, divide every hour or so. As the embryo takes on human characteristics, however, and cells begin to develop specialized functions, the rate of division slows to once every several hours.

Cell division is stimulated by injury. Cells in human skin, for example, may normally divide only once per day, but a cut will stimulate hourly divisions until the damaged tissue is regenerated. It appears that the rate of division is normally held in check by the proximity of other cells.

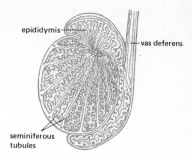

FIGURE 11-6
A cross section of a human testicle.

MEIOSIS

Within the nucleus of every somatic cell of the human body are 23 pairs of chromosomes, or 46 chromosomes in all. Because they contain pairs of chromosomes, somatic cells are said to be **diploid.** During mitosis the chromosomes duplicate, and one of each of the duplicates goes to each daughter nuclei.

Within the male and female reproductive organs, the **testicles** and **ovaries,** a different kind of cell division takes place. **Meiosis** gives rise to **reproductive cells, sperm** and **ova** (eggs), which have only one of each pair of chromosomes. They are said to be **haploid.** Complete meiosis involves two successive divisions, which result in four haploid cells being generated from a single diploid precursor. It is the first division sequence that results in the halving of the chromosome number; the second division is essentially a mitotic one. Each of the divisions of meiosis may be described as occurring in an interphase, prophase, metaphase, anaphase, and telophase; but the following descriptions touch mainly on where the process differs from mitosis. Since the ways in which sperm and eggs are produced are slightly different, spermatogenesis and oogenesis are described separately.

Spermatogenesis — formation of sperm

Consider the interior of a human testicle, where over 100,000,000 sperm per day are produced within several hundred yards of tiny, convoluted **seminiferous tubules** (Figure 11-6). The cells that give rise to sperm, **primary spermatocytes,** line the interior walls of these tubules. In a process involving two nuclear divisions, a primary spermatocyte gives rise to four sperm.

During a prolonged prophase, sometimes lasting several days, chromosomes in a primary spermatocyte shorten and thicken, becoming clearly visible with a light microscope. As in cells developing for mitosis, each chromosome consists of two chromatids joined by a centromere. Since chromosomes are paired at this point they can be designated as 1^a-1^a, 1^b-1^b, 2^a-2^a, 2^b-2^b, etc. For simplicity only two chromosome pairs are illustrated in Figure 11-7.

At this point meiosis begins to differ from mitosis. Chromosome pairs, each consisting of two chromatids joined at a centromere, apparently attract each other and come together. This is **synapsis** (Figure 11-7c). Synaptic pairs of chromosomes do not actually join; rather their centromeres attach, side-by-side, to a single fibril of the spindle (Figure 11-7d). This is different from mitosis, during which each chromosome attaches to a separate fibril. Near the close of prophase, the synaptic pairs are drawn to the center of the cell.

During meiotic anaphase, one chromosome of a synaptic pair, with its chromatids still joined, moves toward one pole. Concurrently its mate moves toward the opposite pole (Figure 11-7e). It is important to remember that *pairs of chromosomes, not chromatids, separate during the first meiotic division.* Further-

FIGURE 11-7
Spermatogenesis. Meiotic divisions of a
diploid primary spermatocyte give rise to
four haploid sperms.

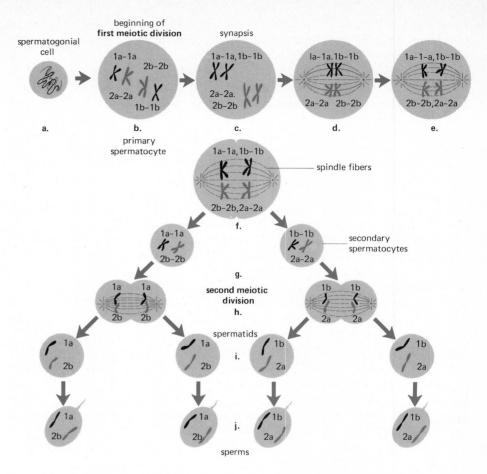

more, separation of one pair of chromosomes during meiotic anaphase does not
seem to influence the way other pairs separate. In the diagram, chromosome
1^a-1^a moves with 2^b-2^b to one pole, and 1^b-1^b joins 2^a-2^a at the other pole. It
could just as easily have been that 1^a-1^a joined with 2^a-2^a, and 1^b-1^b joined with
2^b-2^b. With only two pairs of chromosomes, there are only two possible combinations. However, in humans, with 23 pairs of chromosomes, the number of possible
combinations is immense.

After the chromosomes arrive at the poles, a furrow forms in the primary
spermatocyte near its center, and cell division results in two **secondary spermatocytes** (Figure 11-7g). Following an interphase during which there is no duplication of genetic material (remember that although the chromosomes are unpaired
at this point, they still consist of identical chromatids joined by centromeres) the
secondary spermatocytes undergo *second meiotic divisions*. During metaphase of
this division the centromeres break, and the single-stranded chromosomes move to
opposite poles. Division of two secondary spermatocytes produces four **spermatids.** As they mature, spermatids develop whiplike flagella and become functioning sperm, each with a haploid set of chromosomes (Figure 11-7j).

Oogenesis — formation of ova

A process similar to spermatogenesis occurs in the formation of ova. However,
there are certain differences (Figure 11-8). In human females ova develop one at a
time, at a rate of about one per month (rather than hundreds of millions per day)
near the surfaces of the ovaries. Oogenesis also differs from spermatogenesis in

that division of the cytoplasm of the **primary oocyte** (which is analogous to the primary spermatocyte) is unequal, resulting in a **secondary oocyte** and a smaller **first polar body** (Figure 11-8g). Like secondary spermatocytes, however, both of these cells receive haploid numbers of chromosomes.

The secondary oocyte undergoes a second meiotic division in which the chromatids separate and move to opposite poles. This division results in an **ootid** and a **second polar body.** The ootid matures to form an egg. The polar bodies usually decompose. They are non-functional in any case.

SEXUAL REPRODUCTION

In the preceding sections, formation of human ova and sperm were considered as the models for meiosis. The function of meiosis in all animals and in certain plants is essentially the same; it results in the formation of haploid reproductive cells, or **gametes.** However, there are certain variations.

In some fungi and algae, all gametes are alike. They cannot be recognized as sperm or ova because they are structurally identical. They are known as **isogametes.** Primitive forms of life more commonly reproduce by isogametes than do more advanced species but **heterogametes** have structural differences that make them recognizable as sperm or ova. Most sperm are motile (they swim with lashing movements of flagella) and are smaller than ova, which are nonmotile.

There are many different patterns of gamete formation. In some species they are produced by different individuals, but in other species both male and female gametes are produced by a single individual. Such species are said to be **hermaphroditic.** This is the case with earthworms, two of which can fertilize each other simultaneously (Figure 11-9).

During **fertilization** two gametes unite, thus establishing a diploid set of chromosomes in the fertilized cell, or **zygote,** of a new organism (Figure 11-10). Reproduction involving the union of gametes is termed **sexual reproduction.** Soon after fertilization, the zygote divides by mitosis. Thus, the diploid chromosome number is maintained in the daughter cells. Repeated mitotic divisions may

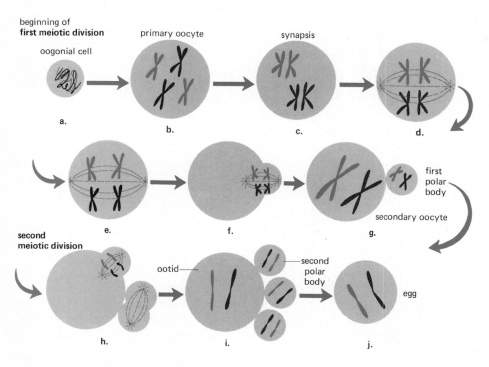

FIGURE 11-8
Oogenesis. Meiotic divisions of a diploid primary oocyte give rise to a haploid egg.

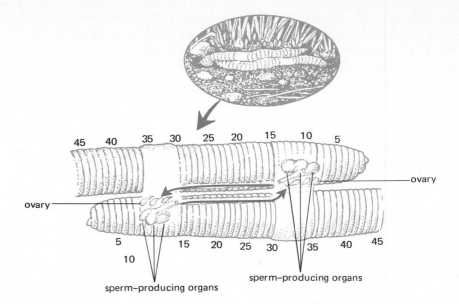

FIGURE 11-9

Reproduction in the earthworm. Even though the earthworm has the reproductive organs of both sexes, it exchanges its sperms for those of another earthworm. The sperms travel from the seminal vesicles of one worm to the seminal receptacles of the other.

ultimately generate trillions of diploid cells in the body of a mature sexually reproducing organism.

REPRODUCTIVE VARIETY

Reproduction does not necessarily have to be sexual. Certain species generate more of their kind **asexually** (without the recombination of genes by the union of gametes). Other species may be sexual or asexual, depending on the phase of their lives or the nature of their surroundings.

For certain single-celled species, reproduction is simply a matter of cell divi-

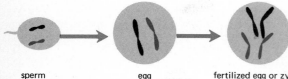

sperm   egg   fertilized egg or zygote

FIGURE 11-10a

Fertilization. The combining of the chromosomes of two haploid gametes restores the diploid condition.

FIGURE 11-10b

Scanning electron microscope photographs taken during fertilization of a sea urchin egg. The egg was mixed with a solution of 3×10^8 sperm per cubic millimeter. From left to right, top to bottom: one second after exposure to the sperm, after 15 seconds, after 30 seconds when a sperm has penetrated the egg, and at 180 seconds. After 180 seconds, a process that detaches all other sperm is complete and development of the organism begins.

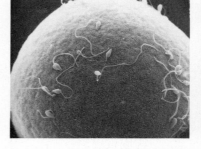

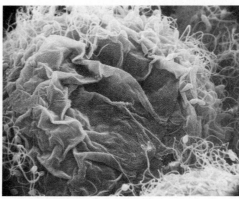
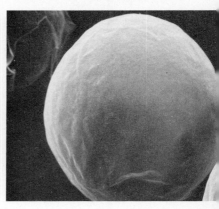

sion. Bacteria, for example, which lack well-defined nuclei reproduce asexually by a primitive form of mitosis, termed **binary fission.** Prior to binary fission the genetic material migrates to the center of the cell and duplicates. Without the formation of a spindle, the cell wall grows inward at the center of the cell, separating the genetic duplicates.

Budding is another type of asexual reproduction, which differs from binary fission in that the resulting cells are of unequal size. A bud develops as a knob or bulge on a mother cell, and the nucleus divides equally by mitosis, one portion remaining in the mother cell and the other entering the bud. The bud receives only a small proportion of the cytoplasm, but it grows under the direction of its full complement of chromosomes. Yeasts and certain other microorganisms reproduce in this fashion (Figure 11-11).

Various organisms such as bacteria, fungi, and yeasts reproduce by forming **spores** during periods of unfavorable environmental conditions. In some organisms spores are reproductive cells formed by a series of divisions within a mother cell (Figure 11-11). Spores are shed by the parent organism, and in some cases are carried passively by the wind, water, or animals. If they are produced by aquatic organisms, spores may have flagella with which they swim. Spores are usually protected by a resistant cell wall and may survive for long periods under such severe environmental conditions as drying or freezing. When the weather becomes more hospitable, a spore germinates to form an active organism.

A given species doesn't always reproduce in the same way. Some bacteria, for example, reproduce by binary fission until their environment becomes unfavorable. Then they switch to the formation of spores. A species that thrives in pond mud, for example, might make the switch during times of drought, when the pond has gone dry.

Some species of microorganisms alternate between phases of asexuality and sexuality. *Chlamydomonas,* for example, is a genus of green algae which alternately produces haploid flagellated forms by mitosis and innactive diploid forms (Figure 11-12). During its flagellate phase, *Chlamydomonas* lives near the surfaces of ponds and other bodies of fresh water. In preparing to reproduce asexually it absorbs its flagella and divides by mitosis into two or more daughter cells, while

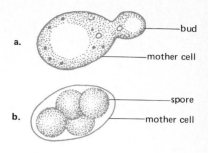

FIGURE 11-11
Yeast cells reproduce rapidly by budding, a form of fission (a). Under certain conditions, they also form spores (b).

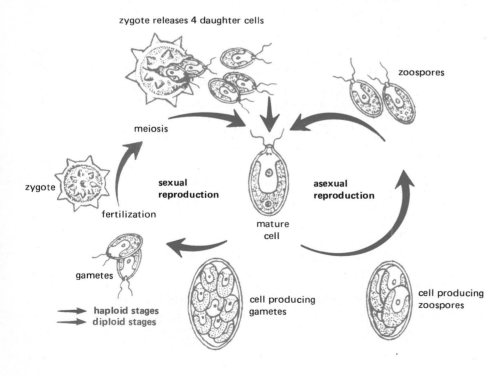

FIGURE 11-12
Reproduction by *Chlamydomonas* may be sexual or asexual. (Illustration is reprinted from *Biological Science*, Second Edition, by William T. Keeton, illustrated by Paula Di Santo Bensadoun, with the permission of W.W. Norton and Company, Inc. Copyright © 1972, 1967 by W.W. Norton and Company, Inc.)

CELLULAR AND ORGANISMIC REPRODUCTION **265**

still encased in the original cell wall. Eventually the cell wall bursts releasing the flagellates, which may repeat the process or go through a sexual phase.

The sex life of *Chlamydomonas* begins similarly to the asexual phase. The flagella are absorbed and successive mitotic divisions produce gametes within the original cell wall. When the cell wall bursts the flagellated isogametes swarm with others formed by other cells. Ultimately they pair off and fuse together, merging their contents. Thus, a diploid number of chromosomes is established within a zygote, which forms a protective coating, and then sinks to the bottom of the pond. There it can sit out a cold winter that the flagellated form could never survive. *Chlamydomonas* differs from most higher organisms in that its diploid embodiment is a resting form, and the haploid embodiment is the active, growing form.

Evolutionists believe that organisms similar to *Chlamydomonas* gave rise to the rest of the plant kingdom. Over the course of evolutionary history the genetic variety endowed to organisms by the fusion of nuclear materials has led to increasingly complex diploid forms, while the haploid stage has remained simple.

Nevertheless, meiosis in certain species of plants produces haploid spores that subsequently develop into complex haploid vegetative forms. Certain species of multicellular algae, molds, and some land plants reproduce in this fashion. A haploid multicellular plant may eventually produce cells specialized as gametes; but *these gametes are produced by mitosis, not meiosis, because the cells that divide to produce the gametes are already haploid.* Two of these gametes unite to form a diploid zygote, which divides mitotically to form a diploid multicellular plant. Eventually this plant produces spores, and the cycle starts over again. Thus, a single species of plant may alternately go through haploid and diploid phases of life.

There is wide variability in the relative importance of the haploid and diploid phases in plant life cycles. In some cases only the zygote is diploid, and the multicellular form is haploid. This is so with filamentous forms of green algae and molds. Other plants have life cycles in which haploid and diploid multicellular phases alternate. *Ulva*, or "sea lettuce," is a species that grows in the intertidal zone whose leaflike haploid and diploid multicellular forms are almost identical (Figure 11-13). The plants with which people are most familiar, the flowering plants, spend most of their lives as diploid multicellular organisms. Their reproductive cycles are considered in detail later in this chapter.

Haploid phases of life are quite common among plants, whereas mature forms of essentially all animals are diploid. Nevertheless, certain animals are haploid. They develop by **parthenogenesis** from unfertilized eggs. Best known among these organisms are male bees, or drones, which develop from eggs laid by a queen bee in huge numbers.

There is immense diversity in modes of reproduction in the living world; what has been described here is but a small sampling. In the following sections of this chapter, plant reproduction will be looked at more closely. Chapter 12 considers the details of human reproduction.

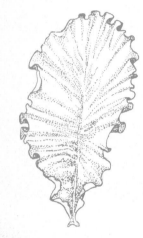

FIGURE 11-13
Ulva is a marine green alga whose haploid and diploid forms are nearly identical.

FIGURE 11-14
Vegetative reproduction in the strawberry. Roots that arise from an unusual place such as stems or leafs are called adventitious.

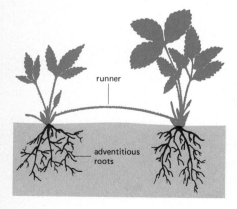

VEGETATIVE REPRODUCTION

A common form of plant reproduction, **vegetative reproduction,** can be seen in a strawberry patch. Certain stems of the strawberry, called runners, grow along the ground from parent plants. Runners form roots at the tips, and can give rise to new plants when they settle on a clear patch of ground (Figure 11-14). Thus, a single parent can give rise to many new plants in one season.

For some plants leaves are organs of vegetative reproduction. *Bryophyllum*, for example, develops fledgeling plants at the edges of their thick, fleshy leaves (Figure 11-15). As they increase in size, the tiny plantlets drop to the ground and take root.

Commercial plant growers use many methods of starting new plants vegetatively. One of the most common techniques is the preparation of **cuttings**, which are portions of stems or roots that are removed and transferred to loose, damp soil or sand. Many house plants can be propagated in the same manner. A cutting made from a stem should include one or two buds or leaves. The lower end of the cutting is buried, and in a few weeks roots begin to emerge from it. New shoots then develop from the buds or the active **meristematic tissue** at the juncture of a leaf and a stem. Many woody plants, including roses and pussy willow, are easily propagated by stem cuttings. Herbaceous plants, such as the geranium, can also be grown from cuttings.

In recent years agriculturists have begun to use plant hormones to insure the success of cuttings. Various growth-regulating hormones, including the **auxins**, initiate cell division and promote root growth. The cuttings are treated by soaking them in a dilute solution of the hormone for 12 to 24 hours or by applying a salve or powder that contains tiny amounts of the hormone. Growth of roots on treated stem cuttings is more rapid if leaves are present to supply sugar for the developing root tissues. By means of hormone treatment, many plants that are difficult to propagate vegetatively can be successfully grown from cuttings. Among these are holly and ornamental conifers, such as spruce and juniper.

Layering is another method of artificially propagating certain plants. Rose stems, for example, can be bent to the ground and covered with soil, leaving only the tip exposed (Figure 11-16). The buried portion will form roots, and growth of the stem will continue from the tip. After the new plant is established, it is cut from the parent bush and set out in a new location. This process is used most widely in propagating climbing roses and other varieties that develop vigorous roots.

Large leaved house plants, such as the rubber plant, can be propagated by **air layering** (Figure 11-17). In this process, a branch is cut about half of the way through and propped open by inserting a match stick. Then it is surrounded with a ball of wet peat moss which is held to the stem with a piece of fabric or a plastic bag tied above and below the incision. In several weeks, when roots have developed, the remaining point of attachment to the base of the plant is cut, and the newly rooted portion is planted.

FIGURE 11-15
Vegetative reproduction in *Bryophyllum*. Detail of the tiny plants growing along the toothed margin of a fleshy *Bryophyllum* leaf can be seen below.

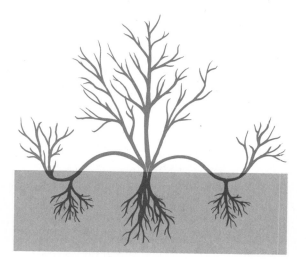

FIGURE 11-16
Simple layering.

Grafting is a common technique for propagating woody plants, especially fruit trees (Figure 11-18). It is the splicing together of two stems. The rooted portion is called the **stock,** and the cutting that is joined to the stock is the **scion.** In general, only closely related plants can be grafted. A scion from an orange tree cannot be grafted to an apple tree, for example. But, many types of apples can be grafted to a single stock apple tree, which has been chosen for its strong root system. Each portion of the tree that develops from a scion bears fruit like the tree from which the scion was taken.

A graft can be successful only if the thin layers of Meristematic tissue beneath the bark, the **cambium layers,** are joined (Figure 11-19). It is also necessary that the graft be made during a season of dormancy, so that the living tissues of the stems can unite before shoots develop. For this reason, grafts are usually prepared during the winter months. The region of the graft is covered with wax or "nurseryman's paint" to prevent drying or invasion of the wound by fungi, insects, or

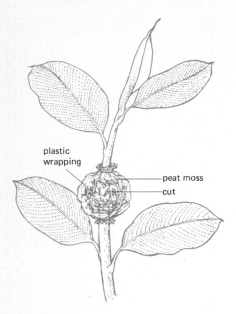

plastic wrapping

peat moss

cut

FIGURE   11-17

In air layering, the stem of the plant is cut about half way through, and the cut is propped open with a match stick. A ball of damp peat moss, about the size of a baseball, is wrapped around the cut and held in position with a piece of plastic wrapping until roots begin to develop (which may take a few weeks). Large house plants such as rubber trees may be rejuvenated by air layering.

FIGURE   11-18

Grafting and budding.

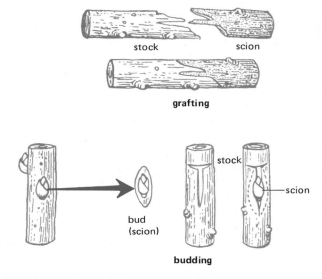

stock   scion

grafting

bud (scion)

stock

scion

budding

other damaging organisms. Come the spring, the stock begins to develop into a branch.

Budding is similar to grafting except that a bud, rather than a branch, is used as the scion. A bud is removed from the tree on which it is growing, together with a small amount of surrounding bark, by cutting at the level of the cambium. Then the bud is united with the stock by slipping it beneath a T-shaped cut in the bark of the stock, thus uniting the cambium layers (Figure 11-18). The incision is wrapped with elastic tape to secure the bud and prevent drying. Budding is usually done in the late summer or fall. When the bud opens and a shoot develops the following spring, the stock is cut off above the point where the bud was inserted. In this way the energies of the plant are concentrated on promoting growth of the graft.

Through grafting, nurserymen have been able to preserve desirable mutant strains. The original Golden Delicious apple, for example, was produced only once from seed, as the result of a fortuitous chance combination of genes. It is unlikely that another tree with the same characteristics will ever be grown from seed again. The original tree was the source of buds, however, from which grafts could be made. Thus, all Golden Delicious apple trees are the same as the original tree, propagated and perpetuated by grafting. The same applies to prized varieties of other fruit trees, shade trees, and roses.

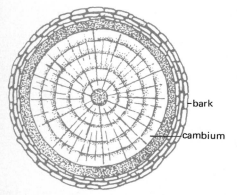

bark

cambium

FIGURE   11-19

A cross section of a woody stem.

SEXUAL REPRODUCTION IN FLOWERING PLANTS

By sexual reproduction, flowering plants bring together new combinations of genes and encase them within seeds. Flowers are reproductive organs, which, over the course of millions of years, have evolved from leaves. There is great variation among flowers. The lily, rose, orchid, and tulip have been bred for their large, colorful blooms. In contrast, the flowers of the oak, elm, maple, corn, wheat, and lawn grasses are usually unnoticed, although they are equally efficient reproductive organs.

A typical flower is diagrammed in Figure 11-20. The sexual parts of flowers aren't the showy petals. The reproductively active parts are the **stamen** and the **pistil.** Each stamen consists of a slender stalk, which supports a knoblike sac called the **anther.** The anther produces **pollen** grains, which carry male gametes. The pistil is usually a vase-shaped organ, the top of which, the **stigma,** produces a sticky substance that catches and holds pollen. The stigma is supported by a slender stalk that swells at the base to form the **ovary.** Inside the ovary are the **ovules,** which will later become seeds. The ovules are attached to the ovary (either at its base, along the inner wall, or on an axis running down the center of the ovary). Depending on the species, there may be from one to several hundred ovules within the ovary.

There is as wide a variation in the reproductive structures of flowers as there is in the colors and shapes of their blooms. Many plants have both stamens and a pistil within the same flower; others, such as the oaks, squash, and corn, have these structures in separate flowers on the same plant. Still other species have separate plants for stamens and pistils. Willows and cottonwoods are of this type.

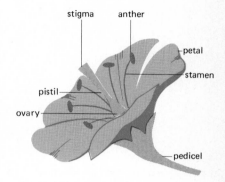

FIGURE 11-20
The floral parts of a complete flower.

Pollen formation

If you examine a cross section of the developing anther of a large stamen, such as that of a lily or a tulip, you can clearly see four chambers, or **pollen sacs** (Figure 11-21). During development of the flower, each pollen sac is filled with cells having large nuclei. These are known as **microspore mother cells,** and they contain the diploid number of chromosomes. As the anther grows, each microspore mother cell undergoes meiotic divisions, forming four **microspores.** They contain the haploid number of chromosomes.

Each microspore develops into a pollen grain. First, the nucleus divides by mitosis, forming two haploid nuclei. One is designated as a **tube nucleus** and the other as a **generative nucleus.** The wall of the microspore thickens and becomes

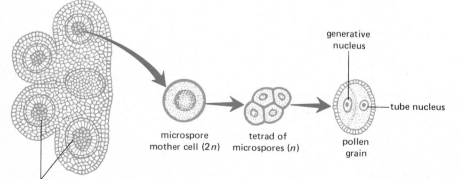

microspore
mother cell (2*n*)

tetrad of
microspores (*n*)

generative
nucleus

tube nucleus

pollen
grain

pollen sacs

FIGURE 11-21
Development of a pollen grain within the pollen sac of an anther (n = haploid, 2n = diploid).

the protective covering of the pollen grain. At about this time, the anther ripens and the walls of the pollen sacs burst open, spewing out the pollen.

Development of the ovule

While pollen grains are forming in the anthers, changes are occurring in the ovary at the base of the pistil. For the sake of simplicity, these changes will be described as they occur in an ovary that contains a single ovule, like the avocado pear. Keep in mind that the ovaries of many flowers contain several ovules.

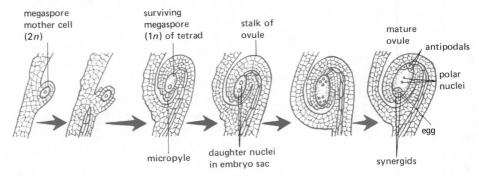

An ovule appears first as a tiny knob on the ovary wall. This swelling contains a single **megaspore mother cell** (Figure 11-22). The megaspore mother cell contains the diploid number of chromosomes. As the ovule grows it is raised from the ovary wall by a short stalk through which nourishment is received. One or two protective layers form around the ovule and enclose it completely except for a tiny pore, the **micropyle,** which is usually on the lower side of the ovule.

The megaspore mother cell undergoes two meiotic divisions, resulting in a row of four haploid **megaspores.** Of the four megaspores, one survives, and the other three disintegrate. The surviving megaspore is usually the one farthest from the micropyle. It rapidly enlarges and forms an oval **embryo sac,** in which further development occurs in the following steps:

1 The megaspore nucleus divides by mitosis, forming two daughter nuclei. Two additional mitotic divisions result in eight haploid nuclei.
2 Four nuclei migrate to each end of the embryo sac.
3 One nucleus from each group of four, known as a **polar nucleus,** migrates to the center of the embryo sac.
4 Each nucleus in the groups of three at either end of the embryo sac is enclosed by a thin membrane.
5 One of the cells nearest the micropyle enlarges and becomes the **egg.** The other two cells in this group are known as **synergids.**
6 The three cells farthest from the micropyle are known as **antipodals.**

The ovule is now ready for fertilization. However, only the egg and the polar nuclei will be involved. The synergids and the antipodals are short-lived and have no apparent function. Before fertilization can occur, a pollen grain must be transferred to the stigma of the pistil by one of the various agents of pollination.

Pollination

Pollination is the transfer of pollen from an anther to a stigma. In some plants, pollen is transferred from anther to stigma in the same flower or to the stigma of another flower on the same plant. This is **self-pollination.** If flowers on two separate plants are involved, the process is called **cross-pollination,** and this requires an outside agent. Among the agents of pollination are insects, wind, and water.

Chief among the insect pollinators are bees. But moths, butterflies, and certain kinds of flies visit flowers regularly and, in so doing, cross-pollinate them. Insects seek the sweet nectar secreted deep within flowers by glands at the base of the petals. The plump hairy body of the bee makes it an ideal agent of pollination (Figure 11-23). In reaching the nectar glands the bee picks up pollen as it rubs against the anthers. These are usually located near the opening of the flower. When the bee moves to the next flower, some of the pollen is sure to rub off on the sticky stigma of the pistil, while a new supply is brushed from the anthers onto the bee's body.

Brightly colored petals and sweet odors aid insects in locating flowers. Stripes along the petals of certain flowers may serve as guides (like the lights along a runway guide an airplane in for a landing).

At least one bird must be included in a discussion of agents of pollination. Tiny hummingbirds feed on the nectar of certain flowers. Their long bills, encasing needle-like tongues, reach down to the nectar glands while the bird hovers over the flower. With its flashing movements, the delicate bill is an instrument of pollination.

The flowers of wind-pollinated plants are usually less striking than the flowers of those pollinated by insects (Figure 11-24). Wind-pollinated flowers generally bloom in dense clusters near the ends of branches. As a rule, petals are lacking, and the flowers seldom have any nectar. Frequently, the stamens are long and produce enormous quantities of pollen. The pistils are also long, and the stigmas are large and often feathery to catch pollen grains that are blown about by the wind. As sufferers of hay fever well know, wind-pollinated plants (including cottonwood, willow, corn, oats, and ragweed) literally fill the air with pollen when their stamens are ripe.

FIGURE 11-23
The bee is an ideal pollinator because of its plump, hairy body.

FIGURE 11-24
The pollen-producing flowers of the giant ragweed. The pollen of this wind-pollinated plant causes hay fever in people who are sensitive to it.

Fertilization

Once a pollen grain lodges on the surface of a stigma, a **pollen tube** forms which penetrates the surface of the stigma by enzymatic action (Figure 11-25). As the

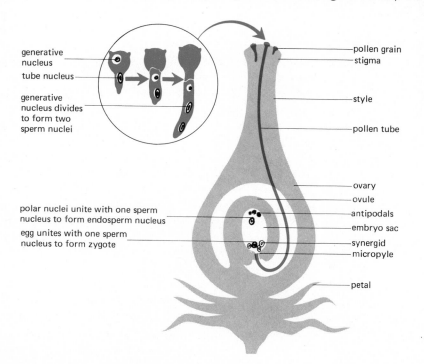

generative nucleus

tube nucleus

generative nucleus divides to form two sperm nuclei

polar nuclei unite with one sperm nucleus to form endosperm nucleus

egg unites with one sperm nucleus to form zygote

pollen grain
stigma

style

pollen tube

ovary
ovule
antipodals
embryo sac
synergid
micropyle

petal

FIGURE 11-25
Growth of the pollen tube and double fertilization.

tube lengthens, it grows through the soft tissue of the style and eventually reaches the micropyle of the ovule. The pollen tube may have to grow only a few millimeters (as in the pea) or over a hundred millimeters (along a corn silk). Rates of growth of the tube vary from a few millimeters a year (in some oaks) to a few millimeters in an hour (in corn).

Recall from the discussion of pollen formation that each pollen grain contains two nuclei: a generative nucleus and a tube nucleus. As the generative nucleus moves into the pollen tube, it divides and forms two **sperm nuclei,** which are male gametes. After passing through the micropyle, the pollen tube "digests" its way through the thin wall of the embryo sac. The tip of the tube ruptures, and the two sperm nuclei enter the embryo sac. Meanwhile, the tube nucleus degenerates.

Within the embryo sac one of the two sperm nuclei fertilizes the egg, forming a zygote. Since both the egg and sperm are haploid, fertilization results in the diploid number of chromosomes within the zygote, just as it does in animals. While the zygote is being formed, the second sperm nucleus unites with *both* polar nuclei to produce an **endosperm nucleus.** Since the polar nuclei and the sperm nucleus each contain the haploid chromosome number, the endosperm nucleus is *triploid* —it contains three of each type of chromosome.

Immediately after this double fertilization (the union of the egg and first sperm nucleus and the fusion of the polar nuclei with the second sperm nucleus) rapid cell division and tissue growth begin within the ovule. The zygote undergoes an orderly development into the embryo plant. Meanwhile, the endosperm nucleus gives rise to a mass of tissue known as the **endosperm** of the seed, which is stored food for the embryonic plant. In some seeds, the nutrients in the endosperm are absorbed by the embryo while the seed is developing. In others the endosperm remains as a part of the seed at maturity.

From flower to fruit and seed

Fertilization brings a sudden end to the roles of most flower parts. As the petals and stamens wither, the plant's energies are turned to the development of the ovules and the surrounding ovary. The ovary becomes the **fruit,** while the ovule(s) become **seed**(s).

Fruits are as variable in structure as flowers. Figures 11-26 and 11-27 show some of the common fruits according to structure. As you can see, a fruit need not be fleshy, like an apple, a peach, or pear. A kernel of corn, an oak nut, and a bean pod are just as much fruits as the fleshy, juicy type. Thus, the biological meaning of the term fruit is quite different from the meaning used in a grocery store.

Seed dispersal

Consider for a moment what would happen if seeds all fell to the ground and started to grow close to the parent plant. In a short time, the parent would be surrounded by seedlings that would compete for a limited supply of nutrients, water, and light. Few, if any, of the offspring would survive to maturity. Nature has avoided this dilemma by giving seeds modes of dispersal.

In some cases seeds are spread by mechanical dispersal. The pods of beans and peas twist as they ripen, due to changes in humidity. This puts a strain on a pod, which bursts open suddenly, spewing its seeds away from the parent plant. When the fruits of the touch-me-not plant become ripe, they open upon the slightest touch and curl rapidly upward, throwing their seeds several feet. In some fruits, including those of the poppy, holes form in the stigma. When the ripened fruit

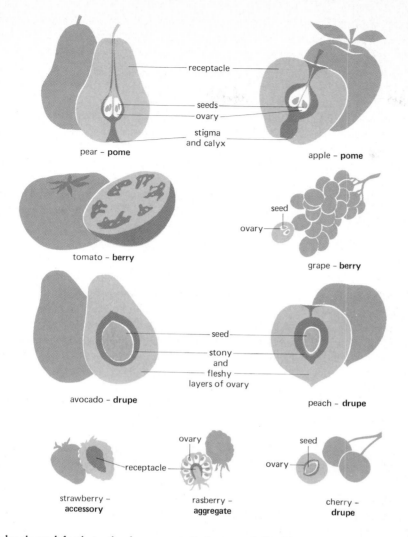

receptacle

seeds
ovary

stigma
and calyx

pear – **pome**

apple – **pome**

tomato – **berry**

seed
ovary

grape – **berry**

seed
stony
and
fleshy
layers of ovary

avocado – **drupe**

peach – **drupe**

receptacle

ovary

seed
ovary

strawberry –
accessory

rasberry –
aggregate

cherry –
drupe

sways back and forth in the breeze on its long and flexible stem, seeds sift out, as if from a salt-shaker.

Animals play a role in the dispersal of certain seeds. The tasty tissues of the apple, grape, and blackberry serve as a sort of biological bribe. Birds and other animals ingest seeds along with fruits. Because the cellulose coatings of the seeds cannot be digested, they pass unharmed through the digestive tracts of the animals, and are distributed in excrement (an excellent fertilizer for a developing seedling).

Animals aid in seed dispersal in another way. Many plants produce fruits with stickers or spines that cause the seed to cling to fur. If you have ever removed

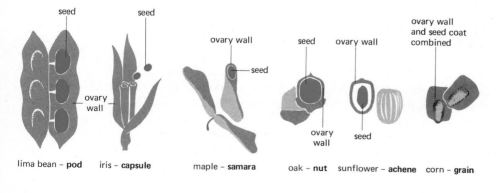

seed

seed

ovary wall

seed

seed

ovary wall

ovary wall
and seed coat
combined

ovary
wall

ovary
wall

seed

lima bean – **pod**   iris – **capsule**   maple – **samara**   oak – **nut**   sunflower – **achene**   corn – **grain**

CELLULAR AND ORGANISMIC REPRODUCTION   **273**

beggarticks, stick-tights, or burdocks from your clothes while hiking in the woods, you have played a part in seed dispersal.

Water is an agent of dispersal for many seeds. The coconut palm, for instance, lives close to the seashore and its large fruits may be caught up in the surf. The thick, stringy husk of the coconut is waterproof; thus it can travel unharmed across several hundred miles of ocean.

Certain grasslike plants, known as sedges, also drop their fruits into water. They, like the coconut palm, are generally found along the shores of oceans or the banks of rivers and streams where their seeds have found a foothold on land.

The wind is another agent of seed dispersal. When the milkweed pod splits open, the wind whisks away its seeds, each of which is tufted with long, loose hairs. The seeds of the dandelion are distributed in a similar fashion. The winged seeds of maple, ash, elm, and pine whirl through the air like tiny helicopters.

Dormancy of seeds

Many seeds go through periods of inactivity before they germinate (sprout). Seeds that are produced in the fall, for example, normally lie dormant throughout the winter but germinate during the following spring or summer. Drought, cold, and heat are all potentially lethal to a delicate young plant; thus, germination doesn't begin until conditions for growth are favorable.

Certain seeds can lie dormant for many years and still remain viable. Lotus seeds discovered among the relics of a past civilization in India, for example, were induced to sprout after an estimated 1000 years of dormancy. On the other hand, some seeds, like the maple, must sprout almost immediately after falling from the tree or they lose their potential for life.

For **annuals** (plants whose life cycles are completed in a year) seeds are the only form in which the plants can survive cold winter months. Their period of dormancy extends from one growing season to the next. For **biennials** (plants whose lives span two years) and **perennials** (plants that survive for more than two years) seed production need not occur every year. Most biennials, in fact, do not produce seeds during their first year of growth.

Whether or not a seed survives dormancy depends to a great extent on environmental conditions. Gardeners know that seeds that are stored in dark, cool, dry places (a basement, for example) have a better chance of surviving than those which are exposed to light, warmth, or moisture. Warmth and moisture may induce germination prior to planting or cause fungus infections to develop. Prolonged exposure to light can kill the embryo.

Germination

For germination, most seeds require moisture, a fairly warm temperature, and oxygen. The conditions for germination, however, vary widely from species to species. Seeds of many water plants, for instance, must be totally submerged in order to sprout. But seeds of most garden plants rot if they are kept too damp.

A temperature between 16°C and 27°C is usually the best for germinating garden seeds. Wild plants, however, may have different temperature optimums, depending on their native climates and life cycles. Maple seeds require cold weather to germinate; they will sprout on a cake of ice.

During germination the cells of a seedling divide rapidly, and this increased activity requires a high rate of respiration. Therefore, the oxygen supply to a seedling is critical. Garden soil should be loose for good ventilation.

Much of the food stored in a seed is starch. As seeds sprout, this starch is changed to sugar through the action of an enzyme known as **amylase,** and the sugar is used by the cells of the growing embryo. This change accounts for the sweet flavor of sprouting seeds and explains why sugar can be extracted from sprouting grain (malt). In the process of making beer this sugar is fermented to produce alcohol. Normally, the sugar goes through the Krebs cycle to make ATP.

A seed consists of an embryonic plant, stored food, and protective seed coats. The stored food nourishes the young plant from the time it starts to grow until it can produce its own nourishment by photosynthesis. In some seeds nutrients are stored in one or more thick "seed leaves," the **cotyledons.** You may have seen cotyledons on the stems of green beans shortly after they have pushed through the soil. They are thick protuberances beneath the first foliage leaves, which last for only a few days before they wither and fall off.

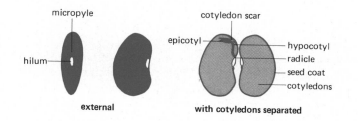

FIGURE 11-28
The structure of a bean seed.

In beans all of a seed's nutrients are stored in the cotyledons (Figure 11-28). In some other plants, however, much of the food is stored in the endosperm. A kernel of corn, for example, has starch and protein in the endosperm, while the cotyledon contains oils and protein.

Seedcoats, which develop from the wall of the ovule, protect the seed and prevent it from drying out. Usually there are two seed coats, but some seeds have only one. The outer coat is generally tougher and thicker than the inner one.

EARLY DEVELOPMENT OF FLOWERING PLANTS

The following sections consider the structures and early development of seeds of the two classes of flowering plants. The bean is a representative dicotyledon, or **dicot** (each bean contains two cotyledons); and corn is a **monocot** (a kernel of corn contains a single cotyledon). Monocots and dicots also differ in other aspects, such as the number of flower parts.

Structure and germination of a bean

A bean is usually more or less kidney-shaped. The hard, smooth outer seed coat (testa) may be white, brown, red, or mottled in color, depending on the species. The **hilum,** an oval scar on the concave side, marks the place where the bean was attached to the wall of the pod. Near one end of the hilum is the tiny micropyle through which the pollen tube grew. The inner seed coat of a bean is a thin, white tissue that adheres to the outer seed coat. Both of these coats develop from the wall of the ovule.

If you soak a dried bean and remove the seed coats, the cotyledons will separate easily. They fill most of the space within the seed coats and are fleshy (not at all leaf-like). Lying between the cotyledons are the other parts of the

FIGURE 11-29
The structure of a kernel of corn.

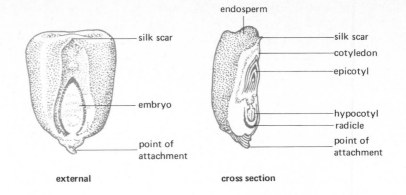

external

cross section

embryonic plant. A fingerlike projection, the **hypocotyl** (Figure 11-28) fits into a protective pocket. At one end of the hypocotyl is the **radicle,** the embryonic root; at the other end is the **epicotyl,** two tiny leaves folded over each other. Between the leaves of the epicotyl is the bud that will later form the plant's terminal (or tip) bud as a shoot develops. The hypocotyl, radicle, and epicotyl develop rapidly as they receive nourishment from the cotyledons during germination.

Figure 11-30 illustrates the stages in the germination of a bean. After the bean has absorbed water and its seed coats have softened, the hypocotyl bursts out. It breaks ground, pulling the cotyledons after it. Once the cotyledons have cleared the surface, the developing stem straightens and the cotyledons fold open, exposing the minute leaves of the epicotyl. These unfold, forming the plant's first foliage leaves. These are true leaves, and, if they are not eaten by garden pests (they are tender and highly desirable to foraging insects), they will remain with the plant throughout its life.

FIGURE 11-30
Germination of a bean seed. The testa is the outer coat of the cotyledon.

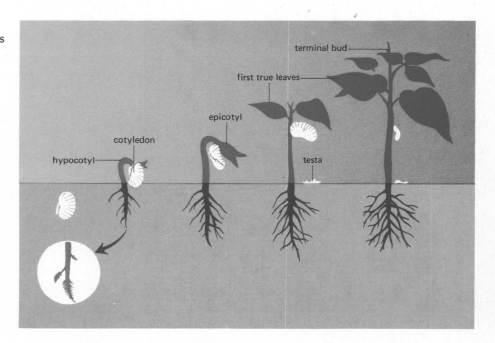

The stem lengthens rapidly, developing more leaves, and the small bud between the epicotyl leaves develops as the terminal bud of the plant. The cotyledons remain attached to the stem for a time, below the true leaves. But, as the nourishment within them is depleted and the plant becomes able to sustain itself by photosynthesis, the cotyledons wither and fall off. Like corn and most other vegetables, beans are annuals; they die at the end of a single growing season, after bearing fruit.

Structure and germination of corn

Each kernel of corn is a complete fruit. Therefore it corresponds to an entire pod of beans rather than to an individual bean seed. However, there is only one seed in each kernel. It completely fills the fruit, the outer coat of the kernel having been formed from the flower's ovary wall. The point of attachment of a kernel to the cob corresponds not to the hilum of a bean, but to the stalk by which a bean pod is attached to the plant. It is the pathway through which the developing fruit receives nourishment.

On one side of a grain of corn near the bottom is a light-colored, oval area that marks the location of the embryo. This is plainly visible through the fruit coat. Near the top of the kernel, on the same side as the embryo, is a tiny point, the **silk scar,** where the stalk of the pistil was attached.

If you cut a kernel of corn lengthwise through the region of the embryo, the internal parts can be clearly seen, especially if the cut surface is treated with a solution of iodine. The endosperm fills much of the seed. It contains starch, which turns purple when it reacts with iodine. "Field corn" stores only starch in the endosperm; whereas "sweet corn" stores sugar in addition to starch. The flavor of sweet corn is implied in its name; the starch in field corn makes it mealy and bland in taste to us.

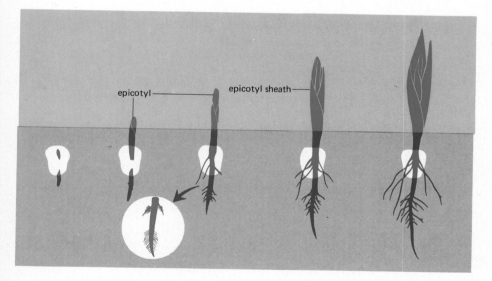

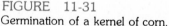

FIGURE 11-31
Germination of a kernel of corn.

The embryo, consisting of a very small hypocotyl, an epicotyl, a cotyledon, and a radicle, lies on one side of the kernel (Figure 11-31). The radicle points downward toward the point of attachment and is surrounded by a protective cap. The epicotyl is also protected by a cap. The leaves of the epicotyl are rolled (not folded as they are in the bean) into a compact spear. This is typical of all monocot seeds. During germination the cotyledon absorbs food from the endosperm and passes it on to the developing embryonic plant, which pushes its sheaf of tightly rolled leaves through the surface of the soil. Once through the surface, the leaves unfurl as the stem grows upward to form the cornstalk.

CHAPTER 11: SUMMARY

1 Cell size varies from about 1 micrometer to over 15 centimeters.

2 Spermatogenesis is the production of sperm cells.

3 Oogenesis is the production of egg cells.

4 Sexual reproduction involves the union of gametes.

5 Some forms of cellular reproduction are budding, binary fission, mitosis and meiosis.

6 Development of certain animals occurs from unfertilized eggs—the process of parthenogenesis.

7 Some plants develop by vegetative (asexual) reproduction.

8 Sexual reproduction in flowering plants involves the movement of pollen from anther to stigma and subsequent fertilization.

9 Seeds are dispersed in a variety of ways, and may lie dormant before they germinate or grow into a plant.

Suggested Readings

Flanagan, Geraldine Lux   *Window into an Egg,* William R. Scott, Inc. New York, 1969—Excellent photographs of the day-by-day development of a chick embryo.

Grant, V.   "The Fertilization of Flowers," *Scientific American,* June, 1961—The adaptive features of flowers and the correlated adaptations of pollinating animals.

Haldane, J. B. S.   "Some Alternatives to Sex," *New Biologist,* volume 19, 1955—A noted biologist discusses reproductive processes not involving sexuality.

Jenkins, Marie M.   *Embryos and How They Develop,* Holiday House, New York, 1975—A survey of embryology, covering developmental processes from initial stages to adulthood.

Ray, Peter   *The Living Plant,* 2nd Ed., Holt, Rinehart and Winston, New York, 1972—Includes a good chapter on plant development.

Salisbury, Frank B.   *The Biology of Flowering,* Natural History Press, New York, 1971—How flowers do it.

SWITCHING SEXES CAN SAVE SPECIES
Jack Horn

Human sex change is a surgical affair, requiring ongoing administration of sex hormones to maintain the switch. For some species, however, sex change is part of the normal course of life.

The ability to change sex in midlife has its advantages, at least evolutionarily. Biologists Robert Warner, D. Ross Robertson and Egbert Leigh, Jr., show how it works in the case of the bluehead wrasse, a tropical fish. Young wrasse come in both sexes and in various color combinations; older wrasse have blue heads and green bodies separated by two black stripes, and are invariably male. All young females turn into males when they reach this final color phase.

What makes this arrangement worthwhile for the bluehead wrasse is its mating style. Nearly all fertilization is done by the larger, older males who set up temporary territories on the reef and wait for females to drop by. These females have a strong preference for the largest males, and usually won't give the small males a tumble even if their first choice is busy.

The largest males sometimes mate as often as 100 times a day; their younger fellows are reduced to snatching the odd chance by hanging around a popular territory. More commonly, young males cluster in groups of hundreds, occasionally descending *en masse* on a passing female for a finny orgy.

With the odds stacked this way, an individual fish's reproductive potential, and therefore the species' chances for survival, is best served by being a female early in life and a male later, when he's bigger and can successfully defend a mating territory. The reason all young bluehead wrasse aren't females, according to the researchers, is a matter of available partners.

When large numbers of females arrive at the spawning site simultaneously, even small males have a good chance of fertilizing one or more, since the larger males are overbooked. In this situation, mating rates for young males and females are about the same—one a day—so being a young male is just as profitable as being a young female. However, among wrasse species that normally live in small groups, young males stand little chance of scoring, and females predominate.

Although sequential hermaphroditism works well for the wrasse, the idea hasn't taken hold among mammals, even in species in which older, bigger males monopolize the females, such as deer, baboons and seals. Warner *et al.* explain that sex change is physiologically difficult in animals who fertilize and carry their young internally or bear large eggs. Further, sex determination in birds and mammals may be too rigid for sex change. Finally, experience, as well as size, plays a role in an older mammal's dominance. A male baboon who misspent his youth as a female wouldn't have this advantage.

MATING, REPRODUCTION, PAIR-BONDING, AND BROOD-TENDING
Wolfgang Wickler

Why do certain animals form lasting pair bonds? Do organisms always mate in order to reproduce? Such questions are considered in the following excerpt from *The Sexual Code.*

Is it true that animals and humans mate in order to reproduce? Or that parents stay together because this facilitates brood-tending? We can determine whether these statements are in fact laws of nature by observing nature. A survey of conditions among very primitive living things tells us about primeval conditions.

Mating among unicellular animals takes two different forms. Either the copulating partners fuse, or they lay themselves one against the other, exchange parts of the cell nucleus, and then separate again. The biological significance of this process is that the genes, which are always slightly different between individuals, are mixed anew, producing new variations. These slightly divergent variants form the basis for further developments. The same applies in the reproductive behavior of higher animals who recombine their genes. With unicellulars, however, it is to be noted that mating has nothing to do with reproduction; on the contrary, when the partners have fused there are only half as many individuals left as before. When unicellulars reproduce, it is without sexual activity, by division into two (fission). We also find reproduction without sexual activity among fairly highly developed animals, who do not simply divide into two. This is called parthenogenesis (i.e., development of ovum without fertilization into new individual). New offspring develop from the ova of the female without the intervention of a male. Many polyps and various small crustacea such as the water flea (*Daphnia*) reproduce in this manner. So mating and reproduction can coexist without any necessary connection. The animals we have just named reproduce without mating as long as the environment is favorable; then they produce as many offspring as they can as fast as they can, to make use of the good nutritional possibilities. Reproduction becomes sexual when living conditions deteriorate, for instance when the available space is exhausted. When unicellular animals or crustacea mate, reproduction ceases for a considerable period of time. That is to say, in this case mating and reproduction, each serve a different function: reproduction serves the extension and survival of the species, while mating serves to increase the number of variations within the species and provides the basis for the development of new types and even of new species.

Sexual reproduction is not always tied to mating among vertebrates either. The male water newt deposits a packet of sperm on the lake or pond bottom and the female comes and picks it up with her cloaca. . . . So mating and sexual reproduction are also independent in a number of cases. Many species of fish who discharge eggs and sperm anywhere and abandon them to their fate nevertheless live in permanent monogamy; examples are many well-known butterfly fish (*Chaetodon*) of the tropical seas, are familiar to all aquarium owners. Since, on the other hand, all animals that mate are certainly not all monogamous, monogamy and mating are also independent. Moreover, both can occur independently of brood-tending. We already find brood-tending among the lower animals that bear offspring anonymously, without contact between the parents. Various starfish tend their brood on the body of the mother; in the case of *Lepasterias,* the offspring develop in the mother's body, as with many shellfish. *Stygiomedusa,* a colored deep-sea jellyfish, gives birth to live offspring ten centimeters long. But the monogamous butterfly fish do not concern themselves with the eggs or larvae they hatch, and some birds who no longer tend their brood, because they have become brood parasites, nevertheless live monogamously, like the African didric cuckoo (*Chrysococcyx caprius*), the cuckoo weaver (*Anomalospiza imberbis*), and probably the black-headed duck (*Heternonetta atricapills*).

What use, then, is monogamy in these cases? It achieves the same end as the fusion for life of the sexual partners among many lower animals (e.g., the double animal—*Diplozoon*—among sucker worms, and the blood fluke—*Schistosomum*—that causes bilharziasis). In fact, it still occurs among vertebrates, for example, deep-sea anglers (*Ceratias*). It means that the sexual partners do not have to search for each other again and again. Any search entails the risk of error, and if animals of different species mistakenly fuse their gametes, they produce hybrids who can be sterile or otherwise affected. More important is the fact that closely related species in particular, who could still bear fertile offspring with one another, are specialized in different ways, with respect to nutrition, habitat, or some other part of their environment. This different specialization allows them to make better use of existing conditions and prevents intraspecific competition. . . .

Such signals are also necessary in sexual reproduction, for if hybrids appear, the species will lose its specialization again. The animals must, therefore, be able to recognize a conspecific sexual partner. Here it is enough if one sex is different from species to species, so that the other sex can choose according to the differences. Among many animals, the female is occupied with brood-tending, so she would put herself and her offspring in great danger if she had bright markings. In these cases the males are more striking, and wear a display dress, differing for every species, while it is the female who chooses. The most attractive example of this phenomenon is in the emphasis on contrast among closely related species inhabiting neighboring territories. In places where the species occur side by side and there is a risk of mistaken identity, the recognition signals, songs, or markings contrast more strongly than at the other ends of the area of distribution where only one species occurs. Living things that do not choose and largely leave the meeting of the gametes to chance suffer from continual hybridization, which spoils any chance of different specializations and the formation of different species. . . . The display dress, which often differs so much from species to species, prevents mistakes in the choice of sexual partner. The more often an individual reproduces, the more often it will have to choose, and the more risk it runs of making a mistake, especially if the partners only meet briefly, copulate, and then go their way again. The longer they stay together, the easier it is for them to notice and correct an initial mistake. If they stay together permanently, they avoid the need for a new choice and the ensuing risk of error.

Comparisons between certain groups of animals have shown that permanently monogamous animals are indeed protected at least as well if not better from choosing the wrong mate than those of their relatives who do not form lasting pair-bonds. This applies even when the latter have evolved extremely conspicuous display dresses, differing according to species, as recognition signals. It is very clear in the case of tropical cichlids and birds of paradise, and it explains why the male and female of non-pair-bonding species have strikingly differentiated colors and clearly distinct display dresses, while the monogamous species do not as a rule. *Monogamy replaces display dress.* It prevents mistakes in mating and the ensuing waste of time and gametes; that is to say, it preserves the characteristics of the species. This is true of monogamy whether or not the parent animals tend their brood. So monogamy is functionally independent of brood-tending; but it can also serve the interests of brood-tending. One could describe the physical fusion of sexual partners as "physical marriage." But in a genuine, permanent marriage, the individuals remain mobile independently of each other; in a sense they grow together in their behavior while recognizing each other as individuals. A transitional step between the two forms is "local marriage" (*Ortsehe*). Here the partners attach themselves to the same locality or nest but not directly to each other. The Californian blind goby (*Typhlogobius californiensis*) spends its whole life in pairs in the channels a burrowing shrimp digs in the seabed; the male or female of the fish will drive off all rivals of the same sex but tolerate any partner of the opposite sex who has chosen the same habitat. One can replace either the male or the female by another at will. Similarly, the stork is more attached to its eyrie than to its partner: the male stork and female stork are "married" not to each other but each to the nest; they are faithful to a place but not to a partner.

Because mating, reproduction, and pair-bonding have entirely different functions and are largely independent from one another in nature does not, of course, mean that they cannot also be related to one another. The more highly developed animals have combined mating and reproduction and made pair-bonding subserve brood-tending. Eventually, brood-tending and mating are made to promote the interests of pair-bonding.

CHAPTER

12

Human Sexual Biology

Objectives

1 Learn the anatomy of the external and internal male and female sex organs.

2 Understand what controls the menstrual cycle.

3 Describe the development of the embryo and fetus.

4 Learn the stages of sexual response.

5 Define cloning.

6 Learn the various methods of contraception, their faults and relative advantages.

7 Describe the sexually related diseases.

HUMAN SEXUALITY

Humans exist at the peak of vertebrate evolution. Like our other physiological systems, our reproductive physiology is not much different from the sexual systems of our closest living evolutionary relatives, the other primates (apes and monkeys). But the fact that we are sufficiently self-aware to be able to control our reproductive capability, sets us far apart from other organisms. Modern technologies of contraception and abortion give us a mode of control over our lives which is probably inconceivable to other species.

Nevertheless, we have not taken complete control of our reproductive nature. In fact, in one way we seem to be losing what control we now possess. Venereal disease (VD) is rampant in our civilization, and though we are capable of killing most of the agents of VD that now exist, they possess amazing capacities of adaptive resistance to the drugs we use to stop them. New strains crop up which are not phased by treatments that are lethal to their predecessors. Certain viral agents of VD are resistent to every available medication.

This chapter considers human sexual biology at many levels — from reproductive physiology to erotic response, to contraception, to venereal disease.

THE MALE REPRODUCTIVE SYSTEM

As mentioned in the discussion of meiosis in Chapter 3, male reproductive cells, **sperm,** are prolifically produced within the tiny, convoluted **seminiferous tubules** of the male gonads, the **testicles.** The testicles, in turn, are suspended from the abdomen in a pouch of pliant flesh, the **scrotum.** By gently squeezing a scrotum, the testicles can be felt to be egg-shaped bodies, about two inches long. They are similar in firmness to a flexed muscle. Within the scrotum testicles are maintained at a temperature a couple of degrees below body temperature. This relatively cool environment is necessary for sperm to develop properly. If held too closely to the abdomen by tight clothes for long periods of time, the testicles may produce relatively inactive sperm.

In addition to producing sperm, the testicles produce the male sex hormone, **testosterone.** Testosterone stimulates sex drive and causes the male body to manifest **secondary sex characteristics,** including facial hair, broad musculature, and a lower pitched voice. (Eighteenth century male *castrato* singers had high voices because they were castrated before puberty.) Testosterone is produced by **interstitial cells,** which are scattered in the connective tissues outside the seminiferous tubules (Figure 11-6). Testosterone enters the bloodstream via the thin-walled capillaries that permeate the connective tissue.

Mature sperm pass from the seminiferous tubules into a collecting duct, the **epididymis** (Figure 11-6). An epididymis would be several yards in extended length, but it is folded back and forth upon itself to form a compact mass adhering to the surface of a testicle. Sperm are stored within the epididymises between ejaculations.

Ultimately, the epididymises straighten and form two ducts, the **vas deferens** (Figure 12-1 left), which rise into the abdomen. Within the abdomen the vas deferens loop over the bladder and form **ampullae,** which are storage vessels into which sperm move from the epididymises immediately prior to orgasm. Toward the ends of the ampullae, sperm are joined by fluid from the paired **seminal vesicles.** The secretion from the seminal vesicles stimulates sperm, activating whiplike tail movements.

After merging with the seminal vesicles, the ampullae enter the prostate gland, which is a muscular structure about the size of a plum. Sperm enter the

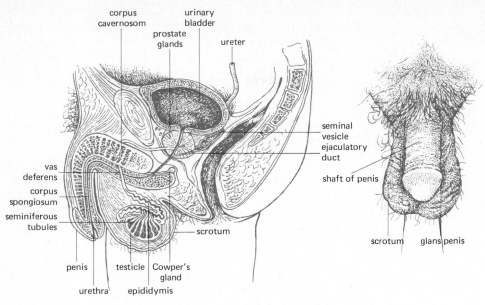

FIGURE 12-1

The male reproductive system (left) and the male genitalia (right).

prostate via two **ejaculatory ducts.** Secretions from the porous tissues within the prostate gland add to the **semen** (sperm plus the fluid bathing them). The prostate produces the bulk of semen, giving it a characteristic odor, milky hue, and slightly gelatinous consistency.

In addition to its secretory functions, the prostate gland serves as a two-way valve, allowing semen or urine to pass through it, depending on what the occasion calls for. During erection a sphincter muscle pinches the **urinary duct** (which joins the bladder to the ejaculatory ducts) closed. At the point of ejaculation, rhythmic contractions of muscles at the base of the penis pump the semen outward along the **urethra,** the single duct through the **penis** which serves in both ejaculation and urination.

In external structure the penis (Figure 12-1 right) has two distinct parts—a smooth acorn-shaped tip, the **glans penis,** and a shaft of thin, loose (in the absence of erection) skin. The glans penis and the portion of the shaft immediately behind the glans penis are the most erotically sensitive portions of the male anatomy. High densities of touch-sensitive and pressure-sensitive nerve endings are embedded close to the surface of the tip of the penis. These nerves set off impulses that reach pleasure centers within the brain via the spinal cord. Since some of the nerve endings within the penis are pressure-sensitive, there is no need for outside stimulation in order for an erection to be somewhat gratifying. However, in order to achieve orgasmic sensations, a certain amount of physical massage is necessary.

The penis contains muscles, blood vessels, nerves and three parallel tubes of spongy tissue (Figure 12-2). Two relatively large tubes, the **corpora cavernosa,** lie above the third the **corpus spongiosum,** which surrounds the urethra. During erotic arousal the three tubes engorge with blood by a rapid influx through the branching arteries permeating the penis. Simultaneously the veins of the penis constrict, and it becomes erect as it swells with trapped blood. Unlike the penis of many other mammals, the human penis does not contain a bone to help erection.

As it leaves the prostate gland and enters the penis, the urethra passes between two pea-sized organs of secretion, the **Cowper's glands.** The clear, sticky fluid that the Cowper's glands secrete into the urethra prior to ejaculation is basic; it neutralizes the acidity of the traces of urine left along the penile urethra, preparing

FIGURE 12-2

The penis.

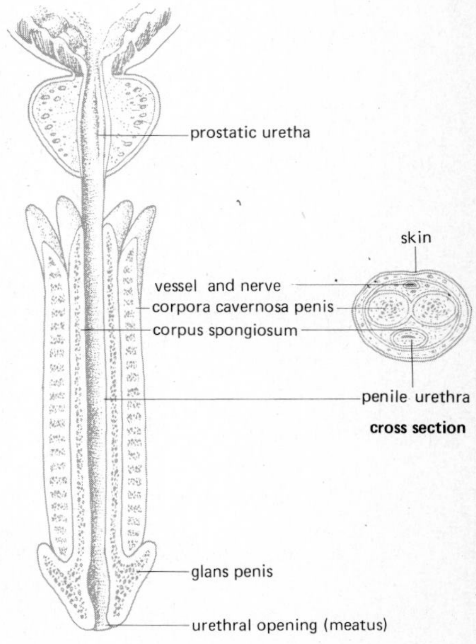

prostatic uretha

skin

vessel and nerve
corpora cavernosa penis
corpus spongiosum

penile urethra

cross section

glans penis

urethral opening (meatus)

the way for the acid-sensitive sperm. For practitioners of the "withdrawal" technique of contraception (**coitus interruptus**) the Cowper's glands have another interesting property. Sperm left within them following ejaculation may survive for many days. Thus, fertilization may be accomplished by one of the tens of thousands of active sperm deposited in the vagina in the few drops of fluid from the Cowper's glands that ooze from the **urethral meatus** (exit of the urethra) prior to ejaculation.

THE FEMALE REPRODUCTIVE SYSTEM

The female **genitalia** (external sex organs) are usually covered by two mounds of flesh, the **labia majora** or major lips (Figure 12-3). At times of erotic arousal, however, the tender inner lips, the **labia minora,** becomes engorged with blood and may protrude from the labia majora.

At the peak of the labia minora is an organ positioned somewhat like a penis. This is the **clitoris,** and its function is similar to the penis in that it is the center of female erotic arousal. Unlike the penis, however, the clitoris serves no direct reproductive function. It does, however, contains *corpora cavernosa* that engorge with blood during erotic arousal, and it also contains high densities of touch-sensitive nerve endings. Rhythmic stimulation of the clitoris sets off waves of pleasure-giving impulses along the neurons that connect the clitoris to the lower spinal cord and ultimately the brain—the most erotic organ.

Within the labia minora, an area termed the **vulva,** the meatus of the urethra lies between the clitoris and the vaginal orifice. Unlike the male urethra, the female urethra serves no reproductive function; it serves only in urination.

The vagina is a collapsible tube of membranous flesh that angles up and back, into the lower abdomen, where it is sandwiched between the bladder and the large intestine (Figure 12-4). During times of erotic arousal, the mucoid tissues lining

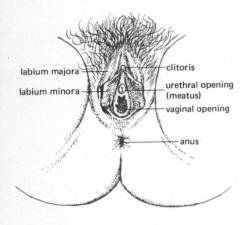

FIGURE 12-3
The female genitalia.

labium majora
labium minora

clitoris
urethral opening (meatus)
vaginal opening
anus

FIGURE 12-4
The female reproductive system.

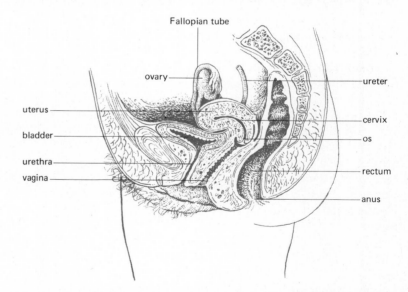

Fallopian tube
ovary
uterus
bladder
urethra
vagina
ureter
cervix
os
rectum
anus

the vagina secrete a lubricating fluid. The orifice and first third of the inner vagina, the **orgasmic platform,** are innervated with pleasure receptors similar to those in the clitoris. The innermost vagina has very few nerve endings, and hence, is relatively insensitive.

Above the vagina a muscular organ, the **uterus,** is suspended by ligaments within the pelvic bone (Figure 12-5). Normally the uterus is about the size and shape of a pear, but its volume increases by twenty-fold when a fetus develops inside it. The small end of the uterus, the **cervix,** protrudes through the roof of the vagina. It can be felt as a bump about the size of the tip of a nose. Through the cervix a pore, the **os,** connects the vagina with the uterine cavity. Prior to pregnancy the diameter of the os is about the same as that of a pencil lead. In order to allow the passage of a baby's head, however, it must increase to 10 cm across.

The female gonads, the **ovaries,** are suspended by ligaments at the sides of the uterus. The **ova** (eggs) that they periodically release are transported to the uterus by **Fallopian tubes.** Fringed funnels of diaphanous tissue at the ends of these tubes cling to the ovaries but are not attached. Currents set up by wafting cilia within the Fallopian tubes push the ova away. Like the testicles, the ovaries are not only producers of reproductive cells, they are also **endocrine glands** (hormone-secreting glands). The ovaries produce the female sex hormones, **estrogen** and **progesterone.**

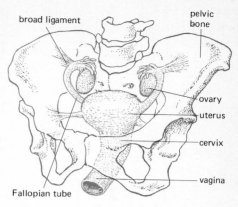

FIGURE 12-5
The female reproductive organs as they are situated within the pelvic bone.

The menstrual cycle

Approximately once per month an ovum develops and is released by one of the ovaries (normally the ovaries alternate in the production of ova). Underlying this process is a cyclical interplay of hormones between the ovaries and a deeply rooted anatomical part of the brain, the **pituitary gland.** The processes of the pituitary gland are partially under control of the brain and partially of hormones secreted from the various other endocrine glands distributed throughout the body. The female sex hormones are but two of the dozen or so hormones that are involved in the processes of the pituitary gland. The pituitary is under the control of another brain area called the hypothalamus which regulates many body functions.

In a controlled response to stimulation by estrogen and progesterone, the pituitary gland secretes hormones, known as **gonadotropins,** which modify the rates of secretion of the sex hormones. With the bloodstream as the messenger medium, a cyclical series of exchanges occurs between the pituitary gland and the ovaries. Concurrently the sex hormones stimulate changes within the uterus. Under stimulation by the sex hormones, the **endometrium** (the delicate inner lining of the uterus) prepares to receive a developing embryo. Thus, the ovaries, pituitary gland, and the uterus are involved in the cycle by which a new human life may begin (Figure 12-6).

Near the surface of an ovary, the mass of cells producing an ovum first forms a **follicle** (Figure 12-7). This portion of the menstrual cycle is initiated by **follicle stimulating hormone** (FSH), which is one of the gonadotropins secreted by the pituitary gland. As the egg reaches maturity, the follicle becomes filled with a fluid containing the sex hormone, estrogen. Ultimately the ovum breaks free of the surface of the ovary, and the follicle that contained it loses its original cuplike shape. The follicle buckles in on itself and turns yellowish in color. At this stage it is known as the **corpus luteum** (Latin for "yellow body").

Meanwhile, the estrogen that had entered the bloodstream from the follicle while the ovum was developing, depresses production of FSH but stimulates production of the second gonadotropin of the pituitary gland, **luteinizing hormone** (LH). At the corpus luteum, LH stimulates production of the second sex hormone, **progesterone.** At this point in the menstrual cycle the concentrations of both estrogen and progesterone in the bloodstream are relatively high. As these hormones course through the capillaries within the inner lining of the uterus, they stimulate the development of the delicate tissues of the endometrium. If fertilization of the ovum occurs and a developing embryo implants within the endometrium, the site of implantation will itself produce estrogen and progesterone. If fertilization does not occur, however, the endometrium breaks down as production of the sex hormones by the corpus luteum falls off. **Menstruation** is the subsequent expulsion of blood and bits of uterine wall tissue through the vagina.

FIGURE 12-6
The relationship of the pituitary gland in the brain to the uterine cycle.

FIGURE 12-7
A representation of a section through the human ovary. In stages 1-4 the ovum is shown maturing. In stage 5 ovulation is shown taking place. In stage 6 the corpus luteum is shown. Mature follicles may be several millimeters in diameter.

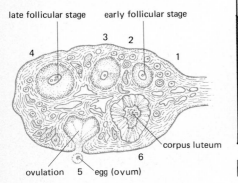

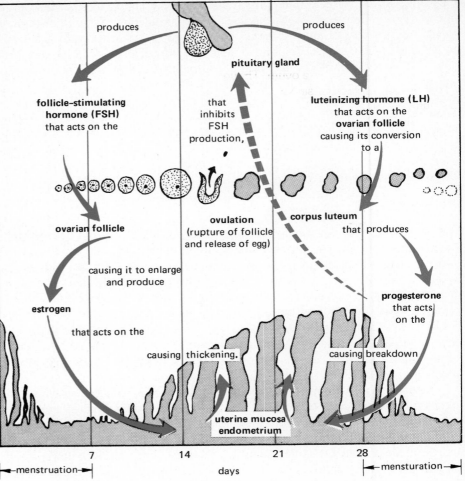

The menstrual cycle comes full swing when the pituitary responds to the depressed levels of estrogen in the bloodstream by stepping up its production of FSH. Under stimulation by FSH a new ovarian follicle begins development. In summary, the menstrual cycle consists of four stages (Figure 12-6):

1 Menstruation   A period of bleeding lasting about five days.

2 Follicle stage   Development of an ovum within an ovarian follicle occurs over a period of ten to fourteen days following menstruation.

3 Ovulation   This is the release of a mature ovum from a follicle.

4 Corpus luteum stage   The changes in the follicle, lasting from ovulation to menstruation—about ten to fourteen days.

The past couple of decades have brought the development and widespread use of contraceptive pills. There are various types, but they are all composed of female sex hormones, and all of them interfere with the menstrual cycle. "The pill," for example (which contains synthetic estrogen and progesterone), prevents the monthly release of an ovum. More is said about "the pill" and the other types of hormonal contraceptives in the section on birth control, after a consideration of the amazing sequence of events that occur after fertilization.

FERTILIZATION

Hundreds of millions of flailing sperm (Figure 12-8) are released by an average ejaculation. Mitochondria within their flagella produce the ATP necessary to power their tortuous swim to the ovum. Though it is a journey of only about 0.4 meter, it is a relative marathon, since sperm are on the order of 10^{-5} (0.00001) meter in length.

Given the opportunity to at least land on the labia minora, which become moist at times of erotic arousal, sperm have a chance to make it to the vagina, thence to the ovum. Thus, **intromission** (insertion of a penis into a vagina) isn't necessary for fertilization. However, if less than twenty million sperm are deposited within the vagina, fertilization is unlikely. The obstacles to fertilization are all but insurmountable.

Most sperm are either lost from the vagina when the penis is withdrawn or they perish in the acidic vaginal secretions. Only a small proportion of them make it through the tiny os, into the uterus. Once within the uterus, conditions are somewhat better; the slightly basic secretions of the endometrium are less harsh than the vaginal fluid. But, only a fraction of the sperm that make it to the uterus ever reach the orifice of the proper Fallopian tube, which is a small pore, a millimeter across. Usually only one ovum is released per month, and since the life spans of sperm and ova are but a few days, a trip up the wrong Fallopian tube is a dead end. If a sperm misses its one chance it degenerates.

The process of fertilization involves many sperm (Figure 12-9a) but the nuclear material from only one of them actually fuses with the nuclear material of the ovum (Figure 12-9b). Hundreds of tiny cells are left clinging to the surface

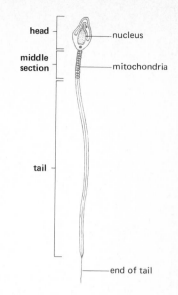

FIGURE 12-8

Anatomy of a sperm cell.

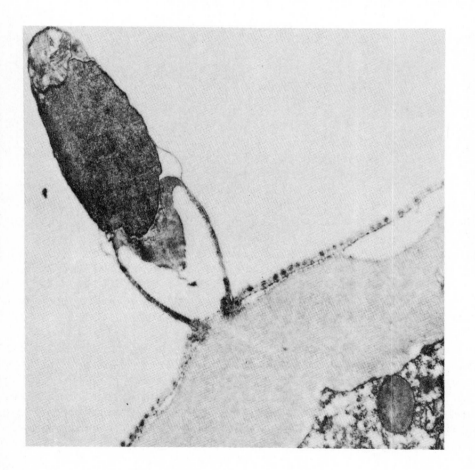

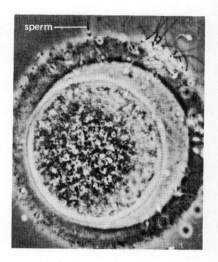

FIGURE 12-9a

A human ovum and sperms at fertilization.

FIGURE 12-9b

Electron microscope photograph of the initial contact of a sperm with an egg.

of an ovum when it ruptures from an ovary, and the hoarde of sperm which meets the ovum in the Fallopian tube prod at this coating and release digestivelike enzymes to break it down. Ultimately one sperm makes contact with the ovum itself. The plasma membranes of the sperm and ovum dissolve at the point of contact, the sperm surges inward, and the genetic materials merge.

EMBRYONIC AND FETAL DEVELOPMENT

As soon as an ovum is penetrated by a sperm, a membrane forms around it. This is the **fertilization membrane,** and it prevents other sperm from entering the newly formed zygote. The genetic constitution of the human-to-be is locked in for life, barring mutations from the outside.

Following the establishment of the diploid number of chromosomes, the zygote begins a series of mitotic divisions, and in a few days it consists of a conglomerate of many cells (Figure 12-10). This rapidly developing cell mass travels to the uterus in from three to five days, during which it is nourished by food stored in the ovum.

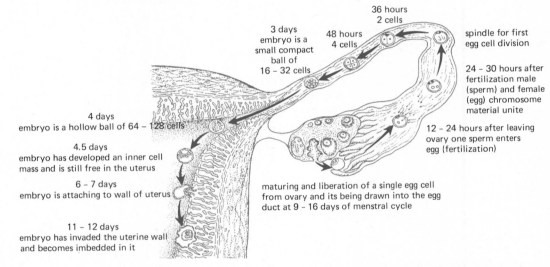

36 hours
2 cells

3 days
embryo is a
small compact
ball of
16 – 32 cells

48 hours
4 cells

spindle for first
egg cell division

24 – 30 hours after
fertilization male
(sperm) and female
(egg) chromosome
material unite

4 days
embryo is a hollow ball of 64 – 128 cells

12 – 24 hours after leaving
ovary one sperm enters
egg (fertilization)

4.5 days
embryo has developed an inner cell
mass and is still free in the uterus

6 – 7 days
embryo is attaching to wall of uterus

maturing and liberation of a single egg cell
from ovary and its being drawn into the egg
duct at 9 – 16 days of menstral cycle

11 – 12 days
embryo has invaded the uterine wall
and becomes imbedded in it

FIGURE   12-10
Fertilization and early embryonic development.

Repeated mitotic divisions result in a sphere of cells, the **blastocyst,** containing a fluid-filled cacity. As the blastocyst is gently wafted along the Fallopian tube, a mass of cells begins to develop on a portion of its inner wall. These cells will become the **embryo,** a term that refers to a developing organism whose body form has not yet acquired the characteristics that make it recognizable as a particular species. For human beings, the embryonic phase last about six to eight weeks following fertilization. The human embryo is practically indistinguishable from the embryos of other vertebrates, suggesting, of course, a similar evolution.

At the end of about six weeks the embryo is only about 15 millimeters long; but it rapidly grows, as human characteristics become obvious. From this time until birth, it is called a **fetus.**

Embryonic development   During the brief span of embryonic development, the spherical blastocyst undergoes an amazing metamorphosis, by which the rudiments of all the tissues of a human body begin to take form.

At about the same time the blastocyst grows into the endometrium (six to seven days following fertilization) the patch of cells within it begins to differentiate. The patch partially lifts away from the inner surface of the blastocyst and forms two **primary germ layers** of cells, the **ectoderm** and the **endoderm** (Figure 12-11a-c). The endoderm is the innermost layer, and the ectoderm is separated from

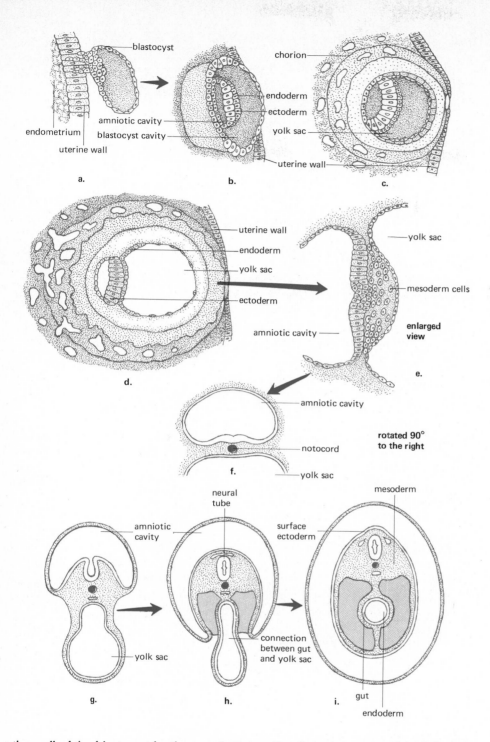

FIGURE 12-11
Early developmental stages of the human embryo, ending in the development of the nervous and digestive systems.

the wall of the blastocyst by the **amniotic cavity.** Over the course of the following weeks the amniotic cavity becomes larger. Eventually it almost completely surrounds the embryo, which develops from the ectoderm, the endoderm, and a third primary germ layer that forms between them, the **mesoderm** (Figure 12-11d,e). The fluid within the amniotic cavity acts as a "shock-absorber," protecting the embryo (and later the fetus) from being jarred. The structures of the body, which eventually develop from the three primary germ layers, are listed in Table 12-1.

Early in embryonic development, the endoderm forms a pouch, called the **yolk sac** (Figure 12-11c). For birds and other animals that hatch from eggs, the

TABLE 12-1   STRUCTURES FORMED FROM SPECIFIC PRIMITIVE GERM LAYERS

ECTODERM
    skin and skin glands
    hair
    most cartilage
    nervous system
    pituitary gland
    lining of mouth to the pharynx
    part of the lining of rectum
    adrenal medulla

MESODERM
    connective tissue
    bone
    most muscles
    kidneys and ducts
    gonads and ducts
    blood, blood vessels, heart, and lymphatic system

ENDODERM
    lining of alimentary canal from pharynx to rectum
    thyroid and parathyroids
    trachea and lungs
    bladder

yolk sac has a significant function; it fills with nutrients that will later be utilized by the embryo when it is cut off from contact with its mother. For human embryos the yolk sac (Figure 12-12) is small and serves little purpose, except that a portion of it becomes pinched off, by a process termed **gastrulation,** and eventually develops into the tissues of the digestive tract (Figure 12-11f-i).

Concurrent with the early stages of differentiation of the primary germ layers, the outer surface of the blastocyst, the **chorion** (Figure 12-11c) interacts with the endometrium to form an organ of exchange between the mother and the embryo. Many small, fingerlike projections from the chorion, called **chorionic villi,** grow into the endometrium. Enzymes produced by the chorionic villi dissolve capillary walls within the endometrium, forming blood channels which ultimately produce capillaries that are extensions of the embryonic circulatory system form. Normally there is no mixing between the bloodstreams of the mother and embryo, but nutrients pass into the embryo and wastes are eliminated through the thin, membranous **placenta,** which forms at the interface of the chorionic villi and endo-

FIGURE   12-12

Later development of the digestive organs.

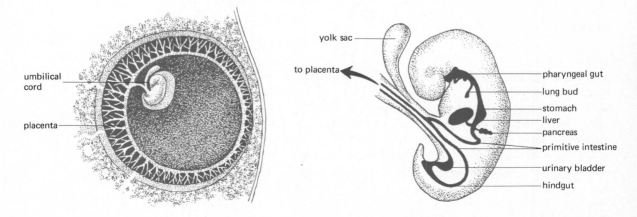

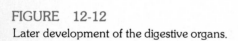

292

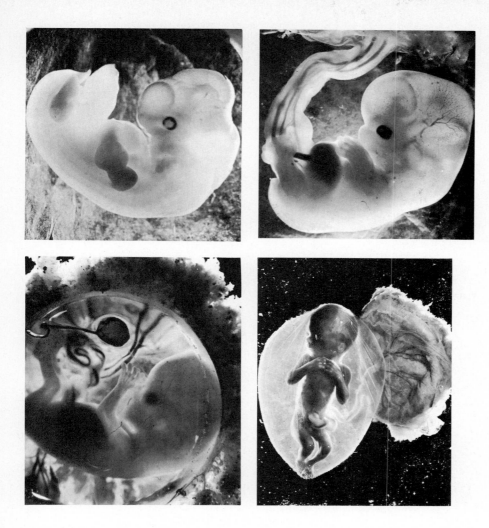

FIGURE 12-13
Development of the human embryo and
fetus. Starting above from left to right,
development has progressed for five
weeks, six and a half weeks, eleven weeks,
and sixteen weeks; note the progressive
development of the hands.

metrium. An **umbilical cord,** containing 2 arteries and 1 vein, joins the embryo
and the placenta (Figure 12-12).

The placenta is not only an organ of exchange between the mother and
embryo; it is also an endocrine gland that produces estrogen and progesterone.
High concentrations of estrogen and progesterone throughout pregnancy interrupt
the normal menstrual cycle. Thus, there is no menstrual bleeding during pregnancy.

Fetal development   By nine weeks after conception most basic human features
have appeared (Figure 12-13). At this stage the early **fetus** is only about 30
millimeters in length. Hands, facial features, and internal structures are recog-
nizably human but lack refinements.

The sexual system is the slowest organ system to develop. Rudiments of sex
organs are usually present by the seventh week of pregnancy, but they don't
become discernible as male or female for another month or so (Figure 12-14 top).
Sex organs develop in a process that involves degeneration of certain structures
and growth of others. Early male and female fetuses both have **Müllerian ducts**
(forerunners of Fallopian tubes) and **Wolffian ducts** (forerunners of sperm ducts).
In female fetuses Müllerian ducts develop while Wolffian ducts degenerate; the
reverse is true for males (Figure 12-14 bottom).

A related process results in the formation of homologous sex organs (which
develop from common embryonic structures). As described in Chapter 2, homologies have been found in the development of anatomical features of different species

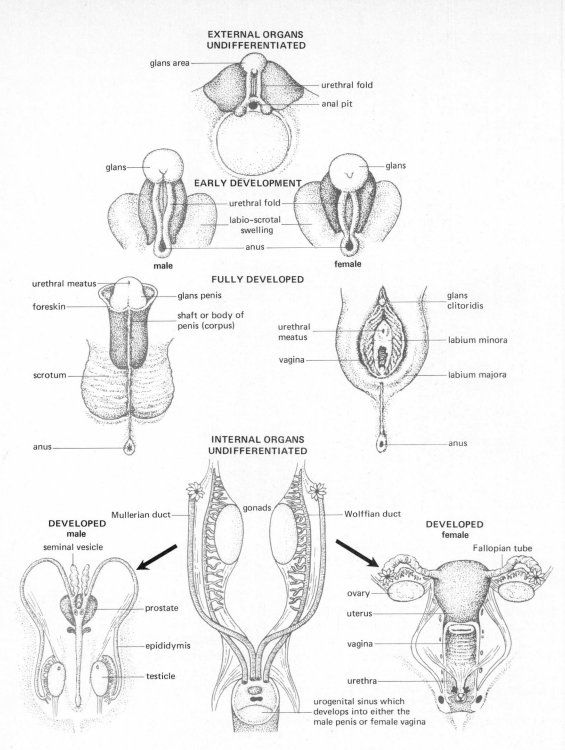

EXTERNAL ORGANS UNDIFFERENTIATED

glans area

urethral fold

anal pit

EARLY DEVELOPMENT

glans

urethral fold

labio-scrotal swelling

anus

glans

male

female

FULLY DEVELOPED

urethral meatus

glans penis

foreskin

shaft or body of penis (corpus)

glans clitoridis

urethral meatus

labium minora

vagina

scrotum

labium majora

anus

anus

INTERNAL ORGANS UNDIFFERENTIATED

Mullerian duct

gonads

Wolffian duct

DEVELOPED male

seminal vesicle

DEVELOPED female

Fallopian tube

ovary

uterus

prostate

vagina

epididymis

urethra

testicle

urogenital sinus which develops into either the male penis or female vagina

FIGURE 12-14

Differentiation of external sex organs (top) and differentiation of internal sex organs (bottom). (Adapted from an original painting by Frank H. Netter, M.D., from *The CIBA Collection of Medical Illustrations,* copyright © by CIBA Pharmaceutical Company, a Division of CIBA-GEIGY Corporation.)

(bird wings and human arms, for example, develop from a common embryonic structure). The same is true of the male and female sexual systems of the same species. For instance, early in the fetal phase, the undifferentiated gonads that may become ovaries or testicles are situated high in the abdomen. In women they shift down to the level of the upper edge of the pelvis. For men, they settle into the scrotum, with the Wolffian ducts trailing behind them. Other female/male homologies include the labia major/scrotum and clitoris/penis (Figure 12-14).

Sexual development

In recent years significant advances have been made in the sciences exploring genetic, biochemical, and physiological interrelationships during early sexual development. Most of the evidence has come from experiments with mammals other than humans (including rats, rabbits, hamsters, monkeys, and dogs). But, the mechanisms of sexual differentiation appear essentially the same for all mammals.

With delicate surgery, gonads have been removed from embryos before sexual differentiation (approximately six weeks after fertilization). Such embryos (even those that were chromosomally male, XY) subsequently developed into anatomical females. In other experiments testosterone has been injected into the bloodstreams of chromosomally female (XX) fetuses, with subsequent masculinization of their anatomies. Female fetuses masculinized in this fashion developed with oversized clitorises.

It has been suggested that a particular disfunction of the interaction of male and female sex hormones in fetal development may lead to homosexuality in adults. What causes the disfunction is not known. Preliminary studies have shown that male homosexuals have lower levels of testosterone than male heterosexuals while female homosexuals have higher levels of testosterone than female heterosexuals. It may be, then, that biochemicals and not the social environment determine our sexual preferences.

Embryonic and fetal development is a very delicate process. It can be upset by a variety of things, including drugs (such as aspirin, antihistamines, and various consciousness-altering drugs), which readily reach the embryo via the placenta. Exposure to X rays, poor nutrition, and infectious diseases can also disrupt fetal development. "German" measles (also known as "three-day" measles) is particularly dangerous. If contracted by a pregnant woman early in pregnancy, her baby may be born blind, deaf, mentally retarded, or crippled. Gonorrhea and syphilis can have similar effects. The first several weeks of pregnancy are the most critical (although the whole process is delicate), since during this period the organs and limbs begin development.

Childbirth

The period of fetal development ends with the birth of the infant approximately forty weeks after fertilization of the ovum (Figure 12-15). The smooth muscles of

FIGURE 12-15
Some 40 weeks after fertilization of an ovum, uterine contractions begin, and a baby is born.

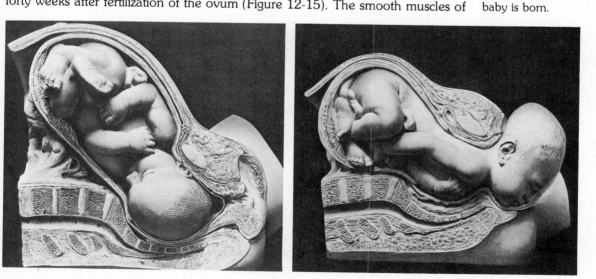

the uterus begin periodic, spontaneous contractions, which are out of control of the mother. An early sign that birth is imminent is the dislodging of the plug of slightly bloody mucus that has been blocking the cervix. Either concurrently or slightly later the amniotic membrane bursts, with the subsequent rush from the vagina of the watery amniotic fluid. Further contractions force the infant from the vagina. Shortly after the child is born, the placenta and remains of the amniotic sac (called "afterbirth") are expelled. At this time, the umbilical cord still attaches the baby to the placenta. The cord is tied and cut, leaving the naval as a scar.

CONTEMPORARY RESEARCH IN DEVELOPMENTAL BIOLOGY

An embryo goes through amazing changes in its development to a free-living organism, and most biologists believe the entire sequence is programmed by the DNA contained in the nucleus of a fertilized egg. One of the great unknowns in contemporary biology is how DNA acts to allow growth and development to proceed as it should. While organs are forming, certain genes controlling development must be active. But, when formation is completed, these genes must be somehow "turned off" and those needed for maintenance "turned on." How this process works, however, is unknown.

While the chemical mechanisms of development are a mystery, biologists have had a good deal of success in mimicking the environment of the process. Nature's place for human development is the uterus, but human embryos have been kept alive (and developing) for a few weeks in the laboratory. The biological problems of developing an artificial uterus in which an infant could develop to the same extent it does within the female body may even be surmountable.

Another line of research in developmental biology has resulted in the potential for producing genetic duplicates of any individual. Since the DNA in every cell of an organism contains all the genetic information necessary to construct its body, it is theoretically possible to develop an organism from any body cell. This has actually been done with certain animals and plants (Figure 12-16). Given the proper chemical and physical environment, a cell from the stalk of a tobacco plant, for instance, can be induced to divide and develop into a mature plant. Almost limitless numbers of genetically identical individuals could be produced by this asexual process of **cloning,** and theoretically, it is feasible for any species, including humans.

EROTIC RESPONSE

Obviously, sexuality is not all reproduction. Nature has imbued sexual intercourse with physical, erotic pleasure. Erogenous sensation is a sort of biological bribe to encourage reproduction. Through techniques of contraception we have been the only species clever enough to circumvent the reproductive process for the sake of pure erotic pleasure.

The past several years have brought an arousal of scientific interest in eroticism as a physical phenomenon. Among contemporary researchers of sexual physiology, William Masters and Virginia Johnson are preeminent. They have analyzed the orgasmic responses of hundreds of people of both sexes and found striking similarities and subtle differences in the sexual responses of males and females. According to Masters and Johnson, the sexual response patterns of both sexes occur in four phases: excitement, plateau, orgasm, and resolution. Of course, every individual experiences eroticism uniquely, and there are not abrupt

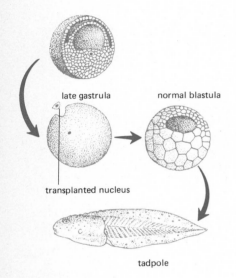

late gastrula

normal blastula

transplanted nucleus

tadpole

FIGURE 12-16
The development of a complete tadpole from a late gastrula cell to which a nucleus has been added. An unfertilized egg (with nucleus removed) is taken from a tadpole, and to it is added a nucleus from an intestine cell of another tadpole. With the proper chemicals added, a normal tadpole develops.

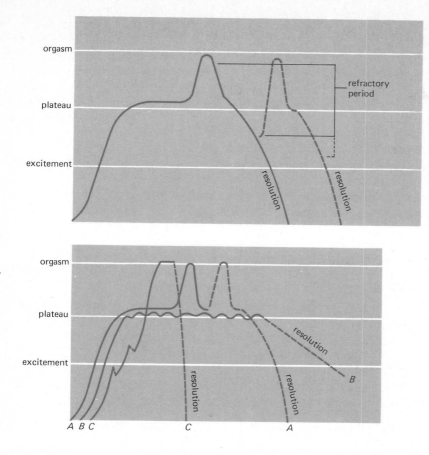

FIGURE 12-17
The male (top) and female (bottom) sexual response cycles. Three possibilities of female response (ABC) are indicated.

shifts between the phases of the sexual response cycle. Nevertheless in general the following characteristics are typical (Figure 12-17).

The female cycle

Excitement   Many parts of the body in addition to the genitals respond to sexual stimulation. **Myotonia,** an increase in muscular tension, is a generalized sexual response involving both voluntary and involuntary muscles. Heart rate and blood pressure increase during the excitement phase, and blood accumulates in certain blood vessels, a phenomenon known as **vasocongestion.**

Vasocongestion occurs in various parts of the body, but particularly the labia minora, clitoris, and vagina. Vasocongestion within the mucoid tissues of the vagina causes a "sweating" reaction, which lubricates the vagina in preparation to receive a penis. Pressure within vasocongested tissues sets off impulses along neurons leading to pleasure centers in the brain, and the caresses of sex play stimulate more frequent impulses. Vasocongestion may also develop in the capillaries just beneath the surface of the skin of the breasts, abdomen, and shoulders, where a **sex flush** (a blush) may appear, particularly on hot days. Vasocongestion within the breasts causes them to enlarge and the nipples to become erect. The clitoral glans, which contains the highest density of nerve endings of the female genitals, also enlarges due to vasocongestion.

Late in the excitement phase, muscular contractions within the pelvis cause shifts of the internal sex organs. The inner two-thirds of the vagina expands (a phenomenon called **tenting**), and the uterus rises up and back toward the spine. The vagina increases about an inch in length and two- to threefold in width.

Plateau   The first part of this phase is mainly a continuation of reactions that originated during the excitement phase. If effective sexual stimulation is maintained, myotonia and vasocongestion will intensify. During the plateau phase the clitoris becomes increasingly sensitive to touch, and it withdraws beneath the clitoral hood. Excessive stimulation of the clitoris during the plateau phase may be irritating; however, what is excessive for one woman may be extremely pleasurable for another.

The expansion that affected the inner two-thirds of the vagina during the excitement phase now extends to the outer one-third (the orgasmic platform). Intense vasocongestion within the orgasmic platform gives it a deep red or burgundy hue. A similar color change is seen in the labia minora.

During the plateau phase the uterus reaches its peak of elevation. As sexual tensions increase and orgasm becomes imminent, further increases in respiration, heartbeat, and blood pressure develop. Twitching and clutching movements of the hands and feet (**carpopedal spasms**) commonly occur during this phase. The plateau is the delicate balance between the peak of excitement and the release of orgasm. Much of the joy of sex is the titillation of the balance at this borderline.

Orgasm   Immediately before orgasm the physiological responses described for earlier phases are at their most intense level. The breasts are enlarged, with nipples erect. If a sex-flush has developed it is at its maximum extent. Only myotonic tension will become even more intense with orgasm.

Orgasm involves pleasurable, rhythmic contractions of the orgasmic platform. It begins with a long contraction, lasting from two to four seconds, and then a series of shorter contractions less than a second apart. The factors that initiate this response have not been identified. They may be neural, hormonal, muscular, or some combination of factors. The contractions of the orgasmic platform are accompanied by uterine contractions of a less definite pattern and often by contractions of the sphincter muscle of the rectum. The frequency of these rectal contractions usually is a function of the intensity of the orgasmic response. No particular reaction has been observed in the breasts or the clitoris during orgasm.

Until recent years it was a common fallacy that orgasm resulting from vaginal stimulation (as in normal intercourse) was physiologically distinct from that resulting from clitoral stimulation alone (as in masturbation). Sigmund Freud initiated the belief that in normal sexual development women shift from clitoral response to vaginal response. The studies by Masters and Johnson, however, refuted this view. They found that the physiological responses from any form of erotic stimulation were essentially the same.

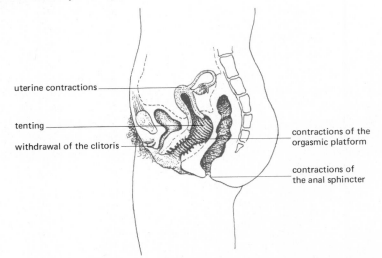

uterine contractions

tenting

withdrawal of the clitoris

contractions of the orgasmic platform

contractions of the anal sphincter

FIGURE   12-18
The physiology of female orgasm.

Resolution   This is a phase of return to the physiological conditions which preceded erotic stimulation. The return is fairly rapid if orgasm has occurred. The average time is about five to ten minutes. Myotonia, sex-flush, and the orgasmic platform subside quickly. The clitoris regains normal size and position. Changes resulting from vasocongestion disappear throughout the body. This process is much slower if plateau levels of tension were reached but orgasm did not occur. In this case, thirty minutes or more may be required. Complete resolution does not occur between cycles if more than one orgasm is attained over a relatively short period. Unlike men, who require a period of rest between orgasms, some women are capable of rearousal to several orgasms in rapid succession.

The female cycle of erotic response is illustrated in Figure 12-8.

The male cycle

Excitement   As in the woman, many parts of the male body besides the genitals respond to sexual stimulation. Breast enlargement is not part of the male cycle, although the nipples may become erect. When nipple erection does occur, it usually begins late in this phase. Other extragenital responses include increased heart rate and blood pressure. Elevated tension (myotonia) in both voluntary and involuntary muscles develops early in this phase.

Of course, the most obvious characteristic of male erotic arousal is erection of the penis. As previously mentioned, the penis becomes erect because of massive influx of blood (vasocongestion) into the spongy tissues within it. Concurrently, tightening of the muscles around the vas deferens draw the testicles upward (they may ultimately be drawn into the abdominal cavity before the response is over). As they rise, the testicles become larger due to vasocongestion. During the excitement phase a few drops of the clear fluid from the Cowper's glands, which contains active sperm, may ooze from the urethral meatus at the tip of the penis.

Plateau   The penis becomes fully erect during the excitement phase, except for the glans penis, which swells still further during plateau. The ridge at the back of the glans penis (which, for most men, is the most erotically sensitive portion of the penis) sometimes develops a purplish hue, due to intense vasocongestion. As heart rate, breathing rate, blood pressure, and myotonia increase, carpopedal spasms and a sex flush may develop.

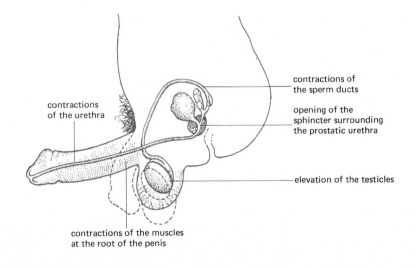

contractions of the urethra

contractions of the sperm ducts

opening of the sphincter surrounding the prostatic urethra

elevation of the testicles

contractions of the muscles at the root of the penis

FIGURE 12-19
The physiology of male orgasm.

Orgasm   With the exception of fluid discharge, the physiological responses of orgasm are essentially the same for men as for women. Orgasm for both sexes involves a general loss of muscular control and pleasurable contractions of the genital area.

Male orgasm is initiated by contractions of the seminal vesicles and vas deferens which compress semen behind a sphincter muscle that surrounds the urethra in the prostate gland. The sensation of this pre-ejaculatory phase, which lasts 2 to 3 seconds, is one of inevitability. The man knows that ejaculation is coming, and he can do nothing to stop it. Ultimately the sphincter opens, and semen is pumped outward by pulsating contractions of the muscles at the base of the penis. The contractions first come at 0.8 second intervals, but soon taper off. After two or three expulsive efforts there are a couple of slower, less forceful ones.

The total body response described for women also occurs in men. Just as women can voluntarily contract certain of their muscles to enhance erotic enjoyment, some men learn to heighten the intensity of orgasm by voluntarily contracting the muscles of the buttocks, abdomen, and anus before and during orgasm.

Orgasm and ejaculation are commonly considered to be inseparable for men, but the former can occur without the latter. Prior to the maturation of the reproductive organs during puberty, boys may experience orgasm without ejaculation.

Resolution   Following orgasm men have a **refractory period,** which begins as the nonexpulsive contractions of the penis muscles subside and lasts until sexual tension is reduced to excitement-phase levels. During this period (which may last less than ten minutes for young men but gradually increases with age) rearousal to orgasm is impossible. As previously mentioned, females have no refractory lapse and are capable of arousal to renewed orgasmic responses without falling below plateau-phase levels of excitation.

Although some men prefer bringing a quick end to erotic play following orgasm, prolonged sensuous (if not orgasmic) pleasures are also possible for them. Sensitivity to the unique sensual desires of one's mate is particularly important during the resolution phase. When a woman might prefer continued stroking of her genitals, for instance, her mate might like having his back massaged. The sensual desires of both partners may not always be satisfied at the same time.

The male cycle of erotic response is illustrated in Figure 12-19.

CONTRACEPTION

The idea of **contraception** (literally, against conception) is not new. People have tried to prevent pregnancy for thousands of years (there is evidence that the Egyptians, Greeks, and Romans practiced it). A highly abbreviated list of things that have been tried as contraceptive agents includes castor beans, parsley, rosemary, thyme, crocodile dung, tea made from gunpowder and foam from a camel's mouth. The last item on the list was popular for several centuries, but there is no reliable record as to how well it worked. Chances are, however, it was far inferior to contemporary concoctions, which are nearly 100 percent effective.

Oral contraceptives

The active ingredients in modern oral contraceptives are female sex hormones, or compounds closely resembling them. The earliest ones included only **progestins** (synthetically produced hormones resembling progesterone). Progestins have two contraceptive effects. Firstly they depress production of LH by the pituitary gland, creating a break in the normal menstrual cycle. Hence, ovulation ceases. Secondly, progestins cause the mucus produced within the uterus to thicken, thereby impeding penetration by sperm. Although there are several progestins, the singular term, progestin is generally used to refer to them.

Contraception is accomplished with progestin, but using it alone causes health problems, including nausea and abnormal uterine bleeding. Today, the so-called "minipill" utilizes only progestin, but many women cannot tolerate it. The most commonly used oral contraceptives of today contain estrogen as well as progestin. Health problems have been alleviated by this adjustment, but they have never been totally solved.

Combination pills (also known as "the pill") contain about 2 milligrams of progestin and a fraction of a milligram of estrogen per dose. Like progestin, estrogen administered out of phase with the normal menstrual cycle inhibits the usual interplay of hormones between the pituitary gland and the ovaries.

Recent British studies have revealed that formation of blood clots in the legs (thrombophlebitis), the lodging of blood clots in the lungs (pulmonary embolism), and heart attack are potentially fatal side effects of oral contraception. One study indicated that about 0.05 percent of women using oral contraceptives develop thrombophlebitis serious enough for hospitalization every year. For women not using oral contraceptives the annual rate is about 0.0005 percent.

According to a study published in the *British Medical Journal* of May, 1975, the threat of heart attack induced by "the pill" increases with age. Women in the 30–39 age bracket who use oral contraceptives were found to have fatal heart attacks at the rate of about 0.005 percent (the rate for nonusers is 0.002 percent). For women of ages 40–44 who use oral contraceptives, the rate of fatal heart attack is over 0.05 percent (versus about 0.01 percent for nonusers).

Apart from circulatory problems, oral contraceptives have been found to have various other ill effects, including breast pains and changes in size of breasts. Pain and tenderness are most common with pills high in estrogen, while high-progestin pills tend to cause enlargement. There has been a great deal of concern over whether oral contraceptives cause breast cancer. To date, statistical studies have indicated that they aggravate existing cases of breast cancer, but they have not been shown to start the disease. The results of long-term testing are still forthcoming at the time of this writing (studies now in progress will be completed early in the 1980's).

Other problems that have been reported with the pill include weight gain, increased susceptibility to vaginitis (vaginal infection), and depressed sex drive (some women, on the other hand, find their sex drives enhanced by the pill). It must be stressed that the responses to oral contraceptives are highly variable; some women have used them for years without noticing detrimental side-effects.

Condoms plus spermicide

If a couple is willing to put up with a bit of distraction from erotic play, there is a contraceptive technique that gives nearly the same protection as oral contraceptives, without the risk. **Condoms** (nipple-ended sheaths of thin, latex rubber which fit snugly over an erect penis) used with **spermicidal foams** (sperm-killing agents injected into the vagina with a plunger device) are nearly 100 percent effective at preventing fertilization (Figures 12-20 and 12-21). Another advantage of the condom/spermicide combination is that it provides the best available protection against venereal disease. Apart from inconvenience the only disadvantage of the condom/spermicide combination is that a small percentage of people have allergic reactions to the spermicide.

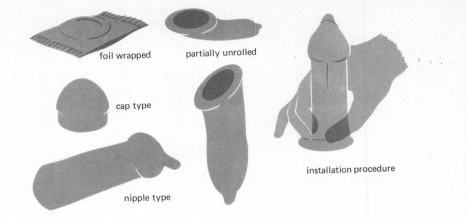

FIGURE 12-20
Condoms.

foil wrapped   partially unrolled

cap type

installation procedure

nipple type

FIGURE 12-21
Using foams, creams and jellies.

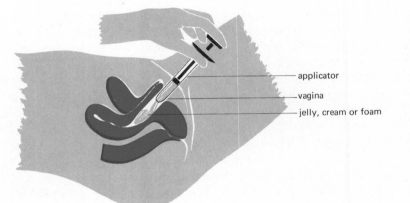

applicator

vagina

jelly, cream or foam

FIGURE 12-22
I.U.D.s (left) and their insertion through
the vagina (right).

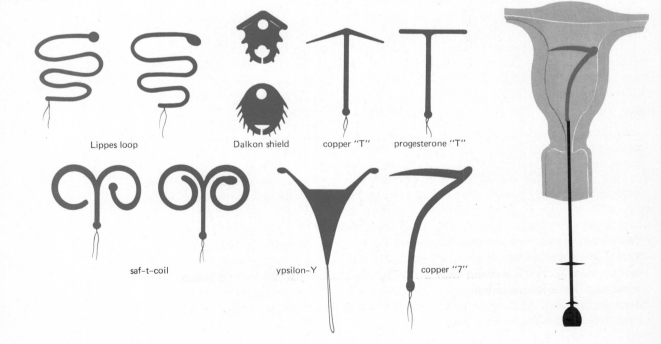

Lippes loop   Dalkon shield   copper "T"   progesterone "T"

saf-t-coil   ypsilon-Y   copper "7"

Intrauterine devices

Intrauterine devices (IUDs) are objects inserted into the uterus via the cervix (Figure 12-22). Precisely how IUDs prevent pregnancy is not known, but they are thought to act more as agents of abortion rather than contraception. Constant irritation of the endometrium by an IUD is thought to prevent implantation by embryos. The failure rate of IUDs is about 5 percent.

Not everyone can use an IUD. About 15 percent of them must be removed because they cause cramping in the abdomen, backache, or unusual bleeding. In general, they are more easily tolerated by women who have already had a baby.

In addition to physical discomfort, there is a potentially dangerous physical side-effects of IUDs. They double the risk of infection of the uterus. In itself a uterine infection is irritating but not permanently damaging. However, if the infection spreads to the Fallopian tubes there is a possibility that the woman will be left permanently sterile. The presence of an IUD hastens the spread of uterine infections to the Fallopian tubes.

The diaphragm

A diaphragm is a thin, flexible rubber cup placed over the cervix to prevent sperm from entering the uterus (Figure 12-23). From Victorian times until the 1960s the diaphragm was the most effective method of contraception available to women (having a failure rate of 15 percent). Even today there is still a good reason for choosing a diaphragm; there is essentially no danger of physical harm in using

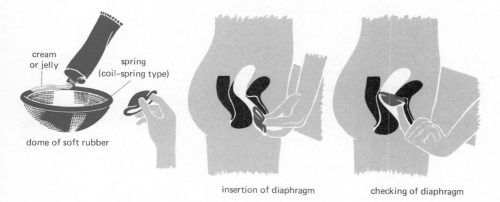

cream or jelly

spring (coil-spring type)

dome of soft rubber

insertion of diaphragm   checking of diaphragm

FIGURE   12-23
The preparation and insertion of a diaphragm.

one. On the other hand, there is a good argument against diaphragms because they are significantly less effective than "the pill," the condom/spermicide combination, or the IUD. When a diaphragm is used with spermicidal cream or jelly, the failure rate is reduced to 3 percent.

Rhythm

Basically rhythm involves refraining from intercourse for a week or so around the time of ovulation, which usually occurs about midway between periods of menstrual bleeding. For a woman with a 28-day menstrual cycle this would mean abstention for a few days before and after the 14th day following each period of menstrual bleeding. There are two methods of estimating the timing of ovulation. One is to keep record of the dates of menstrual bleeding, and the other involves

FIGURE 12-24
Basal body temperature and the menstrual cycle.

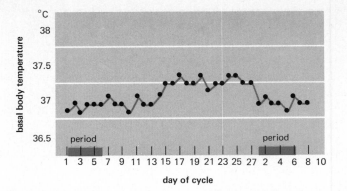

monitoring basal body temperature (Figure 12-24). By taking her oral temperature every morning before becoming active, a woman looks for the rise of a few tenths of a degree which follows ovulation. Data should be accumulated for several months (preferably a year) before beginning to practice rhythm.

Rhythm is a notoriously ineffective method of contraception, particularly for women who have irregular menstrual periods (the failure rate is about 21%). One gynecologist observed that people who depend on rhythm are called parents. Rhythm fails because the timing of ovulation isn't entirely predictable; it may happen more than once per month, and it may be caused by orgasm at a time when it normally wouldn't occur. Although rhythm by itself is a poor method of contraception, it's worth keeping in mind that the most fertile time of the month is generally half way between periods of menstrual bleeding. Particular care in using spermicides with a condom or diaphragm is most important midway between periods of menstrual bleeding.

Withdrawal

Withdrawal of the penis prior to ejaculation takes a man of unusual self-control. The success of it depends on the agility and honesty of the male, since once a woman has committed herself to it, she is usually in no position to do anything about lack of control or a last minute change of plans on the part of her mate. The rate of failure for people depending exclusively on withdrawal is about 25 percent.

Douching

Of all the common methods of contraception, douching is the worst (with a failure rate of 40 percent). It involves attempting to wash sperm out of the vagina with a flow of water. But within a minute after ejaculation, sperm are on their way through the cervix to the uterus—well out of the way of a douche's spray. Moreover, even if a woman is able to jump from bed and sprint to the bathroom in less than a minute, the procedure is self-defeating. The rush of water inevitably forces some sperm up into the cervix at the same time that it washes others out of the vagina.

Table 12-2 summarizes the effectiveness of contraceptive procedures.

Research in contraception

Research in contraception is pushing forward on several fronts. New modes of introducing female sex hormones, for example, are currently under clinical study. Injections of long-lasting hormonal preparations (which are taken up by fat deposits

TABLE 12-2   APPROXIMATE FAILURE RATES OF COMMON METHODS OF BIRTH CONTROL (PREGNANCIES PER 100 WOMEN)

METHOD	IF USED CORRECTLY AND CONSISTENTLY	AVERAGE U.S. EXPERIENCE AMONG 100 WOMEN WHO WANTED NO MORE CHILDREN
Abortion	0	0
Abstinence	0	0
Hysterectomy	.0001	.0001
Tubal Ligation	.04	.04
Vasectomy	.05	.15
Oral Contraceptive (combined)	.34	2-5
I.M. Long-Acting Progestin	.25	5-10
Condom + Spermicidal Agent	Less than 1.0	5
Low Dose Oral Progestin	1-1.5	5-10
IUD	1-3	5
Condom	3	15
Diaphragm (with spermicide)	3	17
Spermicidal Foam	3	22
Withdrawal	12	20-25
Rhythm (Calendar)	13	31
Chance (sexually active)	90	90
Douche	30	40

and slowly released) have proved highly effective in preventing pregnancy. However, the return to fertility after the injections are stopped is slow.

Hormone-releasing devices which are placed in the vagina, cervix, or uterus show more promise than injected preparations. The "vaginal ring," for example, is inserted like a diaphragm and slowly releases progestin. Experiments with hormone-impregnated IUDs are also under way. Since such devices are placed within the reproductive organs, their effect is direct, and the amounts of hormones that must be released are far less than the amounts needed with injections or oral contraceptives.

Because of the problems inherent with using the female sex hormones for contraception (already considered in the discussion of oral contraception), researchers are looking for other chemical methods of preventing pregnancy. Another line of research is considering the possibility of vaccinating women against sperm. Just as immunity to certain diseases develops after injections of weak strains of disease-causing organism, an "immunity" to pregnancy can be achieved through injections of semen. Once immunized, however, some experimental subjects have stayed that way. Presently there are no chemical agents of contraception proven to be as effective and reversible as "the pill."

To this point little has been said about methods of contraception applied to males. There is good reason for this, since the condom is the only presently available agent of contraception for men. The relative lack of contraceptive techniques with which the male must take responsibility has brought criticism on the predominately male researchers of contraception. Indeed, in the past most research has been directed toward developing techniques for women. At present, however, several lines of investigation into male contraception are in progress. One of them is considering using female sex hormones. The chemical mechanism by which estrogen and progestin block fertility in men is similar to the way they work in women—they inhibit secretion of the pituitary hormones, LH and ICSH. The latter is precisely the same hormone as LH, it is just given a different name in men (**interstitial cell stimulating hormone**). Inhibition of gonadotropin secretion makes a break in the hormonal support system of sperm, and the production of them diminishes. Unfortunately, the female sex hormones tend to depress sex drive in men and cause enlargement of the breasts. Testosterone has shown promise as a contraceptive in affecting sperm production, but it also has drawbacks such as a tendency to cause cancer of the prostate gland.

ABORTION

In 1973 the Supreme Court legalized **abortion** (termination of pregnancy by destruction of the fetus) up through the 24th week of pregnancy. Today there are four commonly used techniques of abortion. Which one is used depends upon the state of development of the pregnancy.

The **"morning-after" pill** stops pregnancy with a large dose of estrogen (actually, a series of doses taken over several days) which renders the endometrium inhospitable to settlement by an embryo. The treatment must begin within three days following intercourse. The morning-after pill is nothing to get into the habit of taking. It contains the equivalent of estrogen in a 40-year supply of contraceptive pills, and it hasn't been in use long enough to determine the nature of its long-term effects.

Dilation and curettage is a method utilized during the first three months of pregnancy. In the first phase of the operation the cervix is dilated or expanded (this is commonly accomplished by insertion of a thin, porous rod which swells as it absorbs fluid). Once the cervix is dilated the uterus is scraped with a sharp instrument (curet).

Vacuum aspiration is a newer and somewhat more gentle method of abortion for the first three months of pregnancy. It involves removal of the fetus by suction through a plastic tube which is inserted into the uterus (Figure 12-25).

If pregnancy progresses beyond the first three months, abortion becomes more difficult. Abortion performed after the first three months can be dangerous to the mother, regardless of the method used. A technique is used which induces delivery of the immature fetus. The procedure, **saline injection,** involves the introduction of a concentrated salt solution into the uterus. The consequent osmotic imbalance causes the muscles of the uterus to contract, and the fetus is expelled.

An experimental method of abortion involves administration of hormone-like substances, **prostaglandins,** which induce contractions of the uterine muscles. Prostaglandins are normally produced by several glands within the human body (the prostate gland is the primary producer of them in males). The normal function(s) of them aren't known. Early in the 1970s prostaglandins were cleared for use as abortive agents in England. As yet, the Federal Drug Administration has not approved them for general use in the United States.

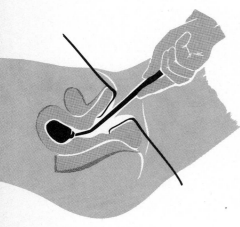

FIGURE 12-25

The vacuum aspiration method of abortion. The suction device is inserted into the cervix.

STERILIZATION

Sterilization is the rendering of the reproductive organs infertile by surgery. The procedure for male sterilization, **vasectomy,** involves severing the vas deferens (Figure 12-26). After small injections of local anesthetic are made into each side of the scrotum, the vas deferens are withdrawn through small incisions at the top of the scrotum. They may be either tied and cut or cauterized. At present it appears that the side effects of vasectomy are mild. However, there have been tentative reports that it disrupts the functioning of the immunological system. Sperm cells, which are still produced by the testicles following vasectomy, eventually deteriorate and are absorbed into the surrounding fluids. Consequently the

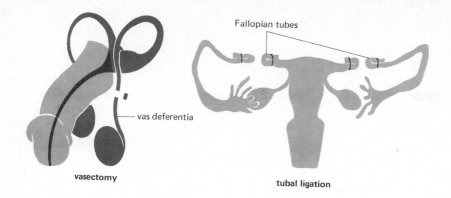

FIGURE 12-26
Sterilization: vasectomy (left) and tubal ligation (right).

Fallopian tubes

vas deferentia

vasectomy

tubal ligation

body reacts as if it were being invaded by disease organisms and produces sperm-destroying antibodies. Some researchers believe that this triggering of the immunological system may lead to auto-immune diseases (becoming sensitive to one's self), which include rheumatoid arthritis. Research aimed at determining whether or not there is any danger is in progress.

The most common technique for female sterilization involves severing the Fallopian tubes (Figure 12-26). This procedure, **tubal ligation,** may be accomplished by approaching the Fallopian tubes via incisions in the abdomen or at the back of the vagina. In the latter case the Fallopian tubes are located with a miniature periscope. As with vasectomy the side effects of female sterilization seem to be negligible.

VENEREAL DISEASES

Venereal is a term derived from the name of the Roman goddess of love, Venus, and is aptly applied to diseases that are transmitted during intimate physical contact. In our promiscuous society **venereal disease** (VD) is a rampant health problem. In 1977 public-health experts estimated that over 3,000,000 North Americans had some form of it. Among these cases the vast majority were of a single disease, gonorrhea. Before considering the characterisitcs and the cures of gonorrhea and the other common venereal disease, it is worth noting that there are certain things that can be done to prevent them. As previously mentioned, the condom/spermicide combination gives a certain amount of protection against VD (as well as being a nearly 95 percent effective contraceptive). Additional protection is afforded by bactericidal soaps, which can be used to wash the penis before intercourse (if the female is in doubt about the health of the male) or after intercourse (if the male is in doubt about the female).

Gonorrhea

Neisseria gonorrhea, the bacterium that causes **gonorrhea,** can infect the moist mucous membranes of the vagina, penis, anus, mouth, and throat (thus, gonorrhea can be passed by oral and anal sexual contact). If untreated the bacteria spread from the penis and vagina upward along the passageways of the genital tract and may produce sterility in both males and females. In addition, the infection may enter the bloodstream and cause arthritis or heart disease.

Gonorrhea in females   Gonorrhea is particularly dangerous to women, since most gonorrheal infections of the female reproductive organs occur without obvious early symptoms. Usually women discover the disease in themselves only through reports from male partners. Thus, if a male discovers he has gonorrhea, he should be quick to inform those with whom he has had intercourse. (A trial in Wyoming in 1976 put teeth in the injunction that a man with VD should inform his sexual partners; a man who had infected a woman and didn't tell her was ordered to pay $1.3 million in damages.)

Gonorrheal infection of the female reproductive organs usually begins in the cervix. A few days following the infecting intercourse, pus is discharged from the cervix. If the pus reaches the vaginal opening (it usually doesn't) it appears yellow-green and is irritating to the vulva. As the infection progresses, some women begin to feel a persistent low backache or pain in the abdomen. In some cases the urethra and anus become infected, and a burning sensation is felt during elimination.

If left untreated, the infection progresses along the membranous lining of uterus and eventually reaches the Fallopian tubes (about two to three months following the original infection). Pus forms within the tubes and eventually leaks through the funnellike ends onto the ovaries and other organs in the lower abdomen. These infections are extremely painful and may be irreversibly damaging to the reproductive system. In some cases the internal reproductive organs are misshapened and pulled out of place by thick growths of scar tissue. A quarter of gonorrhea victims may be sterilized by the complete closing of both Fallopian tubes if not treated properly.

Gonorrhea in males   In males the penile urethra is usually the first area of attack by gonorrhea bacteria, and symptoms are noticed within two weeks after the infecting intercourse. A discharge seeps from the urethral meatus, which, at first, is thin and clear but soon thickens. It may be white, yellow, or yellowish-green in color. Most victims feel a burning sensation in the penis, particularly during urination. The urine may become cloudy with pus and sometimes contains traces of blood.

After about two weeks following their initial appearance, the symptoms of gonorrhea begin to fade and may disappear for a time. Nevertheless, the bacteria continue to penetrate further along the urethra. Eventually the infection reaches the prostate gland, which enlarges and presses into the bladder and rectum. Urination and defecation become painful.

From the prostate gland the infection may spread along the vas deferens and invade the epididymises. This condition, **epididymitis,** is characterized by hard, painful swellings at the bottoms of the testicles. If epididymitis isn't soon treated, scar tissue develops which prevents passage of sperm from the testicles into the vas deferens. Thus, sterility may result.

Transmission   Gonorrhea is most often contracted through sexual intercourse. But the pus is so laden with the disease organism that it can be passed by hand to susceptible tissues (mucous membranes).

Gonorrhea from infected pregnant women used to be a frequent cause of infant blindness. The bacteria were transferred to the baby's eyes as it passed through the birth canal. Today newborn babies are routinely treated with a bactericidal agent (usually silver nitrate) which is administered as eye drops.

Treatment   Penicillin is the first choice for treatment of gonorrhea, despite the recent appearance of strains of gonorrheal bacteria that are resistant to it. Other antibiotics also can be used with success, particularly for those persons who are sensitive or allergic to penicillin. The response to treatment is rapid. Prompt treatment of the early stages almost immediately clears up the discharge from the vagina or penis and kills the infecting bacteria. It is more difficult to treat advanced cases, and it is impossible to repair scarred tissue.

Syphilis

The spiral-shaped bacterium (spirochete) that causes **syphilis,** *Treponema pallidum* (Figure 12-27) does not long survive the drying effects of air, but it will grow prolifically in the warm, moist tissues of the body. The mucous membranes of the genital tract, the rectum, and the mouth are perfect breeding grounds.

Primary syphilis   Syphilis begins when a spirochete enters the body. The infected person may show no sign of the disease for from 10 to 28 days. The first sign is a sore known as a chancre, which is a lump about the size of a dime or smaller, teeming with microscopic spirochetes. It is moist, although there is no discharge, and it is generally painless. A chancre often appears at the site of infection, usually in the genital region—on the shaft of the penis (Figure 12-28) or on the vulva (Figure 12-29). Unfortunately, it also can develop out of sight and never be noticed, deep in the recesses of the vagina, the rectum, or the male urethra. Thus many infected persons, women in particular, never know they have primary syphilis. But, visible or hidden, the chancre is dangerously infectious and readily transmits the spirochete from person to person.

Primary syphilis can easily be cured with antibiotic treatments. Even without treatment, however, the chancre disappears within several weeks, giving the false impression that the infection is over. But, in fact, it has entered the blood, and the spirochetes are being carried to all parts of the body. The primary stage of syphilis has ended, and the secondary stage has begun.

Secondary syphilis   Secondary-stage symptoms appear some time between a few weeks to a year after the appearance of the chancre. The symptoms vary greatly in severity—many people show no symptoms at all while others become disabled. The symptoms may last only a few days or may persist for several months. If lesions (sores) develop, they may heal and leave no scars even without treatment, again lulling the afflicted person into a false sense of security.

Symptoms of secondary syphilis may include a skin rash, the development of small, flat lesions in regions where the skin is moist, whitish patches on the mucous membranes in the mouth and throat, spotty, temporary baldness, and various constitutional symptoms, including general discomfort and uneasiness, low-grade fever, headache, and swollen glands. These symptoms are easily mistaken for those of other diseases. However, an absolute diagnosis is possible with a blood test.

During the secondary stage, syphilis is more contagious than at any other phase of its development. All the lesions are filled with spirochetes, hence any contact with open lesions—even without sexual intercourse—can transmit the disease.

Latent syphilis   The third stage of syphilis is called the latent period because all signs and symptoms of the disease disappear. Essentially the disease is hidden, but it is not gone. Spirochetes actively invade various organs, including the heart and the brain. This phase sometimes lasts only a few months, but it may last for 20 years or until the end of life.

In the latent stage the infected individual appears disease-free and is usually not infectious. There is, however, an important exception; a pregnant woman may transmit the disease to her child. Although there are no overt symptoms of latent syphilis, a blood test during this phase will reveal the disease. The infected person is still a potential transmitter, however. Within the first two years a relapse may occur, with the redevelopment of the highly contagious secondary stage.

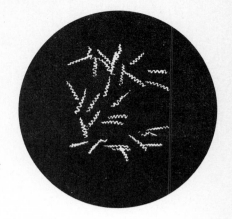

FIGURE   12-27
The bacteria *Treponema pallidum* (length, 10 micrometers).

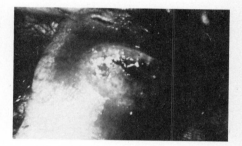

FIGURE   12-28
A chancre of the penis.

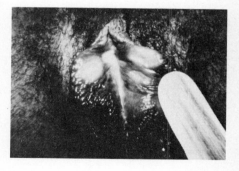

FIGURE   12-29
Multiple primary chancres of the labium minora.

Late syphilis   The late stage of syphilis generally begins 10 to 20 years after the beginning of the latent phase, but can occasionally occur much earlier. For the infected person, it is the most dangerous stage of the disease. The disabling effects of late syphilis depend on which organ or organs the spirochetes settle in during the latent period. In late syphilis, 23 out of every 100 untreated patients become incapacitated. Thirteen of the 23 develop serious cardiovascular disease; many die of severe heart damage or rupture of the aorta. The other 10 have slowly progressive brain or spinal cord damage, eventually leading to blindness, insanity, or crippling.

Congenital syphilis   Syphilis may be transmitted from a pregnant women to her unborn child. If the fetus is exposed to the disease in its fourth month of development, the infection may kill it or cause various disfigurations. If the fetus is infected late in pregnancy it may be born with a case of the disease, but the infection may not be apparent for several months or years after birth, when the child may become disabled as in late syphilis. Treatment of the infected pregnant woman within the first four months of pregnancy halts the spread of the disease in the unborn child.

Treatment of syphilis   Syphilis is easily cured with antibiotics when the treatment is begun in the first two stages or even in the latent phase. Penicillin has been the drug used most often, but other antibiotics may be used when the patient is allergic to penicillin or the bacterial strain does not respond to it. The infection will not recur unless one contracts the disease again from an infected person. An individual can be reinfected over and over again (syphilis and gonorrhea confer no immunity to succeeding infections), and there are no preventive vaccines. Thus, continual caution must be taken to ensure prompt diagnosis and treatment after each possible exposure to syphilis and gonorrhea.

Other "social" diseases

Besides gonorrhea and syphilis various other diseases and infections are transmitted by sexual intercourse. This is not the place for an exhaustive consideration of them, but a few are worth mentioning.

Pubic lice   Commonly called "crabs" (or "crotch crickets") pubic lice are insect infestations. With three pairs of claws, a louse is a fearsome monster when seen under magnification, but it is no larger than the head of a pin. Pubic lice cling to the base of the pubic hairs, pierce the flesh with their mouth parts, and feed from the capillaries. They are usually transmitted during sexual intercourse, but, since they can survive for a day apart from a host, they can be passed via infested bedding or clothing. Delousing creams and shampoos are available without prescription at drugstores.

Cystitis   This is a bacterial infection of the bladder, which is most common in women but may develop in men. Various bacteria can cause cystitis. But a bacteria which is present in the intestines of all healthy people, *Escherichia coli,* is the most common agent. Cystitis usually develops as the result of frequent, vigorous intercourse with a new partner (the old name, "honeymoon disease" is now somewhat out of date) which transfers bacteria from the anus to the urethra. The symptoms include a burning sensation upon urination and urine which is hazy or reddish with blood from the infected bladder. Cystitis can be cured with sulfa drugs or antibiotics.

Genital herpes   This is a viral disease, whose first symptom is blistering on the genitals. For men the penis is most commonly infected. In women, the blisters may develop on the labia minora, clitoris, anus, or cervix. Eventually the blisters rupture forming painful, open sores. Soon after rupturing herpes sores begin to disappear, and within a month they are usually healed. But, the virus is still present, and blistering may sporadically reappear even without sexual contact.

Sometimes genital herpes is misdiagnosed as syphilis and treated with penicillin. But penicillin does not cure it. In fact, there is no known drug that completely eradicates genital herpes (although sulfa creams help clear up the blisters). Once the herpes virus has invaded, the victim might carry it for life.

CHAPTER 12: SUMMARY

1   In the male, sperm are produced in the testicles located in the scrotum.

2   Sperm leave the testicles, mix with seminal fluid from the prostate gland and seminal vesicles and are ejaculated.

3   The development and release of an egg are controlled by the menstrual cycle.

4   The fertilized egg goes through a complicated series of cell divisions and cell movements to form the embryo which develops into the fetus.

5   Cloning is the development of an entire organism from just one or a few cells.

6   Both male and female sexual responses involve several distinct stages.

7   There are many forms of contraception that differ markedly in their effectiveness.

8   Venereal diseases are transmitted primarily by sexual contact.

Suggested Readings

Boston Women's Health Book Collective   *Our Bodies Our Selves,* Simon and Schuster, 1971 — A book by women about female sexuality.

Chase, H. B.   *Sex: The Universal Fact,* Dell, New York, 1965 — The structure and function of sexuality.

Katchadourian, H. A., and D. T. Lunde   *Fundamentals of Human Sexuality,* Holt, Rinehart and Winston, 1975 — A consideration of all aspects of human sexuality.

Segal, Sheldon J.   "The Physiology of Human Reproduction," *Scientific American,* November, 1968 — The physiology of the human reproductive system.

Wilson, Sam, R. Roe, and L. Autrey   *Readings in Human Sexuality,* West Publ. Co., 1975 — Readings on the biology, psychology, and history of human sexuality.

CLONING: THE ETHICAL QUESTION

The reading accompanying Chapter 10 introduced the question of human eugenics. The following article considers the ethics of a method by which eugenics could be accomplished.

The idea of human reproduction by cloning individuals asexually meets with the favor of Dr. Joseph Fletcher, professor of medical ethics at the University of Virginia—if cloning accomplishes the greatest good for the greatest number.

There is nothing sacred about sexual reproduction in his opinion: "It seems to me that laboratory reproduction is more human than conception by ordinary heterosexual intercourse," he says. "It is willed, chosen, purposed and controlled, and surely these are among the traits that distinguish *Homo sapiens* from others in the animal genus."

Cloning is the experimental process of duplicating an animal by inserting the nucleus of one of its cells in an unfertilized egg of another animal. The recipient nucleus, now having all the chromosomes for reproduction of another animal, theoretically proceeds to multiply.

Prof. Robert Sinsheimer, a biologist at the California Institute of Technology, has predicted that cloning in mammals, including man, may be accomplished in as early as a decade. It has already been done in lower life forms such as amphibians.

Fletcher says he can envision several situations in which cloning may be desirable from a social view. It may be necessary to control a child's sex to avoid any one of 50 sex-linked genetic diseases.

"It is entirely possible," he adds, "given our present progressive pollution of the human gene pool through uncontrolled sexual reproduction that we might have to replicate healthy people to compensate for the spread of genetic disease.

He would favor cloning top-grade soldiers and scientists if they were needed to offset a tyrannical power plot by other cloning nations—"a truly science fiction situation, but possible."

Opponents of genetic control view with alarm the production of identical copies of an individual, Fletcher acknowledges, but he points out that personalities are not produced by genes alone. Each is shaped by the environment as well.

THE SUPREME COURT IGNITES A FIERY ABORTION DEBATE

While intellectuals debate whether or not extreme modifications of the reproductive process, such as cloning, would be ethical, the general public is concerned with much less drastic intrusions into the normal course of human reproduction. A large segment of society is morally opposed to abortion, which was legalized by a Supreme Court decision in 1973. A more recent decision by the Court has weakened the original ruling, and "right-to-lifers" are pushing for even more stringent legislation.

Government must not interfere with the right of any woman to have an abortion in the first three months of her pregnancy if she so chooses. That fundamental and sweeping principle was established by the U.S. Supreme Court in 1973. But last week the court seriously compromised and restricted its edict. By a vote of 6 to 3, the Justices held in effect that the only women who can be certain of being able to have a medically safe abortion are those who can afford to pay for one. State and local governments, said the court, can choose whether or not to finance the abortions of needy women for nontherapeutic reasons—conditions that do not endanger their health.

The court's decision aroused once again the fiery moral, religious, medical and political issue that most elected officials would like to dodge. They will have no such chance. The court gives legislators in states and, by implication, the U.S. Congress the duty of deciding whether or not indigent women should be allowed to have elective abortions at public expense.

Cancer of Poverty. The fight will center around the familiar Medicaid program, which is jointly financed by the states and the Federal Government and needs the approval of both levels to work. Medicaid is the main way for welfare recipients to get safe abortions in the U.S.; it accounts for nearly a third of the 1 million or so operations performed annually. The court's ruling is already being hailed as a great victory for the forces determined to move Medicaid out of the abortion business. It removes any constitutional cloud over the decision of 15 states to deny payment to Medicaid patients for nontherapeutic abortions. These states are now free to cut off all nontherapeutic Medicaid abortions.

The dissent from the decision of the majority was angry, even bitter. Justice Harry Blackmun, who wrote the landmark 1973 decision, assailed the ruling as "disingenuous and alarming, almost reminiscent of 'Let them eat cake.'" Departing from Chief Justice Warren Burger, his "Minnesota twin," Blackmun roundly scolded his colleagues: "There is another world 'out there,' the existence of which the court, I suspect, either chooses to ignore or fears to recognize. And so the cancer of poverty will continue to grow." Justice Thurgood Marshall charged that the court's decision would "brutally coerce poor women to bear children," and said that he was "appalled at the ethical bankruptcy of those who preach a 'right to life' that means a bare existence in utter misery for so many." Justice William Brennan claimed the decision "seriously erodes" the principle, set forth by the court in its historic 1973 ruling, that it is unconstitutional for Government to interfere

with a woman's choice whether or not to have an abortion.

The majority opinion, written by Justice Lewis Powell, held that there is nothing in the Social Security Act, which created the Medicaid program in 1965, that requires participating states to provide any specific medical treatment, including abortion.

The court brushed aside claims that providing funds through Medicaid for childbirth but not abortion violates both the equal-protection clause of the Constitution and a woman's right to make her choice unhindered by Government. The Justices conceded that "the state may have made childbirth a more attractive alternative, thereby influencing the woman's decision," but nonetheless ruled that it did not really eliminate her free choice, since privately funded abortions still were possible—even though the woman might find it impossible to afford one.

The court's majority was perhaps on firmest ground in contending that the whole question of public funding of abortion ought to be resolved by legislators rather than by judges. Wrote Powell: "We leave entirely free both the Federal Government and the states, through the normal process of democracy, to provide the desired funding. The issues present policy decisions of the widest concern. They should be resolved by the representatives of the people, not by this court."

Presidential Support. But the "representatives of the people" are just as hesitant to grapple with the issue, one that matches militant right-to-lifers against dedicated pro-abortionists. There is no practical way to compromise the issue. Anti-abortionists feel that life begins at the moment of conception and that abortion is therefore murder; pro-abortionists believe that terminating an unwanted pregnancy is a personal decision that should be left to the conscience of the woman rather than be made unlawful. Paying far more attention to preventing undesired pregnancies is one measure on which the contending sides may agree—but no contraceptive has yet won wide enough acceptance to ensure that kind of solution.

The main battle arena is Washington, where

the anti-abortionists have the tacit support of President Jimmy Carter, who personally opposes abortion and has repeatedly said that he does not think the Federal Government should do anything to encourage it. Almost alone at the Department of HEW, which administers Medicaid, Secretary Joseph Califano shares Carter's view. When he submitted HEW's budget for the fiscal year beginning in October, he urged that no Medicaid funds be used for nontherapeutic abortions.

Killer! Killer! In Congress, the issue cuts confusingly across ideological lines, making it difficult for many liberals from areas with strong Catholic enclaves to vote for abortion. Whatever his own feelings, any legislator in a swing district hesitates to arouse the anger of such an uncompromising group as the anti-abortionists. Their attack can be so personal that New Jersey Congressman Andy Maguire was actually chased through his office building by lobbyists screaming "Killer! Killer!" He nevertheless maintained his pro-abortion stand. It takes courage for a Congresswoman like Maryland's Barbara Mikulski, whose Baltimore area is heavily Catholic, to fight the right-to-lifers as resolutely as she has. "I am a professionally trained social worker," she explains. "I know what the coathanger abortion is all about."

Last year Republican Congressman Henry J. Hyde of suburban Chicago introduced an amendment to the HEW appropriations bill for the current fiscal year that denied funds for abortions not necessary to save the life of the mother. To the surprise of the overconfident pro-abortion forces, it passed both chambers. The restriction failed to go into effect only because its opponents secured a court injunction against the measure in New York while its constitutionality was being tested. In view of last week's decision, the Supreme Court could now lift that injunction. If reinstated, this ban on abortion payments would expire at the end of September, when the fiscal year ends.

The odds now favor the success of a new effort in Congress to place a similar restriction on next year's HEW appropriations. By a vote of 201 to 155, the House two weeks ago approved another Hyde amendment that would even ban abortions deemed necessary to save a mother's life. This week the Senate will start debating the measure. Even many anti-abortionists will join in trying to restore Medicaid for therapeutic abortions. But pro-abortionists seriously doubt that they will be able to persuade their colleagues to fund all abortions for needy women. Massachusetts Senator Ed Brooke, fighting hard to preserve full funding, is incensed at the legislators' mood. "It is really inconceivable that in 1977 the Congress would take this step backward," he protests. "This is nothing but a means test saying who's allowed to have an abortion."

Gaining momentum, the anti-abortionists do not conceal the fact that their real aim is to make all elective abortions illegal again. "I would certainly like to prevent, if I could legally, anybody having an abortion—a rich woman, a middle-class woman or a poor woman," Hyde, a father of four, frankly told the House. The right-to-lifers are even hoping to call a constitutional convention that would adopt an amendment restricting abortions, although there are differences of opinion about how strong to make the measure.

However the fight in Washington turns out, the struggle is likely to continue at state and local levels. Even if federal funds are shut off at their source, the states could theoretically carry on with tax money of their own. It is highly doubtful, however, that many could do so. Since federal funds now cover up to 90% of abortion costs, the state's burden would be heavy.

If the abortion defenders somehow rally to keep the federal funds flowing, their opponents are well prepared to persuade state legislatures to get the abortion programs killed at the lower level. Beyond the 15 states in which they have already succeeded,* the right-to-lifers appear to be on the verge of pushing through other bans in Massachusetts, Illinois, Michigan and Wisconsin. In heavily Catholic Massachusetts, the pro-abortion forces frankly admit they see little

*Connecticut, Idaho, Indiana, Louisiana, Missouri, New Hampshire, New Jersey, New York, North Dakota, Ohio, Pennsylvania, South Carolina, South Dakota, Utah, West Virginia.

chance of checking such a ban and are hoping they can persuade Democratic Governor Michael Dukakis to take the hazardous political step of vetoing such a bill.

Anti-abortion sentiment is so strong in Illinois that the legislature has approved a bill requiring unmarried women under the age of 18 to get their parents' consent before having an abortion, even though the Supreme Court has already declared unconstitutional such a limitation on the mother's freedom of choice. The strong opposition of Michigan Governor William G. Milliken is the main obstacle to passage of a ban on public funding of abortions in his state. While he has not promised to veto such a measure, he has warned that it could produce "a system of back-alley abortions—and that would be wrong, morally wrong."

Pro-abortion leaders had been lulled by the 1973 court decision and polls showing that more than 60% of Americans favor permitting abortions in the first three months of pregnancy. Yet as the battle now centers on the narrower question of whether public money should support abortions, that majority could fade.

The pro-abortionists hope to win over some legislators to their cause by arguing their case on purely practical grounds: it may cost the government only $200 to fund an abortion, but childbirth care for a needy woman may run to $1,000, and welfare payments for the baby could add another $1,000 in the first year alone.

But perhaps the strongest point the pro-abortionists can make is that cutting off government funds will not stop abortions for needy women: they will have them anyway—somehow, somewhere. "It's outrageous, it's crazy," contends Joyce Yaeger of Planned Parenthood in New York. She points out that the $200 fee for an abortion performed in a hospital "is a lot of bread when you're poor. But you can always get $25 to $50 for the knitting-needle crowd of abortionists."

Making Liars. A California public-health official predicts that if government funds are cut off, more welfare mothers will die by going to cut-rate quacks for their abortions. Says she: "I'll send a copy of the first death certificate to Justice Burger."

If their counterattacks on the funding curbs fail, the pro-abortionists expect to try to expand the definitions of what constitutes a therapeutic abortion. They hope that the court will support their contention that a mother's mental as well as physical health can be severely affected by an unwanted childbirth. At the least, they assume, sympathetic doctors will sign papers claiming that the mother's health might be endangered. "We'll just have to educate poor women to do what wealthy women always have done," says Fran Avallone of New Jersey's National Abortion Rights Action League. "The court will make poor women lie, physicians lie and psychiatrists lie."

Legalities and theology aside, the prevalence of unwanted children—and the lengths to which women may go to prevent such births—presents a social problem that legislators, as the court suggests, should face squarely. Whether abortion is an acceptable alternative is a difficult question to resolve but one that should be decided by rational debate, not by which noisy side poses the more immediate threat to a legislator's political survival.

CHAPTER

13

Physiology, Growth and Regulation

Objectives

1 Learn what biologists mean by food and how foods are classified.

2 Understand the process of digestion.

3 Describe circulation and the functions of blood.

4 Learn how wastes are eliminated from organisms.

5 Understand the function of plant and animal hormones and where they originate.

FIGURE 13-1

Examples of foods containing high proportions of the three major groups of organic nutrients: fats, or lipids, proteins, and carbohydrates.

FOODS AND NUTRITION

Food is the solid and liquid matter taken in to maintain existence. A short while after eating, food substances are in the tissues, and activities are being powered by the energy released in the breakdown of some of them. In these chemical reactions, carbohydrates, fats, and proteins react with oxygen to liberate the energy stored in their structures. As these complex compounds are transformed to simpler substances (mainly carbon dioxide and water), energy is transferred to ATP.

Carbohydrates, fats, and proteins are the major organic nutrients (Figure 13-1). Carbohydrates and fats supply energy. Proteins are needed to build tissue, but can also be utilized for energy. The tissue-building value of foods can be seen qualitatively in the growth and health of an animal, while energy value can be measured more quantitatively, as **Calories.**

One food Calorie is the amount of heat required to raise the temperature of 1 kilogram of water (the mass of one liter) 1 centigrade degree. (This amount of heat is 1,000 times greater than the "small" calorie used in measurements by physical scientists.) A piece of apple pie, for example, can be utilized by the body to liberate about 300 Calories. This is approximately equivalent to the amount of energy in one pound of coal or one sixth cup of gasoline. The amount of energy needed by an individual varies with activity, age, and body build. But a daily intake of 2,000 to 3,000 Calories is about average in this country.

Carbohydrates

More than half of the average American diet is **carbohydrate,** but a person's accumulated carbohydrate reserve is generally less than one percent of total weight. This is evidence that carbohydrates are primarily fuel foods, and that they are oxidized rapidly to liberate energy.

Many different kinds of carbohydrates are included in the average diet (Table 13-1). Some are easily digested and are transported to the tissues with little chemical change. Others require more chemical simplification. **Cellulose,** a carbohydrate which is formed in the cell walls of plants, is indigestible, but provides necessary roughage in the diet. In the digestive tract, cellulose stimulates the contractions of muscles which move food through the intestine. All carbohydrates that are digestible reach the body tissues eventually transformed into glucose (dextose).

TABLE 13-1   FOOD SUBSTANCES

SUBSTANCE	ESSENTIAL FOR	SOURCE
A. Inorganic compound		
water	composition of protoplasm, tissue fluid, and blood; dissolving substances	all foods (released during oxidation)
B. Mineral salts		
sodium compounds	blood and other body tissues	table salt, vegetables
calcium compounds	deposition in bones and teeth, heart and nerve action, clotting of blood	milk, whole-grain cereals, vegetables, meats
phosphorus compounds	deposition in bones and teeth; formation of ATP, nucleic acids	milk, whole-grain cereals, vegetables, meats
magnesium	muscle and nerve action	vegetables
potassium compounds	blood and cell activities growth, nerve action	vegetables
iron compounds	formation of red blood cells	leafy vegetables, liver, meats, raisins, prunes
iodine	thyroid gland function	seafoods, water, iodized salt
C. Complex organic substances		
vitamins	regulation of body processes, prevention of deficiency diseases	various foods, especially milk, butter, lean meats, fruits, leafy vegetables; also made synthetically
D. Organic nutrients		
carbohydrates	energy (stored as fat or glycogen) bulk in diet	cereals, bread, pastries, tapioca, fruits, vegetables
fats	energy (stored as fat deposits)	butter, cream, lard, oils, cheese, oleomargarine, nuts, meats
proteins	growth, maintenance, and repair of protoplasm	lean meats, egg, milk, wheat, beans, peas, cheese

Sugars   Many simple sugars are present in the American diet. These include glucose, fructose, and galactose. As discussed in Chapter 5, these sugars are classed as monosaccharides, because they consist of single hexose molecules with the chemical formula $C_6H_{12}O_6$. They are quick-energy sources since they require little or no chemical change before they are absorbed by the blood from the digestive organs.

Sucrose (cane sugar), **lactose** (milk sugar), and **maltose** (malt sugar) are **disaccharides,** composed of two hexose units. They all have the chemical formula $C_{12}H_{22}O_{11}$. These "double-sugars" undergo **hydrolysis** reactions with water and are thereby broken into single sugar molecules for absorption into cells.

Starches   A large part of the carbohydrate portion of the diet is composed of starches, or polysaccharides. They are abundant in cereal grains such as wheat, corn, rye, barley, oats, and rice, in addition to potatoes and tapioca. Starches are composed of large aggregates of sugar units, and they are broken-down to simple sugars during digestion (Figure 13-2a).

Much of the **glucose** received by the blood and transported to the liver is converted temporarily to "animal starch," or **glycogen.** As glucose is oxidized in the body tissues, glycogen is changed back to glucose in the liver, and is released into the bloodstream. In this way, stability is maintained in the level of blood sugar. Were it not for this function of the liver, we would constantly have to eat small quantities of carbohydrates.

Fats

Fats and **oils** (Table 13-1) yield more than twice as much energy per given weight as carbohydrates. However, relative to carbohydrates, fats are broken down slowly during digestion (Figure 13-2c). Common sources of fats include butter, cream, cheese, margarine, shortenings, vegetable oils, and meats.

Body fats are also synthesized from excess carbohydrates. The chief storehouses of body fat are the tissue spaces beneath the skin, the region of the kidneys, and the liver. Excess body fat is detrimental to health, and both the carbohydrate and fat content of the diet should be carefully regulated.

Proteins

As described in Chapter 5, proteins are composed of amino acids linked in chainlike molecules. Because of the immense variability in the order and amounts in which the 21 amino acids can be joined, there is an essentially unlimited variety of proteins.

Proteins consumed as food (Table 13-1) are foreign to the body; they cannot be directly utilized. However, reduced to amino acids during digestion (Figure 13-2b) they supply the units required to form the specific proteins needed by an individual. Growth and repair of body substances depend on protein intake.

Among the most concentrated protein sources are eggs, milk, peanuts, cheese, whole wheat, beans, and corn.

Certain of the amino acid molecules absorbed by the blood aren't used in cellular protein synthesis. These are broken down into two parts by a chemical activity of the liver called **deamination** (removal of amino or nitrogen parts). One part contains carbon and is sent to the tissues as glucose. The other portion, the nitrogen-containing part, is converted to **urea,** a waste product received by the blood and transported to the kidneys for excretion in the urine.

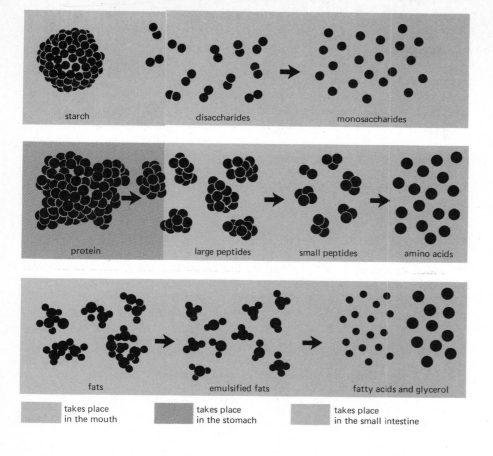

FIGURE 13-2

Phases in the digestion of (from top to bottom) carbohydrates, proteins, and fats, or lipids. The arrows indicate reactions produced by enzymes.

starch   disaccharides   monosaccharides

protein   large peptides   small peptides   amino acids

fats   emulsified fats   fatty acids and glycerol

takes place in the mouth   takes place in the stomach   takes place in the small intestine

Water

Water can be consumed directly, or it may be produced by chemical reactions within the body (Table 13-1). If you recall the discussion of dehydration synthesis in Chapter 5, you will remember that water is liberated in the formation of proteins from amino acids, and starches from sugar molecules. Water is inorganic and does not yield energy to the tissues. Yet, we are more dependent on a constant availability of water than on any other nutrient.

Water composes 60 to 70 percent, by weight, of a human body. Much of it is included in protoplasm and in the spaces between the cells. The fluid part of the blood, called **plasma,** is 91 to 92 percent water. It is essential in the plasma as a solvent for the food and waste products that are transported to and from the body tissues.

Mineral salts

Besides water, mineral salts are the other major inorganic food requirement. Table salt (sodium chloride) is consumed directly, and along with other mineral salts, is also present in food. Since salts are lost in perspiration, persons exposed to excessive heat over long intervals must either increase the salt in their diet or supplement their normal diet with salt tablets.

Animals require calcium and phosphorus in greater abundance than other mineral elements. Calcium is necessary for proper functioning of plasma membranes, and to insure the clotting of blood. Phosphorous is a component of ATP,

DNA, and RNA. Calcium phosphate is a primary component of bones and teeth. Milk is a rich source of both calcium and phosphorous; other sources include whole-grain cereals, meat, and fish. As with all foods, however, it should be remembered that excesses can be just as bad as too little. Excess calcium, for instance, can lead to the buildup of painful deposits in the kidneys, known as "kidney stones." Generally, as people grow older and their bodies reach full development, their calcium and phosphorous requirements diminish.

Iron compounds are essential for the formation of red blood cells. Meats, green vegetables, and certain fruits, including plums and grapes are important sources of iron in the diet. Iodine salts are essential in the formation of the thyroid gland's secretion (Figure 13-3). Iodine is usually found in tap water and table salt. It can also be obtained by eating seafoods.

Vitamins

A **vitamin** (Table 13-1) is an organic compound necessary in small quantities for nutrition. Not all vitamins can be synthesized by all organisms. Thus, these vitamins must be obtained in foods. Vitamins are essential as catalysts; they function together with proteins in enzymes. Vitamins were first designated by letters — A, B, C, and so on. Later, it was discovered that certain vitamins thought to be single substances were actually made up of many different components (for example, the vitamin-B complex). Then such names as B_1, B_2, and so forth were adopted. Today vitamins have names which indicate their chemical composition, although letters are still used as a means of easy and simple references. The functions of some vitamins required by humans are listed in Table 13-2.

Some vitamins can be stored in the human body, while others must be supplied constantly because the excess is excreted in the urine. Vitamin D is produced in the skin during exposure to sunlight. The absolute requirements of vitamins seem to be very small, although, in recent years, some biologists have come to believe that extremely large doses of vitamins may be beneficial. In particular, vitamin C has been acclaimed as a preventitive for the common cold, and some studies support this point of view. As much as a gram of vitamin C daily (as contrasted with a few milligrams of other vitamins) has been suggested to prevent colds. The value of this so-called "mega-vitamin therapy" has been disputed. We do not know, however, what any optimal vitamin doses are.

In some cases too much of certain vitamins have actually been shown to be harmful. Overdoses of vitamins A and D, for example, can lead to serious kidney damage. The possibility of ill effects from large doses of vitamin C is a matter of debate.

THE DIGESTIVE PROCESS

There are two basic reasons why tissues cannot use most foods in the forms in which they are eaten. First, many substances are insoluble in water and cannot enter the circulatory system or pass through cellular membranes. Second, foods are too chemically complex for cells to use. Digestion brings about changes in both of these conditions, with the result that cells can absorb and use the products. Thus, in **digestion** complex foods are broken down into smaller molecules of water-soluble substances that can be used by the body cells.

Digestion is both mechanical and chemical. The mechanical phase involves chewing and a churning action of the walls of the digestive organs (Figure 13-4). Chemical digestion involves several enzymes that catalyze breakdown reactions.

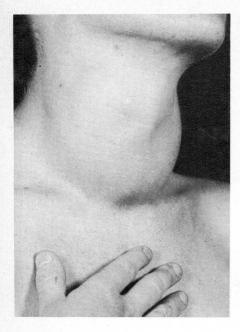

FIGURE 13-3
Simple goiter, an enlargement of the thyroid gland, can be prevented by the addition of iodine to the diet. Certain brands of table salt (NaCl) contain added iodine.

TABLE 13-2   IMPORTANT SOURCES AND FUNCTIONS OF SOME VITAMINS FOR HUMAN BEINGS

VITAMINS	BEST SOURCES	ESSENTIAL FOR	DEFICIENCY SYMPTOMS

vitamin A (oil soluble)	fish-liver oils
	liver and kidney
	green and yellow
	    vegetables
	yellow fruit
	tomatoes
	butter
	egg yolk
		growth
		health of the eyes
		structure and functions
		    of the cells of the skin
		    and mucous membranes
			retarded growth
			night blindness
			susceptibility to infections
			changes in skin and
			    membranes
			defective tooth formation

thiamine (B₁) (water soluble)	seafood
	meat
	soybeans
	milk
	whole grain
	green vegetables
	fowl
		growth
		carbohydrate metabolism
		functioning of the heart,
		    nerves, and muscles
			retarded growth
			loss of appetite and
			    weight
			nerve disorders
			less resistance to fatigue
			faulty digestion
			    (beriberi)

riboflavin (B₂) (water soluble)	meat
	soybeans
	milk
	green vegetables
	eggs
	fowl
	yeast
		growth
		health of the skin and
		    mouth
		carbohydrate metabolism
		functioning of the eyes
			retarded growth
			dimness of vision
			inflammation of the
			    tongue
			premature aging
			intolerance to light

niacin (water soluble)	meat
	fowl
	fish
	peanut butter
	potatoes
	whole grain
	tomatoes
	leafy vegetables
		growth
		carbohydrate metabolism
		functioning of the
		    stomach and intestines
		functioning of the
		    nervous system
			smoothness of the tongue
			skin eruptions
			digestive disturbances
			mental disorders
			    (pellagra)

vitamin B₁₂ (water soluble)	green vegetables
	liver
		preventing pernicious
		    anemia
			a reduction in number of
			    red blood cells

ascorbic acid (C) (water soluble)	citrus fruit
	other fruit
	tomatoes
	leafy vegetables
		growth
		maintaining strength of
		    the blood vessels
		development of teeth
		gum health
			sore gums
			hemorrhages around the
			    bones
			tendency to bruise easily
			    (scurvy)

vitamin D (oil soluble)	fish-liver oil
	liver
	fortified milk
	eggs
	irradiated foods
		growth
		regulating calcium and
		    phosphorus metabolism
		building and maintaining
		    bones, teeth
			soft bones
			poor development of teeth
			dental decay
			    (rickets)

tocopherol (E) (oil soluble)	wheat-germ oil leafy vegetables milk butter	fertility	infertility muscular dystrophy
vitamin K (oil soluble)	green vegetables soybean oil tomatoes	normal clotting of the blood normal liver functions	hemorrhages

FIGURE 13-4
The organs of digestion in the human body.

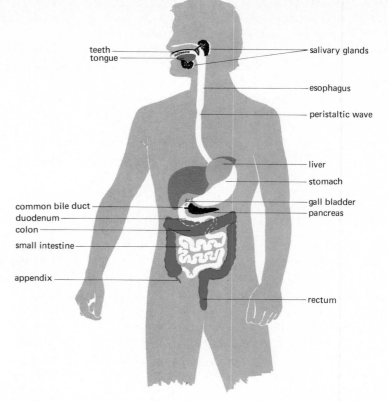

The digestive system includes the organs that form the **alimentary canal,** or digestive tract. It also includes those organs that secrete digestive chemicals into the alimentary canal through tubes, or **ducts.** As foods move through the organs of the alimentary canal, a series of chemical changes occur in the step-by-step process of simplification. Each of these changes requires a specific enzyme secreted by the digestive system.

Digestion in the mouth

Chemical action on food begins in the mouth, where an enzyme in **saliva** begins the hydrolysis of starch. Saliva is a thin, alkaline secretion of the **salivary glands.** It is more than 95 percent water and contains mineral salts, lubricating mucus, and the enzyme **ptyalin** which converts cooked starch to maltose, a disaccharide. It is necessary to cook starchy foods such as potatoes in order to burst the cellulose cell walls. This allows the ptyalin to contact the starch grains. Because of the short time food is in the mouth, starch digestion is seldom completed when food is swallowed. However, ptyalin continues to act in the stomach. Food is carried from the mouth to the stomach by a tube, the **esophagus.** Rhythmic contractions, **peristalsis,** of the esophagus propel the food downward.

The action of gastric fluid

The principal enzyme in **gastric fluid** (the fluid secreted into the stomach) is **pepsin.** It acts on protein, splitting the complex molecules into simpler groups of amino acids, known as **peptones.** This is the first in a series of chemical changes involved in protein digestion. Another component of gastric fluid, **hydrochloric acid,** provides the proper pH for the action of pepsin, dissolves minerals that are insoluble at a neutral pH, and kills many bacteria that enter the stomach with food. It also regulates the action of the pyloric valve, which opens at the completion of stomach digestion and allows food to pass to the small intestine.

Food usually remains in the stomach for two or three hours. During this period, contractions of the stomach muscles churn it. This action separates food particles, and mixes them thoroughly with gastric fluid.

The food passing from the stomach to the small intestine may contain the following: (1) fats (unchanged); (2) sugars (unchanged); (3) the starches that were not acted on by ptyalin; (4) maltose formed by the action of ptyalin; (5) coagulated milk protein called casein; (6) those proteins that were unchanged by the pepsin of the gastric fluid; and (7) aggregates of amino acids formed from pepsin acting on protein.

Functions of the liver and bile

In receiving glucose from the blood and changing it to glycogen, the liver serves as a chemical "storehouse" for holding reserve carbohydrates in the form of glycogen. In acting on excess amino acids not needed by the body and forming urea, the liver is an organ of excretion.

As a digestive gland, the liver secretes **bile,** which acts on food in the small intestine. In the formation of bile, the liver plays a part in using what might otherwise be discarded as waste. Part of the bile is formed of hemoglobin from worn out red blood cells. From the liver, bile moves to the **gall bladder** where it is stored and concentrated before secretion into the alimentary canal. Bile has at least two functions in digestion. First, it aids enzymatic action by breaking globules of fat into smaller, more easily acted on droplets. This process is termed **emulsification.** Second, it activates an enzyme from the pancreas, **lipase,** which breaks down lipids.

The role of the pancreas in digestion

Pancreatic fluid contains enzymes (trypsin, amylase, and lipase) which act on the three major classes of nutrients. **Trypsin** continues the breakdown of proteins that began in the stomach by changing peptones into still simpler amino acid groups called **peptides.** Peptides are not the final product of protein digestion; one additional step is necessary to form the amino acids used in protein synthesis by the cells. **Amylase** continues with the action of the ptyalin by changing starch into maltose. **Lipase** splits fat into **fatty acids** and **glycerol,** both of which can be absorbed by the body cells.

Digestion in the small intestine

The fluid secreted by the intestinal glands is highly alkaline and contains four principal enzymes: protease, maltase, lactase, and sucrase. **Protease** completes protein digestion by decomposing peptides into amino acids. **Maltase** splits the disaccharide maltose into two molecules of the monosaccharide glucose. **Lactase** acts similarly on lactose, or milk sugar, changing it to glucose and galactose. **Sucrase** acts on sucrose, breaking it down to the simple sugars glucose and fructose. The suffix "ase" (meaning separation) indicates an enzyme.

Thus, with the combined action of bile, pancreatic fluid, and intestinal fluid in the small intestine, all three classes of foods are completely digested. Soluble substances, including simple sugars, fatty acids, glycerol, and amino acids, leave the digestive system and enter the blood and lymph.

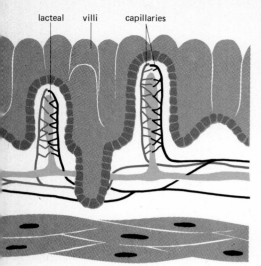

lacteal   villi   capillaries

muscle of intestinal wall

FIGURE   13-5

The absorption surface of the small intestine is greatly increased by the villi.

Absorption in the small intestine

Magnification shows that the lining of the small intestine is a mat of tiny, fingerlike projections called **villi** (Figure 13-5). They are so numerous that they give a velvety appearance to the inside of an intestine. Within the villi are capillaries and branching lymph vessels called **lacteals.** The villi bring blood and lymph close to the digested food and increase the surface area of the intestine enormously so that nutrients can diffuse more easily into the intestine. Diffusion is increased further by a constant swaying motion of the villi throughout the intestine.

Glycerol and fatty acids enter the villi and are carried away by the lymph. They eventually reach the general circulation and travel to cells by diffusion from the circulatory system. Monosaccharides and amino acids enter the blood vessels of the villi, and are carried to the liver.

Water absorption in the large intestine

The large intestine receives a watery mass of undigestible food bulk from the small intestine. As it slowly moves along, much of the water is absorbed and taken into the tissues. The remaining intestinal content, the **feces,** becomes more solid as its water is absorbed. The feces pass into the rectum, from which they are eventually eliminated through the **anus.**

Table 13-3 summarizes the processes that occur along the digestive tract.

TABLE 13-3   SUMMARY OF DIGESTIVE PROCESS

PLACE OF DIGESTION	GLANDS	SECRETION	ENZYMES	DIGESTIVE ACTIVITY
mouth	salivary	saliva	ptyalin	changes starch to maltose, lubricates
	mucous	mucus		lubricates
esophagus	mucous	mucus		lubricates
stomach	gastric	gastric fluid	pepsin	changes proteins to peptones and proteoses
		hydrochloric acid		activates pepsin; dissolves minerals; kills bacteria
	mucous	mucus		lubricates
small intestine	liver	bile		emulsifies fats; activates lipase
	pancreas	pancreatic fluid	trypsin	changes proteins and peptones to peptides
			amylase	changes starch to maltose
			lipase	changes fats to fatty acids and glycerin
	intestinal glands	intestinal fluid	protease	changes peptides to amino acids
			maltase	changes maltose to glucose
			lactase	changes lactose to glucose and galactose
			sucrase	changes sucrose to glucose and fructose
	mucous	mucus		lubricates
large intestine (colon)	mucous	mucus		lubricates, stores and releases fecal matter

326

THE CIRCULATORY SYSTEM

The products of digestion enter the body's fluids through the walls of the digestive tract and are dispersed throughout the body. The flow of nutritive fluids, waste materials, and water in living organisms is **circulation.** Within the animal kingdom as we saw are a wide variety of circulatory systems.

The sponges, for example, accomplish circulation by simply pumping seawater through their bodies (Figure 13-6). The seawater supplies each cell with oxygen and washes wastes away. The fluids within humans and other land-dwelling organisms are strikingly similar to seawater. This is not surprising since our earliest ancestors were probably sea-dwelling organisms. Over the course of evolutionary history, circulatory fluids have been internalized.

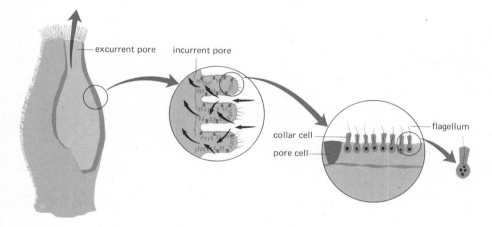

FIGURE 13-6

Water is continually drawn into the sponge by the flagella of the collar cells. It passes through small pores into the cavity of the sponge and out through the excurrent pore.

In humans, as well as other higher organisms where the volumes involved are large, circulatory fluids must be moved with a pump—the heart. (See Figure 13-7 for a summary of the circulatory system.) If the heart stops working, a person's cells are in the same predicament in which a sponge would be if it were thrown up on the beach.

Blood—a fluid tissue

Blood is the transporting medium for nutritive substances and wastes. An average mature person has about 12 pints of blood, which composes about nine percent of the body weight. It consists of a fluid portion, the **plasma,** and the suspended blood cells (Figure 13-8). The plasma is a straw-colored liquid of which nine-tenths is water.

Proteins in plasma give it a sticky consistency. One of them, **fibrinogen,** is essential in the clotting of blood. Another, **serum globulin,** gives rise to anti-bodies that provide immunity to certain diseases. These and other proteins in the plasma also function to give the blood a thickness, or **viscosity,** that aids in maintaining pressure within the vessels.

Besides proteins, the following materials are also present in plasma:

- Inorganic minerals   These give plasma a salt content of approximately 1 percent (that of seawater is about 3 percent). Among the mineral salts are positive ions (including calcium, sodium, magnesium, and potassium) and negative ions (including carbonate, phosphate, and chloride). They are absolutely essential for the proper functioning

FIGURE 13-7

CIRCULATORY SYSTEM

(flow of nutritive fluids, waste materials and water)

COMPONENT	DESCRIPTION	FUNCTION
blood	about 12 pints of fluid tissue consisting of plasma (fluid) and blood cells (solid)	carries nutrients to cells, removes cell wastes and water, carries oxygen to tissues, equalizes body temperature, distributes ductless gland secretions
pulmonary circulation	involves right side of heart, lung capillaries and pulmonary veins	blood from body is pumped from the right ventricle through the pulmonary arteries to lungs; blood discharges carbon dioxide and water, picks up oxygen and returns to the left atrium through pulmonary veins
systemic circulation	left side of heart, aorta, arteries, capillaries and veins leading to *venae cavae,* plus several shorter circulations	oxygenated blood pumped from left ventricle through the aorta to arteries and capillaries and returned through venules and veins to venae cavae, then to the right atrium
coronary circulation	aorta, coronary arteries and veins	blood flows through the heart muscle, coronary arteries and veins; every heartbeat depends on free flow through these *coronary vessels*
renal circulation	renal artery, kidney capillaries, renal veins	blood nourishes kidneys, discharges water, salts, nitrogenous cell wastes
portal circulation	veins leading from spleen, stomach, pancreas, small intestine and colon	large veins of portal circulation unite to form portal vein which enters liver; blood from digestive organs brings digested food and water; blood with food for body tissues flows from liver in hepatic veins which empty into inferior vena cava

Reprinted from MODERN BIOLOGY, © 1977, by Otto and Towle.

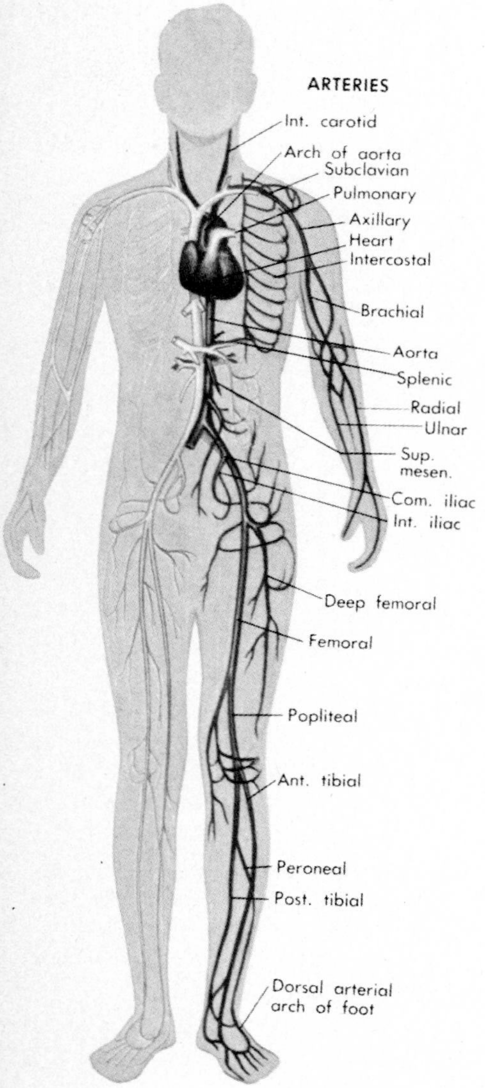
ARTERIES
Int. carotid
Arch of aorta
Subclavian
Pulmonary
Axillary
Heart
Intercostal
Brachial
Aorta
Splenic
Radial
Ulnar
Sup. mesen.
Com. iliac
Int. iliac
Deep femoral
Femoral
Popliteal
Ant. tibial
Peroneal
Post. tibial
Dorsal arterial arch of foot

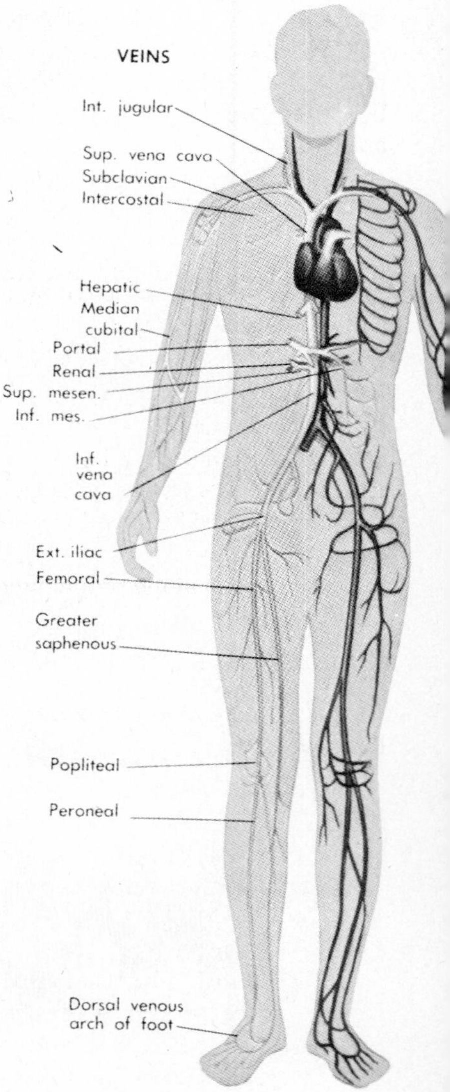
VEINS
Int. jugular
Sup. vena cava
Subclavian
Intercostal
Hepatic
Median cubital
Portal
Renal
Sup. mesen.
Inf. mes.
Inf. vena cava
Ext. iliac
Femoral
Greater saphenous
Popliteal
Peroneal
Dorsal venous arch of foot

of body tissues. Without calcium compounds, for example, blood wouldn't clot and bones wouldn't develop.

- Digested foods   These are present in plasma in the form of glucose, fatty acids, glycerol, and amino acids. They are transported to the liver and other body tissues.

- Nitrogen-containing wastes   These result from protein degradation, and travel in the plasma to the organs of excretion, the kidneys.

The suspended components of blood

Red blood cells (red corpuscles), **white blood cells** (white corpuscles), and **platelets** are suspended in blood (Figure 13-9). The red blood cells are disk-shaped with concave faces. Sometimes they travel in rows, like stacks of coins, or they may separate and float individually. They are so small that ten million of them could be spread out in one square inch. Yet, they are so numerous that the red blood cells from one person would cover an area of 3,500 square yards. The pigment in them is the complex iron-containing protein, **hemoglobin.** It gives blood its red color and carries oxygen throughout the body.

Red blood cells are produced by the **red marrow** of the insides of certain bones, including the ribs, vertebrae, and skull. In children, they are also produced by the ends of long bones. During their development, red blood cells are colorless and have large nuclei. Normally, by the time they are released into the bloodstream, they have lost the nuclei and have accumulated hemoglobin.

The life span of a red blood cell is usually between 20 and 120 days. When they cease functioning, they are removed by the liver or the spleen. The liver utilizes hemoglobin in forming bile, and cycles substances vital in the formation of new red blood cells back into the blood.

The pigment hemoglobin is an unusual iron-containing protein within red blood cells. The chemical properties of hemoglobin are such that it can rapidly take on oxygen from air, and it can just as readily let loose of it in the body's tissues.

Undoubtedly you have seen the rusting hulk of a wrecked car. Rust is iron which has chemically reacted with oxygen from the air. It is a product of the **oxidation** of iron. The iron of hemoglobin combines with oxygen in the lungs, but there is an important difference between this union and the process of rusting. The iron in rust doesn't easily give up its oxygen. The iron in hemoglobin, however, lets loose of oxygen at the proper time and place in the body. In the tissues, oxygen is given up, and carbon dioxide that has formed during respiration combines with hemoglobin. The carbon dioxide is carried to the lungs, via the red blood cells, where it is released, as oxygen is absorbed.

Most white blood cells are larger than red blood cells and differ from them in three ways:

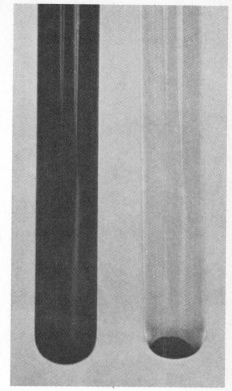

FIGURE   13-8

Fresh blood in a test tube and blood in a test tube after settling and centrifugation. Notice how little suspended matter is in blood.

red corpuscles

platelets

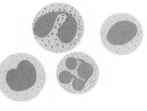

white corpuscles

FIGURE   13-9

The suspended components of blood include red corpuscles, or blood cells (size, 8 micrometers), white corpuscles, or blood cells (size, 16 micrometers), and platelets (size, 1 micrometer).

- White blood cells have nuclei.
- White blood cells do not contain hemoglobin; they are nearly colorless.
- Some white blood cells are capable of **amoeboid movement**—they can move about by extending bulges of protoplasm, like the amoeba.

White blood cells are less numerous than the red cells, the ratio being about one white cell to every 600 red cells. They are formed in several places, including the red bone marrow, the spleen, and the lymph glands. White blood cells are, in general, more short-lived than red cells; they usually exist only a few days.

White blood cells are able to ooze through capillary walls into intercellular spaces. Here they engulf solid materials, including bacteria, and thus are an important defense against infection. When an infection develops in the tissues, the white-cell count may go from 8,000 to more than 25,000 per cubic millimeter. They collect by amoeboid movement in an area of infection and destroy bacteria by engulfing them. The thick, yellowish remains of dead bacteria, white blood cells and tissue fluid is **pus.**

Platelets are irregularly shaped, colorless bodies. They are much smaller than the red blood cells, with a count of about 300,000 per cubic millimeter of blood. Platelets are thought to be formed in red bone marrow. They function in clotting. An overview of the functions of blood as a transporting medium is given in Table 13-4.

TABLE 13-4   BLOOD AS A TRANSPORTING MEDIUM

TRANSPORTATION OF	FROM	TO	FOR THE PURPOSE OF
digested food	digestive organs and liver	tissues	growth and repair of cells, supplying energy, and regulating life processes
cell wastes	active tissues	lungs, kidneys, and skin	excretion
water	digestive organs	kidneys, skin, and lungs	excretion and equalization of body fluids
oxygen	lungs	tissues	oxidation
heat	tissues	skin	equalization of the body temperature
secretions	ductless glands	various organs, glands	regulation of body activities

Clotting

When blood begins to ooze from a break in the skin, a series of clotting reactions soon starts (Figure 13-10). First, the platelets disintegrate and release an activating substance, **thromboplastin,** which reacts with a protein in blood, **prothrombin,** and calcium to form **thrombin.** This, in turn, reacts with another blood protein, **fibrinogen,** to form **fibrin.** The tiny threads of fibrin congeal in a network that traps blood cells, thus forming a clot, and preventing further escape of blood. A scab forms as the clot dries, and healing occurs as the edges of the wound grow toward the center. The process can be summarized as:

1   thromboplastin + prothrombin + calcium → thrombin
2   thrombin + fibrinogen → fibrin

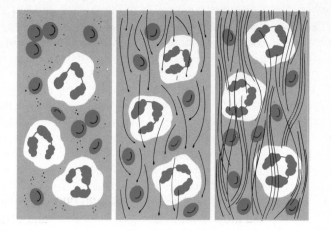

If blood vessels are broken under the skin, a discolored area, or a bruise, may appear as clotting occurs. Gradually, the clotted blood is absorbed, and the color of the bruise changes and finally disappears.

Blood transfusions

Conditions like severe **hemorrhage** (bleeding), wound shock, and certain illnesses may require blood transfusions. If whole blood is used, the patient receives both the necessary plasma and blood cells from a donor. The blood of the donor, however, must be matched with that of the patient. This is done by adding a drop of a test serum to a drop of the donor's blood. If the red cells **agglutinate,** or clump together, the bloods are incompatible. If, however, the red cells remain in suspension, the samples are compatible, and it is possible to make a transfusion.

The Rh factor

One of the ways in which blood types may be non-compatible is in their Rh factors. The Rh factor was first discovered in the rhesus monkey (the designation, "Rh," is derived from the first two letters of the monkey's name). It is an inherited difference in the proteins on the surfaces of red blood cells. Among the human population of the United States about 85 percent of the people have the Rh proteins on their red blood cells. They are said to be Rh-positive. The remaining 15 percent do not have the Rh proteins (more precisely, they possess them in insignificant amounts), and they are said to be Rh-negative.

An antigen/antibody reaction causes problems when an Rh-negative individual receives a transfusion of Rh-positive blood. **Antibodies** are proteins produced within higher organisms which destroy foreign substances called **antigens.** Antibodies are normally produced to ward off disease carrying microorganisms but the Rh antibody causes the red blood cells of Rh-negative blood to agglutinate and break up (Figure 13-11). Complications from the Rh factor occur with childbearing in about one in three hundred births. An antigen/antibody reaction may occur when the mother is Rh-negative, the father is Rh-positive, and the child inherits the Rh-positive factor from the father.

The bloodstreams of an expectant woman and the child she is bearing normally do not mix, although they come close to each other within the **placenta.** Sometimes, slight ruptures develop in the placenta through which blood from the child may seep into the mother's bloodstream and vice versa. If the child's blood

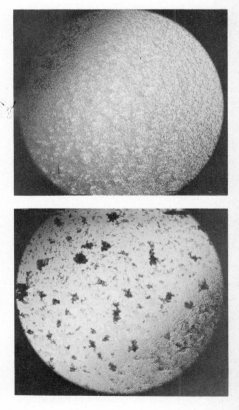

is Rh-positive and the mother's blood is Rh-negative, the mother will begin to produce the Rh antibody.

If the exchange of blood is slight, there are usually no problems with the first pregnancy. However, if seepage occurs again when the woman is pregnant, with a second Rh-positive child, the antibody in the mother's blood (which has remained from the first pregnancy) enters the fetus's circulatory system and causes agglutination of its red blood cells. Occasionally the child dies before birth. This problem has been eliminated with a recent technique. It involves the injection of small amounts of Rh-positive antibody into the mother soon after the birth of the first child. The injected antibodies destroy the Rh positive fetal blood cells that leak through the placenta into the mother during a second pregnancy. This prevents the mother from producing antibody. Thus, a second Rh-positive child that is conceived is safe because the mother is not producing antibody toward it.

A-B-O blood types

In addition to the Rh factor, people may inherit one or two of another group of antigens on their red blood cells. These have been termed antigen-A and antigen-B. People whose red blood cells contain antigen-A are said to have type-A blood, and those who have antigen-B possess type-B blood. People who inherit both antigens are said to be type-AB, and those who inherit neither antigen are termed type-O. Thus, there are four possible A-B-O blood types. (See Chapter 10.)

It is a curious property of the inheritance of A-B-O blood types that antibodies as well as antigens are inherited. People are born with antibodies in their blood plasma which will act against the antigens their red blood cells do *not* contain. For instance, a person with type-A blood has plasma which contains anti-B antibodies, and someone with type-O blood has plasma with both anti-A and anti-B antibodies. This is one of the few cases in which the human body normally synthesizes antibodies against antigens to which it has not yet been exposed. The anti-A and anti-B antibodies react with A and B antigens in essentially the same way that the Rh antibody reacts with the Rh antigen; they cause agglutination of red blood cells.

Because the antibodies present in the plasma of blood of one type cause the red blood cells of blood of other types to agglutinate, it is best when seeking a donor for a blood transfusion to find someone who has the same blood type as the patient. It has been found, however, that when such a donor is not available blood of another type may be used, provided that the plasma of the patient and the red blood cells of the donor are compatible. This is possible because, unless the transfusion is to be a massive one or is to be made very rapidly, the donor's plasma is sufficiently diluted during transfusion so that little or no agglutination occurs.

This means that type-O blood can be given to anyone because its red blood cells have no antigens and are compatible with the plasma of any patient. Type-O blood, therefore, is called the "universal donor." But, type-O patient can receive

TABLE 13-5   CHARACTERISTICS OF A-B-O BLOOD TYPES

	TRANSFUSION RELATIONSHIPS		ANTIGEN AND ANTIBODY CONTENTS	
BLOOD TYPE	CAN DONATE BLOOD TO TYPE(S)	CAN RECEIVE BLOOD FROM TYPE(S)	ANTIGENS ON RED BLOOD CELLS	ANTIBODIES IN PLASMA
O	O, A, B, AB	O	None	anti-A, anti-B
A	A, AB	O, A	A	anti-B
B	B, AB	O, B	B	anti-A
AB	AB	O, A, B, AB	A, B	None

transfusions only from type-O donors because their plasma contains both anti-A and anti-B and is not compatible with the red blood cells of any other class of donor. Conversely, people with type-AB blood, whose plasma contains no antibodies, are "universal recipients" but cannot act as donors for any except type-AB patients. Table 13-5 summarizes the A-B-O transfusion relationships.

More is said about antigen/antibody reactions in Chapter 14. For now, however, we will turn to the structure and function of the circulatory system.

THE STRUCTURE OF THE HEART

The heart is a cone-shaped, muscular organ situated under the breastbone and between the lungs in a sac called the **pericardium.** It usually lies a little to the left of the midline of the chest cavity, with its point extending downward and to the left between the fifth and sixth ribs. Since the beat is strongest near the tip, many people have the mistaken idea that the entire heart is on the left side.

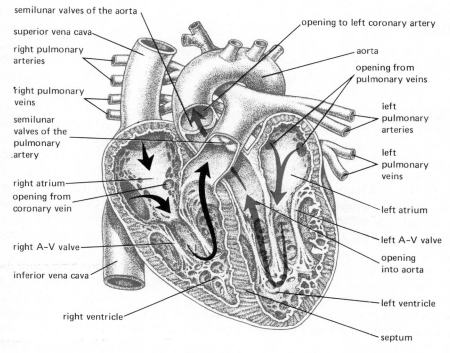

semilunar valves of the aorta
superior vena cava
right pulmonary arteries
right pulmonary veins
semilunar valves of the pulmonary artery
right atrium
opening from coronary vein
right A–V valve
inferior vena cava
right ventricle

opening to left coronary artery
aorta
opening from pulmonary veins
left pulmonary arteries
left pulmonary veins
left atrium
left A–V valve
opening into aorta
left ventricle
septum

FIGURE 13-12
The human heart. Note the location of the valves to the heart chambers and blood vessels. The arrows indicate the direction of blood flow.

The heart is separated into two halves by a wall called the **septum** (Figure 13-12). Each half is composed of two chambers, a relatively thin-walled **atrium** and a thick, muscular **ventricle.** The two atria (plural of atrium) act as reservoirs for the blood entering the heart. Both contract at the same time, filling the two ventricles rapidly. Next, the thick, muscular walls of the ventricles contract, forcing the blood out through the great arteries.

Flow of blood from the ventricles and maintenance of pressure in the arteries between beats require two sets of one-way heart valves. The valves between the atria and ventricles are called the **atrioventricular valves,** or **a-v valves.** They are flaps of tissue anchored to the floor of the ventricles. Blood passes freely through these valves into the ventricles. The valves cannot be opened from the lower side, however, because of the tendons anchoring them. Thus, blood is unable to flow backward into the atria during contraction of the ventricles. Other valves, called the **semilunar valves,** or **s-l valves,** are located at the opening of the arteries. These cuplike valves are opened by the force of blood passing from the ventricles into the arteries, and they prevent blood from returning to the ventricles.

Circulation within the heart

Blood enters the right atrium of the heart by way of the **superior vena cava** and the **inferior vena cava.** The former carries blood from the head and upper parts of the body, and the latter returns blood from the lower body regions (Figure 13-12). From the right atrium, blood passes through the right a-v valve into the right ventricle. When the ventricle contracts, blood is forced through a set of s-l valves into the **pulmonary artery,** which carries it to the lungs. After the blood has passed through the lungs, it is returned to the heart through the right and left **pulmonary veins.** These vessels open into the left atrium, from which the blood passes through the left a-v valve into the left ventricle. From here, blood passes out the **aorta** and is distributed to all parts of the body.

Although the chambers of the heart are continuously filled with blood, its muscle layers are too thick to be nourished by blood. Thus, there is an internal circulatory system consisting of the **coronary arteries.** The right and left coronary arteries branch off from the aorta just above the heart. They curve downward around each side, sending off smaller vessels that penetrate the heart muscle.

The heart beat

A complete cycle of heart activity, or beat, consists of two phases. During the first phase, **systole,** the ventricles contract and force blood into the arteries. During the second phase, **diastole,** the ventricles relax and receive blood from the atria. If you have ever listened to your heartbeat with a stethoscope, you know that there are two distinct alternating sounds. One is the sound of the contraction of the muscles of the ventricles and the closing of the a-v valves during systole. The other is the closing of the s-l valves at the bases of the arteries during diastole.

At rest the heart of an average adult beats about 70 times per minute. During strenuous work, the rate may be as high as 180 beats per minute. With the body at rest, the heart pumps about 10½ pints of blood per minute. An average adult has about 12 pints of blood, so all of it makes a complete circuit in slightly over a minute. However, mild exercise, such as walking, speeds the heart output to about 20 pints per minute, and strenuous exercise may increase it to as much as 42 pints per minute. The heart is an incredibly efficient pump, whose Herculean task is all the more amazing since it weighs only one-fourth of a pound.

The blood vessels

Blood moves in a system of vessels of varying sizes. **Arteries** carry blood away from the heart, while **veins** bring it back for oxygen. **Capillaries** are the minute blood vessels where arteries blend into veins (Figure 13-13). At the capillaries, substances are transferred between the blood and tissues.

Arteries have elastic, muscular walls and smooth linings. Because of their elasticity, they can expand and absorb part of the pressure resulting from contraction of the ventricles at systole. The pressure in the aorta leading from the left ventricle is greater than that in the pulmonary artery pumped by the smaller right ventricle. If the aorta were cut, blood would spurt out in a stream of six feet or more. When the ventricles contract, arterial pressure is greatest and is called **systolic pressure.** The elasticity of the artery walls maintains part of this pressure while the ventricles are at rest. This is the time of lowest pressure in the arteries, or **diastolic pressure.** The bulge in an artery wall caused by systolic pressure can be felt in the wrist or any part of the body where an artery is near the surface.

FIGURE 13-13

Three types of blood vessels.

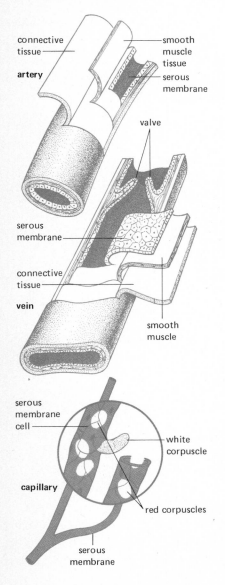

connective tissue

artery

smooth muscle tissue

serous membrane

valve

serous membrane

connective tissue

vein

smooth muscle

serous membrane cell

white corpuscle

capillary

red corpuscles

serous membrane

The capillaries

As the arteries penetrate the tissues, they branch into smaller and smaller vessels, which finally become capillaries. Capillary walls are only one-cell layer thick; they are only slightly greater in diameter than the red blood cells which they carry. Red blood cells must pass through the capillaries in single file, and may even be pressed out of shape by the capillary walls.

Dissolved nutrients, waste products, and gases diffuse through the thin walls of capillaries and in and out of the tissue spaces. Tiny openings in the walls are penetrated by white blood cells as they leave the bloodstream and enter the tissue spaces. Also, part of the plasma diffuses from the blood and becomes tissue fluid. Thus, all the vital relationships between the blood and the tissues occur in the capillaries and not in arteries and veins.

Vein structure and function

On leaving an organ, capillaries unite to form veins. Veins carry dark red blood, which contains little oxygen. Yellow pigments in the skin cause veins to appear bluish in color. The walls of veins are thinner and less muscular than those of arteries, which carry bright red, oxygenated blood. Many of the larger veins have cuplike valves that prevent the backward flow of blood.

Veins have no pulse wave, and the blood pressure within them is much lower than that in arteries. Blood pressure resulting from heart action is almost completely lost as blood passes through the capillaries. Blood from the head may return to the heart with the aid of gravity, but in the body regions below the level of the heart, other factors are required. Venous flow from these regions is aided by the working muscles and by respiration movements.

Circulation throughout the body

Our four-chambered heart is really a double pump in which the two sides work in unison. Each side pumps blood through a major division of the circulatory system (Figure 13-14). The right side of the heart receives dark, deoxygenated blood from the body and pumps it through the arteries of the **pulmonary circulation.** The large pulmonary artery, extending from the right ventricle, sends a branch to each lung. These arteries in turn branch within the lungs, forming a vast number of smaller vessels. Here the blood discharges carbon dioxide and water and receives oxygen. Oxygenated blood, now bright scarlet in color, leaves the lungs and returns to the left atrium of the heart through the pulmonary veins.

Oxygenated blood passes through the left chambers of the heart and out the aorta under great pressure. The blood is now in the **systemic circulation,** which supplies the body tissues. This extensive circulation includes all of the arteries that branch from the aorta, the capillaries that penetrate the body tissues, and the vast number of veins that lead to the two vena cava. The systemic circulation also includes several subsystems that supply or drain certain organs of the body.

The **coronary circulation,** referred to in the discussion of the heart muscle, supplies the heart itself. This short but vital circulation begins at the aorta and ends where the coronary veins empty into the right atrium. Every beat of the heart depends on the free flow of blood through the coronary vessels.

The **renal circulation** starts where a **renal artery** branches from the aorta to each kidney. It includes the capillaries that penetrate the kidney tissue and the **renal veins,** which return blood from the kidneys to the inferior vena cava. Blood

FIGURE 13-14

A representation of the various circulations in the human body. Blood leaves the heart via arteries and returns by veins.

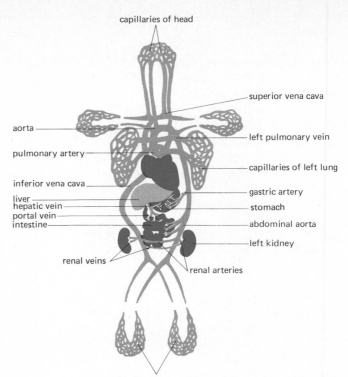

capillaries of head

superior vena cava

aorta

left pulmonary vein

pulmonary artery

capillaries of left lung

inferior vena cava

gastric artery

liver
hepatic vein
portal vein
intestine

stomach

abdominal aorta

left kidney

renal veins

renal arteries

capillaries of legs and feet

on this route nourishes the kidneys and discharges water, salts, and nitrogenous cell wastes. Thus, even though it is low in oxygen content, blood in the renal veins is the purest in the body.

The **portal circulation** includes an extensive system of veins that lead from the spleen, stomach, pancreas, small intestine, and colon. These large veins of the portal circulation unite to form the **portal vein,** which enters the liver. Blood flowing from the digestive organs transports digested food and water to the liver. Blood laden with nutrients for the body tissues flows from the liver in the **hepatic veins,** which in turn empty into the inferior vena cava, completing the systemic cycle.

Return of tissue fluid to the circulatory system

The tissue fluid, lymph, bathes the cells, and is collected in minute vessels. Small lymph vessels join one another to become larger ones in the same way in which capillaries join to form veins. **Lymph nodes,** like beads on a necklace, are enlargements in the lymph vessels, called **lymphatics** (Figure 13-15). Here white corpuscles collect and destroy bacteria. The **lymphatic system** returns some of the tissue fluid brought by blood capillaries back to the veins.

The greatest concentrations of lymph nodes are in the neck, the armpit, the bend of the arm, and the groin. When there is an infection in the hand or arm, the lymph nodes of the armpit may swell and become painful. Both the **tonsils** and **adenoids** in the throat are masses of lymphatic tissue that often become inflamed during childhood and have to be removed surgically.

The lymph of the right side of the head, neck and right arm enters a large vessel named the **right lymphatic duct.** This vessel returns the lymph to the blood by opening into the **right subclavian vein.** The lymphatics from the rest of the body drain into the **thoracic duct,** which in turn empties into the **left subclavian vein.**

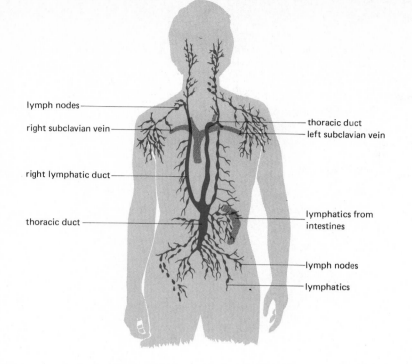

FIGURE 13-15
The lymphatic system returns tissue fluid to the blood stream.

lymph nodes

right subclavian vein

thoracic duct
left subclavian vein

right lymphatic duct

thoracic duct

lymphatics from intestines

lymph nodes

lymphatics

In the walls of the larger lymph vessels are valves to control the flow. These valves are similar in structure to those found in the veins. The return of lymph to the bloodstream is aided by contractions of the body muscles. Thus, the flow of lymph increases with body activity.

THE EXCRETORY SYSTEM

The oxidation of foods in metabolism produces waste products that are removed by **excretion.** In protein metabolism, waste products result from the separation of the carbon and nitrogen parts of amino acids before oxidation of the carbon part. Other waste products result from the synthesis of proteins from amino acids during growth processes. These nonprotein **nitrogenous wastes** include **urea** and **uric acid.**

Any great accumulation of wastes in the tissues, especially nonprotein nitrogens, causes rapid tissue poisoning, starvation, and eventually suffocation. Tissues filled with waste products can absorb neither food nor oxygen. Fever, convulsions, coma, and death are inevitable if metabolic wastes aren't eliminated from the tissues.

Further complications arise if mineral acids and salts accumulate in the body. Their presence disrupts delicate acid-base balances and also upsets the osmotic relationships between blood, lymph, and the tissues. When excess salts are held in the tissues, water accumulates, causes swelling, and a weight gain.

One-celled organisms and animals like the sponge and jellyfish discharge their cell wastes directly into their marine environment. However, when many millions of cells form an organism, as in higher animals, the removal of cell waste products becomes a complicated process involving many organs. Each cell discharges its waste materials into the tissue fluid, which in turn reaches the bloodstream. The blood transports the cell wastes to excretory organs such as the kidneys and skin for elimination.

The kidneys—the principal excretory organs

The **kidneys** are bean-shaped organs, about the size of clenched fists. They lie on either side of the spine, in the small of the back. Layers of fat form a protective covering around them.

A cross-section of a kidney reveals several regions (Figure 13-16). The firm outer covering that composes about one-third of the kidney tissue is the **cortex.** The inner two-thirds, or **medulla,** contains conical protrusions which extend into a sac-like cavity, the **pelvis** of the kidney. The pelvis, in turn, leads into a long, narrow tube called the **ureter.** The two ureters (one for each kidney) empty into the **urinary bladder.**

FIGURE 13-16

The kidney is an efficient filtering organ.

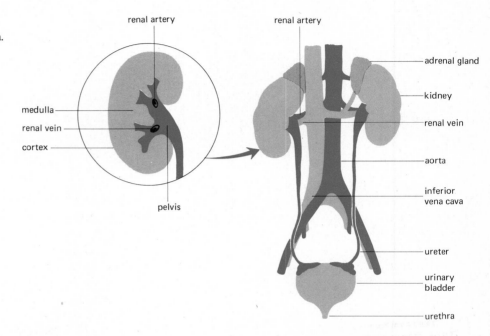

Each kidney contains over a million tiny filters called **nephrons** (Figure 13-17). Their function is to control the chemical composition of blood and urine. Each nephron consists of a small, cup-shaped structure called a **Bowman's capsule.** A tiny, winding **convoluted tubule** comes from each capsule. It loops toward the pelvis, then returns to the cortex where it joins the **collecting tubule,** which receives fluid from many nephrons. The collecting tubule carries urine to the renal pelvis which as we saw becomes the ureter.

Blood enters each kidney through a large **renal artery,** which branches directly from the aorta. In the kidney, the renal artery branches and rebranches to form a maze of smaller vessels which penetrate all areas of the cortex. Each of these vessels ends in a coiled knoblike mass of capillaries, the **glomerulus,** which fills the cup of a Bowman's capsule. In the first stage of waste removal, fluids diffuse from the bloodstream and enter the Bowman's capsule. In the second stage, valuable substances which left in excess in the first stage are returned to the blood.

Removal of fluids from the bloodstream takes place in the coiled capillaries of the glomeruli. Here, water, nitrogenous wastes, glucose, and mineral salts pass through the walls of the capillaries into the surrounding capsule. This solution resembles blood plasma without the proteins. Complete loss of this much water, glucose, and minerals would be fatal. However, after the fluid leaves the capsule it passes a network of capillaries, which reabsorb vital materials. Only nitrogenous

338

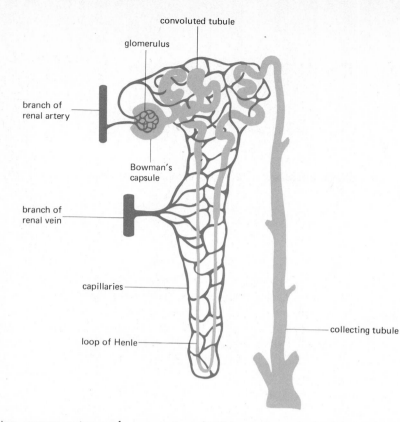

convoluted tubule

glomerulus

branch of
renal artery

Bowman's
capsule

branch of
renal vein

capillaries

loop of Henle

collecting tubule

FIGURE 13-17
The human nephron. For description, see text. (Adapted from *The Kidney: Structure and Function in Health and Disease* by Homer W. Smith. Copyright © 1951 by Oxford University Press, Inc. Reprinted by permission.)

wastes, excess water, and excess mineral salts pass through the tubules to the pelvis of the kidney as **urine.**

For every 100 milliliters of fluid that pass from the blood into the capsules, 99 milliliters are reabsorbed. The urine passes from the pelvis of each kidney through the ureters to the urinary bladder. Contractions of the bladder expels the urine through the **urethra.** Blood leaves the kidneys through the **renal veins** and returns to the general circulation by way of the inferior vena cava. The two kidneys have tremendous reserve power. If one is removed, its mate becomes enlarged and assumes a near normal function of two kidneys.

Other organs of excretion

During **expiration** (a synonym for exhalation) the lungs give off carbon dioxide and water vapor—the end products of the Krebs cycle and the electron transport chain. The excretory function of the **liver** in forming urea has been discussed earlier. The bile formed in the liver and stored in the **gall bladder** is also a waste-containing substance, in that is is produced from dysfunctional red blood cells.

The **large intestine** removes undigested food. This, however, is not cell excretion in the strict sense, since the food refuse collected there has never actually been absorbed into the tissues.

VARIED FUNCTIONS OF THE SKIN

The skin supplements the kidneys in the excretion of water, salts, and some urea, in the form of **perspiration.** But, this fluid is more important in regulating body temperature than it is as an excretory substance. Heat is absorbed during the change from liquid to vapor. Thus, as perspiration evaporates (vaporizes) from the

FIGURE 13-18

The skin, an organ of varied functions.

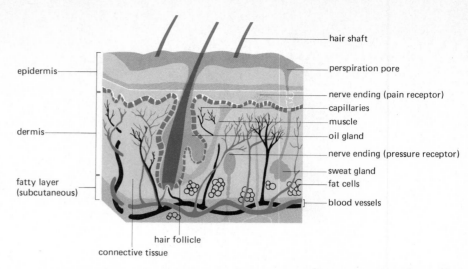

- hair shaft
- perspiration pore
- nerve ending (pain receptor)
- capillaries
- muscle
- oil gland
- nerve ending (pressure receptor)
- sweat gland
- fat cells
- blood vessels

epidermis

dermis

fatty layer (subcutaneous)

hair follicle

connective tissue

body surface, heat is withdrawn from the outer tissues. In a sense, therefore, skin is a radiator. The capillaries within it bear heated blood from the tissues.

As the body temperature rises, the capillaries dilate (expand), the skin becomes flushed with blood (our faces become reddened) and heat is rapidly conducted to the surface. At the same time, sweat is secreted more rapidly. This increases the rate of evaporation and, hence, the rate of heat loss.

In addition to its function as an organ of excretion and a regulator of body temperature, the skin has several other functions. The outermost layer, the **epidermis** (Figure 13-18) serves primarily to protect the active tissues beneath it. It is rubbed off constantly, but active cells in the lower layers continuously regenerate it. Friction and pressure on the epidermis stimulate cell division and may produce a **callus** more than a hundred cells thick. Hair and nails are special outgrowths of the epidermis. The **dermis** lies under the epidermis. It is a thick, active layer, composed of tough, fibrous tissue, richly supplied with blood and lymph vessels, nerves, sweat glands, and oil glands. The following are among the varied functions of the skin:

- Protection of the body from mechanical injury and bacterial invasion (Figure 13-19).
- Protection of the inner tissues from drying. The skin, aided by oil glands, is nearly waterproof. Little water passes through it, except through the pores.
- Location of the nerve receptors that respond to touch, pressure, pain, and temperature.
- Excretion of wastes in perspiration.
- Control of the loss of body heat through the evporation of perspiration.

FIGURE 13-19

An electron micrograph of human skin. Ever-present bacteria (size, a few micrometers) lie in its folds.

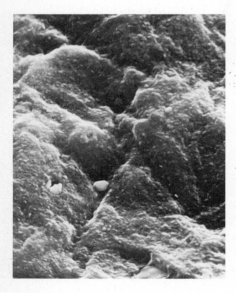

THE DUCTLESS GLANDS

You are already familiar with certain glands, such as the salivary glands of the mouth and gastric glands of the stomach. These pour secretions into the digestive tract through the tubes, or ducts. The **ductless glands** give off their secretions directly into the bloodstream. With blood as a transporting medium, these substances are distributed throughout the body. Ductless glands are also called **endocrine glands** (Figure 13-20).

The secretions of the endocrine glands, **hormones,** are effective in very small amounts. They are complex compounds of various types formed from sub-

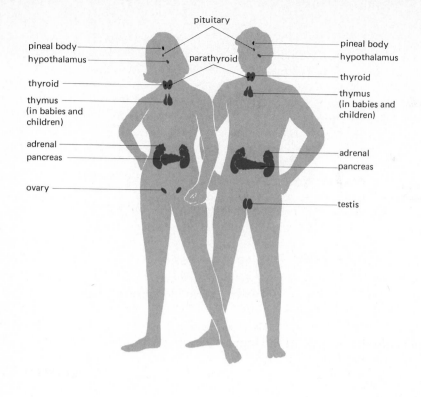

FIGURE 13-20
The locations of endocrine glands.

stances taken from the blood. Hormones regulate the activities of many body processes, including growth and physical development. They are also involved in the flux of human emotions.

The circulatory system is vital to the endocrine system, both in supplying the raw materials for building hormones and in delivering the finished product. For the most part, the endocrine glands are small, but their sizes do not reflect their vital influences on the body.

The thyroid gland

The **thyroid gland** is a relatively large endocrine gland, which lies close to the body surface in the neck (Figure 13-21). It consists of two lobes at the sides of the juncture of the **larynx** (voice box) and **trachea** (wind pipe). Four smaller glands, the **parathyroid glands,** whose function is described later, are embedded within the thyroid glands. The lobes are joined by a narrow isthmus across the front surface of the trachea. The complete thyroid gland somewhat resembles a butterfly with spread wings. One of the hormones of the thyroid gland, **thyroxine,** contains the highest concentration of iodine of any substance in the body.

The hormones of the thyroid gland are involved in various processes of the body, particularly those involved in growth and respiration. Overactivity of the thyroid gland causes a condition known as **hyperthyroidism,** marked by an increased rate of respiration and high body temperature. The rate of heart action is increased, and blood pressure is elevated. Sweating when the body should be cool and extreme nervousness and irritability are also symptoms.

One treatment for hyperthyroidism is surgical removal of a portion of the thyroid gland. A more recently developed treatment involves ingestion of radio-

FIGURE 13-21
The lobes of the thyroid gland lie on either side of the traceha. Note also the positions of the four parathyroid glands, embedded in the back of the thyroid gland.

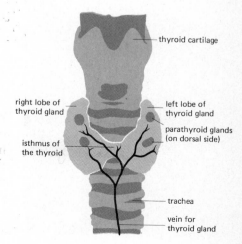

active iodine. This is picked up by the gland just as ordinary iodine would be. In bombarding the gland with radioactivity, it destroys some of the thyroid tissue, as surgery formerly did. Since most of the body's iodine is concentrated in the thyroid, tissues elsewhere in the body are thought to be unharmed.

Underactivity of the thyroid gland is known as **hypothyroidism.** This condition is characterized by symptoms opposite to those of hyperthyroidism. The rate of respiration is depressed, and activity of the nervous system is reduced. Heart action decreases and in many cases the heart enlarges. If left untreated hypothyroidism can lead to stunted growth and mental retardation, but it can be treated with thyroid extract from sheep.

If the thyroid gland is defective during infancy, **cretinism** can result. This condition is characterized by severely stunted physical and mental development. If the cretin passes from infancy to childhood without thyroid extract treatment, the dwarfism and mental deficiency can never be corrected.

Iodine deficiency is the major cause of enlargement of the thyroid gland, known as **simple goiter.** This condition is rare along the seacoast, where people eat large amounts of seafoods. It is most common in mountainous regions, where the iodine content of the soil is low. The addition of iodine compounds to table salt and to the water supply in certain regions are adequate preventive measures.

Since metabolism and oxygen consumption are affected by thyroid activity, the basal metabolic rate gives an indication of thyroid health. More accurate tests, determine the rate at which iodine is taken into the thyroid gland. The amount of iodine in the blood may be measured, but the most precise tests use radioactive iodine. A small amount of radioactive iodine (insufficient to cause radioactive damage) is given to the patient orally. Later, at successive intervals a Geiger counter is placed over the thyroid gland (Figure 13-22). The rate of uptake of the radioactive iodine indicates the activity of the gland.

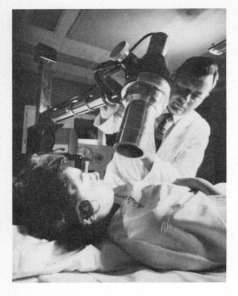

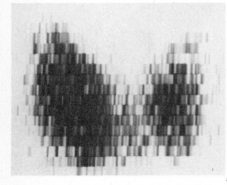

FIGURE 13-22
Measurement of uptake of radioactive iodine in the thyroid gland of a patient. The counter in the upper picture records the areas of concentration of iodine in the gland. A photograph of the record is shown in the lower picture.

The parathyroid glands

The **parathyroids** are four small glands embedded in the back of the thyroid, two in each lobe (Figure 13-21). Their secretion, **calcitonin,** regulates the amount of calcium in the blood. Bone growth, muscle tone, and normal nervous activity are absolutely dependent on a constant, stable calcium balance.

The pituitary gland and the hypothalamus

The **pituitary,** a gland about the size of an acorn, lies at the base of the brain with the hypothalamus next to it (Figure 13-23a). The **hypothalamus** directly influences the pituitary with its secretion of proteins that cause the release of hormones by the pituitary.

The pituitary consists of two lobes, anterior and posterior (Figure 13-23b). The **anterior lobe** secretes several hormones. One of these, **somatotrophin hormone,** or **growth stimulating hormone,** regulates the growth of the skeleton. Other secretions of the anterior lobe, the **gonadotropins,** influence the activities of the sex organs, bringing on the sweeping developmental changes that occur during puberty. **Thyroid stimulating hormone** affects the activities of the thyroid gland, and **ACTH (adrenocorticotropic hormone)** stimulates the outer part, or cortex, of the adrenal glands. ACTH has been used in the treatment of leukemia and, more successfully, in the treatment of arthritis. Good results in the treatment of asthma and other allergies with ACTH have also been reported.

The **posterior lobe** of the pituitary gland produces two hormones, oxytocin and vasopressin. **Oxytocin** helps regulate blood pressure and causes milk release in nursing. Because it also stimulates contractions by muscles of the uterus, oxytocin is sometimes administered during childbirth. **Vasopressin** controls water reabsorption in the kidneys. A deficiency of vasopressin causes a condition called **diabetes insipidus,** in which too much water is eliminated. This disease should not be confused with "true diabetes," which will be discussed in connection with the pancreas.

The most frequent disorder of the pituitary gland involves somatotropin hormone. If too much of this hormone is produced during the growing years, a person becomes abnormally tall. If oversecretion occurs during adult life, the bones of the face and hands thicken, but don't grow in length. Also, the organs and the soft tissues enlarge tremendously. This condition is known as **acromegaly.**

Somatotrophin hormone deficiency results in a pituitary dwarf, or **midget.** Such individuals are perfectly proportioned people in miniature. They are quite different from thyroid dwarfs (who have oversized heads) in that they are not mentally retarded.

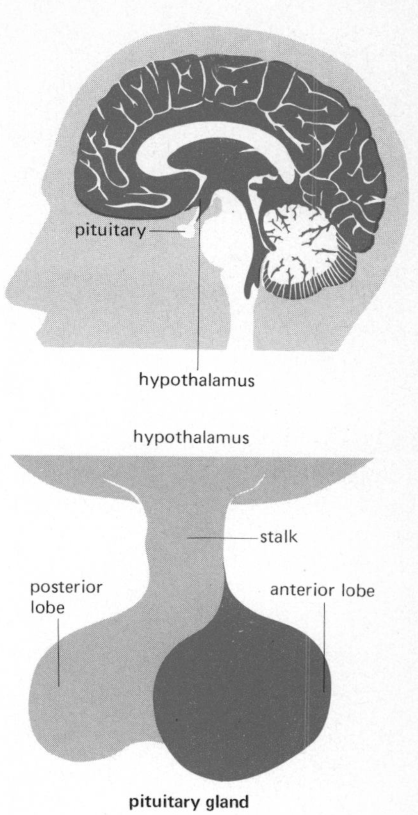

The adrenals—glands of emergency

The **adrenal glands,** are located on top of each kidney (Figure 13-24). They are composed of an outer region, the **cortex,** and an inner part, the **medulla.** The adrenal medulla is dispensable; the adrenal cortex is absolutely essential for life. The adrenal cortex secretes hormones called **corticosteroids.** These hormones are responsible for the control of certain phases of carbohydrate, fat, and protein metabolism as well as of the salt and water balance in the body. The adrenal cortex also produces hormones that control the production of some types of white blood cells and the structure of connective tissue.

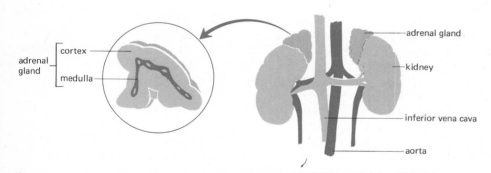

The medulla of the adrenal glands secretes the hormone, **epinephrine** (also called **adrenalin**). The adrenal glands have been called the glands of emergency because of the action of this hormone. Many people have performed superhuman feats of strength during periods of anger or fright, stimulated by a rush of epinephrine. This "strength of desperation" results from the following series of rapid changes in body activity:

- The person becomes pale because of constriction of the blood vessels in the skin. The rapid movement of blood from the body surfaces reduces loss of blood if there is a surface wound. It also increases the blood supply to the muscles, brain, heart, and other vital organs.
- The blood pressure rises, because of constriction of surface blood vessels.

FIGURE 13-23

The pituitary gland is suspended from the base of the brain (top). An enlarged view of the pituitary gland (bottom).

FIGURE 13-24

The position of the adrenal glands on the kidneys.

PHYSIOLOGY, GROWTH, AND REGULATION **343**

- Heart action increases circulation of nutrients.
- The liver releases some of its stored sugar so that ATP can be generated to supply energy for increased body activity.

The pancreas

The production of pancreatic fluid in connection with digestion, which has already been discussed, is only part of the function of the **pancreas** (Figure 13-25). Special groups of cells within the pancreas called **islets of Langerhans,** secrete the hormone **insulin,** which enables glucose to diffuse from the blood and enter the liver where it is stored and released on demand.

FIGURE 13-25
The pancreas is both an endocrine gland and a digestive gland located in the curvature of the duodenum.

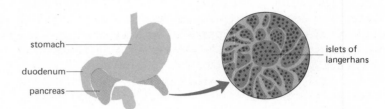

stomach

duodenum

pancreas

islets of langerhans

A person who lacks insulin cannot metabolize sugar efficiently. Thus, the body is deprived of energy and sugar collects in the blood. As the blood sugar rises, some of it is excreted in the urine. This is a symptom of *diabetes mellitus*. Diabetes isn't just a malfunction of the pancreas. The pituitary, thyroid, and adrenal glands, as well as the liver, are known to play roles in the disease. Excess body weight also increases the severity of this condition. Diabetes mellitus is definitely hereditary. If it runs in the family, regular periodic checkups of the level of blood sugar should be made. Once discovered, diabetes can usually be controlled. In fact, if treatment is begun in the disease's early stages, the patient can lead an essentially normal life.

Production of excess insulin results in a condition called **hypoglycemia** (low blood sugar). Excess insulin in the blood causes sugar that should be delivered to the cells to be stored in the liver. Extreme fatigue is the most obvious symptom of hypoglycemia. It is treated by administering glucose.

The ovaries and testicles

In addition to producing reproductive cells (eggs and sperm) the ovaries and testicles are also endocrine glands. Glandular cells in the ovaries secrete the female hormones, **estrogen** and **progesterone;** analogous cells in the testicles produce the male hormone, **testosterone.**

The production of testosterone isn't limited to the testicles. It is also secreted by the cortex of the adrenal glands in females as well as males, and it stimulates sex drive in people of both sexes. Normally estrogen prevents the masculinizing effects of testosterone from occurring in women. However, if the estrogen secretion by the ovaries is relatively low, a female may develop what are considered masculine characteristics, such as extensive body hair. Similarly, low production of testosterone by the testicles of a male can result in what are considered feminine characteristics, including a high-pitched voice. Thus, different individuals may show varying degrees of the male and female **secondary sex characteristics** (physical traits unrelated to reproduction, the development of which is controlled by the sex hormones). The bright plumage and large comb of a rooster are secondary sex characteristics. So are the antlers of a male deer.

For humans the secondary sex characteristics rapidly develop during a phase called **puberty,** which usually begins early in the second decade of life, when the ovaries and testicles rapidly increase their outpourings of sex hormones. Several distinct changes mark the onset of puberty. A pubescent boy's voice, for example, cracks and then deepens. His beard appears, together with a general increase in body hair. The chest broadens and rapid growth of the long bones add to his height. As a girl matures, her breasts develop and hips broaden. Menstruation begins.

The physical changes of puberty are accompanied by sweeping emotional changes. Aggression, for instance, seems to be related to testosterone. In most animals, adult males are more aggressive than females. Nursing females are the notable exception. Destruction of the testicles, **castration,** has been used for thousands of years to modify the behavior of domestic animals. If the male is castrated when it is young, it is not only sterilized; it is also more docile, and it stores more fat. For these reasons, livestock breeders use castration to increase the weight and gentleness of cattle. When testosterone is administered to a male castrate, aggressive behavior is restored. Similarly, when it is given to a female, aggressive tendencies become more pronounced.

Are there genes for aggression? The synthesis of testosterone (all hormones) requires enzymes which, being proteins, are coded for by DNA. Therefore, some of us may have genes producing more testosterone and more aggression than we would like.

Prostaglandins

The prostaglandins are a recently discovered class of hormone-like compounds formed throughout the body. They are secreted by various animal tissues, and have several effects. Some prostaglandins lower blood pressure; others raise it. Certain endocrine glands are regulated by them. The male prostate gland secretes prostaglandins into semen, which causes it to stimulate contractions of the female uterus to aid in sperm movement.

Although the mechanism(s) by which prostaglandins work aren't precisely known, they somehow influence the permeabilities of cell membranes. Curiously, aspirin seems to work on pain by inhibiting the action of prostaglandins.

MECHANISMS OF HORMONAL ACTION

The effects of hormonal secretions have been known for many years, but it has only been in the past decade that biologists have begun to discover the chemical level of their action. In the 1960's, it was discovered that the same compound that stores and transmits life's energy, ATP, was also involved in hormonal effects. As described in Chapter 7, ATP (adenosine triphosphate) binds energy in a series of three linked phosphate units. Removal of two of these phosphates generates AMP (adenosine monophosphate). According to a recently proposed **two-messenger model,** hormones (the first messengers) stimulate the conversation of ATP to a form of AMP called **cyclic AMP** (the second messenger).

Hormones are distributed more or less everywhere throughout the body's tissues by the bloodstream. However, hormones are very specific in their effects —only certain cells react in certain specific ways to a given hormone. The question naturally arises: How does a cell "know" which hormones to respond to? The two-messenger model answers this question by assuming that different types of cells have different hormone receptor sites. When a hormone activates this site, ATP is converted to cyclic AMP within the cell membrane. Cyclic AMP, in turn,

FIGURE 13-26

A representation of the mechanism of hormonal activity. When activated by a specific hormone, a receptor site catalyzes the conversion of ATP to cyclic AMP. Cyclic AMP in turn influences chemical reactions within the cell to bring about its characteristic response to the hormone.

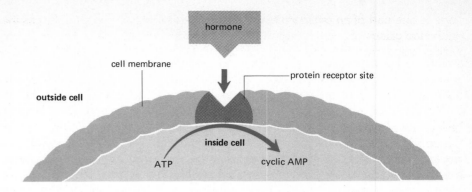

activates the cell's chemical response. Thus, hormones are extracellular messengers, while cyclic AMP is an intracellular messenger (Figure 13-26).

When the role of cyclic AMP as a "second messenger" was first revealed, some researchers envisioned it as a possible universal component of hormonal action. More recent research indicates, however, that it is only a mediator of hormones which are proteins or peptides. Some hormones such as the sex hormones are of a class of compounds known as steroids, and these are thought to penetrate directly into the cytoplasm of cells where they affect RNA synthesis.

One line of research into the functions of cyclic AMP indicates that it may be linked to the mental disorder, **manic-depression.** In the manic state (characterized by great elation) high levels of cyclic AMP have been found in the urine of afflicted individuals. In the depressed state low levels of the compound are usually found in the same people. It is not yet clear whether this fluctuation is a cause or an effect of the illness, and this, of course, is the crucial question.

Dynamic balance in the endocrine glands

The endocrine glands are closely interrelated. The hormonal secretions of one gland may affect one or more other glands. Besides the influence of glands on one another, there are two other factors operating to produce the delicate checks-and-balances system in body chemistry.

By the first of these processes, called **feedback,** the accumulation of a substance in the blood automatically depresses the process by which that substance is released into the blood. For example, the hormone, calcitonin which is produced by the parathyroid glands, regulates the level of calcium in the body. The concentration of calcium, in turn, regulates the production of calcitonin. When the calcium level in the body drops, the secretion of calcitonin increases to restore the calcium level. When the proper level is reached the calcium influences the parathyroids to decrease their secretion. This same kind of feedback occurs in the other glands and between various glands. In this way, a balanced state is automatically maintained in a body that is functioning normally.

The endocrine glands are also affected by the activity of the nervous system, which acts as a monitor of both internal and external conditions. The adrenal medulla, for example, may be stimulated to produce epinephrine as the need is signaled by the nervous system. Nervous control and feedback are further examples of homeostatic mechanisms operating in the body to maintain a steady state in the face of constantly changing conditions.

HORMONES AND PLANT GROWTH

With all the other similarities of plant and animal biochemistry, it isn't surprising that plants also have hormone systems. As in animals, plant hormones are secretions in one part of an organism that influence cells in other areas, and, just as the endocrine glands of animals respond to internal and external influences, so do the hormone systems of plants.

The auxins

One group of plant hormones, the **auxins,** act primarily to regulate cell growth and the development of flowers and fruit. Certain auxins stimulate growth while others depress it. The principal natural auxin produced in plants is **IAA (indoleacetic acid)**. It is secreted in largest amounts in growing and developing regions of plant organs. These include the tips of shoots and roots, as well as developing leaves, flowers, and fruits. From these regions, it is distributed to other parts of the plant via tube-like vascular tissue (xylem and phloem).

Different plant tissues react differently to auxins. High auxin concentration seems to inhibit stem growth, while lower concentrations act as a stimulus. Root tissues are stimulated by much lower auxin concentrations than stem tissues.

The effects of auxin activity are easily seen in the shoots of oats. The primary leaf of the oat and other grasses is enclosed in a protective sheath, or **coleoptile.** If the tip of a coleoptile is removed, as shown in Figure 13-27, growth of the shoot ceases. If the tip is replaced, elongation resumes. This indicates that auxins secreted in the tip diffuse downward, causing cells to grow lengthwise. This assumption is supported by placing a coleoptile tip on an agar (a substance somewhat like Jello) block into which auxins have been absorbed. Placing the agar block on the decapitated coleoptile causes elongation, just as replacing the tip does.

FIGURE 13-27
Demonstration of the presence and action of auxin in the tip of an oat coleoptile.

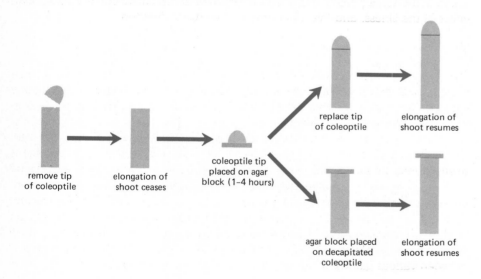

remove tip of coleoptile

elongation of shoot ceases

coleoptile tip placed on agar block (1–4 hours)

replace tip of coleoptile

elongation of shoot resumes

agar block placed on decapitated coleoptile

elongation of shoot resumes

Agriculturalists have found several ways to utilize auxins to ease crop production. They can be used, for instance, to kill dandelions and other broad-leaved weeds. The most common of this family of weed killers is a synthetic auxin, known as 2,4-D, which, in low concentrations (around 0.1 percent) kills broad-leaved plants without injuring grasses. Other auxin preparations are used to prevent fruit from dropping. This permits the harvesting of an entire fruit crop at one time.

Auxins have also been used to stimulate the development of fruit without seeds. Certain auxins cause plants to bear fruit without pollination. By using them, growers have been able to produce seedless watermelons, cucumbers, and tomatoes. Fruit development without pollination occurs naturally in the banana and navel orange, both seedless fruits.

FIGURE 13-28
A flight of four United States Air Force planes spray a Viet Cong jungle position in South Vietnam with a defoliating liquid.

FIGURE 13-29
Effect of gibberellin on spinach plant growth. A normal and a giant plant given gibberellin.

As with many powerful and useful scientific insights, knowledge of plant auxins has had its tragic applications. One auxin, 2,4,5-T, was used by the United States armed forces in Vietnam. It was sprayed over vast areas to destroy vegetation. The long-term effect of this defoliation was the creation of huge ecological wastelands (Figure 13-28). An even more alarming side-effect of 2,4,5-T is that it has been found to cause birth defects in mammals.

Gibberellins and cytokinins

Another group of growth-regulating substances, the **gibberellins,** are similar to auxins in promoting cell growth in plants. Gibberellins were first discovered in Japan in 1926, when a connection was established between a fungus (Gibberella fujikuroi) and the "foolish seedling disease" of rice. Infected plants became greatly elongated due to a substance secreted by the fungus. In 1935, it was isolated as a crystalline compound and named *gibberellin,* after the fungus. For many years, biologists believed that gibberellin was produced only by the fungus. Later, they discovered that several gibberellins are natural products in higher plants.

Besides promoting cell elongation, gibberellins influence growth and development in other ways. They stimulate flowering as well as fruit development. Evidence also indicates that they promote activity of the cambium in woody plants. In many of their influences, gibberellins may be associated with auxins.

Interesting results are obtained when gibberellin is applied artificially to growing plants. Giant plants result from abnormal cell elongation (Figure 13-29). Rapid growth may be accompanied by early flowering.

Cytokinins are less familiar regulating substances secreted in plants. While auxins and gibberellins promote cell enlargement and development, cytokinins influence cell division. They stimulate leaf growth and the development of buds.

LIGHT AND PLANT GROWTH

Light affects plant growth in several ways. As the source of energy for photosynthesis, it determines the amount of food produced. The amount of food avail-

FIGURE 13-30
Compare the plant on the left, grown in a greenhouse under normal light conditions, with that on the right, grown in a growth chamber under controlled environmental conditions.

able to the tissues, in turn, influences the rate of cell division and the growth of all organs of the plant (Figure 13-30).

Lack of light seems to stimulate elongation of the stem but prevents normal growth and expansion of leaves. Thus, stems of plants grown in the dark are weak and spindly, with leaves that are widely spaced and poorly developed.

On the other hand, reduced light (as opposed to heavy shade or darkness) may be a growth stimulus. In a moderate degree of shade, the rate of transpiration (water loss) is reduced more than the rate of photosynthesis. This results in an increase in the water content of the growing tissues without a comparable decrease in the rate of foodmaking. Thus, stems grow rapidly, and large leaves are produced.

The duration of light exposure also has direct effects on plant growth and reproduction. The changing responses of plants to varying periods of light and darkness is **photoperiodism.** As you know, the number of hours of daylight and darkness varies with the seasons. From December 21 until June 21, the hours of daylight increase in the Northern Hemisphere. Between June 21 and December 21, the days become shorter. Studies of plant reproduction have revealed a striking relationship between flowering and the length of days and nights. Many plants flower in the spring or in the fall, when the days are short and the nights are long. These ''short-day'' plants include such spring-blooming flowers, shrubs, and trees as the daffodil, tulip, crocus, forsythia, and dogwood. Among the short-day plants which bloom in the late summer and fall are chrysanthemums (Figure 13-31). goldenrods, ragweeds, and poinsettias. It isn't uncommon for short-day plants, including the forsythia, to bloom in the spring and again in the fall, when the hours of daylight, or **photoperiod,** are equal in length.

Other flowering plants grow **vegetatively** (that is, their non-flowering parts grow) when the photoperiod is short, and bloom later in the season, when the days are longer. These ''long-day'' plants flower in the late spring and early summer. Among them are the iris, hollyhock, clover, and such garden vegetables as the beet and radish.

FIGURE 13-31

Photoperiodism in a short-day plant. The blooming chrysanthemum was kept alternately in light for eight hours and in dark for sixteen hours. The one on the right received much longer periods of light relative to dark and has no flowers.

A third group of plants seems to be influenced only slightly, if at all, by the length of days and nights. Flowering starts after a period of vegetative growth, regardless of the length of the photoperiod. Among these "neutral day" plants are the nasturtium, marigold, zinnia, snapdragon, carnation, tomato, and garden bean.

Commercial flower and vegetable growers make extensive use of artificial regulation of photoperiods. They have found that the number of hours of darkness and light must be carefully regulated to optimize flower production.

TEMPERATURE AND PLANT GROWTH

Temperature fluctuations also have influences on plant growth and reproduction. In most plants, temperatures affect photosynthesis, translocation, respiration, and transpiration. Between the hot and cold extremes, at which plants die, there is an optimum range. It varies with species, but for most plants it is between 50° and 100° F.

DORMANCY IN PLANTS

In many regions of the world, there are seasons when the water supply, temperature, and/or light conditions aren't favorable for active plant growth. During these periods, many plants enter a period of inactivity, or **dormancy. Woody perennials** remain alive above the ground, with their living tissues protected by bark or the scales of winter buds. **Herbaceous perennials** die to the ground, with only the roots or underground stems surviving through the dormant period. Several species grow and flower, then enter a period of dormancy even though conditions are still favorable for growth. For example, daffodils bloom in the spring, then grow for a time. By summer, the leaves have dried, and only the bulb remains alive. Annual plants do not survive, but the seeds that are produced during a growing season preserve the species for the next years.

A plant hormone, **abscisic acid,** has been found that causes dormancy (by delaying growth) in certain species (Figure 13-32). As with many other phenomenon in both plants and animals, dormancy appears to be brought on by an interplay of hormonal and environmental influences.

FIGURE 13-32
Abscisic acid treatments have delayed budding in white ash cuttings. Left to right: control, 0.4 ppm, 2.0 ppm, 10.0 ppm abscisic acid after 22 days of treatment.

350

CHAPTER 13: SUMMARY

1 Food is the assortment of chemicals taken in by plants and animals from which energy is extracted and cellular structures are built.

2 Food consists of carbohydrates, fats, proteins, vitamins, and minerals.

3 Digestion is the process by which food is utilized by an organism.

4 Circulation is the flow of nutritive fluids and waste materials in an organism, with blood the transporting medium in higher animals.

5 Unused food and metabolic by-products are excreted by all cells in a variety of ways.

6 The endocrine glands are responsible in vertebrates for the secretion of a variety of hormones.

7 Plants also have hormones used in growth and regulation.

Suggested Readings

Asimov, Isaac   *How Did We Find Out about Vitamins?*, Walker & Co., New York, 1974 — Includes case studies of several vitamin deficiency diseases.

Galston, Arthur W. and Peter V. Davies   *Control Mechanisms in Plant Development*, Prentice-Hall, Englewood Cliffs, New Jersey, 1970 — A pleasant and readable introduction to plant hormones.

McMinn, R. M. H.   *The Human Gut*, Oxford Biology Reader, Oxford University Press, London, 1974 — A basic description of human digestion.

Muir, A. R.   *The Mammalian Heart*, Oxford Biology Reader, Oxford University Press, London, 1971 — The anatomy and physiology of the mammalian heart.

Riedman, Sarah R.   *Hormones: How They Work*, Abelard-Schuman Ltd., New York, 1973 — The functions of the various human hormones.

Simeons, A. J. W., MD.   *Food: Facts, Foibles, and Fables*, Funk and Wagnalls, New York, 1968 — An historical look at the world's eating habits.

Sussman, Maurice   *Developmental Biology*, Prentice-Hall, Englewood Cliffs, New Jersey, 1973 — A look at how many kinds of plants and animals develop.

An Excerpt from THE GREAT VITAMIN HOAX
M. Daniel Tatkon

Vitamins are essential to a balanced diet, but by taking massive doses of concentrated vitamins, some people do themselves more harm than good. Certain vitamins are water-soluble, and amounts which are ingested in excess of requirements are eliminated in the urine. Other vitamins are fat-soluble, and rather than being eliminated, excess amounts of these accumulate within the tissues of the body. Among the vitamins that are stored in the body are A, D, E, and B_{12}. Extreme excesses of two of these, A and D, have toxic side effects. The following reading considers the ill effects of too much vitamin A, of which the daily allowance recommended by the federal government is 5000 IU (International Units). Enough vitamin A can be obtained in a well-balanced diet without supplements.

Hypervitaminosis A, the condition that results from an overdose of vitamin A, is revealed by the following symptoms: loss of appetite, intense itching, loss of hair, dryness or roughness of the skin, fissuring at the angles of the lips, bone pain, fatigue, enlargement of the liver and spleen, increased intracranial pressure (headaches), frequent vomiting, swellings in the forearms, feet, and shanks, and irritability. In infants, a swelling of the fontanel can be observed and in adults hydrocephaly (an increase of the fluid within the head and subsequent enlargement) may develop.

Hypervitaminosis A is treated primarily by cutting off any further use of the vitamin. In most cases, the damage caused by an overdose can be corrected; however in a report entitled "Focal Retardation and Arrestment of Growth of Bones Due to Vitamin A Intoxication," which appeared in the *Journal of the American Medical Association* on December 8, 1962, Dr. Charles N. Pease reports a case involving seven children, three of whom were permanently harmed by overdoses of vitamin A given to them by zealous mothers. The damage: one leg developed shorter than another. The other four infants were taken off the heavy doses of vitamin A in time and bone damage was prevented. In one case a mother had given her child three teaspoonfuls of vitamins A and D instead of three drops; in another, a mother gave her child one teaspoonful of oleum percomoph daily until the child was one, and then proceeded to give it two teaspoonfuls daily; another mother gave her child 10 drops of oleum percomoph when the infant was one month, then increased it to one teaspoonful daily. When the infant was nine months old, she started using a multiple vitamin preparation three times a day in addition to the oleum percomoph.

Adolescents and adults are also victims of overdoses, either prescribed or self-medicated. Acne is one condition treated by large doses of vitamin A and there are several cases in which teenagers have suffered from a toxic condition because of irresponsible dermatologists, druggists, or friends. In one case a seventeen-year-old boy was taking 200,000 IU of vitamin A daily; in another instance a fourteen-year-old boy had been given a prescription of 50,000 IU daily that he then increased to 300,000 IU daily; and another case reported three teenage girls who had taken up to 200,000 IU a day of vitamin A. All were severely affected.

In 1961 Dr. William H. Stimson wrote an article for the August 24 issue of the *New England Journal of Medicine* entitled "Vitamin A Intoxication in Adults." He described six adults and their particular toxic conditions. One patient had taken 600,000 IU of vitamin A as a result of a radio ad for dry throat and colds, two had taken it for acne, and one for eczema. Dr. Joaquin Soler-Bechara and Dr. John L. Soscia reported another case of hypervitaminosis A in the October 1963 issue of the *Archives of Internal Medicine*. In this instance their patient had been taking a multivitamin substance and individual tablets of vitamin A to improve her vision.

There are complications in the diagnosis of hypervitaminosis A, one being that there is a latent period of anywhere from six to ten months before symptoms develop. Because of this and since this toxic condition involves many of the human systems, hypervitaminosis A is many times misdiagnosed, even to such extents as diagnosing the condition to be a brain tumor, meningitis, or some other equally serious condition. As Drs. Soler-Bechara and Soscia point out in their report: "Dermatologic changes are a common occurrence after excessive intake of vitamin A. Loss of body hair with thinning of eyebrows and lashes, pruritus, and dry skin are described in the case report. The skin manifestations resemble those of vitamin A deficiency.*

These are not the only cases of the dangers of overdose, but they do illustrate how and why hypervitaminosis A occurs and emphasize the danger of self-diagnosis and self-medication, so readily encouraged by our vitamin salesmen. Hypervitaminosis A has occurred in an adult even when the dose was only 50,000 IU taken daily over a period of eighteen months. Although generally much larger quantities of this vitamin have been ingested before the toxic condition appears, it is still no justification for placing ourselves or our children in jeopardy by self-determining the medication for our daughter's acne or for our own tired eyes.

*Soler-Bechara and John L. Soscia, "Chronic Hypervitaminosis A," in *Archives of Internal Medicine.* Vol. 12 (October 1963), pp. 462–466.

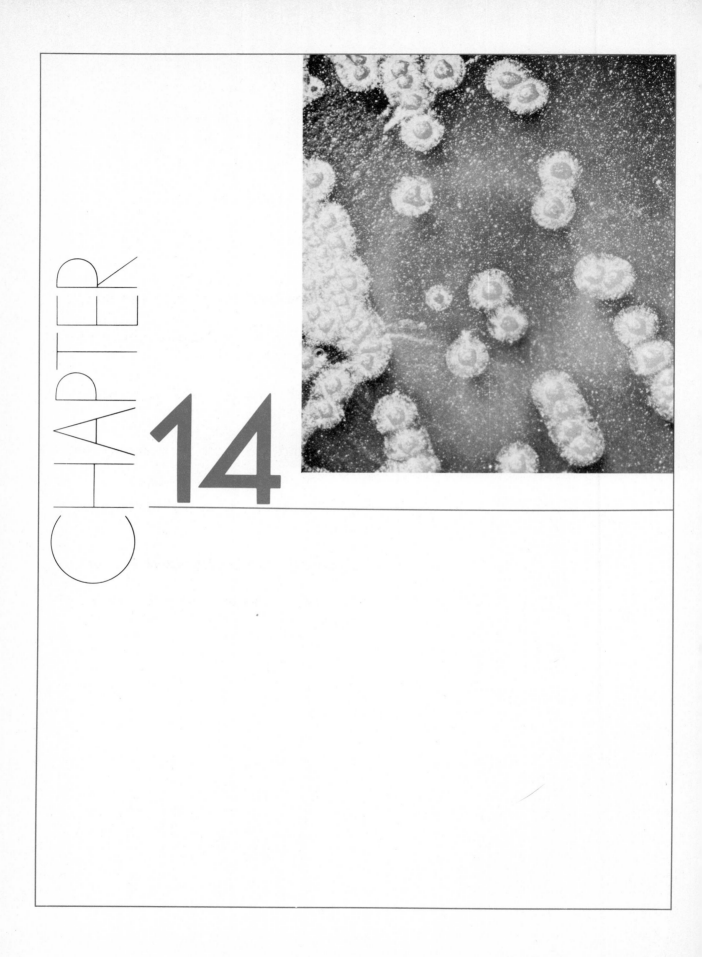

CHAPTER 14

Disease and Aging

Objectives

1 Learn how microorganisms cause disease.

2 Describe how cells and organs fight off infection.

3 Distinguish between viral and bacterial life cycles.

4 Understand the mode of action of antibiotics.

5 Describe hardening of the arteries and its causes.

6 Understand the possible causes of cancer.

7 Be able to theorize about the causes of aging.

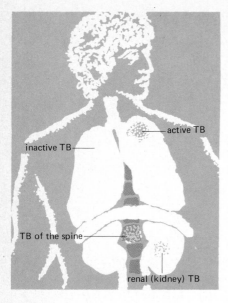

FIGURE 14-1

Common sites of tuberculosis infection. The lungs are the sites of approximately 93 per cent of all TB cases. It can also attack the nervous system and parts of the skeletal system. (From *Life and Health.* Copyright © 1972 by Ziff-Davis Publishing Company. Reprinted by permission of CRM Books, a Division of Random House, Inc.)

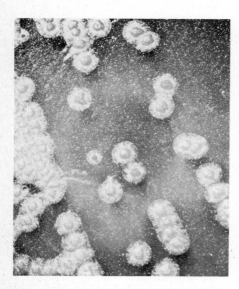

FIGURE 14-2

Photograph through a microscope of *Clostridium botulinus* bacteria (size, about 5 micrometers) which cause botulism.

HUMAN DEGENERATIVE PROCESSES

All organisms are subject to infection by agents of disease. Even the simplest of bacteria may be taken over by viruses, and complex organisms are susceptible to a wide variety of agents of disease. Apart from diseases caused by foreign species, an organism is subject to spontaneous degeneration and ultimately death.

Certain diseases have been discussed in preceding chapters. Parasitic diseases were considered in the context of ecology, and genetic/hormonal malfunctions have been considered in various chapters. This chapter considers diseases caused by microorganisms, the mysterious degenerative disease cancer, heart disease, and finally the ultimate degenerative process, aging.

HOW MICROORGANISMS CAUSE DISEASE

Microorganisms can damage the body in several ways. Tuberculosis, for instance, is a bacterial infection that results in tissue destruction. The primary infection of tuberculosis is usually in the lungs, although other organs may be affected (Figure 14-1). As the microorganisms multiply in lung tissue, they destroy cells and produce lesions (ruptures in tissues) from which blood seeps into air passages. In advanced stages of tuberculosis, patients often hemorrhage in their lungs and cough up blood. Tissue destruction characterizes many infections besides tuberculosis. Some *Streptococci*, for example, destroy blood cells and typhoid bacteria destroy cells of the intestine wall.

Often, bacteria form poisonous protein substances, **exotoxins,** that diffuse from the bacterial cells into the host's body tissues. They may enter the bloodstream and reach body tissues far removed from the site of the infection. For example, exotoxin from a tetanus infection in a foot can be carried to all parts of the body. If it reaches the jaw muscles it causes spasms and rigid paralysis—hence, the name "lockjaw."

Certain bacteria which live in foods produce exotoxins that cause food poisoning. The most deadly type of food poisoning is **botulism,** caused by a relative of the tetanus bacteria (Figure 14-2). Botulism bacteria form spores, which can contaminate meat, vegetables, and other foods prior to canning. If the foods are improperly sterilized, the spores germinate into cells that release deadly exotoxins. Symptoms of botulism appear within twelve to thirty-six hours after contaminated food is eaten. They include double vision, weakness, and paralysis that spreads from the neck to other parts of the body. Death may result from paralysis of muscles involved in breathing or from heart failure. About 65 percent of the cases of botulism are fatal. Since the exotoxins are destroyed by heat, botulism can be prevented by thoroughly cooking canned meats and vegetables, especially those that are home-canned. Cans containing spoiled food often bulge because of gases generated by growing bacteria. The food in them should never be used.

Certain bacteria form toxins that remain within them until they die and disintegrate. These are **endotoxins,** and their ultimate effects are similar to exotoxins. Among the endotoxin diseases are typhoid fever, tuberculosis, cholera, bubonic plague, and bacterial dysentery.

Structural defenses against disease

The human body normally harbors an enormous number of bacteria and other microorganisms that cause no harm, but there is a constant threat of invasion by organisms that are capable of causing disease. To prevent uncontrolled infection, the body has several defenses that function at various stages of bacterial attack.

The skin, if unbroken, is nearly impermeable to bacteria. In addition, salts and various fatty acids in perspiration destroy bacteria and thereby help make the skin an even more effective defense against infection. However, natural openings in the skin, such as pores and hair follicles, may allow entry of microorganisms.

The mouth, digestive tract, respiratory passages, and genital tract are lined with **mucous membranes,** which are a second line of defense against bacterial infection. **Mucus** is a slimy substance composed of carbohydrates bound to proteins that traps microorganisms. Cells of the mucous membranes in the respiratory tract have tiny, hairlike projections (cilia) that sweep bacteria and other foreign particles upward toward the throat, where they stimulate a cough. They are spewed outwards, suspended on tiny droplets of mucus. Irritation of the membranes of the nasal passages results in sneezing.

Tears, secreted by tear glands, protect the eyes from infection by washing bacteria and other foreign particles into the tear ducts, which empty into the nasal passages. Both tears and mucus contain enzymes known as **lysozymes,** which prevent infection by destroying the cell walls of bacteria.

The acid secretion of the stomach is another effective body defense. Millions of bacteria enter the stomach in food, but few can survive the hydrochloric acid secreted by glands in the stomach wall. Furthermore, bacterial populations which are normally present in the digestive system interfere with the growth of invading organisms and thus protect the host. The importance of these intestinal bacteria becomes obvious when they are accidentally destroyed by antibiotics. Before they can multiply and reestablish their normal population size, foreign species can take over and cause intestinal infections. Thus, antibiotics taken to control one disease can accidentally lead to another.

Cellular defenses against disease

Once bacteria have passed through the "mechanical" defenses of the skin and mucous membranes, they are met by cells which are specialized for disease resistance. **Phagocytic cells** engulf bacteria and dissolve them with enzymes, including lysozymes (Figure 14-3). Among the phagocytic cells are the white blood cells. These are capable of amoeboid movement and can escape from the blood and lymph vessels by squeezing through the vessel walls at the points of contact between cells. They migrate through tissue fluids to the site of a local infection. Here many of them form a wall around invading organisms and begin to engulf them. Debris of the battle—blood serum, digested bacteria, and degenerated phagocytic cells—is left as pus.

In local infections (which are limited to a small area, rather than spread throughout the body) tissues usually swell and turn red. This inflammation results from increased flow of blood to the region of the infection. Blood vessels enlarge, and lymph seeps into the tissue spaces. The lymph carries bacteria and phagocytic cells that have engulfed bacteria to lymph nodes, where they are filtered out.

During the struggle between invading microorganisms and the cellular defenses of the body, the body temperature often rises. **Fever** is a reaction that inhibits the growth of many bacteria, but not necessarily the ones causing the infection. It is not completely clear how a fever develops, although it is believed that foreign bacteria and viruses interfere with the temperature regulatory system in the hypothalamus of the brain. A prolonged high fever can permanently damage certain tissues, particularly brain cells, by deactivating certain enzymes.

FIGURE 14-3

Phagocytosis. This series of photographs, taken at ten-second intervals, shows a human white blood cell, or leucocyte, engulfing a chain of *Bacillus megaterium* bacteria.

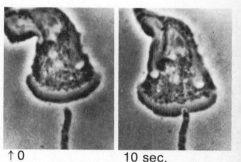
↑ 0   10 sec.

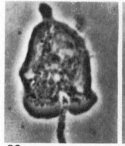
20 sec.   30 sec.

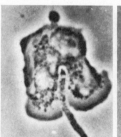
40 sec.   50 sec.

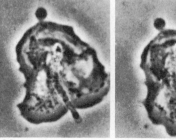
60 sec.   70 sec.

The defenses described so far are non-specific — that is they act on any invading microorganisms. However, the body also produces highly specific chemical defenses against disease. Certain white blood cells, those of the type known as **lymphocytes,** give rise to specialized cells called **plasma cells,** which play a central role in immunological reactions. Plasma cells respond to the presence of certain kinds of foreign substances, or **antigens,** which are usually (but not always) proteins, by making **antibodies,** proteins that destroy or inactivate the antigens. Each type of antibody is usually highly specific; it will inactivate only the antigen that stimulated its synthesis. It is not the virus, bacterium, or fungus itself which acts as the antigen. Rather, it is usually some protein molecule on its surface that triggers the antibody response.

Antibodies are believed to be produced primarily in the lymph nodes, spleen, and thymus gland, from which they enter the blood and lymph. Antibody production begins within a few hours after invasion by the microorganism. Within a few days, antibodies begin to enter the bloodstream and continue to increase in quantity for three or four weeks. This period usually marks the highest level of their production. Antibodies may remain in the bloodstream for many weeks or even years. The level of a given antibody in the blood declines slowly, and a second exposure to the corresponding antigen speeds up production of the antibody in the lymph nodes and spleen. An organism reacts faster with greater antibody synthesis the second time it is invaded by a given antigen; this phenomenon has been termed **immunological memory.**

VIRUSES AND DISEASE

We have noted that the most basic scientific questions are the ones which are furthest from being answered. For the question: What is life? we listed several characteristics, but pointed out that a hard and fast definition was not justified by our present knowledge. In Chapter 3 we considered the structure and function of the major groups of organisms. We purposely left out one group which most biologists consider to be nonliving, but which certainly possesses characteristics closer to life than other nonliving matter. In discovering viruses, biologists have found units of matter at the very borderline of the living and nonliving.

Viruses are tiny particles composed of protein and nucleic acid and because of their simplicity, they are unlike any other living material. In themselves, they are not actually living. They only "live" in close associations with host organisms.

Regardless of whether viruses are classified as living or nonliving, it is appropriate to deal with them in a text on biology. If they are nonliving, they are unique, because their influence on cell activities is different from that of any other nonliving material. If they are living, they are certainly the most basic organisms known.

What are viruses?

At the mention of the word "virus," you probably think of disease, and certainly there are many viral diseases. Polio, small pox, chicken pox, influenza, rabies, and the common cold are well known viral infections. Many cancer researchers believe that viruses are at the root of human cancer. But biologists are finding that few, if any, organisms are free of viruses, and many viruses cause little apparent damage.

Virus particles lack the complexity of cells. A virus has no nucleus, no cytoplasm, and no surrounding membrane. It is larger than a molecule, yet much smaller than the smallest cell (Figure 14-4).

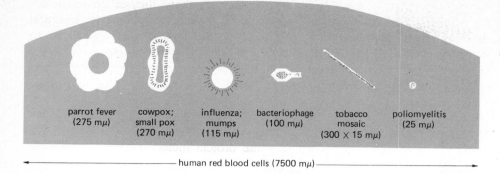

parrot fever (275 mμ)   cowpox; small pox (270 mμ)   influenza; mumps (115 mμ)   bacteriophage (100 mμ)   tobacco mosaic (300 $\times$ 15 mμ)   poliomyelitis (25 mμ)

←————— human red blood cells (7500 mμ) —————→

Virus particles are all but invisible under even the highest magnification of a light microscope. For this reason, before the invention of the electron microscope, little could be determined about their structures. Along with revolutionizing cell biology, the electron microscope brought viruses into view.

Viruses are of a wide variety of shapes. Some are needle-like rods, while others are spherical, cubical or brick-shaped. Still others have an oval or many-sided head and a slender tail.

The discovery of viruses

Scientists studied viral diseases long before the existence of viruses was known. Dr. Edward Jenner performed the first vaccination against smallpox in 1796, when he transferred a virus-containing fluid from a cowpox sore on the hand of a dairymaid to a scratch on the arm of an eight-year-old boy. About a century later, Louis Pasteur discovered that the rabies infection centered in the brain and spinal cord. He did it by transferring the disease to a healthy animal with an injection of infected brain and spinal-cord substance. While both Jenner and Pasteur made significant medical discoveries, neither of them knew the nature of the infectious agents with which they were working. Nineteenth and early twentieth century researchers described viruses as "contagious fluids," destructive chemical substances," and "destructive enzymes."

One such investigator was Dimitri Iwanowski, a Russian biologist, who worked with diseased tobacco plants in the 1890's. These plants were infected with a virus disease known as tobacco mosaic. The term **mosaic** refers to a curious pattern of light green and yellow areas that appears as leaf tissues are destroyed by the virus. As the disease progresses, the leaves become stunted and wrinkled (Figure 14-5).

Iwanowski squeezed fluid from the infected leaves and rubbed it onto the leaves of healthy plants. Soon the healthy leaves developed the mosaic pattern. He repeated his experiment, but this time he passed the fluid through a filter with pores small enough to trap all bacteria. Microscopic examination revealed no bacteria or other visible bodies that might cause disease. Nevertheless, healthy plants inoculated with the filtered leaf fluid soon developed mosaic disease.

In Iwanowski's day, bacteria were thought to be the smallest agents of disease, and he assumed that the infectious "juice" contained an invisible bacterial poison that had passed through the filter. Six years later, the Dutch botanist Martinus Beijerinck repeated Iwanowski's experiments. He concluded that an invisible agent, smaller than the smallest bacterium, must be present in the infected "juice." He named this unknown agent *virus*, a Latin word meaning "poison."

In an effort to isolate the virus in 1935, Dr. Wendell Stanley ground more than a ton of diseased tobacco leaves and extracted the juice. From this extract, he

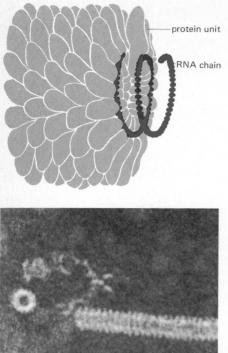

protein unit

RNA chain

obtained about a spoonful of needlelike crystals. These crystals could be stored in a bottle in an apparently lifeless state. Yet when they were suspended in water and rubbed on a tobacco leaf, they produced the mosic disease.

The composition of viruses

In recent years, scientists have determined the chemical compositions of several viruses, including tobacco mosaic and polio. They have found the simplest virus particles to consist of a protein coat encasing a core of nucleic acid. In some viruses, including plant viruses, the core is RNA. In others, it is DNA. The simplest viruses are giant molecules of **nucleoprotein,** similar in some ways to the nucleoproteins in the nuclei of cells. More complex viruses also contain carbohydrates, lipids, metals, and other substances but in small quantities.

At a magnification of 60,000×, the tobacco mosic virus appears as a spiral of rod-shaped bodies. The protein coat, composing 95 percent of the virus, consists of more than 2,000 protein units. A core of RNA composing 5 percent of the virus is sheathed within the protein coat (Figure 14-6).

Viral reproduction

A virus particle can be active only in direct association with the content of a living host cell. Removed from a cell, a virus ceases all apparent activity, but it still retains its ability to infect a cell.

A virus cannot reproduce independently. That is, it cannot duplicate its own structure in the way in which cells self-reproduce by fission. *In order to reproduce, a virus must invade a host cell and assume control of the cell's metabolic processes.* Biologists aren't sure how this is accomplished. Furthermore, different kinds of viruses may alter the chemical activities of cells in different ways. One hypothesis suggests that an invading virus alters the enzyme pattern that normally regulates protein synthesis and growth of the cell. In this way, the virus uses the biochemical system of the cell to form virus particles rather than normal cell constituents. In a sense, a virus is a gene without a "home" until it invades a cell.

Just as genes mutate, so do viruses. More than 50 mutant strains of the tobacco mosaic virus have been discovered. These mutants differ in virulence (potency) and in the symptoms they produce in the host plant.

FIGURE 14-6
Tobacco mosaic virus. Compare the schematic drawing (above) with the electron micrograph (below).

Viral specificity

Virus/host relationships are usually highly specific. A given bacterial virus, for instance, may only be capable of invading a single species of bacteria. Similarly, a plant virus may be specific for the cells of the flower petals of a particular kind of plant (Figure 14-7). A specific human or animal virus may exist only in the cells of the skin. Another may invade only the respiratory organs.

Some viruses are even more specific. Polio viruses attack only the cells of one kind of nerve in the brain and spinal cord. Similarly, mumps is an infection of only one pair of salivary glands; the virus never invades the other two pairs.

FIGURE 14-7
A Rembrandt tulip, showing the "broken" coloration that results from a virus infection.

Bacteriophages

Much of our knowledge of viruses has come from investigations of the bacterial viruses, which are referred to as **bacteriophages.** Phages are shaped somewhat

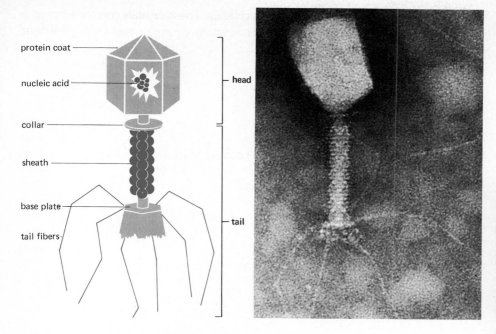

protein coat

nucleic acid

collar

sheath

base plate

tail fibers

head

tail

FIGURE 14-8
A bacterial virus, or phage. Compare the schematic drawing (left) with the electron micrograph (right). Length of virus is 130nm.

FIGURE 14-9
An electron micrograph of an "exploded" T₂ bacteriophage. Its central core of DNA occurs as a single strand of length about 50 micrometers.

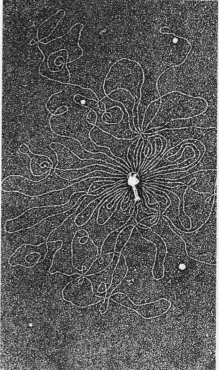

like tadpoles, with a round or many-sided head and a slender tail with attachment fibers (Figure 14-8) that stick to the host. The coat of a phage is composed of protein, while the core is usually DNA (Figure 14-9), but sometimes RNA.

Early researchers were able to see the gross effects of phages on bacterial colonies (Figure 14-10), but it remained for the electron microscope to reveal what actually happened to individual bacteria. The result of viral invasion is often the disintegration, or **lysis** of a bacterium. A phage that produces lysis goes through a **lytic cycle.** The lytic cycle is outlined below and diagrammed in Figure 14-11:

- A normal uninfected bacterium (1).
- A phage has hooked itself, tail down, to the cell wall of the bacterium. An enzyme in the tail of the phage is dissolving an opening in the bacterial wall (2).
- Having formed an opening in the wall, the tail contracts and injects the DNA or RNA of the phage core into the bacterial cell. The empty protein coat remains outside (3).
- Within a few minutes, the phage DNA appears near the DNA of the host cell. The phage takes over control and the chemical system of the bacterium is used to synthesize phage DNA and protein molecules. The bacterium has become a virus "factory" (4).
- Soon the bacterium contains hundreds of phage particles (5).
- The bacterial cell ruptures, releasing its phage content to attack other bacterial cells (6).

This entire cycle, from the entry of the phage DNA to the bursting of the bacterial cell, takes about forty minutes (Figure 14-12). Since each phage can multiply several hundred times within a lytic cycle, a colony of millions of bacteria can be destroyed by phage within a few hours.

Human viruses

Many familiar human diseases are caused by viruses. In most of them, the virus invades only specific tissues while other tissues are left unaffected. Symptoms of

FIGURE 14-10
A bacterial culture in a dish spotted with areas of viral infection. The spots are where the viruses have destroyed the bacteria.

FIGURE 14-11

The lytic cycle of destruction caused by a virulent (infectious) phage.

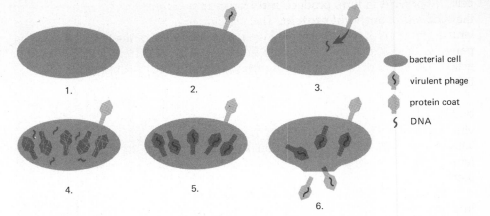

bacterial cell

virulent phage

protein coat

DNA

1. 2. 3.

4. 5. 6.

FIGURE 14-12

An electron microscope photograph of T₄ bacteriophages (small particles) destroying the bacterium *E. coli*.

FIGURE 14-13

Human body cells before (top) and after (bottom) infection with measles virus.

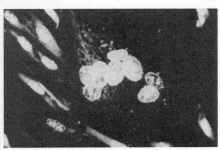

viral infections are as different as the viruses that cause them. Generally, however, a virus infection results in the disruption of metabolic processes and damage or destruction of tissue cells. A lasting immunity remains after recovery from many virus infections. The common cold and influenza are notable exceptions, however.

The following are some of the better-known virus diseases that affect human beings: smallpox, cowpox, chicken pox, shingles, cold sores, fever blisters, warts, influenza, measles, German measles, also known as three-day measles (Figure 14-13), virus pneumonia, the common cold, parrot fever, yellow fever, infectious hepatitis, infectious mononucleosis, and mumps. As described later in this chapter, viruses have also been linked to the most mysterious of diseases, cancer.

The body's defenses against viruses

Until a couple of decades ago the body's chemical defenses against viral diseases were thought to be limited to antigen/antibody reactions. However, in 1957 two doctors, Alick Isaacs and Jean Lindenmann, found that cells invaded by viruses produce a protein which seemed to interfere with viral action. They called the protein **interferon.** Further studies have indicated that interferon itself is not the antiviral agent. The action involves a second protein synthesized in a cell as a result of stimulation by interferon.

Within a few hours after the onset of virus infection, the interferon system is in operation. First, the presence of the virus stimulates a cell to synthesize interferon, which is released into the intercellular spaces and taken into the nuclei of other cells. Their DNA in turn, produce **interferon messenger RNA,** which codes for the synthesis of **antiviral protein.** This protein protects the cell by preventing the virus from modifying the protein synthesizing machinery of the cell to form virus particles. Thus, the virus is prevented from multiplying in the protected cells. The actions of interferon and antiviral proteins are intracellular (within cells) and continue for a period of one to three weeks.

About three days after the onset of a virus infection, specific antiviral antibodies begin to appear in the blood serum. This is a second line of defense against viral infection and is extracellular (occurring outside the cell). Antibodies combine with virus particles, making them noninfectious. White blood cells aid in this defense. They do this by engulfing the incapacitated virus particles and destroying them.

Thus, the interferon and antibody systems provide a two-pronged attack. Interferon acts rapidly and for a short time. It is nonspecific—that is, effective against any virus. Antibodies are produced more slowly but are formed for a much longer period of time (in some cases for life). They are specific in that one kind of antibody acts only on one kind of virus.

IMMUNITY

Resistance of the body against infections is called **immunity.** Certain kinds of immunity are inborn, while others are acquired during the lifetime of an individual.

For the most part, humans have inborn immunity to diseases that affect other animals and plants, because conditions in the human body don't support the growth and activity of the infectious organisms that cause these diseases. Such inherited characteristics as body structure, body temperature, and biochemical makeup are responsible for this resistance. The general immunity of a given species to pathogens of other species is called **species immunity.** There are, however, several notable exceptions to species immunity. Tuberculosis and undulant fever may be transmitted to human beings in milk from diseased cattle. Anthrax may be picked up by us by contact with lesions in the skin of infected sheep, cattle, horses, and other animals.

Acquired immunity may be active or passive, depending on how it is established. **Active immunity** involves the production of antibodies within an individual, and **passive immunity** involves transferral of antibodies from an outside source.

Active immunity develops naturally during recovery from certain infectious diseases, including diphtheria, scarlet fever, measles, and mumps. During the infection the body produces specific antibodies against the pathogenic organisms or their products. The antibody production may continue after recovery, resulting in permanent active immunity. But following some diseases like colds, influenza, and syphilis, the immunity may only last a week or so. Active immunity may also be acquired artificially by preparations known as **vaccines** that contain dead or weakened disease organisms. The body is thereby stimulated to form antibodies without actually having to suffer the symptoms and dangers of a full-blown case of the disease. Immunological memory will then protect the person.

Passive immunity is acquired artificially by transferring antibodies that have been produced in animals or other people into an individual. While the introduction of antibodies from an outside source provides immediate protection against an infection, it is only a temporary measure. In most cases the antibodies are eliminated by the body within a few months.

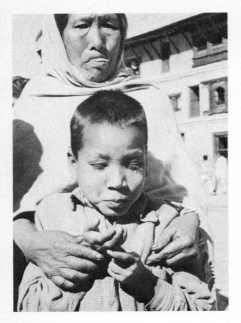

FIGURE 14-14

The face of an unvaccinated child scarred by an attack of smallpox.

Temporary passive immunity may also be acquired naturally in an unborn infant through the transfer of antibodies from the mother's blood across the placenta. Other antibodies may be transferred with milk to a nursing baby. Immunity thus transferred usually lasts from six months to a year.

A summary of the types of immunity, how they are established, and their duration is outlined in Table 14-1. The following sections will consider the historical development of immune therapy as an illustration of how scientists work.

Edward Jenner's risk

In the history of biology there have been a few memorable occasions when researchers have taken chances on risky human experimentation and succeeded. Perhaps the most fruitful risk of all time was taken by an obscure British country doctor. His discovery not only laid the groundwork for subsequent studies of the immune system, it also saved countless human lives.

It was during one of the most dreadful smallpox epidemics in England's history that Dr. Edward Jenner made a discovery that was to revolutionize medicine. Smallpox (Figure 14-14) took its greatest toll in cities, and Jenner noticed that the disease seldom struck people who lived in rural areas and worked around cattle. Most farmers and dairy workers had contracted a similar but relatively mild disease, cowpox, and had recovered with nothing more serious than a pustule (pus filled pimple) that left a scar. Were they immune to smallpox? If so, why not deliberately infect people with cowpox to protect them from smallpox?

In 1796 Jenner had a chance to test his theory. A young boy was offered by his mother as an experimental subject, and Jenner took him to a dairymaid, Sarah Nelmes, who had cowpox. Jenner made two shallow cuts, each about an inch long, on the child's arm and inoculated the cuts with matter taken from a cowpox sore. A pustule developed, formed a scab, and healed, leaving only a scar. Was the boy now immune to smallpox? There was only one way to find out. That was to inoculate him with smallpox. (Conceivably Jenner could have answered this question statistically by inoculating many people with cowpox, then waiting to see if any of them caught smallpox by chance; but there undoubtedly was a lack of people willing to try it). Two months after the cowpox inoculation, Jenner repeated the procedure with matter from a smallpox pustule. During the following weeks he watched for signs of the disease, but they did not appear. Several months later, he repeated the smallpox inoculation. Again, smallpox didn't develop. The boy was immune.

Following his daring success, Jenner wrote a paper explaining the method of **vaccination** (from the Latin word for cow). At first, he was ridiculed by his medical peers as well as the general public. Anti-vaccination campaigns were organized. Some people, however, were willing to try it, and the success was undeniable. Gradually vaccination was accepted, and today is part of growing up for children throughout most of the world.

Pasteur's work with weakened strains

About eighty years after Jenner's first vaccination, Louis Pasteur began his prolific work, which led to immunization against several diseases. Among Pasteur's most outstanding achievements was the development of a vaccine for anthrax, a bacterial intestinal disease. It was made from a culture of bacteria that were rendered less infectious after being isolated from the blood of infected animals.

TABLE 14-1   TYPES OF IMMUNITY

TYPE		HOW ESTABLISHED	DURATION
INBORN species		through inherited anatomical, physio-logical, and chemical characteristics	permanent
ACQUIRED active	natural	by experiencing an infection during which contact with microorganisms or their products stimulates antibody production	usually lasting or permanent
	artificial	by injecting vaccines, toxoids, or other weakened microorganisms	usually from several years to per-manent; booster shots may be necessary
passive	natural	by transfer of antibodies from mother to infant through the placenta prior to birth or in milk after birth	from 6 months to 1 year
	artificial	by injecting a serum containing anti-bodies	usually from 2 or 3 weeks to several months

To test the vaccine, Pasteur selected forty-eight healthy animals (mostly sheep) and divided them into two groups. He gave the animals in one group injections containing a few drops of anthrax vaccine. Twelve days later, he gave the same animals a second injection of the vaccine. Fourteen days later, he gave all forty-eight animals injections of active anthrax bacteria. All of the vaccinated animals lived, while most of the others died.

Subsequent to his work with anthrax, Pasteur used a similar technique to create a vaccine for rabies. The convulsions, insatiable thirst, and paralysis that climaxed a rabies infection led Pasteur to believe that it was centered in the nervous system. Microscopic examinations of the brain tissues of victims, however, didn't reveal any microorganisms. (It is known today that rabies is caused by a virus which is far too small to be seen with the microscopes available to Pasteur.) Nevertheless, Pasteur found he could transmit rabies by injecting infected brain tissue from a rabid dog (or other mammal) into a healthy one. He repeated the inoculations with rabbits and discovered that the disease became more severe as it was passed from one animal to another. If spinal-cord tissue taken from an infected dog or rabbit was dried for a few weeks, however, it lost its potency.

The discovery that the virus weakened with aging and drying led to experiments to find if immunity to rabies could be produced. Pasteur began by injecting fourteen-day-old brain tissue from a rabid animal into a healthy dog; the next day he injected the same dog with thirteen-day-old material (which was only slightly stronger). The injections were continued day after day until the dog was given full-strength rabies virus in the fourteenth injection. The animal suffered no lasting ill-effects. Thus, the series of injections with material of increasing strength had produced immunity to rabies. In 1885, Pasteur successfully used a similar sequential vaccination on a child who had been bitten by a rabid dog.

Von Behring's new approach to immunization

Historically, diphtheria was one of the worst epidemic killers, especially among children. Diphtheria bacteria grow in a thick, grayish-white membrane on the back of the throat. As the membrane spreads, it may block the opening to the lungs and cause death by strangulation. Furthermore, the bacteria produce highly poisonous exotoxins that are absorbed through infected tissues into the blood. These toxins often cause severe damage to the heart, nervous system, and other organs.

A drug which gives immunity to diphtheria was first developed only about fifty years ago, and involved the work of several scientists. One found the rod-shaped bacteria which caused the disease growing in the throats of patients. Another cultured the bacteria and developed a stain suitable to highlight it for microscopic studies. Much of the credit for the cure, however, belongs to Emil von Behring, a German bacteriologist.

Von Behring was puzzled by the fact that even though diphtheria organisms remained in the throat, the effects of the disease appeared in far removed organs. When an extract of the medium on which the bacteria were grown was injected into guinea pigs, the symptoms of diptheria occurred, even though no bacteria were present. This was evidence that some poisonous secretion (an exotoxin) was involved.

It was while doing such experiments that Von Behring noticed that guinea pigs and rabbits could be used only once, because they developed immunity to the disease during the first exposure. Could this immunity be transferred to animals that had never been given doses of toxin? To answer this question, Von Behring took a blood sample from immune animals and injected it into other animals. Exposure to diphtheria toxin now showed that they were immune. Von Behring named the substance that had produced immunity **antitoxin.** By transferring antibodies (antitoxin) that had developed in one organism to another, Von Behring had produced passive immunity in the recipient.

Sheep were used in the production of diphtheria antitoxin, and after extensive testing of it in guinea pigs, it was used in the Children's Hospital in Berlin. Von Behring found, however, that immunity resulting from injections of sheep antitoxin lasted only a few weeks. Apparently it was slowly destroyed in the human bloodstream (the basic problem with passive immunity).

Von Behring reasoned that if children could be made to produce their own antitoxin, the resulting active immunity would be as long lasting as if they had had diphtheria and recovered from it. To give diphtheria toxin alone, however, would be nearly as dangerous as inoculating them with the disease itself, so he began experiments using mixtures of toxin and antitoxin. World War I prevented Von Behring from finishing his work, but a successful toxin-antitoxin was soon produced in the United States by other researchers.

Toxin-antitoxin was widely used to produce immunity to diphtheria until it was discovered that some people had a serious reaction to it. This problem has been solved by using a **toxoid,** a substance chemically similar to the toxin and usually derived from it, but lacking its poisonous quality. The toxoid is made by treating the toxin with heat or with chemicals. The toxoid is similar enough in structure to the toxin so that, in response to it, the host manufactures an effective antitoxin.

CHEMOTHERAPY

Paralleling the development of vaccines, the battle against infectious diseases has been proceeding on another front. In **chemotherapy** synthetic chemical compounds are used to stop diseases. Unlike vaccines, chemotherapeutic agents don't act in conjunction with a person's natural immunity (the antibody-producing system). They function in other ways to destroy disease-causing organisms.

Like many other sciences, chemotherapy has long roots in the past. In every ancient civilization potions, powders, and plant and animal extracts have been attributed curative properties. Some ancient remedies were highly effective. Early Peruvians, for example successfully used the bark of the chinchona tree against malaria.

However, it was only after the basic discoveries of Louis Pasteur, Robert Koch, and other medical researchers of the nineteenth century that a systematic search for chemicals could be launched by knowledge of the basic principles of infectious disease. The first successes came early in the twentieth century, primarily through the work of the German scientist, Paul Ehrlich, who spent many years searching for a drug that would cure syphilis. After hundreds of unsuccessful attempts, he finally succeeded. His 606th drug was an arsenic-containing compound, salvarsan, which was subsequently used in treating syphilis before the discovery of penicillin.

Following Ehrlich's success, other medical researchers began experimenting with chemicals for the treatment of disease. In 1932 another German scientist, Gerhard Domagk, discovered that a red dye, prontosil, had germ-killing powers. Soon after his discovery, he tried it on his own daughter, who was dying of a bacterial infection. Prontosil halted the infection and saved the child's life.

Subsequent studies of prontosil showed that its germ-killing powers were concentrated in only a portion of the molecule. This was isolated and called sulfanilamide. It was the first of an important family known as the sulfa drugs, which have been successfully used in treating many diseases, including pneumonia and gonorrhea. Today there are many different sulfa drugs, but they all work on the same principle of **competitive inhibition.** The sulfa drugs are similar in structure to molecules needed by bacteria, and they "confuse" a bacteria's chemical system. When they are utilized in place of the proper substance they bring a halt to bacterial metabolic processes.

ANTIBIOTIC THERAPY

Chemical substances called antibiotics are now commonly used in the treatment of disease. Antibiotics are products of microorganisms, and in this respect they differ from chemotherapeutic agents, which are people-made. The essence of the effects of antibiotics was summed up by Louis Pasteur, who observed that two different microbes cultured together often inhibited each other's growth. "Life hinders life," concluded Pasteur. More recently it has been said that "bugs produce drugs that kill bugs." Antibiotics are chemicals synthesized by microorganisms that inhibit other microorganisms that are competing for the same ecological niche.

Penicillin

The "wonder drug" of World War II, penicillin, was the first of the antibiotics. It was discovered accidentally in 1920 by Sir Alexander Fleming, a Scottish physician and bacteriologist. Fleming was working with cultures of *staphylococcus* bacteria which cause boils and abscesses and noticed that some of the cultures contained fluffy masses of mold. Bacteria had not grown in the immediate vicinity of the mold which was identified as *Penicillium notatum* (Figure 14-15), a relative of the green and white mold often found on oranges. Fleming surmised that the mold had secreted an antibacterial chemical, which he called penicillin. Since the mold

FIGURE 14-15
The mold *Penicillium notatum* growing on a Petri-dish culture. The antibiotic penicillin is obtained from this mold as well as from other species of the same genus.

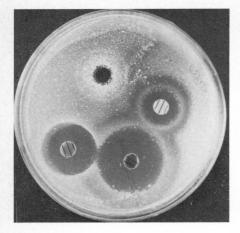

FIGURE 14-16

Determining the sensitivity of a micro-organism to various antibiotics. Each of the four paper disks has been soaked in a different antibiotic and placed in a bacterial culture. The size of the *zone of inhibition* reflects the relative effectiveness of each antibiotic. (Walter Dawn.)

FIGURE 14-17

The effect of penicillin on *Staphylococcus aureus* as seen by scanning electron microscopy. Photo on left shows a control not exposed to the antibiotic, and photo on right shows the result of exposure of *S. aureus* to penicillin for two and one half hours. That cell-wall synthesis is interfered with is obvious in this micrograph.

and the *staphylococci* were competing for the same food supply, it seemed that the mold's secretion was an adaptation that destroyed the bacterial competitor.

In the opening days of World War II, a group of Oxford University researchers began a search for antibacterial substances that would be useful in combating wound infections (Figure 14-16). Their attention turned to Fleming's work, and, in co-operation with him, they developed penicillin and thoroughly tested it.

Today a form of penicillin ten times as powerful as Fleming's is available in unlimited quantity and at low cost. Mutant strains of *Penicillium notatum*, produced by exposure to X rays, yield a far more potent drug than earlier strains. Biochemists have even produced penicillin synthetically.

Penicillin destroys bacteria by interfering with cell wall formation (Figure 14-17). As with sulfa drugs, however, penicillin does not work against viruses. The simple structure of the viruses saves them. Because they lack a cell wall, they are unaffected by drugs like penicillin. However, some antibiotics like the tetracyclines are effective against some types of large viruses.

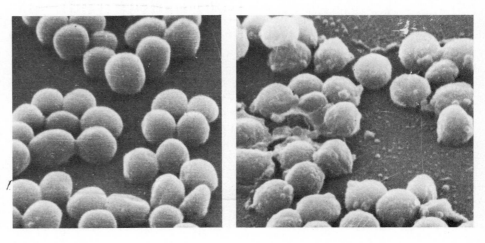

Streptomycin

Dr. Selman Waksman was fascinated by the cleansing powers of soil. He discovered that when the bodies of diseased organisms were buried, the microorganisms sometime disappeared. Waksman traced this phenomenon to a compound formed by a soil bacteria, *Streptomyces griseus*. As an antibiotic, streptomycin has been found to be particularly effective against tuberculosis. It is also partially effective for whooping cough, some forms of pneumonia, dysentery, gonorrhea, and syphilis.

Characteristics of antibiotics

An ideal antibiotic is effective against disease organisms, but does a minimal amount of damage to the host organism. It is difficult to develop the ideal antibiotic (Figure 14-18). Most people have mild negative reactions to antibiotics, and, for some people, serious ill-effects prohibit the use of certain ones. A severe allergic skin reaction to penicillin, for instance, is not uncommon. Highly allergic people may react to antibiotics, or foods, or insect bites with a state of collapse known as **anaphylactic shock**, which can be fatal.

Some antibiotics when taken by mouth destroy the normal bacterial population of the intestines. Intestinal bacteria usually aid digestion and help to keep pathogenic microorganisms in check. Diarrhea and other gastrointestinal dis-

FIGURE 14-18
Various steps in antibiotic production. From left to right, testing activity against specific bacteria, culturing antibiotic organisms on agar, a transfer device used to inoculate large vats with antibiotic organisms, extraction and purification of antibiotics from the producing organisms.

turbances usually follow the destruction of the normal population. Living *lactobacilli*, like those used in making fermented milk products (such as yogurt) help to restore the normal bacterial population if it is low and correct digestive disturbances.

A second reason for sparing use of antibiotics arises from the process of natural selection. Disease-causing organisms vary somewhat in their characteristics (as members of any species do); certain of them may have adaptations that give them resistance to a particular antibiotic. Repeated use of the antibiotic kills off the more susceptible bacteria, while allowing the resistant ones to reproduce. Penicillin, for example, was effective against *staphylococci* for many years. But, resistant strains of *staphylococci* have become widespread and threaten to make penicillin obsolete in treating them. Gonorrhea is becoming an ever greater threat for the same reason.

HEART DISEASE

The number one killer of people in the United States is heart disease—more accurately called **cardiovascular disease**—involving the heart and blood vessels. While there are many manifestations of cardiovascular disease, heart attacks and strokes are the best known to the public.

A **stroke** results when a blood vessel breaks in the brain or when the blood flow to the brain is blocked. Both produce brain damage because the brain cells die due to a lack of oxygen. The extent of damage will depend, of course, on what brain cells and how many die during a stroke. A **heart attack** occurs for essentially the same reason—oxygen (in the blood) fails to reach portions of the heart and heart cells die. Blood fails to reach the heart when the arteries supplying it become blocked. A blood clot from another part of the body may lodge in an artery and block the flow of blood to the heart, or the block may come from **arteriosclerosis**—hardening of the arteries.

Figure 14-19 shows what a hardened artery looks like. It has a buildup of smooth muscle cells, lipids and connective tissues that serve to block blood flow. The development of these so called **plaques** is the subject of much study. Cigarette smoking, high blood pressure, and high levels of cholesterol (a lipid) in the blood are clearly related to plaque formation. Some researchers claim that psychological stress leads to hardening of the arteries, and it may, but the evidence is not yet that strong. Since the plaque is composed partly of cholesterol, a large supply of it in the blood would certainly facilitate plaque formation. However, eating large amounts of high-cholesterol foods does not necessarily lead to high blood levels of it, since our own individual metabolism helps determine to what extent cholesterol will be deposited in arteries—this is a hereditary factor. High blood pressure may irritate arterial cell linings with abnormally high forces on the artery and this may induce arterial cells to grow abnormally and form plaques. Similarly, cigarette smoke may irritate the cells and induce cell growth.

Recently, a virus has been suggested as the cause of heart disease. The irritation of the arterial cells may either activate a virus already present or allow a virus to enter the injured cells. The virus could cause the formation of a **benign** (noncancerous) tumor which grows and blocks the artery.

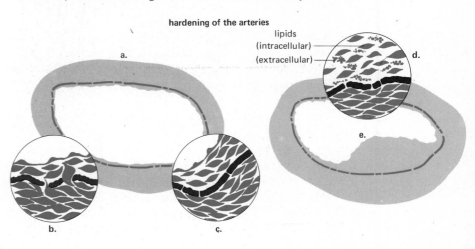

FIGURE 14-19

Hardening of the arteries. A normal artery in cross section (a). Higher magnification of the cells in the artery (b). Higher magnification of the cells in the artery as arteriosclerosis begins to develop; lipids are beginning to accumulate in the cells (c). Further progression of arteriosclerosis as lipids accumulate in the cells and extracellularly (d). The buildup of smooth muscle and connective tissue that blocks part of the artery (e). The dark band in each drawing is the main inner-lining of the artery.

CANCER

Cancer is characterized by the uncontrolled growth of cells and the formation of **tumors,** a mass of tissue that forms when cells that normally perform integral functions in an organism multiply independently. Cancer cells do not necessarily grow faster than other cells, but their growth is unrestrained.

Tumors take nourishment from normal cells and no longer contribute to the functional activity of the tissue or organ from which they descended. Cancer cells may also invade neighboring blood vessels and lymphatic channels, from which they can be swept into the bloodstream or the lymph fluid and be carried to distant parts of the body. Once they settle, they again multiply rapidly and unrestrictedly, forming another tumor. By this process of **metastasis,** cancer may spread throughout the body (Figure 14-20).

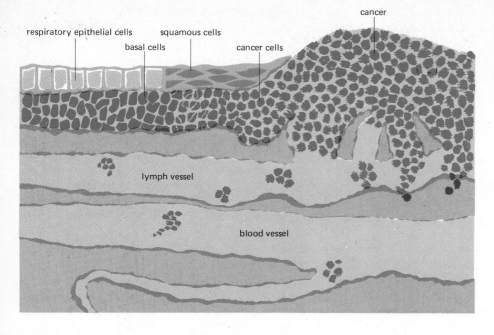

respiratory epithelial cells   squamous cells   cancer cells   cancer
basal cells   cancer cells
lymph vessel
blood vessel

FIGURE 14-20
Cancer cells spreading by metastasis.
(From *Life and Health*. Copyright © 1972
by Ziff-Davis Publishing Company. Re-
printed by permission of CRM Books, a
Division of Random House, Inc.)

Heredity, environment, and cancer

As in any disease, cancer depends on both the environment and the genetic constitution of the host. A clear example of genetic and environmental interplay is found in skin cancer. A large proportion of cases of skin cancer are due to excess exposure to sunlight. People with light skin (particularly if they are freckled) are about ten times more likely to develop skin cancer after prolonged exposure to sunlight than are people with heavy pigmentation, such as Blacks. Nevertheless, American Blacks are as susceptibles as Whites to most other forms of cancer.

Human beings of all races, colors, and environmental habits have been known to develop cancer of one type or another. But certain populations are particularly likely to get certain kinds of cancer. For example, as compared with the United States, stomach cancer is more frequent in Scandinavia and Japan. On the other hand, cancer of the breast and prostate are one-fourth as frequent in Japan as in the United States. Such factors as diet, environmental pollution and occupation are thought to play roles in the differences in the occurrence of a given type of cancer, but, at present, the extent of their influences are unknown.

Causes of cancer

There are many known factors which cause or aggravate cancer, and for a long time these diverse agents seemed unrelated. Today scientists are trying to find a unifying hypothesis—an explanation that will account for the fact that such different agents as cigarette smoke and X rays can produce uncontrolled cellular growth. Certain studies have indicated that viruses may lie at the root of the disease. Other research indicates that cancer results from a defect in the body's immune system or cellular control processes.

Chemical Agents   Various chemical agents are known to cause cancer. People whose work brings them into prolonged contact with certain petroleum byproducts, for example, take a higher than average risk of contracting skin cancer. Workers who breathe the dusts of asbestos and chromium compounds are prone to contracting lung cancer. The **carcinogenic** characteristics of tobacco smoke are infamous. Some researchers claim that pollutants in foods, water, air, and elsewhere in the environment account for most cancers.

Large doses of estrogen, one of the female sex hormones, readily trigger cancers of the breast, uterus, and testicles in laboratory-tested animals. In humans estrogen is known to aggravate existing cases of breast cancer, but whether or not it causes the disease is unknown. Evidence to date points to low cancer risks when estrogen is taken in contraceptive pills or as hormone-replacement therapy for postmenopausal women, but long-term studies are still in progress.

Chemicals that induce cancer trigger complex interactions within cells. The nature of the interactions is not yet known, but the final outcome is presumably a permanent change in the genetic elements that regulate cell growth. Such changes are transmitted to daughter cells during mitosis, and a population of cancer cells becomes established within the body.

Radiation   High-energy radiation has been implicated in several kinds of cancer. Ultraviolet radiation, for example, is the component of sunlight which causes skin cancer, and X rays have been found to increase the occurrence of leukemia. Radiation from atomic explosions also causes leukemia, as was evidenced by the high incidence of the disease among the survivors of the bombing of Hiroshima.

The chain of events between exposure to radiation and onset of cancer has not been definitely established. Radiation is known to increase the rate of genetic mutations in cells and may disrupt the genes coding for cellular control. On the other hand, radiation might make cells more sensitive to a cancer-inducing chemical agent or activate a cancer-inducing virus.

The Virus Theory   According to the virus theory, the various other agents of cancer simply open the door for a latent or hidden cancer-inducing virus that transforms normal body cells into wildly growing parasites that destroy surrounding tissues. As described earlier in this chapter, the primary component of any virus is nucleic acid, and viruses take over host cells by converting the genetic machinery to produce virus particles rather than normal cell constituents. Cancer viruses function no differently. The insertion of virus DNA into a host cell may disrupt the genetic code sequences that control cell division. It is also known that certain noncancer viruses such as cold sore viruses can infect cells but remain hidden or latent for a period of time until they are activated by ultraviolet radiation or chemical agents. It has been suggested that we may "catch" a cancer virus but it needs to be activated by some agent before it actually causes cancer. Viruses have been proved to be the cause of some types of cancer for some species but not for humans. Obviously, we cannot inject humans with suspected cancer viruses and wait for results.

The Defective Immune Mechanism Theory   Another theory proposes a relationship between cancer and the immune mechanism of the body. According to this theory, most people develop cancer cells but they are destroyed or repressed by normal immunological defenses. Occasionally such defenses become defective (possibly under the influence of a chemical agent, virus or radiation) and cancer flares up. It must be stressed that the viral and immune involvements in cancer need not be mutually exclusive; it is conceivable that either one may play a supportive role to the other, although a diminished immune response is probably a result of cancer, not a cause of it.

Fetal Genes   A quite different cause of cancer that has been suggested is that genes which are normally shut off are somehow turned on and their presence interferes with normal cellular control processes. It is thought that in the mature human, 80 percent of the genes may be "turned off" since these genes only functioned in the process of maturation. Most of these non-functioning were fetal genes— genes that enabled the zygote to reach the fetus stage. A characteristic of fetal cells is their sustained growth, somewhat primitive metabolism and ability to move from place to place. These are also characteristics of cancer cells. It was hypothesized, therefore, that cancer results when an agent (such as viruses, X rays, or chemicals) turns on these fetal genes and their gene products interfere with normal cellular growth. There is some evidence that fetal gene products exist in cancer cells but this does not prove that fetal genes cause cancer.

AGING

Hydras, which are freshwater relatives of the jellyfish and therefore multicellular, might live forever (Figure 14-21). Cell division in a hydra occurs in a growth zone just below the tentacles. As new cells are produced, old ones are pushed downward toward the foot or out along the tentacles. Eventually old cells are sloughed off from the foot or the tips of the tentacles. The entire process, from the moment a new cell is generated until it is lost, takes about a month and a half, in which time an entirely new set of cells is generated. Some biologists believe that if a hydra did not become a meal for a fish or other predator, it could be immortal.

Humans and other complex organisms lack the hydra's potential for longevity. Following birth, human development to maturity occurs over a span of 20 years or so, after which are about 10 years of peak health. Subsequently the body begins to deteriorate. In the middle 30's muscles gradually become less firm, and joints which were once supple become stiff. The internal organs—such as the heart, kidneys, and lungs—become less efficient. Mental powers diminish for some.

What is it that causes the changes of aging?

Degeneration without cell division

There is no pat answer to the question of aging, but the fact that organisms which constantly renew their cells seem to live indefinitely suggests that at least part of the answer lies in the process of cell division. In humans cell division is limited. Some cells continue to divide but the cells of the nervous system, for example, undergo very little division at all from a time soon after birth. But, why should this lead to aging?

Even though a cell does not divide, it is not static. The materials of cells are constantly renewed. Studies with amino acids tagged with radioisotopes, for example, have revealed that the structural proteins within cells are continually torn down and rebuilt. However, there is one important component of cells which is not replicated without cell division. This is DNA.

According to one popular hypothesis (which deals only with those cells that do not divide frequently) aging begins when the molecules of DNA begin to deteriorate. Perhaps radiation or toxic chemical substances accumulating in the body over the years cause certain nucleotides to decompose. If this is the case, the proteins coded for by the faulty DNA would have abnormalities in their sequences of amino acids. Thus, a particular brain enzyme, for instance, would be formed with a slightly different structure than it should have. It might still perform its function, but it would not be quite as efficient as the original form of the enzyme.

Over the years more and more errors in the structure of DNA would lead to more and more errors in the structures of proteins, and metabolic processes would consequently become increasingly disrupted. When this hypothetical process goes

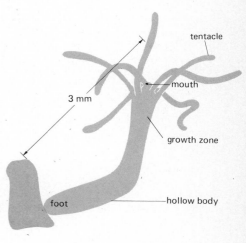

FIGURE 14-21
The hydra has a growth zone just below the mouth. Cells multiply in this zone and then move away in the direction of all tentacles and the foot, in such a manner as to cause successive displacement of older cells. These cells eventually are sloughed off from foot and tentacles.

far enough to disrupt the functions of critical organs, death occurs. Support for this sequence of aging comes from observations of organisms such as the hydra whose cells are periodically replaced and who show few aging effects.

The process of cell division is a rejuvenating process. When a cell divides, both daughter cells seem young. Why repeated cell divisions preserve the structure of DNA is unknown, but apparently they do. Further evidence that the lack of cell division leads to aging comes from the fact that those tissues in which little or no division occurs are the first to age. The highly specialized tissues of the muscular, circulatory, and nervous systems divide infrequently or not at all; and loss of muscle tone, circulatory problems, and mental deterioration are the most obvious signs of aging.

A comparison of the aging of the male and female reproductive systems gives a striking example of the effects of cell division. If a male stays sexually active, viable sperm are produced by his testicles from the time he becomes reproductively mature until he dies of old age. The sperm of an old man are indistinguishable from those of a young man. A female, however, is born with about a half million immature eggs within her ovaries. During the course of her fertile years the eggs mature at the rate of about one per month, but no new ones are produced. Consequently the DNA within any given egg is as old as the woman herself, and, as she ages into her 30s and beyond, her eggs deteriorate. Infants born to women over age 30 are more likely to have genetic defects than infants of younger women.

Other factors in aging

It is possible that aging is not merely the result of degeneration, but that it is somehow genetically preprogrammed. Changes in the secretion of certain hormones necessary for the maintainence of healthy tissues, for example, may occur according to a predetermined genetic schedule. Production of the female sex hormone, estrogen, is drastically cut back around the age of 50 (during menopause) with consequent physical repercussions including some degeneration of bones and the end of fertility. Different species have different life spans. Since their DNA codon sequences are also different, DNA may determine maximum life spans.

Another characteristic of aging is loss of the body's defenses against disease. As a person grows older, the white blood cells which once engulfed invading microorganisms lose their effectiveness somehow. The same is true for the antibody-producing immune system. Moreover, white blood cells and antibodies seem to lose their ability to tell good cells from bad ones. They begin to attack the body's own tissues rather than invaders from outside. Thus, the body advances its own destruction.

CHAPTER 14: SUMMARY

1   An infection is caused by an invading organism, while a disease is the malfunctioning of some part of an organism that may be caused by an invading organism, metabolic or genetic problem.

2   There are many defenses against infection and disease.

3   Viruses are the smallest "organisms" to cause infection.

4   Immunization against foreign substances that can be harmful is a very effective form of cellular defense.

5 Antibiotics interfere with the metabolism of invading organisms.

6 Hardening of the arteries may be caused by a variety of events.

7 Cancer has many potential causative agents.

8 Aging may be built into organisms as a natural process.

Suggested Readings

Boettcher, Helmuth   *Wonder Drugs: A History of Antibiotics,* J. B. Lippincott Co., Philadelphia, 1963 — How antibiotics were developed.

Cooper, Max D.   "The Development of the Immune System," *Scientific American,* November, 1974 — A detailed study of the diverse cells that defend the body against foreign substances.

Kohn, Robert R.   *Principles of Mammalian Aging,* Prentice-Hall, Englewood Cliffs, New Jersey, 1971 — Facts and theories about how we all age.

"Medical Ethics and Human Subjects," *Science News,* July 14, 1973 — The ethics of medical research involving human subjects.

"Out on a Lumph with Burt Bacteria" from *Go to Health,* CRM, Inc., 1962 — A cartoon representation of the functions of the immune system.

Silverstein, Alvin and Virginia Silverstein   *Cancer,* The John Day Co., New York, 1972 — This book considers the various types of cancer, gives the symptoms, and explains the possible causes and treatments.

White, John W.   "Acupuncture: the World's Oldest System of Medicine," *Psychic,* July, 1972 — The story of this ancient Chinese medical technique.

MAMMOGRAM MUDDLE

Modern technology has been a mixed blessing. Insecticides intended to destroy pests have wiped out populations of harmless species, and the search for fossil fuels to power our comfortable but energy-dependent way of life has led to the spoiling of natural environments. Medical technology, which has grown tremendously in the past few decades, also has its drawbacks. The reading "Mammogram Muddle" discusses how the X-ray technique commonly used to ferret out breast tumors may actually cause them to develop. But, as described in the story "Tuning in to Breast Tumors," there is a technological solution to the problem.

More than 250,000 women 35 years old and older have taken part over the past three years in a breast-cancer detection program conducted by the American Cancer Society and the National Cancer Institute. Because any cancerous tumors they may have are detected early, say the sponsors, these women presumably will have a lower-than-average rate of mortality from breast cancer, which will kill some 32,000 in the U.S. this year. Last week the screening program became the center of a major medical storm stirred by a group of doctors who warned that X rays used in the screening might actually increase the risk of breast cancer.

Low Dose. Besides having her breasts examined manually and photographed by a heat-sensitive technique called thermography,* every participant in the screening program is annually subjected to mammography, or breast X rays. Although only an extremely low dose of radiation is required, a team of scientists under the leadership of Dr. Lester Breslow, a U.C.L.A. epidemiologist, nonetheless argues that it may well be enough to cause cancer. Mammography, Breslow insists, is "a striking example of a

*Which can sometimes detect tumors because their temperature is higher than that of surrounding breast tissue.

situation where the very disease may be caused by the technology."

As evidence, Breslow cited a seven-year breast-cancer detection program, involving 62,000 women, undertaken in the 1960s by New York's Health Insurance Plan (H.I.P.). Analysis of the H.I.P. statistics showed that while mammography was of significant value in women over 50, the screening program did not reduce the mortality rate in those under that age. Breslow also noted studies showing increased breast-cancer rates among women exposed to higher radiation levels—those subjected to X rays in treatment of tuberculosis, patients receiving radiotherapy for acute breast infection, and survivors of A-blasts in Japan. Extrapolating from these data, he concluded that "there is no absolutely safe dose" for X rays and urged prompt discontinuation of mammography in routine screening of women under 50.

To allay fears of women alerted by press accounts of Breslow's criticism, the National Cancer Institute (NCI) hastily called a meeting in Bethesda, Md., last week. The directors of the screening program noted that mammography techniques have improved considerably since the H.I.P. study began 12½ years ago and that the radiation doses now used have been reduced to about a third of their old level. More important, they said that about two-thirds of the cases detected were in an early, curable stage—and only about half these cancers could have been detected without X rays. Said Dr. Philip Strax, director of the New York detection center: "The real risk is in *not* doing mammography." Added Dr. Barbara Ward of Boise, Idaho: "These reports are doing more damage than good by scaring women away."

Whether mass mammography will continue is to be decided in the next weeks, after further study by the National Cancer Institute that will include a poll of women Government workers in Bethesda. Asked how he would advise a patient if he were still in medical practice, Dr. Guy

R. Newell, NCI's deputy director, said that he would have no hesitation recommending mammography for any woman over 50. "For a woman under 50," he added, "I would tell her that there is a risk attached to the X-ray technique, a small risk that she might get breast cancer 15 to 30 years from now. But I would also state that by then there is a good chance there will be better treatment and a possible cure."

TUNING IN TO BREAST TUMORS

"Did I understand that you cooked my breast with microwaves?" the woman angrily asked Dr. Norman Sadowsky, chief radiologist at Boston's Faulkner Hospital. Sadowsky reassured her that he had not. Yet her concern is typical of the initial response to the hospital's breast-cancer detection program. To help in the all-important early discovery of a disease that has reached epidemic levels in the U.S. (90,000 cases a year), Faulkner radiologists are using microwaves to spot breast cancers.

Microwaves, though they are being employed for everything from sending telephone messages to cooking steaks, would seem to be a highly unlikely medical tool. Like other electromagnetic radiation—notably X rays—they damage tissue at high enough energies. But the Faulkner microwaves are perfectly safe. Reason: the radiation involved is emitted not by the detector, as in conventional breast X rays (mammography), but by the body itself.

The idea comes from M.I.T. astrophysicist Alan Barrett, who decided that the same electronic wizardry that was enabling him to tune in to microwaves from free-floating molecules in interstellar space could have a down-to-earth application. If they were reduced in size, he reasoned, the sensitive antennas could even pick up the weak microwave (or heat) emissions from a tumor.

Because of its rapid rate of growth and increased blood supply, a tumor is hotter than normal tissue and hence gives off more radiant energy. Thermography, or heat scanning, concentrates on looking for infra-red radiation to find tumors. But such waves are rapidly absorbed by bodily tissue; thus tumors that lie any distance below the skin's surface cannot be readily picked up by infra-red sensors. By contrast, microwaves—which are much longer and more penetrating—can locate tumors up to 10 cm. (4 in.) below the surface.

Not much larger than a stethoscope and used somewhat like it, the little antenna built by Barrett and an M.I.T. colleague, Philip Myers, is placed against nine different sites on the breast and held at each for about 10 seconds. If one spot turns out to be significantly hotter than a comparable area on the other breast, the supervising radiologist is alerted and can make other checks for a tumor, including X rays.

About 70% accurate, the gadget is admittedly less precise than mammography (90%) and only on a par statistically with infra-red thermography. But since there is no radiation risk and no need for a skilled X-ray interpreter to make an initial judgment, Sadowsky points out, the microwave detector could at the very least be used for pre-screening women—especially those under 35 who are ordinarily not encouraged to have mammograms unless they have a family history of breast cancer or symptoms of the disease.

CHAPTER

15

Neurobiology

Objectives

1 Develop an understanding of how nerves are put together.

2 Describe the production of a nerve impulse.

3 Know what happens at a synapse.

4 Understand how muscles contract.

5 Learn how our senses function.

6 Describe the structure and function of the human brain.

7 Learn how drugs affect the nervous system.

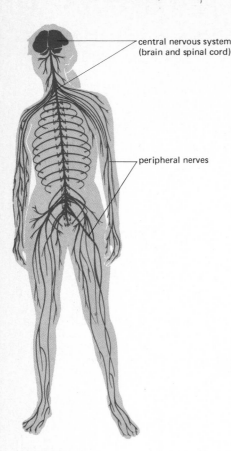

FIGURE 15-1

The central and peripheral nervous systems. For clarity, only a few of the branches from the major nerve trunks and nerves are included.

FIGURE 15-2

Nerve net system in Hydra.

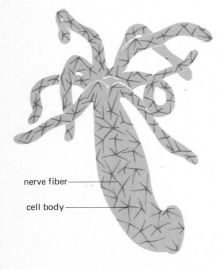

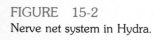

nerve fiber

cell body

THE NERVOUS SYSTEM

So far, living organisms have been portrayed as entities whose sole purposes are to stay alive and reproduce. But, the very fact that your attention is directed toward this page and that you can choose to continue reading or turn your thoughts elsewhere indicates that life is much more than mere survival. In humankind, evolution has led to a highly self-aware species, capable of thinking, reasoning, and titillating its senses. We can ponder the world about us and direct our muscular activities to structure our environments to suit our needs. We have created a disciplined mode of thought, science, to satisfy our curiosity about the physical and biological world.

The nervous system is the core of humankind's amazing uniqueness. It is the center of our awareness, and it gives us capabilities to ponder and to create that are shared by no other animals. Certainly other animals have nervous systems, and in basic ways all nervous systems are similar. But in complexity and capacity, ours is unparalleled.

In **vertebrate animals** (animals with backbones), the nervous system is classified as having three interrelated and overlapping parts. The **central nervous system** consists of the brain and spinal cord, which communicate with all parts of the body via the **peripheral nervous system** (Figure 15-1). The **autonomic nervous system** includes parts of the central and peripheral systems. It regulates the functions of the internal organs, and normally operates without conscious control.

Only a few years ago, most biologists believed that the autonomic nervous system was purely involuntary. It was thought, for example, that it was impossible to regulate the rate of heartbeat while at rest. But, practitioners of the ancient Indian techniques of yoga have proven to western scientists that people can have amazing controls over internal body functions. Through force of will and highly refined nervous control yogis have stopped their heartbeats for three or four minutes at a time.

Neurons

The basic units of the nervous system in all species are **neurons.** These specialized cells have a **cell body,** which contains a nucleus and cytoplasm, and from which project one or more **nerve fibers. Nerve impulses** travel along these fibers. In some of the lower animals, such as the hydra, impulses can travel in either direction (Figure 15-2), but in humans neural transmission occurs only in one direction.

In humans and other higher animals, fibers called **dendrites** receive stimuli from other cells and conduct an impulse toward the cell body. There may be several dendrites for any given neuron. Within our brains as many as 100,000 dendrites per neuron have been found. An **axon** conducts an impulse away from the cell body, and usually there is only one axon per neuron (Figure 15-3). The cell body incorporates all the impulses coming in along the dendrites and "decides" whether or not to send out an impulse along the axon.

In mature vertebrates there is no cell division by neurons, so those that are destroyed are lost forever. Furthermore, neurons die during one's lifetime. This loss may explain the diminished sensory responses of old age that we all are aware of.

Neural pathways

The simplest type of neural pathway in humans is the **reflex arc,** and the simplest reflex arc involves only two neurons. A **sensory neuron** transmits an impulse from

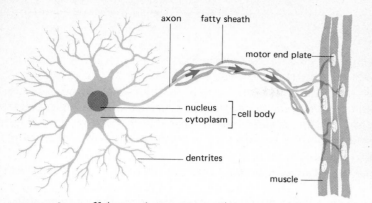

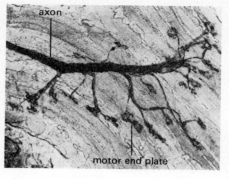

a **receptor cell** (a touch-sensitive cell, for instance) to a **motor neuron,** which, in turn, activates a muscle cell (Figure 15-3).

One reflex arc is the familiar knee-jerk, often checked in a physical examination (Figure 15-4). In this reflex, touch receptors are activated by the doctor's hammer, and they stimulate a sensory neuron, which sends an impulse to the spinal cord.

At left, the structure of a typical motor neuron. The axon of a real motor neuron is considerably longer than shown in proportion to the rest of the neuron. Arrows indicate the flow of the nerve impulse outward to a muscle. At right, a photograph of a motor neuron through a light microscope.

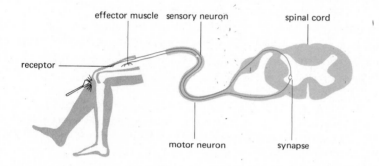

Here the impulse is transferred across a narrow junction called a **synapse** to a motor neuron, whose axon carries the message to muscles in the leg. The muscles contract, producing the jerk.

The two-neuron pathway is the simplest one found in humans, but such simplicity is the exception rather than the rule. Often a reflex arc includes one or more interneurons (Figure 15-5). Further complexities arise from the interconnection of neural pathways. For example, interconnections of the knee-jerk reflex with neural pathways leading to the brain can make you aware that your knee is being struck and allow you to modify your reactions by conscious control. Neural pathways traveling from the brain to the leg muscles could either increase or inhibit your reflex response.

FIGURE 15-4

Diagram of the knee-jerk reflex arc. (Illustration is reprinted from *Biological Science,* Second Edition, by William T. Keeton, illustrated by Paula Di Santo Bensadoun, with the permission of W.W. Norton and Company, Inc. Copyright © 1972, 1967 by W.W. Norton and Company, Inc.)

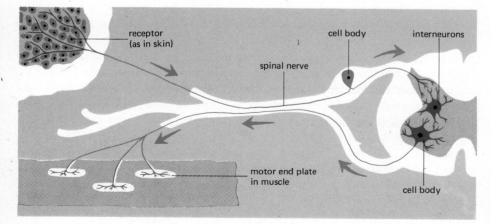

FIGURE 15-5

The path of the impulse from the receptor to the effector (the muscle).

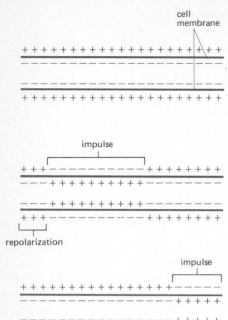

cell
membrane

impulse

repolarization

impulse

repolarization

FIGURE 15-6

The impulse travels along the neuron and can be measured as a change in electrical charge. From top to bottom: resting neuron; impulse moving to the right; original polarity being restored.

Neural impulses

Transmission of neural impulses can be roughly compared with the transmission of an electric current along a wire. There is, however, a basic difference. Electricity involves the movement of electrons, but neurons utilize the movement of ions. The fluid outside a resting neuron (a neuron which is not conducting an impulse) is positively charged while the inside of the neuron is negatively charged. This difference, or **polarization,** is caused by unequal concentrations of ions within the neuron and the fluid that surrounds it.

Potassium ions (K^+) and sodium (Na^+) ions are the primary charge carriers involved in the nerve impulse. Na^+ is highly concentrated outside the neuron (giving the outside a net plus charge). K^+ is highly concentrated inside the neuron, but there are also far more negative ions here (giving the inside a net minus charge). At the membrane of a neuron, energy must be expended to maintain this imbalance of charge.

When a nerve impulse is initiated by light hitting the retina of the eye or when some other receptor cell is stimulated by an outside stimulus, the net charges within a neuron are momentarily reversed (Figure 15-6). Na^+ diffuses into the neuron making the inside more positive than it was before, and with the loss of Na^+ from the outside, the outside becomes less positive in charge. Since charges are measured relative to each other, it is said that the inside of the neuron is now positive with respect to the outside, which is less positive and which is called negative by convention.

A fraction of a second after the Na^+ movement, K^+ diffuses from inside the cell to the outside. Since the inside is losing positive charges (the K^+), it is becoming negative with respect to the outside, which is gaining the K^+ ions. Therefore, the neuron is relatively negative again inside as it was in the resting state. However, the K^+ now is mostly outside and Na^+ is mostly inside — just the opposite of where we started. The process of active transport, which requires ATP energy, "pumps" the Na^+ and K^+ back where they belong (Na^+ to the outside, K^+ to the inside) and the neuron is then ready for another stimulus. This process is shown in Figure 15-7. It can occur in less than 1/250 second. If we wish to move our arm, for instance, we (in some totally unknown way) "will" Na^+ and K^+ to move into the appropriate neurons to initiate a nerve impulse to travel to our arm muscles!

A nerve impulse travels on the surface of a neuron, dendrite or axon because the membrane area immediately adjacent to the area that just reversed charges is stimulated to reverse charges also. The impulse moves along like a row of falling dominos. The development of the nerve impulse and its movement throughout any nervous system can be simply explained just by the movements of sodium and potassium ions, across the nerve cell membrane, plus activity at the synapse.

Activity at the synapse

When an impulse reaches the end of an axon, it may either stimulate the dendrites of an adjacent neuron at a synapse, or activate a muscle cell. In either case, the message is communicated by a chemical substance. This **transmitter substance** is contained in **synaptic vesicles** on the *presynaptic* side at the end of an axon (Figure 15-8). When an impulse arrives at the synaptic vesicles, it causes the release of the transmitter, which then diffuses to the dendrites of an adjacent neuron at the *postsynaptic* side, or to a muscle cell. If the transmitter substance arrives at the dendrites of a neuron, an impulse is initiated, which moves to its cell body and thence to another axon and so on. On the postsynaptic side, the neurotransmitter changes the membrane permeability of the neuron and this allows Na^+ to flow in to reinitiate a nerve impulse.

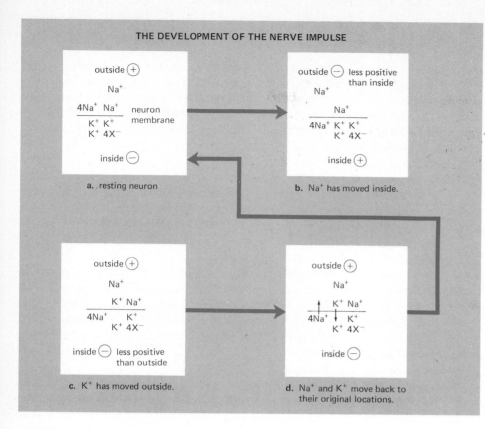

THE DEVELOPMENT OF THE NERVE IMPULSE

outside $\oplus$

Na⁺

$\dfrac{4Na^+\ Na^+}{K^+\ K^+}$ neuron
$K^+\ 4X^-$ membrane

inside $\ominus$

a. resting neuron

outside $\ominus$ less positive than inside

Na⁺

$\dfrac{Na^+}{4Na^+\ K^+\ K^+}$
$K^+\ 4X^-$

inside $\oplus$

b. Na⁺ has moved inside.

outside $\oplus$

Na⁺

$\dfrac{K^+\ Na^+}{4Na^+\quad K^+}$
$K^+\ 4X^-$

inside $\ominus$ less positive than outside

c. K⁺ has moved outside.

outside $\oplus$

Na⁺

$\dfrac{K^+\ Na^+}{4Na^+\ \downarrow\ K^+}$
$K^+\ 4X^-$

inside $\ominus$

d. Na⁺ and K⁺ move back to their original locations.

FIGURE 15-7

The development of the nerve impulse: (a) resting neuron; (b) Na⁺ has moved inside the neuron; (c) K⁺ has moved outside the neuron; (d) Na⁺ and K⁺ move back to their original locations. X⁻ indicates a negatively charged ion.

Because synaptic vesicles are contained only on the presypnaptic side, impulses cannot travel backwards across the synapse. Experiments have shown that if an isolated neuron is stimulated in the middle (by an electric shock, for instance), an impulse travels in both directions from the point of stimulation, but only continues if it is on the presynaptic side and stops as soon as it hits the postsynaptic side.

Between many synapses, the transmitter substance is a compound called *acetylcholine.* Several other substances are also transmitters. *Adrenaline,* for instance, operates in the autonomic nervous system.

A biochemical theory of the cause of schizophrenia involves the neurotransmitter noradrenaline, a form of adrenaline also called norepinephrine (NE). NE is synthesized from dopamine by an enzyme called dopamine hydroxylase and released into the synapse. When this enzyme becomes faulty for some reason, dopamine is not converted to NE and the dopamine is released into the synapse. Dopamine is converted to six-hydroxydopamine there, and it is thought that this is a toxic compound that destroys the synapses and produces schizophrenia. Since a gene produces the dopamine hydroxylase enzyme this gives us a genetic basis for schizophrenia. The psychological environment, via its input into the brain, may in some way turn off this gene or some internal cellular event independent of the psychological input may turn it off. This is only a theory now, but points out the important fact that mental illness may develop as a metabolic disease.

After a transmitter has completed its role of stimulation, it must be destroyed immediately. If it is not, its presence could continue to initiate impulses indefinitely. Acetylcholine is destroyed by an enzyme called *acetylcholinesterase.* Some insecticides and nerve gases work by inhibiting this enzyme's action. The insect dies because its nervous system runs wild with extraneous impulses. Unfortunately these poisons cause the same reactions in humans. The impulses that move across a synapse must also be of a certain intensity or they won't pass. The synapse "decides" in some way which impulses will pass.

FIGURE 15-8

An electron micrograph of the synaptic area of a neuron. The synapse distance between the pre- and post-synaptic neuron membranes is 200 Å.

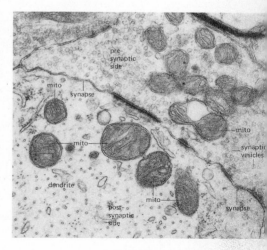

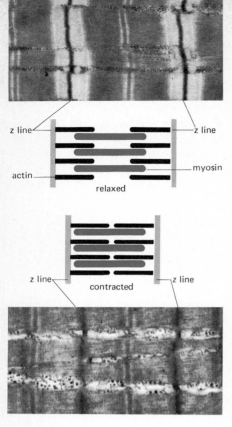

FIGURE 15-9

A representation of the current concept of muscle contraction. Compare and contrast the relaxed muscle tissue (top) with the contracted muscle tissue (bottom). The Z lines have come closer together during contraction.

FIGURE 15-10

An illustration of the role of cross bridges between myosin and actin filaments in muscle. At the left, the cross bridges are part of the thick myosin filaments. At the right, the cross bridges move toward the actin (left), hook onto a site (center), and contract, pulling the actin (right). (Illustration is reprinted from *Biological Science*, Second Edition, by William T. Keeton, illustrated by Paula Di Santo Bensadoun, with the permission of W.W. Norton and Company, Inc. Copyright © 1972, 1967 by W.W. Norton and Company, Inc.)

MUSCLE CONTRACTION

An impulse that reaches a muscle cell has much the same initial effect as one that reaches a neuron. A transmitter substance (acetylcholine) is released by nerve impulse stimulation at the **motor end plate** (where the nerve and muscle meet), causing the muscle's membrane to change its permeability. Up to this point, the initiation of a muscle contraction is quite similar to the mechanism of transmission of neural impulse.

Muscle cells, however, have an additional chemical mechanism that causes them to contract. When the membrane is changed by acetylcholine, calcium ions (Ca^{++}) diffuse from a specialized portion within a muscle cell called the **sarcoplasmic reticulum.** These ions bind to a muscle protein called troponin and the troponin-Ca^{++} complex moves away from another complex composed of the proteins **actin** and **myosin.**

Actin and myosin are long filamentous proteins laid side by side within muscle cells (Figure 15-9). With the troponin removed, actin and myosin are free to move relative to each other (contraction) with ATP utilized in the process. Actually, ATP is bonded to the myosin molecule, and ATPase (which breaks down ATP) is part of the actin molecule. When a muscle is at rest, the troponin molecule keeps the ATP and ATPase apart, but when troponin is removed by the Ca^{++} ions, the ATP of myosin is broken down by the ATPase of actin. Actin and myosin then slide together, causing contraction.

It is believed that the mechanism of contraction involves actual ATP projections from myosin molecules which join with the ATPase sites on actin molecules (Figure 15-10). Once stimulation ceases, the contraction process reverses as the Ca^{++} ions are actively transported back to the sacroplasmic reticulum. The muscle is then ready to contract again if stimulated. The entire process, from the arrival of the nerve impulse at the neuromuscular junction to the sliding together of the filaments, takes a small fraction of a second.

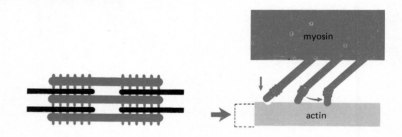

Body musculature

Grouped in bundles, muscle cells perform all the body's movements, from the slow, rhythmic contractions of the small intestine to the batting of an eye. There are about 400 different muscles in the human body, making up about one-half of the body weight. The three main types of muscles are **smooth muscles, skeletal muscles,** and **cardiac muscles** which are described in Figure 15-11.

NEUROLOGICAL CORRELATES IN PLANTS

Traditionally one of the major distinctions between plants and animals has been the lack of nerve and muscle systems in plants. Certainly, these are areas of major structural differences between the two kingdoms. However, at the chemical level, biologists are finding some surprising similarities.

MUSCULAR SYSTEM

MUSCLE TYPES	DESCRIPTION	FUNCTION
smooth muscle	each cell is an elongated spindle containing one nucleus; found in stomach and intestinal walls	contracts in waves to churn food or pass it along through the digestive tract
	found in artery walls	during stress, muscles constrict walls which raises blood pressure
	involuntary	not controlled at will
skeletal muscle	each fiber is a long cylinder with tapering ends, contains many nuclei; fibers are shorter than most muscles, bound in small bundles by connective tissue sheaths; heavier sheath encloses entire muscle which is attached by inelastic tendons to bones; arranged in pairs—	voluntary movement such as walking, running, grasping, lifting, bending
	flexors	bend joints
	extensors	straighten them
	voluntary	controlled at will
cardiac muscle	found only in the heart, cells from branching, interlacing network; plasma membranes separate cardiac muscle cells	muscles contract, squeezing heart chambers and forcing blood out through vessels
	involuntary	not controlled at will

Reprinted from MODERN BIOLOGY © 1977 by Otto and Towle.

Frontalis
Temporalis
Orbicularis oculi
Masseter
Sternocleido-mastoid
Trapezius
Deltoid
Pectoralis major
Biceps
Latissimus dorsi
Flexor carpi radialis
Palmaris longus
Tensor fascia lata
Rectus abdominis (beneath rectus sheath)
Iliotibial band
Vastus lateralis
Peroneus longus
Tibialis anterior
Extensor digitorum longus
Lateral malleolus (fibula)
digitorum tendons
Extensor
Soleus
Tibia
Medial malleolus (tibia)
Gracilis
Sartorius
Vastus medialis
Gastrocnemius
Soleus
Orbicularis oris
Triceps
Serratus anterior
Ext. oblique
Brachioradialis
Extensor carpi radialis longus
Iliopsoas
Pectineus
Adductor longus
Rectus femoris
Tendon of quadriceps femoris muscle group
Patella
Patellar ligament
Sup. extensor retinaculum
Inf. extensor retinaculum
Extensor hallucis longus tendon

FIGURE 15-12

Venus' fly trap, a carnivorous plant. Size of traps are about 10mm.

Venus' fly trap, a carnivorous plant, for example, is thought to have contractile proteins quite similar to actin and myosin. Thus, at the biochemical level, movements of this plant show a correlation to muscular contraction.

The discovery in plants of transmitter substances similar to acetylcholine and adrenalin in animals has indicated another neurological correlate. As in animals, these transmitters may affect permeabilities of cell membranes.

More surprising is the finding that certain plants, including *Mimosa* and the carnivorous Venus' fly trap transmit nerve-like impulses (Figure 15-12) when stimulated. These impulses function, as in animals, to rapidly transfer information from one part of the organism to another. The impulses move more slowly in plants, however.

SENSATION

Neurons transmit sensory stimuli from the organs of perception (skin, mouth, nose, ears, and eyes) to the brain. The initial input may be touch, flavors, odors, sound, and light, as well as pain and temperature. The ability to analyze *and* reflect upon sensory input is the amazing gift of humankind.

The sensations of the skin

The special sensory organs called **receptors** receive all input from the environment we live in. Some receptors are many-celled, others consist of only one specialized cell, and still others are the bare nerve endings (dendrites) themselves. Each receptor is suited to receive only a particular type of stimulus. Some of them respond to touch, while others receive stimuli of pressure, pain, heat, or cold (Figure 15-13).

FIGURE 15-13

The five types of receptors found in the skin.

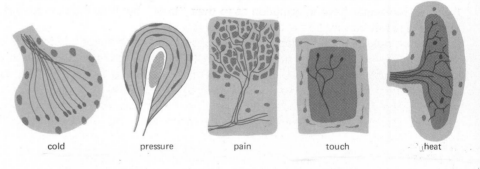

cold   pressure   pain   touch   heat

Normally, no one receptor reacts to more than one stimulus, and thus the five sensations of the skin are distinct. The *pain receptor* is an exception. It is a bare dendrite that will react to mechanical, thermal, electrical, or chemical stimuli. The sensation of pain is a protective device that signals a threat of injury to the body. Pain receptors are distributed throughout the skin.

The sensory nerves of the skin are distributed unevenly throughout the skin, and lie at different depths. For instance, if you move the point of your pencil over your skin very lightly, you stimulate only the nerves of *touch*. The receptors for touch are close to the surface of the skin, often near hair follicles. The fingertips, the forehead, and the tip of the tongue also have high densities of touch receptors.

Receptors that respond to *pressure* lie deeper in the skin. If you press the pencil point against your skin, you feel both pressure and touch. Since the nerves are deeper, a pressure stimulus must be stronger than a touch stimulus. Upon first consideration, you may think that there is no difference between touch and pressure. But the fact that you can distinguish the mere touching of an object from a firm grip on it indicates that separate nerves are involved.

Heat and cold stimulate different receptors. This is an interesting protective adaptation of the body. Actually, cold is not an active condition. Cold results from a reduction in heat energy. If both high and low temperatures stimulated a single receptor, we would be unable to differentiate between the two, and would be unable to react appropriately to either. However, since some receptors are stimulated by heat and others by the absence of it, we are constantly aware of and can react to both conditions.

Although we are always aware of our surroundings on some level, the sensory receptors of the skin adapt (or habituate) to stimuli of a constant intensity. For example, you aren't always aware of the touch of your clothes on your body, and numbness slowly replaces pain. This habituation of your senses is liberating in that your consciousness is freed to deal with more immediate stimulations.

Plants also adapt to stimuli of constant intensity. When rain begins to fall, the leaflets of the sensitive plant close (Figure 16-18), but if the rain continues, the leaflets reopen—they have habituated to the stimulus. If, however, they are touched, they will close again. They have adapted only to the constant patter of the rain.

The sense of taste

Taste results from the chemical stimulation of certain nerve endings. Since nearly all animals prefer some foods to others, we must assume that they can distinguish different chemical substances. The sense of taste in humans is centered in the **taste buds** of the tongue. These flask-shaped structures, containing groups of nerve endings, lie in the front area of the tongue, along its sides, and near the back. Foods mixed with saliva and mucus enter the pores of the taste buds and the molecules of food stimulate their hairlike nerve endings (Figure 15-14). The shapes of the food molecules have information as to their "taste" but how the nervous system obtains this information is not known.

Surprisingly, all the subtleties of flavor arise from combinations of four basic taste sensations: *sour, sweet, salty,* and *bitter.* Taste buds are distributed unevenly over the surface of the tongue (Figure 15-15). Those sensitive to sweet flavors, for instance, are concentrated at the tip. You may have noticed that candy tastes sweeter when you suck it, rather than chew it far back in the mouth. The tip of the tongue is also sensitive to salty flavors. Sour substances stimulate the sides, and bitter sensations are detected at the back of the tongue. If a substance is both bitter and sweet, you sense the sweetness first, then the bitterness. Certain spices, such as pepper, irritate the entire tongue, producing a burning sensation.

Much of the sensation we call taste is really smell. When you chew foods, vapors enter the inner openings of the nose and reach nerve endings of smell. If the external nasal openings are plugged, foods seem to lack the flavor we associate with them. Onion and apples for instance, can have an almost identical sweet flavor. You probably have noticed the loss of what you thought was taste sensation when you had a head cold and temporarily lost your sense of smell.

The sense of smell

Like taste, smell results from the chemical stimulation of nerves, except that odors are in the form of gases, which stimulate the **olfactory nerves** (Figure 15-16). Just as touch receptors become habituated after a time, so do the receptors of odors.

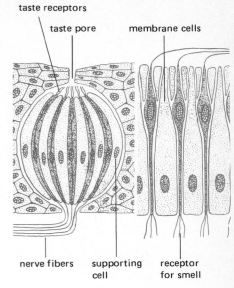

FIGURE 15-14

The receptor of taste (left) and smell (right), as they appear under a microscope.

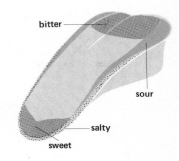

FIGURE 15-15

The distribution of the four basic taste receptors on the human tongue.

FIGURE 15-16

The surface of the inner wall of the nose.

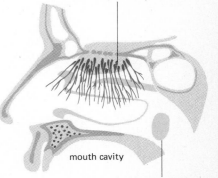

It is believed that there are only a few kinds of odor receptors, combinations of which can produce all sensations of smell. If this is the case, however, it is not yet clear what the basic odors are. As with taste, odors have a molecular structure which has "smell" information. The molecules must fit into specific molecular receptors which trigger specific neural impulses that the brain interprets as odors.

The functions of the human ear

Our ears are beautifully delicate mechanisms for detecting sound waves (Figure 15-17). The **external ear** opens into an **auditory canal** embedded in the skull. The canal is closed at its inner end by the **eardrum.** Behind the eardrum is the **middle ear,** which connects with the throat through the **Eustachian tube.** This connection equalizes the air pressure in the middle ear with the atmosphere.

FIGURE 15-17
Structure of the human ear.

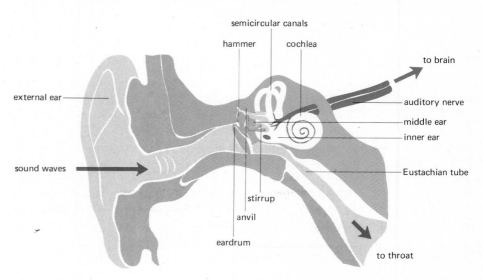

When the connection between the middle ear and throat is blocked by a cold in the Eustachian tube, the pressures outside and inside do not equalize. For this reason, divers and fliers cannot work when they have bad colds. The external pressure increases during a dive and with a blocked Eustachian tube, the middle-ear pressure would not be equalized. The difference might burst the eardrum. The flier's situation would be the reverse, although the eardrum could still burst: the pressure would be less outside than in the middle ear.

Three tiny bones, the *hammer* (malleus), *anvil* (incus), and *stirrup* (stapes) (so called because of their shapes) form a chain across the middle ear. They extend from the inner face of the eardrum to a similar membrane that covers the opening to the **inner ear.**

The inner ear has two parts, the **cochlea** and the **semicircular canals.** The cochlea is a spiral passage resembling a snail shell. It is filled with a liquid and lined with nerve endings that receive sound impressions. The **auditory nerve** leads from the cochlea to the brain. The semicircular canals consist of three loop-shaped tubes, each at right angles to the other. They function in maintaining a sense of balance.

Hearing   An object vibrating in air produces regions in which the air molecules are squeezed together (compressions) and regions in which they are farther apart (rarefactions). Waves of compressions and rarefactions travel outwards from a sound source (Figure 15-18) whether it be a tuning fork or a voice. Upon reaching

FIGURE 15-18
Areas of compressed air travel outward as waves from the tuning fork. The brain perceives the waves as sound.

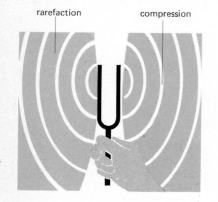

the ear, they cause the eardrum to vibrate. This, in turn, causes the hammer, anvil, and stirrup bones of the middle ear to vibrate, which sets the fluid in the cochlea in motion. Vibration of the fluid, in turn, stimulates the nerve endings in the cochlea. Different sounds would, of course, cause different patterns and affect different nerve endings. Impulses travel through the auditory nerve to the brain where the sensation of sound is perceived. If the auditory region of the brain ceases to function, a person cannot hear, even though his ear mechanisms receive vibrations normally. Similarly, no sounds can be perceived if the ear itself is not functioning properly. Damage to the receptor cells in the cochlea is something that cannot be reversed. However, when hearing loss invokes the three tiny bones, microsurgical procedures can often be used to restore hearing.

Balance

Our sense of orientation, or balance, is centered in the semicircular canals of the inner ear. These canals lie in three planes, at right angles to one another (Figure 15-17).

The semicircular canals contain neural receptors and a fluid similar to that of the cochlea. When the head changes position, the liquid flows in the canals, stimulating the receptors. Impulses traveling from the receptors to the brain communicate the position of the head. Since the canals lie in three planes, any change in position of the head moves the fluid in one or more of them. If you spin around rapidly, the fluid is forced to one end, and impulses travel to the brain. When you stop spinning, the fluid rushes back the other way, giving you the sensation of twirling in the opposite direction, so you feel dizzy. Regular, rhythmic motions produce unpleasant sensations that involve the whole body. These sensations are called **motion sickness**. Disease of the semicircular canals results in temporary or permanent dizziness and loss of equilibrium.

Sight and the human eye

The normal eye is spherical and slightly flattened from front to back (Figure 15-19). The wall of the eyeball is composed of three distinct layers. The outer, **sclerotic layer** is the white of the eye. It bulges and becomes transparent in the front, a portion of the eye called the **cornea.**

The middle, **choroid layer** is richly supplied with blood vessels. It completely encloses the eye except in front, where there is an opening, the **pupil.** Around the pupil, the choroid contains pigmented cells, which absorb excess light, preventing it from reaching the inside of the eye and damaging it. This is the **iris.** It may have a variety of colors. Change in size of the circular pupil is accomplished by muscles in the iris. This adjustment, according to the intensity of light, is an automatic reflex. When the light is reduced, the pupil becomes large, or *dilates*. In bright light it becomes small.

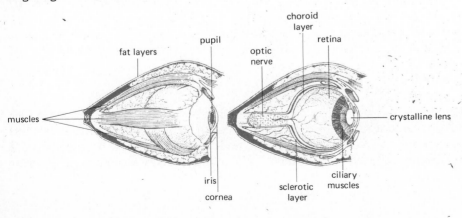

FIGURE 15-19

The human eye. The muscle and socket are shown at the left; the various internal structures can be seen in the cutaway diagram at the right.

Eye doctors use drops of a muscle relaxant to block the automatic iris reflex so that a bright beam of light can be shined in for an examination of the interior. This is the only place in the body where the blood vessels can actually be directly seen. Such examinations give information not only about eyes, but also about the condition of the circulatory system. It's possible, for example to detect hardening of the arteries in this fashion.

A convex, **crystalline lens** lies behind the pupil. The lens is supported by the **ciliary muscles** fastened to the choroid layer. Contraction of these muscles changes the shape of the lens, and in this way, light rays are focused on the retina. Nearsightedness results when the image is focused in front of the retina, and farsightedness develops if the image is focused behind the retina.

The space between the lens and the cornea is filled with a thin, watery substance, the **aqueous humor.** A thicker, jellylike, transparent substance, the **vitreous humor,** fills the interior of the eyeball. It keeps the eyeball firm.

The eye rests in its socket against layers of fat that serve as cushions. Movements of the eyeball are accomplished by pairs of muscles that attach to its sides and extend back into the socket. The sclerotic layer is supplied with nerve endings that register pain when a foreign object touches it. The eye is further protected by its location deep in the recesses of the eye socket, by bony ridges, by the eyelids, and by the tear glands that keep its surface moist. Tears wash over the eye and drain into the tear duct in the lower corner of the eye socket, which leads to the nasal cavity. Because tears contain an antibacterial enzyme, they are mildly antiseptic.

The Structure of the Retina   The inner layer of the eye is the most complicated and delicate of the eye layers. This layer, the **retina,** is 300 micrometers thick. Yet, it is composed of seven layers of cells, receptors, ganglia (large groups of neurons), and nerve fibers. The function of the structures of the eye is to focus light on the retina. The specialized receptors that are stimulated by light are called **photoreceptors** and are of two types, **cones** and **rods** (Figure 15-20). The photoreceptors lie deep in the retina, pointing toward the back surface of the eyeball. There are over 100 million of them in a human eye. Light causes impulses from the cones and rods to travel through a series of short nerves with brushlike endings to ganglia near the front part of the retina. About a million nerve fibers lead from the ganglia over the surface of the retina to the **optic nerve.** There are no

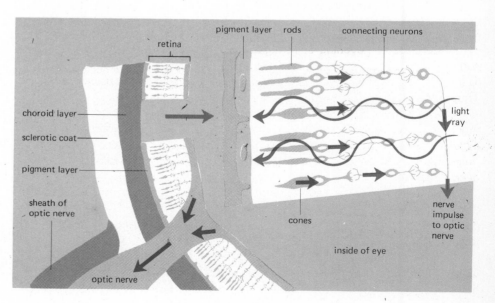

FIGURE   15-20

The structure of the back of the eye. The shapes and arrangement of the rods and cones are shown in greater detail in the enlargement. Short arrows are nerve impulses, long arrows are light rays.

rods or cones at the point where the end of the optic nerve joins the retina. Since there are no visual receptors at this point, it is called the **blind-spot.**

An optic nerve extends from the back of each eyeball to the vision center in the brain. Some of the fibers cross as they lead to the brain. This means that some of the impulses from your right eye go to the left side of the brain, while the rest go to the right side. Thus, what you see with each eye is interpreted in both sides of the brain.

Vision   The cones of the retina are sensitive to bright light and are responsible for color vision. Light rays pass through the cornea, aqueous humor, pupil, lens, and the vitreous humor to the retina. The lens focuses the rays on a small, sensitive portion of the retina called the **fovea,** where the cones are especially dense. Cones outside the fovea register vision only indistinctly. The function of the eye has been compared to that of a camera (Figure 15-21).

There are actually three different kinds of cones, each having a pigment which responds to a different range of colors. Combinations of responses produce the sensations of all the colors of the spectrum. The impressions of colors take place in the brain rather than in the retina, however, as can be demonstrated by looking through a green filter with one eye and a red filter with the other. What is seen appears yellow. Since neither retina received yellow light, it must be assumed that the impression of yellow was created in the brain.

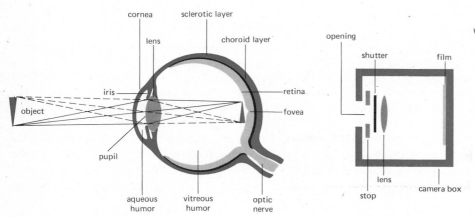

FIGURE 15-21
A comparison of the functioning of the human eye with that of a camera.

During the late evening or at night, the light is too reduced to stimulate the cones, but is sufficient to stimulate the rods. However, while the rods are able to act under conditions of reduced light, they are unable to distinguish along colors. Thus colors become less distinct with decreased illumination.

Rods produce a protein called **rhodopsin,** which is necessary for proper functioning. Bright light destorys rhodopsin, and prevents the rods from functioning. This explains why when you leave a bright room at night, you are temporarily "night-blind." As rhodopsin is restored, the rods begin to work, and objects become visible.

You may have noticed that at night you can see something "out of the corner of your eye," but when you attempt to focus on it, it disappears. This is because the portion of the retina surrounding the fovea has a higher density of rods than the fovea itself.

The human eye contains fewer rods than many animal eyes, so our night vision is relatively poor. The dog, cat, deer, and owl see well at night because they have many rods. The owl, however, lacks cones and therefore is "day-blind."

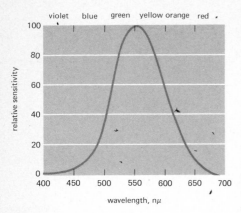

FIGURE 15-22

The relative sensitivity of the human eye at different wavelengths for normal levels of illumination.

Furthermore, they have a reflective coating of cells behind the retina that reflects light which has passed through the retina back to the retina again for sensory processing. This is why when you shine a light on a dog or cat, its eyes shine.

The electromagnetic spectrum, and the limits of our senses

If asked which sense we valued most highly, most of us would probably choose sight. Yet, for all that sight reveals to us, it is, in a way, a very narrow view of our surroundings. As described in Chapter 7 visible light is but a small portion of the **electromagnetic spectrum.** Like all electromagnetic radiation, visible light is propagated through space in waves at a speed of 186,284 miles per second. It is known that each color is characterized by a **wavelength** (the distance between peaks of a wave) and **frequency** (the number of peaks which pass a point every second). Our eyes are most sensitive to the yellow-greens (Figure 15-22); hence, green boards and yellow chalk.

What a small portion of the electromagnetic spectrum we can sense! You need only turn on a radio to realize that we are constantly immersed in radiation of which we are never directly aware. One can only wonder what our impression of reality would be if we could sense a broader range of the electromagnetic spectrum. The question also arises: Are there physical phenomena occurring about us of which we have no knowledge whatsover because of our sensory limitations?

THE BRAIN AND SPINAL CORD

The brain receives and integrates information from the organs of sensation and directs various body processes. Among other things, it is the center of muscular coordination, sense integration, thought, memory, and emotion (Figure 15-23).

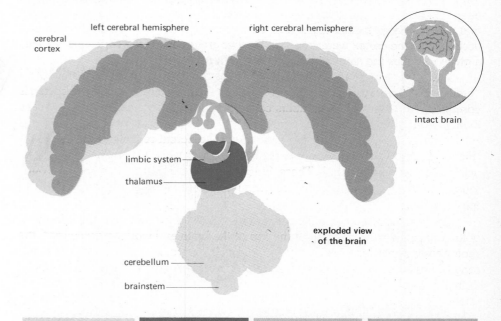

FIGURE 15-23

A diagram of the human brain.

muscular coordination

The cerebellum functions in coordination. It sends messages to the cerebral cortex for "awareness" and also to the various muscles of the body.

sense integration

A mass of gray matter called the thalamus works with the cerebral cortex to integrate messages from the sense organs such as the eyes and ears.

thought and memory

The gray matter making up the cerebral cortex is the area of memory. Sensations are registered here and voluntary actions are begun.

emotion

This area, buried deep within the brain is involved in emotional reactions and control.

Although its function is basic to human awareness, of all the body's organs, we know the least about the brain.

The average human brain weighs about three pounds and fills the cranial cavity. It may have 10^{11} neurons (counting 1 per second would take 30,000 years to count them all) while a nematode for comparison has about 10^2 neurons. The brain is composed of soft nervous tissue and is covered by three protective membranes, together known as the **meninges.** The meninges are richly supplied with blood vessels that carry nutrients and oxygen to the brain cells. Inflammation of them is a serious disorder known as *meningitis.* The space between the inner two meninges is filled with a clear liquid, the **cerebrospinal fluid,** which, as the name implies, is also found around the spinal cord. The composition of cerebrospinal fluid is similar to blood plasma, and its function is similar in that it brings nutrients to the brain. The meninges and cerebrospinal fluid cushion the delicate tissues of the brain and spinal cord, but occasionally, their protection isn't enough. A *concussion* is a brain bruise that results from a severe blow to the head.

The cerebrum—center of highest mentality

The region of the brain known as the **cerebrum** is proportionately larger in humankind than in any other animal (Figure 15-24). Over the course of vertebrate evolution the cerebrum has developed from a small pair of lobes which were originally concerned only with the sense of smell. The cerebrums of modern fish are still little more than this. In the evolutionary sequence from fish to amphibians, to reptiles, to mammals, the cerebrum has become progressively more concerned with integrating all sensory perceptions and ultimately thought processes. Conscious control of muscular response is also centered in the cerebrum.

The human cerebrum consists of two halves, or **hemispheres,** joined by tough connecting fibers and nerve tracts. The outer surface, or **cerebral cortex,** is deeply folded in irregular wrinkles and furrows. These convolutions give the cerebral cortex a large surface area (and, hence, more space for mental processes to occur than if the cortex were smooth). Deeper groves divide the cortex into **lobes;** a given lobe has the same function in every human brain.

The cerebral cortex is composed of billions of neurons and is frequently called "gray matter" because of the color of its cells. The cerebrum within the cortex is composed of masses of nerve fibers covered by insulating sheaths; it is known as "white matter."

Different mental activities are controlled by different regions of the cerebrum (Figure 15-25). The **frontal lobes,** for example, are centers of emotion, judgment, will power, and self-control. These functions aren't totally limited to the frontal lobes, however. As with most higher mental processes, several regions of the brain are involved.

The **motor areas** of the cerebral cortex are concerned with the control of voluntary movement. Starting at the top of the brain and working downward, the motor areas control the muscles of the legs, trunk, arms, shoulders, neck, face, and tongue. Other areas of the cerebral cortex are **sensory areas.** Vision, for example, is interpreted in the sensory areas termed **occipital lobes.** If these lobes are destroyed, vision is lost, even though the eyes themselves might be functioning perfectly. It is an indication of the degree to which evolution has led to dependence on the cerebral cortex that total blindness occurs when a human's occipital lobes are lost. For a frog which has suffered analogous damage to the cerebral cortex, there is no change in vision. Most mammals which have lost their cerebral cortexes retain at least the ability to distinguish between light and dark.

Communication between the sensory and motor areas occurs via **association areas** of the cerebral cortex. The association areas are zones in which sensa-

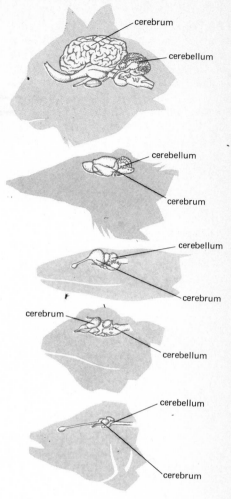

FIGURE 15-24

Comparative sensory and motor areas of the cerebral cortex in various animals. Note the progressive enlargement of the cerebellum as the vertebrates have evolved. (From *Biology Today.* Copyright © 1972 by CRM Books. Reprinted by permission of CRM Books, a Division of Random House, Inc.)

FIGURE 15-25

The upper drawing shows the control areas of the cerebrum. The motor areas are involved with the origin of action and the sensory areas with the interpretation of senses. The lower drawing shows the relative proportions of brain areas controlling human sensory responses. (Lower drawing from *Biology Today*. Copyright © 1972 by CRM Books. Reprinted by permission of CRM Books, a Division of Random House, Inc.)

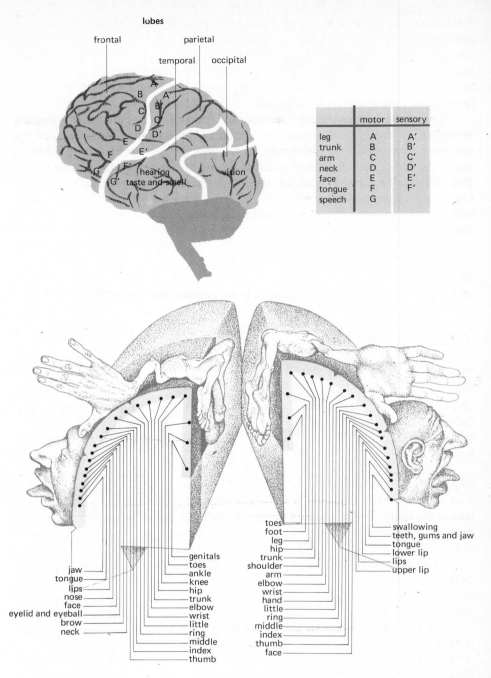

	motor	sensory
leg	A	A'
trunk	B	B'
arm	C	C'
neck	D	D'
face	E	E'
tongue	F	F'
speech	G	

tions are interpreted and courses of action are decided upon. As such the association areas are the centers of the most human of all mental characteristics — complex thought. Association areas account for the largest proportion of the human cerebral cortex.

Limits of Our Knowledge: Learning and Memory   In describing the functions of the cerebrum, one can talk about the structures of cells, electrical properties of membranes, and chemical changes; but, we're still very much in the dark in our understanding of our own brains. Learning and memory — perhaps the most distinctive human mental activities — remain particularly inscrutable.

How, for instance, do we learn to recognize objects in our environment — how are familiar objects, friends, and numbers "coded" in our brains? More significantly, how do we learn to manipulate objects and concepts? How do we learn to add and subtract? How can a creative mathematician see relationships in numbers and abstract variables which no one before her has ever seen? No one yet knows. We do know that changes in proteins, RNA, and ion flow accompany learning in animals. But, it is difficult to delineate between the chemical activities involved in learning and other metabolic functions.

One of the most promising theories of contemporary brain research is that the recording of an impression involves a change throughout a large area of the cerebrum, rather than a development at a specific point. According to this theory, many memories could be stored within the same zone. Other researchers maintain that memories are more localized—that a particular set of neurons and/or biochemical compounds are involved. RNA and protein, for instance, are being considered as possible mediators of memories since they have been shown to increase during learning. Perhaps some combination of these theories is closer to the truth.

In reflecting on the frontiers of human achievement, one cannot help but feel that the search for an understanding of the miraculous powers of the mind is among the most exciting things happening. To know how we know is an awesome goal, and research toward that goal appears to be coming into its classic age.

The cerebellum

The **cerebellum** lies beneath, and toward the back of, the cerebrum. Like the cerebrum, it is composed of two hemispheres, but its convolutions are shallower and more regular. The surface of the cerebellum is composed of gray matter. Its inner structure is largely white matter, although it contains some areas of gray matter. Bundles of nerve fibers connect the cerebellum with the rest of the nervous system.

In a sense, the cerebellum acts as an assistant to the cerebrum in controlling muscular activity. Nervous impulses do not originate in it, and its main function is motor coordination. However, it somehow helps to coordinate the muscular activities of the body. The cerebellum functions further in strengthening impulses to the muscles. This action is a little like a relay station picking up a weak radio or television signal and amplifying it before rebroadcasting.

Another function of the cerebellum is to maintain muscle tone. The cerebellum cannot originate a muscular contraction, but it causes the muscles to remain in a state of partial contraction. You are not aware of this, because the cerebellum operates below the level of consciousness.

The cerebellum functions also in the maintenance of balance. In this activity, it is assisted by impulses from the eyes and from the organs of equilibrium of the inner ears. Impulses from both these organs inform the cerebellum of the body's position relative to its surroundings.

The brain stem

Nerve fibers from the cerebrum and cerebellum enter the brain stem via an enlargement at the base of the brain (Figure 15-26). The lowest portion of the stem, the **medulla oblongata,** is located at the base of the skull and protrudes from it slightly where it joins the spinal cord.

There are twelve pairs of **cranial nerves** connected to the brain. These are part of the peripheral nervous system and act as direct connections with certain organs of the body. One pair, for example, connects with the eyes. Other cranial nerves join with the lungs, heart, and abdominal organs. Through these nerves the

FIGURE 15-26
A longitudinal section of the brain, showing the regions and the meninges.

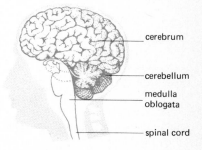

cerebrum

cerebellum

medulla
oblogata

spinal cord

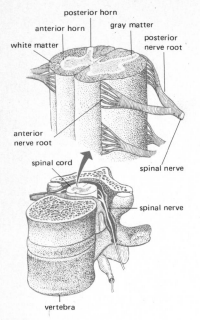

FIGURE 15-27

The structures of the spinal cord and their relationship to the vertebral column. The white matter is located in the outer region of the spinal cord.

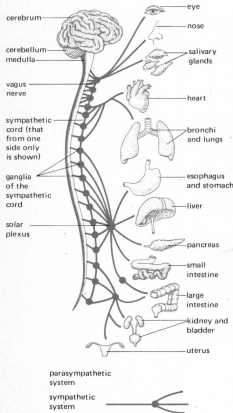

FIGURE 15-28

The autonomic nervous system regulates the internal organs of the body.

medulla oblongata influences the activity of the internal organs. Heartbeat, muscular contractions of the walls of the digestive organs, secretion in the glands, and other automatic activities are regulated by the medulla oblongata.

The spinal cord and spinal nerves

The **spinal cord** extends from the medulla oblongata within the bony arch of vertebrae which compose the spinal column. The outer layer of the spinal cord is white matter, made up of nerve fibers covered by sheaths. Neurons composing the gray matter lie inside the white matter in a shape like that of a butterfly with spread wings (Figure 15-27).

Thirty-one pairs of **spinal nerves** branch off the cord between the vertebrae of the spine. Along with the cranial nerves, they form the peripheral nervous system. One member of a pair goes to the right side of the body. Its mate goes to the left side. Spinal nerve branches begin in the neck and continue all the way along the cord. These large cables are mixed nerves. Of their many fibers, some are sensory fibers that carry impulses into the spinal cord, while others are motor fibers that transport impulses away from it.

If the spinal cord were cut, all parts of the body controlled by nerves below the point of severance would be totally paralyzed. Such an injury might be compared to cutting the main cable leading to a telephone exchange.

THE AUTONOMIC NERVOUS SYSTEM

The *autonomic nervous system* (Figure 15-28) is usually involuntary—its controls of the internal organs are normally not consciously directed. As mentioned in the introduction to this chapter, however, some people have developed amazing abilities to alter its regular functions.

There are two subdivisions of the autonomic system, the *sympathetic* and *parasympathetic* systems. The **sympathetic system** includes two rows of nerve cords which lie on either side of the spinal column. Each cord has ganglia, which contain the cell bodies of neurons. The largest of the sympathetic ganglia is the solar plexus, located just below the diaphragm. Another is near the heart, a third is in the lower part of the abdomen, and a fourth is in the neck. Fibers from the sympathetic nerve cords enter the spinal cord and connect with the brain, as well as with each other.

The principal nerve of the **parasympathetic system** is the *vagus nerve*, a cranial nerve which extends from the medulla oblongata, through the neck, to the chest and abdomen. The parasympathetic system acts in concordance with sympathetic nerves in maintaining harmony in the internal organs. These two systems exert "checks-and-balances" on each other, illustrated by the fact that the sympathetic nerves act to speed up heart action, while the vagus nerve slows it down. Together, the sympathetic and parasympathetic systems regulate the heart, secretions of the ductless glands, constriction of the arteries, actions of the muscles of the stomach and small intestine, and other functions of the internal organs.

ARTIFICIAL STIMULATION OF THE NERVOUS SYSTEM—DRUGS

The nervous system is electrical, and biologists have found that artificially applied electric currents can have some amazing effects on nervous behavior. Electrodes (very fine electric wires) have been implanted in human brains (Figure 15-29). When electricity is applied to them, they stimulate impulses within nearby neurons.

Researchers working with cats and several other animals have found that aggressive behavior can be elicited or prevented by electrical stimulation of specific parts of the brain. Changes in autonomic functions, including the rate of heart beat, the constriction of blood vessels, and pupil dilation can be affected by electrical stimulation also.

In addition to electricity, biologists have used various chemicals to elicit nervous reactions. For example, when the concentration of calcium is artificially raised in the hypothalamus, rats that have just eaten until they are full start voraciously eating again. In cats, body temperature can be raised or lowered by altering the ratio of sodium ions to calcium ions in the hypothalamus.

Many of us often make drastic chemical changes in the biochemistry of our nervous systems. Sometimes we do it unwittingly. Work by one researcher, Ben Feingold, indicates that a diet high in artificial flavors and colorings (included in processed foods such as hot dogs, ice cream, and ready-to-eat cereals) aggravates childhood *hyperkinesis*. This behavioral disorder is characterized by excessive physical activity and an inability to concentrate and learn.

Chemical behavior modification can be accidental, but often it is quite purposeful. Consciousness altering drugs are as old as civilization, and in our age of blossoming biochemical technology, new ones are turning up all the time.

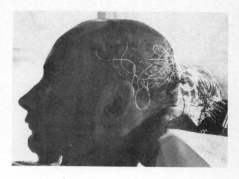

FIGURE 15-29

Brain stimulation in humans has been used successfully to relieve unpleasant symptoms.

Consciousness-altering drugs

Humankind is a drug-taking species. From before the early Egyptians, people have used chemicals to cure their ills and to modify consciousness. Some of our most amazing inventiveness has been expressed in the search for drugs. Chewing dried toad skins, for example, is an old Chinese remedy for curing bleeding gums. (A few years ago, a biochemist discovered that toad skins contain large amounts of *norepinephrine,* a chemical which constricts blood vessels.)

Chemicals that alter consciousness are of particular biological and medical interest because people tend to use them over and over again. Thus, any adverse side effects they may have are aggravated by repeated use. Most drugs, if taken in excess, will cause body damage. The sulfa drugs, for instance, are effective in curing some bacterial diseases, but will cause kidney impairment if they are overused. People sometimes over-use medicinal drugs (including aspirin, vitamins, laxatives, and antibiotics), which can have adverse physical effects. And chemicals which affect the mind can become so compulsively needed that they dominate the psyche as well as impair the health of their users. Questions concerning consciousness altering drugs are pertinent to many of us. Be it coffee or cocaine, a large segment of our society regularly uses one or more drugs to change their approach to life. The risks of drug taking are hard to measure objectively, or to balance against their benefits. Recently scientists have discovered naturally occurring opiates in the brain! They are proteins called **endorphins.** They are secreted by the pituitary to inhibit pain in the body but under just what conditions isn't known. Furthermore, if secreted at an abnormally high level, endorphins may alter consciousness by creating a schizophrenic-like state.

A variety of drugs

One of the frequent effects of using drugs to alter perception, mood and behavior is that one reaches a state in which he or she feels compelled to use them regularly. This is psychological dependence. When deprived of the drug, the habituated person becomes restless and irritable. It has been pointed out, however, that the same symptoms can characterize the loss of a close friend or the cancellation of a favorite TV program. Thus psychological dependency on drugs, such as caffeine, tobacco, alcohol, or marijuana, is a phenomenon that can only be considered on an individual level.

A common characteristic of persistent drug use is that the taker develops **tolerance**—the body becomes adapted to a drug from regular use. As the result of tolerance the user needs progressively greater amounts to achieve the same psychological effects. Alcohol, for instance, is a drug for which people quickly develop a tolerance.

The state of **addiction** is marked by a physical need for a drug. Besides being psychologically dependent upon, or physically tolerant of, a drug, an addict must continue using it, or experience **withdrawal illness.** The *delirium tremens* (D.T.'s) attack of an addicted alcoholic is the most serious withdrawal disease known. Characterized by convulsions, hallucinations, nausea, and vomiting, *delirium tremens* can end in death.

Depressants   Fortunately, although almost anything can bring about psychological dependence, acute physical addiction occurs primarily with the class of drugs known as the **depressants.** These include *alcohol, narcotics,* and *barbiturates*. The narcotics carry with them the quickest rate of addiction, followed by alcohol and barbiturates.

FIGURE   15-30
Member of an opium-growing tribe of North Thailand tending his crops.

Narcotics   *Opium* is a juice extracted from the opium poppy (Figure 15-30). The drugs derived from it, including *heroin, morphine,* and *codeine* (as well as similar synthetic drugs, such as *methadone* and *demerol*) are collectively termed the **narcotics.** They are widely used both medically and illicitly.

Morphine is the primary ingredient of opium, and is often used for medical purposes. Although heroin is derived from morphine, it is barred from medical use in the United States but is the most popular of illicitly used opiates. By weight, pure heroin is about two times more potent than morphine, but "street" heroin averages only about four percent pure. Nevertheless, it is expensive: $300,000 per kilogram.

The opiates are primarily depressants of the central nervous system; they have a high potential for dependency, tolerance, and addiction. The usual starting dose is less than five milligrams, but tolerance can increase the user's needs within a matter of months to doses of 1000 milligrams. Heroin addiction is an expensive habit, costing the addict 50 dollars or more a day.

The initial effects of the opiates are relief of tension and anxiety, decrease in physical drive, and drowsiness. Some users experience a sense of euphoria, but others, particularly if they aren't anxious or in pain, experience distinct unpleasantness. Therapeutically, the opiates are effective in reducing pain, sedating, controlling diarrhea, and suppressing coughs. The biochemical mechanism by which they produce these effects, however, is not known.

Chronic use of opiates often leads to addiction, and the symptoms of withdrawal are excruciating (although generally less intense than with alcohol or the barbiturates). At the peak of withdrawal, the addict experiences severe intestinal pains and diarrhea. The back muscles, bones, and extremities all ache constantly. The symptoms reach a climax within twenty-four hours, and are mostly over within two days.

Death from overdoses of heroin is shockingly common. In New York City alone, about 1000 people die yearly from black-market heroin, whose concentration is highly variable and unpredictable. Although many heroin overdoses are accidental, suicide is common among addicts.

Short of overdose, there appears to be no permanent body damage that results from the opiates. On the other hand, the user may experience the reversible symptoms of loss of appetite, constipation, and impotence. Furthermore, skin abscesses, serum hepatitis, infections, and other medical disorders often result from using unsterilized needles.

Alcohol   Alcohol is similar to the narcotics in that it is a central nervous system depressant. The alcohol in beverages is *ethyl alcohol,* or ethanol. As discussed in Chapter 7, it is produced by the action of yeasts on sugars during anaerobic metabolism. Fruits and grains are commonly used as sources of sugar. The term "grain alcohol" stems from the use of a mash of ground corn, rye, barley, wheat, or other cereal grains.

Alcohol starts to enter the blood within two minutes after it is swallowed, and is rapidly delivered to the tissues, where it is absorbed by the cells. This absorption is most rapid when there is no food in the stomach. In the cells, oxidation begins immediately, and large amounts of heat are released. Within an average person, alcohol is oxidized at the rate of approximately one ounce in three hours.

Since an enzyme is responsible for breaking down alcohol in the body, the gene for this enzyme may be turned on to a different extent in different people. Thus there is a genetic component as to how much alcohol one can drink before being effected by it. The rate of oxidation does not depend on the energy needs of the body, and the excess heat produced raises the temperature of the blood. This, in turn, stimulates the heat-control center in the brain, which responds by causing increased circulation to the skin and a reddened face. There the heat radiates from the body. Since the receptors of heat are in the skin, the rush of blood there gives a false impression of warmth. Actually, the internal organs are being deprived of an adequate blood supply.

Not all the alcohol in the body is oxidized. Part is released into the lungs as vapor, causing an alcoholic breath odor. Some reaches the skin, and is added to perspiration in the sweat glands. Part passes into the kidneys and is carried out of the body in the urine.

In the brain, alcohol has its first effects on the cerebral cortex. Its anesthetic influence on centers of emotional control results in the release of inhibitions. Among the first symptoms of intoxication are loss of judgment, will power, and self-control. Extremes of emotion may be expressed as elation and/or depression. As the effects of alcohol progress, the vision and speech areas of the cerebrum become involved. There is blurred or double vision, inability to judge distance, and slurred speech.

Alcohol's effects on the cerebellum result in loss of muscular coordination. A highly intoxicated person becomes dizzy when standing, and if able to walk at all, does so with a clumsy, staggering gait. In the final stages of drunkenness, a person becomes completely helpless. The cerebral cortex becomes so impaired that the drunken person slips into unconsciousness. The skin becomes pale, cold, and clammy. Heart action, digestive action, and respiration slow down. Possibly the victim dies. It is not yet known precisely how alcohol acts on the brain cells. But its depressant activity suggests that it must interfere with synaptic transmission in some way by slowing the transmission of nerve impulses.

The disease called **alcoholism** is characterized by chronic dependence (not necessarily addiction) on alcohol. In about one case in ten, alcoholism progresses to the stage of **alcohol psychosis,** an acute form of mental illness. The cause of alcohol psychosis is not fully understood. Part of the condition may be caused by the toxic effect of alcohol on nerve tissue. This would account for general deterioration of parts of the brain.

Victims of alcohol psychosis become confused to the point that they may not be able to recognize members of their own family. This confusion is accompanied by terrifying hallucinations, mostly involving visual horror, and uncontrollable trembling. In some cases of alcohol psychosis, the victim suffers a loss of memory of recent events, and may turn to inventing stories to fill the gaps. The treatment of alcohol psychosis involves not only psychotherapy, but also rigid regulation of the diet and vitamin supplement (particularly vitamins of the B-complex, which are essential for normal nervous activity, and are often depleted in the alcoholic).

Barbiturates   Next to alcohol, the most widely used group of depressants are the **barbiturates.** They are synthetic drugs (including *Nembutal, Seconal,* and *phenobarbital*) used to induce sleep, to reduce emotional anxiety, and to relieve pain. They are commonly called sleeping pills. As with alcohol, barbiturates cause impairment of thinking and motor functioning. Low doses cause most people to become drowsy, but tolerance develops rapidly.

If the user keeps daily barbiturate consumption at or below 300 milligrams, he or she probably will not become physically addicted, but may suffer from a variety of side effects including nausea, dizziness, and headaches. A daily consumption over 500 milligrams often leads to physical addiction.

Barbiturate addicts who are deprived of their drug undergo withdrawal symptoms more severe than those associated with opiates. The symptoms begin with nervousness, trembling, and weakness. If untreated, they develop into generalized epileptic-type seizures. The most severe symptoms of untreated withdrawal last about four days, and can be fatal.

Like an alcoholic, a heavy barbiturate user usually cannot function adequately in daily activities. He or she often becomes confused, obstinate, and irritable. This behavior is in marked contrast to that of the narcotics addict, who usually remains passive. Many accidental deaths result from the combined use of barbiturates and alcohol. They have an additive effect more potent than either drug taken singly. Barbiturates are thought to bind to the postsynaptic membrane of neurons. In so doing they prevent the neurotransmitters from binding there and decrease membrane permeability for Na^+ ions. The result is a slowdown of nerve impulses.

Stimulants   Among the broad class of **stimulant drugs** are *nicotine, caffeine, amphetamines,* and *cocaine.* They all act to "speed" the central nervous system, and heavy users of them frequently exhibit hyperactivity. Unlike the depressants, caffeine and cocaine are not addictive (nicotine and amphetamines are mildly so), but psychological dependency and tolerance commonly result from using them. The amphetamines are thought to affect the presynaptic side of the neurons by increasing the membrane permeability for neurotransmitters. With more neurotransmitters available in the synapse, there is an increase in the speed of nerve impulses.

Tobacco—the Nation's Biggest Habit   More than fifty million people in the United States use tobacco in some form. The great majority of these are cigarette smokers.

In 1964, the Public Health Service published a report on the effects of smoking based on experiments with animals, clinical and autopsy studies in humans, and studies of the occurrence of disease in smokers as compared to nonsmokers. The conclusions were a devastating medical condemnation of smoking. In subsequent years, however, the daily per capita consumption of tobacco has remained much the same. Included in the report were these findings:

Tissue damage. Secretions of lung tissue from thousands of smokers have been examined after their deaths. Even in individuals who did not die of cancer, abnormal cells were found in the lungs. Enlarged and ruptured alveoli and thickened blood vessels were found. In the trachea and bronchi, cilia and the protective cells of the mucosa had been destroyed. These structures normally clean and lubricate the respiratory tract and help to prevent infection.

Elevated death rate. For the purpose of this study, the number of deaths among a large sample of nonsmokers was compared with the number of deaths among a similar sample of smokers. The greater number of deaths from certain diseases among smokers was particularly marked. For instance, there were 1000 percent more deaths from lung cancer and 500 percent more from chronic bronchitis and the degenerative lung disease, emphysema. The death rate was also considerably higher for cancer of the tongue, larynx, and esophagus, for peptic ulcer, and for circulatory diseases.

Elevation of the death rate with an increase in the amount smoked. In general, the greater the number of cigarettes smoked, the higher the death rate. For men who smoke less than ten cigarettes a day, the death rate is about 57 percent higher than it is for nonsmokers. For those who smoke forty or more, it is 218 percent higher. The same kind of relationship exists between the number of years of smoking and the death rate (Figure 15-31).

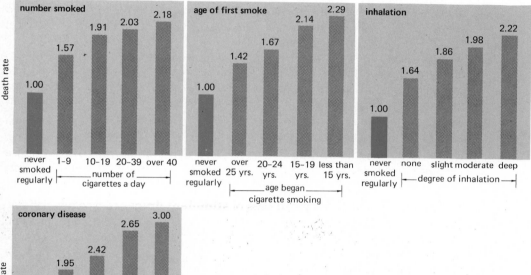

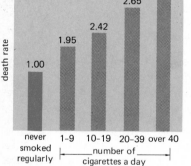

FIGURE 15-31

These graphs show ratios of death rates in the age group 40-49. The death rate for nonsmokers is given in each case as 1.00; the other numbers are ratios to this base number. (Adapted from a report by E.C. Hammond.)

To put the figures on lung cancer in another way, 95 percent of the victims of lung cancer are heavy smokers, and only one-half of 1 percent are nonsmokers. Lip cancer has been traced to irritation from pipes and cigars. Lung and lip cancer, however, do not account for a large percentage of deaths in this country. Heart and circulatory diseases are the primary causes of death in the United States, and the death rate for these diseases is at least 95 percent higher among smokers than among nonsmokers (Figure 15-31).

It is quite possible that heart disease and smoking are sometimes related only indirectly. It has been suggested, for example, that the same psychological or physiological characteristics that predispose someone to smoking may also cause heart disease. For example, a person may feel stressed and smoke to lessen this feeling. But at the same time, the stress causes heart disease. Nevertheless, smoking does place some direct strains on the circulatory system. Since functioning of the lungs is impaired by smoking, for example, the heart must work harder to utilize the remaining air sacs. The heart's job is further complicated because nicotine causes constriction of blood vessels in the extremities. A normal heart may be able to stand these strains, but they are certainly dangerous aggravations for someone with a weak heart.

Researchers have identified over 1,200 different chemicals in tobacco smoke. Among these, **nicotine,** is the primary consciousness-altering drug. It produces dependency and tolerance, and some physical addiction. Nicotine mimics some of the effects of the transmitter acetylcholine enhancing synaptic transmission in low doses and blocking it at higher concentrations. Its effects, therefore, can be both stimulatory and depressing. It is known to stimulate sensory receptors in the skin, depress *diuretic action* (discharge of urine), stimulate heart rate, and alternately stimulate and depress salivation.

A recent theory of how smoking causes cancer proposes that cigarette smoke is radioactive at a very low level. Many years of smoking would build up the radioactivity in the lungs and cancer would ultimately be produced. It has been found that the high phosphate fertilizers used on tobacco have abnormally high radium (a radioactive metal) levels. By radioactive decay, radium turns into a radioactive gas (radon) which is absorbed by the tobacco plants. The radon in turn decays into the element polonium (also radioactive). It is the polonium that ends up in tobacco smoke.

Caffeine   **Caffeine** is the main active ingredient in coffee, tea, cola drinks, and non-prescription preparations for overcoming fatigue, such as "No-Doz." It is a potent stimulant of the central nervous system, but the low doses normally produce only mild side effects. Of all the popularly used consciousness altering drugs, caffeine has the most wide social acceptance.

The caffeine contained in a single cup of coffee (or two cups of tea) has roughly the same stimulatory effect as the nicotine in one cigarette. Many users claim that coffee stimulates thinking, and promotes more sustained physical and intellectual activity. Others have found that the coffee habit leads to disjointed hyperactivity. Like all drugs which affect the nervous system, caffeine can easily be overdone. Physical effects of excessive caffeine include insomnia, irritability, aggravation of peptic ulcers, and high blood pressure. Since coffee is highly acidic, it disrupts the normal pH of the digestive system.

Amphetamines   Among the stimulants, the **amphetamines** have the greatest potential for psychological and physical damage. All amphetamines are synthetically produced. Among them are *methedrine, benzedrine,* and *dexedrine.* Collectively termed "speed," they chemically resemble the hormone adrenalin, which is released by the adrenal glands at times of fright. Like adrenalin, the amphetamines give a sudden burst of energy as would be needed for a "flight or fight" response.

In low doses, amphetamines produce a general increase in alertness, a sense of wellbeing, and decreased feelings of fatigue. They suppress appetite and increase heart rate and blood pressure. When taken for prolonged periods, increased tolerance develops, and larger doses are required to balance the effect. Habitual use of amphetamines can lead to highly paranoid behavior, manifested by unfounded suspiciousness, hostility, hallucinations, and delusions of persecution. Amphetamines are usually taken in pill form, but they are also injected.

Perhaps the worst effect of amphetamine overindulgence is that users frequently turn to barbiturates or heroin in order to come down from a "high." A hideous cycle can develop in which the victim oscillates between the paranoid delusions of the amphetamines, and the compelling physical addiction of depressants.

Cocaine   Legally, **cocaine** is classified with the opium derivatives as a narcotic. But it is produced from a different plant, and its effects are quite different from heroin and the other true narcotics. Although cocaine is similar to the opiates in that it relieves pain, it by no means induces sleep or stupor as do the opiates. When sniffed or injected, cocaine produces much the same effects as the amphetamines. Like the amphetamines, overdoses of injected cocaine can stimulate the central nervous system to the point that death results from heart or respiratory failure.

Cocaine is derived from the leaves of the coca shrub of South America, and was the first local anesthetic to be introduced into medical and dental practice. In the early days of "Coca Cola" production, this soft drink contained cocaine. But by 1906, the Pure Food and Drug Law forced the company to remove the cocaine. The cola part of the name was taken from the kola nut, which contains caffeine, an ingredient later added to the drink. When used in dentistry, cocaine is injected into the gums and has a numbing effect. When a finely crushed powder of cocaine is sniffed, it generally produces euphoria and a rush of energy. Among the negative side effects that are sometimes experienced are headache, anxiety, dizziness, and tissue damage. In those who sniff it regularly, the partition between the nostrils deteriorates. Little or no tolerance develops to cocaine, and withdrawal symptoms are minimal. But, as with amphetamines, excessive use can lead to paranoid reactions.

Marijuana   **Marijuana** is the dried leaves and flowering parts of the hemp plant (*Cannabis sativa*). Its resins contain a behavior-altering chemical called *tetrahydrocannabinol* (THC) (Figure 15-32). Marijuana cannot be classified as either a stimulant or depressant of the nervous system. Its effects vary widely depending on the concentration of THC, the psychological state of the user, and the social setting where it is used. Marijuana has been shown to increase levels of a neurotransmitter called *serotonin* but what this then does to the nervous system is not known. Hash, which is a concentrated form of pollen and resin, has the highest THC content, followed by the flowering parts and leaves.

Apparently THC has various effects on the brain, but it is not clear exactly what they are. It may be that THC itself is not as potent as one or more of the products of its metabolic decomposition.

Under certain circumstances, some people find that marijuana produces a heightened craving for sensory stimulation. It may enhance the pleasures of food, music, and erotic activity. Under the same circumstances other individuals might just become weary. Some behavioral researchers claim to have found that marijuana develops various negative personality traits, including laziness, excessive introspection, and paranoid fantasy. Their studies have been countered by other researchers who claim to have found no such personality changes. Since reactions to marijuana vary so widely from person to person, it is doubtful that generalizations about the effects of marijuana on personality are justifiable. Nevertheless, for most people, marijuana hampers short-term memory and disrupts the sense of balance somewhat. On the other hand, activities requiring quick reflexive responses are generally performed somewhat more precisely under the influence of small amounts of THC.

FIGURE 15-32
Tetrahydrocannabinol (THC), the active ingredient of marijuana.

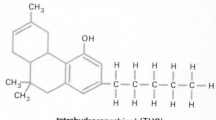

tetrahydrocannabinol (THC)

Recent research has indicated that high levels of marijuana inhibit the production of testosterone and sperm—effects which are ironic in light of the stimulating effect which the drug is purported to have on sex drive. The extent to which marijuana effects reproductive potency isn't known. But, once heavy use of it ceases, production of sperm and testosterone gradually return to normal. Marijuana isn't addictive, and habitual use of it may result in a puzzling reverse tolerance reaction. Progressively smaller amounts of it are required to produce the same effects.

Tranquilizers   **Tranquilizers** (including *Librium, Valium, Chlorpromazine,* and *Thorazine*) reduce hyperactivity and nervous excitability. In this sense, tranquilizers are depressants and physical dependence upon them occurs.

The biochemistry of the action of tranquilizers is unknown, but their function is similar to the depressants at the synapse. They are used extensively by doctors, particularly for calming violent psychotics. In severe cases of emotional disturbance, they act to reduce hallucinations and dampen over-responsiveness to external and internal stimuli.

Although tranquilizers suppress anxiety and irritability, they also cause sleepiness and mental confusion. Although, their side-effects aren't as disabling as those of the barbiturates, tranquilizers can have toxic side-effects, which include skin rashes, epileptic-like seizures, sensitivity to light, and general blunting of the emotions.

Psychedelics   Within this poorly defined group of drugs called **psychedelics** are substances produced by several plants, as well as a growing number of synthetic compounds. *Mescaline* (from the peyote cactus) and *psilocybin* (from the psilocybe mushroom) are the best known plant-derived psychedelics. They also can both be synthesized in the lab. Among the purely synthetic psychedelics are LSD (lysergic acid diethylamide), DMT (dimethyltryptamine), DET (diethyltryptamine) and DOM (dimethoxy-methylamphetamine), which was brought on the drug scene as STP. Even the nutmeg contains psychedelics.

The variable effects of the psychedelics are reflected in the proliferate names for them, including *psychotomimetics* (psychosis-mimickers) and *hallucinogens* (hallucination-producers). The term used here, *psychedelic,* can be roughly defined as mind-manifesting or mind-expanding, but the effects of these drugs defy verbal descriptions. They can transform mental activity to aberrations rarely, if ever, experienced without their use. Some psychologists have suggested a close resemblance in their effects to the psychosis *schizophrenia.*

A minute dose of around 200 micrograms of LSD first causes dizziness, weakness, and nausea. Dilation of the pupils reveals its effect on the autonomic nervous system. Feelings progress through a bizarre flow of mental experiences and emotional states. Sensory perceptions can be transformed to the point that sight and/or hallucination pulsates and convolutes with sounds from outside as well as inside the body. Impressions that seem beautifully harmonious at one point can be shattered by an instantaneous inward revelation or the ringing of the telephone. A psychedelic experience can be deeply pleasurable and/or profoundly frightening.

At the biochemical level, some evidence suggests that LSD blocks the activity of certain nerve cells which form serotonin, a transmitter substance in the brain. LSD is thought to bind to the postsynaptic membrane and prevent serotonin from binding there. Serotonin can act as an inhibitory neurotransmitter, producing nerve impulses that inhibit certain actions. If this inhibitor is in turn inhibited by LSD, the net effect is for LSD to stimulate the nervous system. Other evidence indicates that LSD "fools" nerve cells into reacting to it rather than serotonin. In chemical structure LSD and most of the other psychedelics resemble serotonin (Figure 15-33).

Psychedelics have been reputed to have a variety of long-lasting effects. Many users of LSD have experienced "flashback," during which the sensations of the drug experience were mimicked, months after the last time they took it. Several cases of long-term psychosis have also been reported as caused or aggravated by the psychedelics. They, as well as the other popular illegal drugs, have recently been investigated to determine if they cause birth defects or chromosome damage in users. Experiments with animals have produced conflicting results and observations of people are inevitably complicated by the possibility that other factors in life style and drug are actually to blame. At present, therefore, there is no conclusive evidence linking psychedelics to birth defects or chromosome damage when used irregularly and not in combination with other drugs.

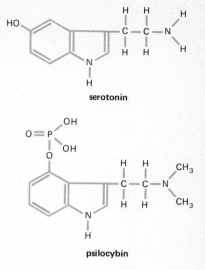

serotonin

psilocybin

FIGURE 15-33
The similarity in chemical structure of serotonin (a neurotransmitter) and psilocybin (a psychedelic) is indicated.

CHAPTER 15: SUMMARY

1 Neurons are the basic units of all nervous systems.

2 Neural impulses transmit information from one place to another.

3 The neural impulse is caused by movements of sodium and potassium across neuronal membranes.

4 A neural impulse travels across a synapse because of the release of a neurotransmitter substance like acetylocholine or adrenalin.

5 Muscle contraction is caused by the action of calcium on specific muscle proteins actin, myosin and troponin.

6 All our senses receive physical stimuli from the environment and translate them into neural impulses for interpretation by the brain.

7 Our brain is divided anatomically into several regions each having its own functions, yet communicating with all the other regions.

8 Most drugs that affect the nervous system do so at the synaptic region.

Suggested Readings

Begbie, G. Hugh   *Seeing and the Eye*, Natural History Press, Garden City, New York, 1969—The physics of light, anatomy of the eye, and the function of the brain in vision.

Bullock, Theodore H.   *Introduction to Nervous Systems*, W. H. Freeman and Co., San Francisco, California, 1977—A beautifully done book on how nervous systems work.

Cohen, Carolyn   "The protein switch of muscle contraction," *Scientific American*, November, 1975—The latest theory of how muscles work.

Girdano, Dorothy Dusek and Daniel A. Girdano   *Drugs—A Factual Account*, Addison-Wesley, Reading, Maine, 1973—Presents a wide variety of information concerning such drugs as alcohol, marijuana, LSD, amphetamines, barbiturates, sedatives, opiates, and nonprescription drugs.

Heimer, Lennart   "Pathways in the Brain," *Scientific American*, July, 1971—Considers the intricate pathways of neural impulses in the brain.

An Excerpt from THE DRAGONS OF EDEN
Carl Sagan

The human brain is the most sophisticated known storehouse of information. Yet we are almost totally in the dark in our understanding of memory. The following excerpt from Carl Sagan's book on the evolution of human intelligence considers some of the experiments that have been directed toward discovering the nature of memory.

What is the information content of the brain? Let us consider two opposite and extreme poles of opinion on brain function. In one view, the brain, or at least its outer layers, the cerebral cortex, is *equipotent:* any part of it may substitute for any other part, and there is no localization of function. In the other view, the brain is completely hard-wired: specific cognitive functions are localized in particular places in the brain. Computer design suggests that the truth lies somewhere between these two extremes. On the one hand, any nonmystical view of brain function must connect physiology with anatomy; particular brain functions must be tied to particular neural patterns or other brain architecture. On the other hand, to assure accuracy and protect against accident we would expect natural selection to have evolved substantial redundancy in brain function. This is also to be expected from the evolutionary path that it is most likely the brain followed.

The redundancy of memory storage was clearly demonstrated by Karl Lashley, a Harvard psychoneurologist, who surgically removed (extirpated) significant fractions of the cerebral cortex of rats without noticeably affecting their recollection of previously learned behavior on how to run mazes. From such experiments it is clear that the same memory must be localized in many different places in the brain, and we now know that some memories are funneled between the left and right cerebral hemispheres by a conduit called the corpus callosum.

Lashley also reported no apparent change in the general behavior of a rat when significant fractions — say, 10 percent — of its brain were removed. But no one asked the rat its opinion. To investigate this question properly would require a detailed study of rat social, foraging, and predator-evasion behavior. There are many conceivable behavioral changes resulting from such extirpations that might not be immediately obvious to the casual scientist but that might be of considerable significance to the rat — such as the amount of post-extirpation interest an attractive rat of the opposite sex now elicits, or the degree of disinterest now evinced by the presence of a stalking cat.*

It is sometimes argued that cuts or lesions in significant parts of the cerebral cortex in humans — as by bilateral prefrontal lobotomy or by an accident — have little effect on behavior. But some sorts of human behavior are not very apparent from the outside, or even from the inside. There are human perceptions and activities that may occur only rarely, such as creativity. The association of ideas involved in acts — even small ones — of creative genius seems to imply substantial investments of brain resources. These creative acts indeed characterize our entire civilization and mankind as a species. Yet in many people they occur only rarely, and their absence may be missed by neither the brain-damaged subject nor the inquiring physician.

While substantial redundancy in brain function is inevitable, the strong equipotent hypothesis is almost certainly wrong, and most contemporary neurophysiologists have rejected it. On the other hand, a weaker equipotent hypothesis — holding, for example, that memory is a function of the cerebral cortex as a whole — is not so readily dismissable, although it is testable. . . .

There is a popular contention that half or

---

*Incidentally, as a test of the influence of animated cartoons on American life, try rereading this paragraph with the word "rat" replaced everywhere by "mouse," and see if your sympathy for the surgically invaded and misunderstood beast suddenly increases.

more of the brain is unused. From an evolutionary point of view this would be quite extraordinary: why should it have evolved if it had no function? But actually the statement is made on very little evidence. Again, it is deduced from the finding that many lesions of the brain, generally of the cerebral cortex, have no apparent effect on behavior. This view does not take into account (1) the possibility of redundant function; and (2) the fact that some human behavior is subtle. For example, lesions in the right hemisphere of the cerebral cortex may lead to impairments in thought and action, but in the nonverbal realm, which is, by definition, difficult for the patient or the physician to describe.

There is also considerable evidence for localization of brain function. Specific brain sites below the cerebral cortex have been found to be concerned with appetite, balance, thermal regulation, the circulation of the blood, precision movements and breathing. A classic study on higher brain function is the work of the Canadian neurosurgeon, Wilder Penfield, on the electrical stimulation of various parts of the cerebral cortex, generally in attempts to relieve symptoms of a disease such as psychomotor epilepsy. Patients reported a snatch of memory, a smell from the past, a sound or color trace—all elicited by a small electrical current at a particular site in the brain.

In a typical case, a patient might hear an orchestral composition in full detail when current flowed through Penfield's electrode to the patient's cortex, exposed after a craniotomy. If Penfield indicated to the patient—who typically is fully conscious during such procedures—that he was stimulating the cortex when he was not, invariably the patient would report no memory trace at that moment. But when, without notice, a current would flow through the electrode into the cortex, a memory trace would begin or continue. A patient might report a feeling tone, or a sense of familiarity, or a full retrieval of an experience of many years previous playing back in his mind, simultaneously but in no conflict with his awareness of being in an operating room conversing with a physician. While some patients described these flashbacks as "little dreams," they contained none of the characteristic symbolism of dream material. These experiences have been reported almost exclusively by epileptics, and it is possible, although it has by no means been demonstrated, that non-epileptics are, under similar circumstances, subject to comparable perceptual reminiscences.

In one case of electrical stimulation of the occipital lobe, which is concerned with vision, the patient reported seeing a fluttering butterfly of such compelling reality that he stretched out his hand from the operating table to catch it. In an identical experiment performed on an ape, the animal peered intently, as if at an object before him, made a swift catching motion with his right hand, and then examined, in apparent bewilderment, his empty fist.

Painless electrical stimulation of at least some human cerebral cortices elicits cascades of memories of particular events. But removal of the brain tissue in contact with the electrode does not erase the memory. It is difficult to resist the conclusion that at least in humans memories are stored somewhere in the cerebral cortex, waiting for the brain to retrieve them by electrical impulses—which, of course, are ordinarily generated within the brain itself.

If memory is a function of the cerebral cortex as a whole—a kind of dynamic reverberation or electrical standing wave pattern of the constituent parts, rather than stored statically in separate brain components—this would explain the survival of memory after significant brain damage. The evidence, however, points in the other direction: In experiments performed by the American neurophysiologist Ralph Gerard at the University of Michigan, hamsters were taught to run a simple maze and then chilled almost to the freezing point in a refrigerator, a kind of induced hibernation. The temperatures were so low that all detectable electrical activity in the animals' brains ceased. If the dynamic view of memory were true, the experiment should have wiped out all memory of successful maze-running. Instead, after thawing, the hamsters remembered. Memory seems to be localized in specific sites in the brain, and the survival of memories after massive brain lesions must be the result of redundant storage of static memory traces in various locales. . . .

But the evidence for localization of function is now much stronger even than this. In an elegant set of experiments, David Hubel of Harvard Medical School discovered the existence of networks of particular brain cells that respond selectively to lines perceived by the eye in different orientations. There are cells for horizontal, and cells for vertical, and cells for diagonal, each of which is stimulated only if lines of the appropriate orientation are perceived. At least some beginnings of abstract thought have thereby been traced to the cells of the brain.

CHAPTER 16

Behavior

Objectives

1 Learn why animals and plants have cycles of behavior.

2 Learn what cycles in the environment influence behavior.

3 See what behaviors are cyclic.

4 Describe what influences migrations.

5 Understand tropisms.

6 Describe the basis of plant movements.

7 Relate theology to evolution and show how social behavior evolved.

8 Contrast and compare instinct and learning

9 Describe imprinting.

10 Learn about forms of animal communication.

FIGURE 16-1
The male stickleback builds the nest (left), "sews" it together by secreting a binding thread (center), and then drives the pregnant female into the nest by attacking her and biting her tail (right).

BEHAVIOR—A PUZZLING PART OF LIFE

Behavior is the most readily observable, yet most puzzling of biological phenomena. Consider, for example, the mating behavior of the stickleback fish (Figure 16-1). At the onset of the sequence of events of the stickleback's mating ritual, a physiological change in the male causes his color to change. It is as if nature had given him a bright flag to wave to indicate his reproductive preparedness. Following the color change he establishes a territory, builds a nest, and waits.

The appearance of a female whose abdomen is distended with eggs stimulates the male to swim in a zigzag pattern toward the nest. By performing this courtship dance and sometimes applying coercion (such as nipping his mate's tail) the male entices the female to enter the nest. Once she has entered, he prods her with his snout, and she releases her eggs. Once the eggs have been deposited, the male suddenly becomes antagonistic toward his mate; he chases her from the nest. Alone with his hoard of eggs, the male releases sperm through a pore in his abdomen to fertilize them.

The chain of events of the reproductive ritual of the stickleback is the same for all members of the species. It is an instinctive behavioral pattern, apparently coded within every individual's genes. Researchers of animal behavior have concluded that much of what animals do is genetically rooted. Genes are thought to somehow stimulate the neuromuscular system (perhaps via hormones) to perform specific functions. However, an organism's behavior must make sense in the context of its surroundings. Thus, environmental influences also play a part in eliciting behavioral responses. Plants do not have neuromuscular systems and their responses to environmental stimuli are not as varied as animal responses. Nevertheless, plants do exhibit certain types of behavior.

BEHAVIOR AND THE CYCLE OF DAY AND NIGHT

In a sense it is impossible to separate the effects of genes from the effects of environments on animal behavior. Over the course of evolutionary history, environments have been sources of pressures of natural selection which have directed evolution. Hence, environments have been forces in determining which genes affecting behavior remain in the gene pool of a species. Apart from the evolutionary influences of environments, however, are short term environmental influences which affect behavior on a day to day basis. Such influences are the subject of the following sections.

412

The cycle of night and day

Perhaps the most significant short term environmental influence on behavior is the cycle of night and day. Within the 24-hour cycle, most animals have periods of activity and periods of rest. Species that are most active during the day are said to be **diurnal,** while those that are active at night are termed **nocturnal.**

In a wooded area, a summer morning may find the birds, chipmunks, and ground squirrels searching for food among the trees as deer browse in the meadows. Before noon the activity of the forest diminishes, but as evening approaches, many of the diurnal animals renew their activities for a short time. Dragonflies and bats dart over streams and ponds searching for the diurnal and nocturnal insects that are active in this transition period. The sounds of the cicadas and birds gradually fade as the crickets begin their activity. Later foxes, raccoons, skunks, and owls begin their nocturnal search for food.

The desert environment, with its extreme temperature fluctuations between the heat of the day and the cool of the night, has a sharp division between diurnal and nocturnal activities. In the early morning hours, birds feed on insects and seeds, and jack rabbits come out of their burrows to look for food. Snakes, hawks, vultures, and ground squirrels are awake and active. Since the temperature may exceed 120°F in areas directly exposed to sun, noontime finds many desert creatures huddled in the shade at the bases of cacti, sagebrush, and creosote bushes. Rodents take to their relatively cool burrows, where the humidity may be more than twice that of the above ground atmosphere. During the night, the Gila monster and the rattlesnake prey upon the many small nocturanl desert mammals. The bobcat, coyote, fox, and owl are some of the larger nocturnal desert carnivores.

In contrast to the desert are the polar regions, where very little activity occurs on frigid nights, especially during winter, when darkness lasts for months at a time. Penguins and polar bears become active during the prolonged days of summer.

In the vast regions of treeless plains of the tundra, where the ground remains frozen most of the year, a shallow layer of earth at the surface thaws briefly enough to allow lichens and some grasses to grow. A few warm-blooded animals such as the caribou, musk ox, Arctic hare, fox, and lemming spend the daylight hours seeking food.

In equatorial regions, the lengths of the day and night are nearly equal and the temperature is less variable. Light, therefore, is a more important factor in determining the periods of activity of equatorial organisms than temperature. In the equatorial forests, both nocturnal and diurnal animals are very numerous.

Day-night rhythms are also found in the oceans, where periodic vertical migrations have been discovered. Vast numbers of shrimp are found near the surface at night, where they feed on smaller organisms of the plankton. During the day, the shrimp swim to a lower level; they may be found three hundred feet below the surface. Small carnivorous fishes follow the daily movements of the shrimp.

Diurnal and nocturnal behavior of plants

Plants lack neuromuscular systems and are, therefore, incapable of complex movements comparable to those involved in animal behavior. Nevertheless, the relatively nonmobile lives of plants are punctuated by certain day-night rhythms. Photosynthetic manufacture of sugars, for instance, proceeds rapidly during daylight hours, but slows as night falls. When plants bloom, the flowers may open and close at regular times of the day. The petals of the poppy, for example, open in the morning and close at night. Certain cacti, on the other hand, open their blossoms only at night; they depend on nocturnal insects for pollination.

Circadian rhythms

In the cycles of the activities of many animals, the alternation of night and day is an important influence. But, behaviorists have found that alternating periods of light and darkness are not always necessary for some daily rhythms. In the nocturnal white-footed mouse, for example, daily behavior is not a simple matter of activity in the dark and rest in the light. When a mouse is kept in constant darkness, its waking and sleeping periods continue as they would as if it were exposed to normal periods of daylight and darkness. Thus, the animal has the capacity to "remember" its rhythm and is not entirely dependent on the external stimulus of light fluctuations. **Endogenous rhythm** is the name applied to those rhythms that persist in the absence of external cues from the environment, that is, in those cases where there are constant environmental conditions (Figure 16-2).

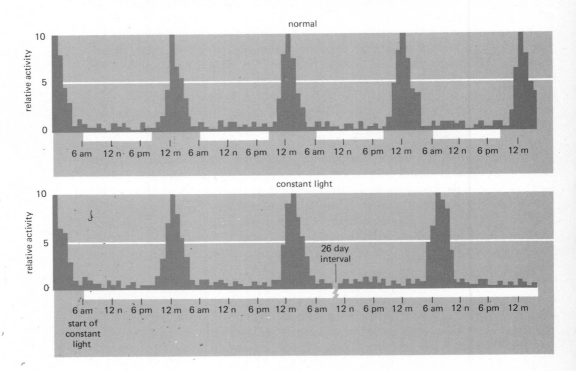

FIGURE 16-2

Rhythmic animal activity. At the top, the activity rhythm under alternate light and dark cycles. At the bottom, the same activity rhythm but in constant light. Note that the animals' activity rhythm continues in constant light without any external light-dark cues.

Even plants in a darkened room have 24-hour cycles of activity. Their leaves and flowers rhythmically open and close. Although different species of plants and animals may show slight variations in the durations of their cycles, most seem to run their courses in from twenty-two to twenty-six hours. These cycles are called **circadian rhythms** and probably exist in all kingdoms except the Monera.

Circadian rhythms have been found in all mammals that have been studied, including humans. People who have been isolated from light-dark cues show 24-hour cycles, not only in sleeping and waking, but also in concentrations of hormones in the blood, the rate of cell division, heartbeat, and excretion. Abilities to perform intellectual tasks also vary with a 24-hour periodicity. Human subjects kept in darkness for long periods settle into a circadian rhythm that differs from the length of a day by only an hour or two. For most people, the period is greater than

24 hours, and after several days they are out of phase with the external world. Interestingly, for a small fraction of test subjects, a 48-hour cycle establishes itself. There is no obvious reason for this doubled periodicity.

Biologists have searched throughout the body for a mechanism that regulates periodicity (a "biological clock"), but have little to show for their efforts. Particular attention in mammals has been focused on the adrenal, hypothalamus, pineal, and pituitary glands. In hamsters, a hormone secreted with a circadian rhythm from the adrenal gland controls certain cyclical body functions (including the rate of storage of glycogen in the liver). These cycles stop if the adrenal glands are removed, but others continue. Clearly, the adrenal glands are not the "master" biological clock.

It has been theorized that rather than a specific organ controlling rhythmic body processes, timing occurs at the level of biochemical systems. Rhythmic fluctuations in the production of certain proteins have been found in many organisms (Figure 16-3). Since proteins are coded for by DNA, there may be some mechanism by which DNA is rhythmically transcribed. The obvious question is: What generates the rhythms?, but there is no clear answer. Perhaps it lies in the maze of feedback systems of hormonal secretions, in which case the biological clock would be everywhere at once but nowhere in particular. One cycle involving

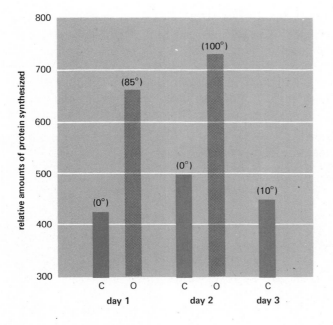

FIGURE 16-3
Rhythm in plant leaflet movement and protein synthesis. Numbers in parentheses are degrees of leaflet opening (0° is complete leaflet closure, 180° is complete leaflet opening). The heights of the vertical bars indicate the amounts of protein synthesized by the leaflets during their closing cycle (C) and during their opening cycle (O). The graph indicates that more protein is synthesized during opening than closing, and that there is a rhythm in the synthesis patterns.

a feedback system of hormones, the month-long female menstrual cycle, was described in Chapter 12.

Sleep

One of the most obvious (but least understood) rhythms of life is the cycle of sleep and awakeness. In humans sleep has somewhat of a circadian rhythm, but after prolonged periods without an awareness of day and night, sleep patterns become sporadic. Some biologists suggest either that sleep is needed so that toxic substances produced while awake can be destroyed or that during sleep, compounds needed for the waking state may be synthesized. It is known that while sleeping more protein, ATP, and neurotransmitters (acetylcholine and serotonin) are synthesized in the brain than during the waking hours. Nevertheless, chemical phenomena are not the only reasons for the need for sleep. Many people, for example, use sleep as an escape from their waking lives. Dreams may be necessary also.

Not all animals sleep equally soundly. In general, the best sleepers are either predators (such as cats) or animals that have safe hiding places (such as moles). Animals that are particularly susceptible to predation (such as rabbits and sheep) are generally poor sleepers.

SEASONAL COMMUNITY CHANGES

The cycle of day and night is not the only environmental rhythm. The yearly fluctuations of the seasons bring several changes, and temperature is one of the most significant (Figure 16-4). Animals meet the problems of seasonal temperature changes in various ways. When winter brings snow and freezing temperatures, an organism must be able to adjust or move away.

The first freezing nights of autumn take a heavy toll of insect life. Most adult insects have completed their life cycles before the beginning of autumn and are killed by the first frosts. There are, however, a few notable exceptions, including the morning cloak butterfly, which finds protection from the winter in a hollow tree or crevice.

The survival of most insect species is guaranteed by a yearly cycle of reproduction. Many moths spend winter as pupae within silk-insulated cocoons (Figure 16-5). Grasshoppers, crickets, and cicadas lay eggs in the ground or in the bark of trees. Stoneflies, mayflies, and dragonflies spend the winter as immature forms (nymphs), sheltered beneath the frozen surface of a pond or stream.

The honeybee finds protection in numbers. During the winter months bees feed on the honey that has been stored in the spring and summer, and they remain

FIGURE 16-4
Plants living here are adapted to the seasonal changes. The animals have adaptations for survival in the various conditions or they move away for the harsh winter season. (Walter Dawn.)

FIGURE 16-5
The rhythm of development. From left to right: Caterpillar of the Cecropia moth feeding on the leaves of the Elderberry bush; Cecropia caterpillar spinning cocoon; The Cecropia larva, after having encased itself in a silken shroud, in which it passes through the pupa stage of life; Cecropia moth shortly after emerging from the cocoon.

416

active in their hives. On a winter day the temperature generated by the active colony within the hive may be 50°F higher than the outside temperature.

Animals like the eastern cottontail rabbit, the whitetailed deer, the cardinal, and the bluejay are permanent residents of their regions. During extremely cold weather, they find protection in woods and thickets, but when snow is on the ground, foodgetting becomes a serious problem. Home animal feeding stations can be a tremendous help to these species (Figure 16-6).

Ground squirrels, chipmunks, woodchucks, and many reptiles and amphibians undergo true **hibernation** during cold weather. For these animals, the rate of body metabolism drops greatly. Heart action and respiration decrease, and they lose consciousness. Greatly reduced activity lowers energy requirements to a minimum. An animal undergoing true hibernation cannot resume activity until the temperature of the environment increases and the body processes speed up.

Some animals periodically enter a state of **dormancy.** This is a less drastic curtailment of body activities than hibernation. The bear, for example finds a hollow log, a cave, or some other protected location and lives on stored fat during its winter sleep. Although metabolism slows down, a normal body temperature is maintained, and the bear can be awakened. It may even leave its shelter on a mild winter day. The skunk, raccoon, and opossum have similar dormant periods.

The ill effects of excessively hot weather can also be avoided by dormancy. A frog, for example, may escape the high temperatures of midsummer by burrowing into the cool mud at the bottom of a pond to lower its rate of metabolism. The box turtle escapes the heat by burying itself in a pile of leaves. The period of withdrawal may be several days or several weeks.

FIGURE 16-6
Tree sparrow at a feeder.

Migration

Many animals move with seasonal fluctuations in climatic conditions. These migratory journeys often cover thousands of miles. Some animals make migratory trips because of seasonal changes in availability of food. Their motivation may also be to relocate in a place where they can produce their young under the most favorable conditions.

The bighorn sheep spends it summers in the high meadows near the summits of the Rocky Mountains. As winter approaches, it moves down into the protection of the forests on the mountain slopes. The Olympic elk has a similar migratory pattern. Through the summer months, herds of them browse in the high altitude of the mountains on the Olympic Peninsula in Washington. During the winter these herds move to the more protected mountain valleys and nearby plains. With the coming of spring, the herds move back up the slopes in long, singlefile processions.

Among the most remarkable migrations is that of the fur seal (Figure 16-7). During the winter, females, young males, and pups roam the waters of the Pacific Ocean to as far south as California. The older males winter in the cold waters near Alaska and the Aleutian Islands. With the approach of the breeding season in the spring, a migration begins to the Pribilof Islands, north of the Aleutians. The males arrive several weeks before the females and battle for territory. The females and young seals start their journey of three thousand miles in the spring and arrive at the Pribilofs in June. A herd of fifty or more females gathers around each male. Pups from the past year's breeding are born almost immediately, and within a week, breeding occurs again. During the mating season the Pribilofs resound with the boisterous barking of the seals. But, soon thereafter, the rugged, rocky islands fall silent when the females and their young leave for southern waters, and the males return to the Aleutians.

FIGURE 16-7
Northern fur seals during summer on St. Paul Island, Alaska.

FIGURE 16-8
Monarch butterflies resting during migration.

A Butterfly Migration   Among the insects, the monarch butterfly is one of the few which migrates. Its seasonal journey is nearly as long as that of the fur seal. In the latter part of the summer, the monarchs gather by the thousands in northern Canada and begin their long flight southward. Some of them travel to the Gulf states to spend the winter, while others travel along the Pacific Coast. At some time between the middle of October and the first of November, tens of thousands of them arrive on the Monterey Peninsula in the small town of Pacific Grove, California. Here they seek shelter in a specific grove of pines where they hang from the branches in such large numbers that the trees appear to be bright orange (Figure 16-8). They stay there in a state of semihibernation until the winter is over. On warm sunny days throughout the fall and winter, they become active and fly about local gardens gathering nectar.

In March the return flight northward begins. Along the way, the monarchs lay eggs on milkweed plants, after which some of the butterflies die. However, after the eggs hatch, and the larvae pupate, the young butterflies continue the migration, laying eggs along the way. By early summer the northern-most point of migration has been reached.

In late summer the monarchs begin gathering for their southward journey to the same locality and the same trees in which their ancestors have wintered. Although their migration is attributed to instinct (genetic programming) the factors causing it aren't known, nor is it known how they find their way.

Bird Migration   Many species of birds fly northward in the spring, to nest and raise their young. It isn't known why some species leave abundant food and warmth in the tropics to migrate to breeding grounds in the north. More logical is the reverse migration which begins before lakes and ponds freeze over and food is blanketed with snow.

Some species make their migratory flights at night; others fly during the day. The V-shaped patterns of geese in the twilight hours, and the daylight flights of

The reproductive cycle of the common garden snail *(Helix pamatia)*. Snails mating. (Hans Pfletschinger from Peter Arnold.)

Snails laying eggs. (Hans Pfletschinger from Peter Arnold.)

Snail young in cave. (Hans Pfletschinger from Peter Arnold.)

Snail young leaving cave. (Hans Pfletschinger from Peter Arnold.)

The temperate grassland biome. Reconstituted prairie at the University of Wisconsin Arboretum. Few if any stands of natural prairie have been left untouched by man. However, this stand, developed in Madison, Wisconsin, is probably quite typical of the natural prairie. (J. C. Cavender.)

The northern coniferous forest (taiga) biome. Winter in the taiga of interior Alaska. (Charlie Ott/Photo Researchers.)

The savannah biome, Serengeti plains, Tanzania. The flat-topped acacia trees and the broad expanses of grassland are typical of the East African savannah. In total numbers and in variety of species, the ungulate population of the African savannas is unequalled anywhere in the world. (E. C. Williams.)

The chaparral biome, a typical example from Southern California, is characterized by abundant winter rainfall and dry summers. The climax vegetation consists of shrubs and small trees with hard, thick, evergreen leaves. Similar communities are also found on the shores of the Mediterranean and this biome type is often called the Mediterranean biome. (Dennis Brokaw.)

The desert biome, the Sonora Desert of northwest Mexico. Among the plants seen here are the organ pipe cactus, mesquite, and brittle bush. (J. C. Cavender.)

The tropical rain forest biome. A giant mora tree with its buttressed trunk in the rain forest on the island of Trinidad. (J. B. Thurston.)

The temperate deciduous forest biome. Midsummer in a typical deciduous forest of eastern United States. (E. R. Degginger.)

The tundra biome, Mt. McKinley Park, Alaska. (John Lewis Stage/Photo Researchers.)·

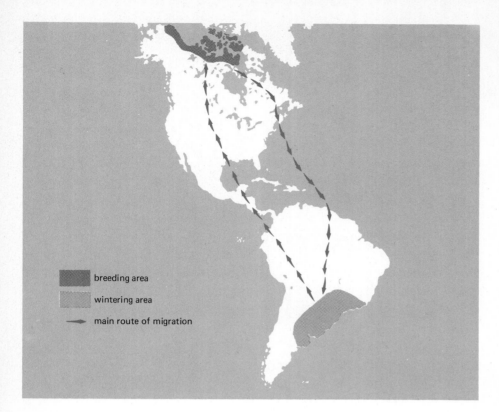

FIGURE 16-9
The migratory route of the golden plover. Flying more than 8,000 miles in a single migratory flight, this bird breeds in northern Canada during the summer, and in the fall flies to South America, where it spends the winter.

breeding area

wintering area

main route of migration

thousands of red-winged blackbirds and grackles are familiar sights during spring and fall. Some species migrate slowly. Feeding along the way, they may average only twenty to thirty miles a day. Others are marvels of speed and endurance. The ruddy turnstone (a shore bird) travels each autumn from Alaska to Hawaii in a single flight. The golden plover makes a biannual journey between Canada and South America, a distance of more than eight thousand miles (Figure 16-9).

Not much is known about the genes that interact with the environment to govern the time and route of migration, but any given species follows the same route year after year and can be expected to arrive at a certain point within a few weeks of the same time each season. Some species travel northward along one route and southward by an entirely different one.

How do migratory birds know where they are going? Perhaps older birds remember, but this does not account for unescorted flights of one-year olds. How migratory birds recognize their destination is not known. But, various techniques of navigation have been discovered. Certain species which fly during the day sense their position relative to the sun, and for night flights some birds are guided by the stars. Investigators in Europe have studied caged birds in a planetarium (in which star patterns are projected upon the inside of a dome). Under a fall star pattern, warblers were found to orient themselves toward their winter home, and under a spring pattern they faced in the direction of their summer home.

For at least some birds, there is a sense of direction relative to the earth's magnetic field. If small magnets are attached to certain species (thus disrupting their impression of the earth's magnetic field), they lose their sense of direction for a while but then regain it when they switch to another navigational aid.

Most migratory birds have more than one method for determining their courses. Prominent landmarks and the direction of prevailing winds are sometimes used in combination with the sun, the stars, or the earth's magnetic field in finding the way between homelands.

FIGURE   16-10
High tide (left), and low tide (right).

Lunar and tidal rhythms

Lunar rhythms affect organisms at several levels. The most obvious is seen along the coasts of continents, where the gravitational attraction of the moon causes the rise and fall of tides (Figure 16-10). Tidal ranges vary from place to place on earth's surface, depending on latitude and the shape of the coastline. In the Bay of Fundy, between Maine and Nova Scotia, fifty-foot tides have been measured. In most places, however, the water level fluctuates only a few feet. Alternately submerged and exposed, the intertidal zone is an area of constant, rhythmic change. But, although the changes at the interface between land and sea are extreme, many organisms thrive there. Mussels and goose-neck barnacles cling in tight clusters, and seaweed forms dense mats of green, red, and brown.

A species of small, prolific fish, the grunion, has geared its reproductive cycle to the flux of the tides. In the spring and summer they spawn on the sandy beaches of California. Somehow they sense the days of the highest tides of the month and ride the waves onto the shore. The females work their way into the damp sand and release their eggs, while the males swarm about them, depositing sperm (Figure 16-11). The prospective parents return to the ocean on the next set of waves (providing they are not snatched up by grunion fishermen), and the eggs are left to incubate.

FIGURE   16-11
The spawning of grunion.

Periodicity

Rhythmic changes pervade life. Some of life's cycles oscillate in a fraction of a second. The hair-like locomotor cilia of certain one-celled organisms, for instance, beat at the rate of twenty times per second. Other cycles, including reproductive habits of birds, butterflies, and bears run their courses in a single year. The periodicity in the blooming of certain tropical bamboo plants can be every 100 years or so.

The reasons for some of life's cycles seem obvious, while others make no obvious sense. It is quite possbile that life responds to cosmic rhythms of which we are yet unaware.

BACTERIAL BEHAVIOR

Even organisms at the lowest level of the evolutionary scheme of life exhibit behavioral responses. Bacteria have been found to be attracted by certain chemicals. If, for instance, the amino acid aspartate is introduced through a capillary tube into a culture of *Escherichia coli*, the bacteria will swim toward the zone of highest concentration with lashing movements of their flagella (Figure 16-12).

It is hypothesized that chemical-sensitive portions of the bacterial membrane, **chemoreceptors,** detect aspartate and initiate directed movements of its flagella. As yet the exact mechanism of chemoreception in bacteria is essentially unknown. Perhaps there are changes in molecular arrangements within the membrane that reach all flagella (the base of a flagellum is in close association with the membrane). Tests with various compounds have indicated that there are at least five different chemoreceptors in *Escherichia coli*—an amazing number considering this bacterium is only a few micrometers long.

PLANT TROPISMS—RESPONSES TO ENVIRONMENTAL STIMULI

A growth response that is produced by an environmental stimulus is called a **tropism.** Such external stimuli include light, gravity, and certain chemicals. A tropic response may be *positive*, toward the stimulus, or *negative*, away from the stimulus. While tropic responses are purely automatic growth reactions, they usually adapt the plant to an environmental condition and are therefore an advantage over the short term.

A tropism familiar to anyone who raises house plants is the growth response to light known as **phototropism.** You have probably noticed a plant growing on a window sill bending toward the light (Figure 16-13). This is a positive tropism, since the plant seeks the light. The bending of the stem is a result of uneven elongation of cells. Those on the shaded side, away from the light, elongate more rapidly than those on the bright side. Light somehow reduces the rate at which the plant growth-stimulating hormones, called auxins, are formed; consequently, growth occurs more rapidly on the shaded side, and the plant bends toward the

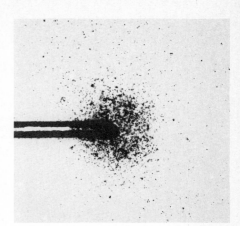

FIGURE 16-12
Photograph through a microscope showing the attraction of *E. coli* bacteria to aspartic acid, an amino acid. The glass tube on the left (diameter 25 micrometers) contains a high concentration of aspartic acid.

FIGURE 16-13
Phototropism. The plant's stem is directed toward the light source and the leaves are at nearly right angles to the direction of the light to maximize the leaf area exposed to the light for photosynthetic purposes.

light. Roots, on the other hand, are negative in their phototropic response (Figure 16-14). Root cells react oppositely from stem cells to changes in auxin concentrations brought on by uneven lighting; hence they grow away from light.

The stems and roots of plants also respond in opposite ways to gravity; that is, they have opposite **geotropic responses.** The root response is positive, or toward gravity, while the shoot response is negatively geotropic, or away from gravity. If a seed is planted upside down the root will grow downward and the shoot will grow upward, even in the absence of light.

The mechanism of geotropic response isn't precisely known, but one line of research indicates that special starch granules in plant cells play a role in the detection of gravity. These **statoliths** (Figure 16-15) move in the cells relative to the position of the plant, but always fall downward.

FIGURE 16-14
An explanation of the role of auxins in controlling geotropic responses. Auxin accumulates on the lower side of the plant. Because cells on the upper side of the stem fail to elongate normally, the stem grows upward. Normal lengthening of the cells on the upper side of the root, however, causes the root to grow downward.

FIGURE 16-15
Photograph through a microscope of statoliths (black dots) in a root tip after 0, 10, 22, and 42 minutes after being horizontal. Note that the root is starting to grow downward after the statoliths fall.

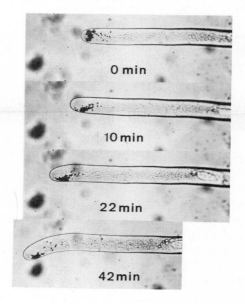

0 min

10 min

22 min

42 min

FIGURE 16-16
These tendrils of wild grape are coiled because of thigmotropism. Upon touch they can be observed to coil after a few minutes.

Besides phototropism and geotropism, plants have several other tropic responses. The tendrils of climbing plants, for instance, wrap around objects they touch (Figure 16-16). This response to contact is **thigmotropism,** and the coiling happens in less than a minute for some species, including the Passion Flower. Adrenaline, which functions as a neurotransmitter in animals, has been found in tendrils and may play some role in their movement.

Nastic movements in plants

Many of the external stimuli that cause tropic responses in plants also trigger **nastic movements.** In nastic movements, however, the response is independent of the direction of the stimulus. Nastic movements can be seen in the closing of flowers or the folding of leaves at certain times of the day or under varying conditions of temperature and humidity. Nastic movements differ from tropic responses in several ways:

- Tropic responses are *growth movements,* while nastic responses are *turgor movements* (regulated by water pressure).
- A nastic movement is not a directed response, as there is no directed movement either toward or away from a stimulus.
- Generally, tropic responses are more permanent than nastic movements.

The opening and closing of flowers are nastic movements, and the Morning Glory gives a beautiful example. During the day, its flowers are open. But, with the approach of evening and the cooling of the air temperature, the flowers close and remain that way all night. With the coming of daylight, they open again. The dandelion and tulip show similar responses (Figure 16-17). The response of flowers of the night-blooming cereus, a cactus, is the opposite. They remain closed

FIGURE 16-17
Nastic movement in tulips: flowers during the night (left), and the same flowers during the day.

throughout the day but open at night. Leaves also show nastic, "sleep" movements. The clover and wood sorrel, are good examples. During the day, their leaflets are open and spread horizontally, but with the approach of night they fold downward and close.

Two other stimuli which cause nastic movements are temperature and touch. The leaves of the rhododendron, a broad-leafed evergreen shrub, roll inward during cold weather and unroll as the temperature rises. The best example of a nastic response to touch is the reaction of the delicate *Mimosa* (sensitive plant). This sensitive plant has compound leaves, with numerous small leaflets that are normally opened horizontally. If a leaf is touched or jarred, the plant immediately reacts by folding its leaflets. Within a few seconds, the leaves droop from their stems (Figure 16-18). This reaction rapidly spreads outward from the point of contact. After having folded, the leaflets gradually reopen to their original horizontal

FIGURE 16-18
Nastic responses to touch in the sensitive plant: normal position of the leaves and leaflets (left), and position of the same leaves and leaflets after they had responded to a touch stimulus.

positions, and become firm as turgor pressure is slowly reestablished. In this remarkable response, the touch stimulus is received by the leaflets and transmitted through the cell fluids in the leaflet and its supporting stem. At the base of the leaflets is a group of cells that holds the leaflet upright. A sudden loss of turgor in these cells relaxes their support, and they sag. A similar turgor change causes the leaves

FIGURE 16-19

The Venus' flytrap is a spectacular insect-catching plant, native to a coastal strip in the Carolinas. If an insect alights on the shiny surface of the outstretched flytrap leaf, it is almost certain to touch the short trigger hairs and, within a second, the leaf closes. The initial movement is rapid enough to capture a large insect, but small ones still have a chance to escape between the stiff projections along the leaf margin; then the leaf will open again in a few hours. When a large victim is captured, the leaf intermeshes the projections. Digestive juices are secreted, and nitrogenous products of disintegration are absorbed. Then the leaf may open again. Indigestible residues are exposed, dry out, and blow away, leaving the trap ready for another victim. Gradually the leaf opens wide, improving the chance that a breeze will clear it of debris.

FIGURE 16-20

Ritualized passing of food during the courtship of *Empis*. The male presents a balloonlike spinning package that contains a fly.

of the Venus' flytrap to snap shut on insects (Figure 16-19). Both *Mimosa* and the Venus' flytrap have also been found to contain neurotransmitter-like chemicals which may be involved in their rapid movements.

ETHOLOGY AND EVOLUTION

Ethology is the study of behavior. It is one of the youngest of the biological sciences, having come into its own only in the past few decades, as biologists have come to realize that behavior as well as anatomy and physiology are subject to the forces of natural selection. Apparently, genes exert influences on the behavioral patterns of organisms, but the mechanisms by which genetic information is translated to behavioral characteristics are unknown. Nevertheless, observers of animals in their natural environments have found traces of evolution's workings.

Flies of the *Empis* family display an amazing example of the effects of natural selection on a behavioral pattern. The males of certain species of Empis flies always present the females with silken spheres before copulation (Figure 16-20). This seems rather unusual, and the basis for it would probably never have been guessed were it not for other species of the same family, still living today, that exhibit various stages of development of this "courtship." In several species in which the male doesn't give anything to the female, she sometimes captures and devours him while he is attempting to mate. In other, slightly more advanced species, the male captures prey, presents it to the female, and then copulates with her while she is busy eating. The male of a still more advanced species captures prey, wraps it in a ball of silk, and offers it to the female. Upon accepting it she may or may not eat the prey, but in either case the male can safely mate with her. In the most advanced species, the male presents his mate with a token sphere of silk, containing no food. At this stage, the sphere has replaced food, becoming part of a display that functions in making the female more receptive to copulation.

During the evolutionary development of the Empis flies' courtship ritual, the pressure of natural selection seems to have favored males which present the most diverting objects to their prospective mates. The behavior of the females, however, is mysterious. It is understandable why they would be receptive to food, but it isn't clear why a silken sphere would be equally diverting. Nevertheless, the ritual of accepting the token seems to have been etched into the females' inheritance.

Inheritance and learning in behavior

Instinctive behaviors are automatic. Instinctive or innate behavior is inherited, and is manifested even when an animal is raised without contact with other members of its species. The sucking responses of a newborn human, for example, are made by any normal infant without training. A young wolf will howl and pounce in

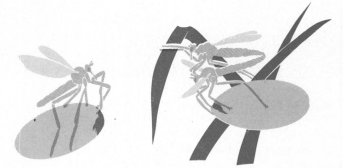

a characteristic manner, even if it has never seen another of its kind. For certain species of birds, complex patterns of mating and nesting are instinctive; they are performed by individuals reared in isolation, with no chance to learn from others of their species.

Unlike instinctive behavioral patterns, **learned behaviors** are developed and modified as a result of experiences. Although the ability to develop learned behaviors must be inherited, experiences are needed as guides. Humans are, by far, the most gifted learners. Our potentials to learn spoken and written languages, for example, are unparalleled. Nevertheless, we are not alone in our ability to learn. Anyone who has ever taught a dog to "shake hands" knows this. But beyond this, learning has been shown to occur in all kinds of animals from flatworms on up, possibly even in protozoa.

The relationship between inheritance and learning is not always obvious. For example, if individuals of certain species of finches are raised in isolation, they will eventually give recognizable finch calls, but will not sing a normal finch song (which is a complex specific series of calls). One would be tempted to conclude from this that the call is inherited, while the song is learned. As the British ethologist W. H. Thorpe has found, however, inheritance plays a subtle role in learning the song.

Thorpe demonstrated that young European Chaffinches, raised in isolation, quickly learned to sing when a recording of the song of their species was played to them. However, when recordings of other species with similar songs were played, the young birds did not respond. Apparently Chaffinches inherit not only the ability to sing, but also the ability to recognize the specific song of their own species. This conclusion was reinforced by experiments in which Thorpe played Chaffinch songs backwards to young birds. To human ears the sounds were nothing like the songs, but the birds still responded to them. At some level deep in their hereditary makeup, they recognized the song.

Imprinting

Imprinting is a behavioral phenomenon, in which heredity and learning both play a role. Konrad Lorenz, a Noble prize-winning ethologist, was among the first to study imprinting. In studying waterfowl, Lorenz found that in the first hours of life, a gosling or duckling which is separated from its mother will dutifully follow almost any moving object, including a wagon, a rubber ball, or a human being (Figure 16-21).

In nature, the first moving object with which a young bird is likely to be in contact is its mother. Since it is of utmost importance that a newborn in the wild stay close to its mother, it is not too surprising that a duckling would be born with a tendency to follow a female duck. It is surprising, however, that the young of many species will, in the absence of the mother, follow almost any moving object. Apparently, newborns inherit the tendency to form attachments with (that is, to be imprinted by) moving objects. This need not be more specific, since, in the natural course of events, the mother either makes sure that she is the object which is followed or, because since she is the source of milk, her offspring will follow her.

Ethologists originally believed that imprinting could occur only during highly restricted time periods ("critical periods"), but this view has changed. It is now known that some attachment can occur somewhat before or after the critical period. Nevertheless, a feature of behavioral development that has become particularly clear in studies of imprinting is that once animals are familiar with a given set of environmental surroundings, they are likely to avoid new stimuli. Indeed, one of the factors that appears to limit the critical period of imprinting is

FIGURE 16-21
Konrad Lorenz, "mother" to his imprinted geese.

the accumulation of sufficient experience to make new stimuli apparent. It is as if the animals build up an internal representation of certain features of their environment, then avoid stimuli that do not match this framework. It could be that this is the onset of the emotional state we know as fear.

Communication

Ethologists have discovered some amazing modes of communication within societies of animals. Bird songs, trails of chemicals left by insects, and the brilliant display of the male peacock are a few of the behavioral traits that transfer information within communities of animals.

In a classic piece of research, an Austrian zoologist, Karl von Frisch, discovered that honeybees communicate through "dances." Within a colony there are three types of honeybees, including one egg-producing *queen,* a few male *drones,* and many *workers,* which are females with undeveloped genitals. Among the workers are a few scouts which hunt for new sources of nectar. If a scout finds a patch of flowers, it eats nectar and pollen and returns to the hive. There it performs a *waggle dance* which informs the other workers of the direction and distance of the flowers.

While rhythmically twitching its tail, the scout walks through a figure-8 on the surface of the hive (Figure 16-22). The distance to the nectar is indicated by the vigor of the dance—the slower the twitching, the further the flowers. The direc-

FIGURE 16-22

Shown from left to right: The dancing bee makes the diameter of the "tail-waggling dance" upward on the vertical surface of the comb. This indicates to the other bees that a source of nectar is located in a direction toward the sun. The second dance indicates that the nectar is located in a direction away from the sun. The third dance indicates the nectar is 60° from the sun.

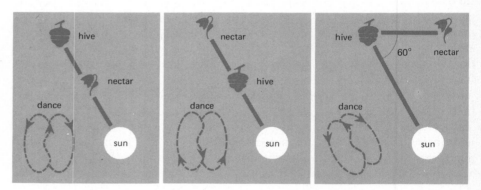

tion is indicated relative to the sun. If, in relation to the hive, the food source is directly toward or away from the sun, the center line of the figure-8 is made perpendicular to the horizon. The dancing scout indicates a direction toward the sun if she walks upwards (away from gravity) along the line. Forming the center line downwards indicates that the food source is directly away from the sun. If the flowers aren't on a direct line with the sun and the hive, the dance is done at an angle to the horizon. If, for example, the flowers are sixty degrees to the left of the sun's direction, the center line is made sixty degrees to the left of the vertical, and so on. Notice that this is all the more remarkable because the dancing bee is transposing the horizontal angle between the direction of the sun and that of the flowers to a vertical angle with respect to gravity on the side of the hive.

Communication among social insects isn't limited to visual displays. Other signals, particularly chemical, are used to transmit a variety of messages. A species may possess a repertoire of different chemical signals, or **pheromones,** produced by various glands. They serve such functions as marking trails, inducing mating, and warning of danger.

Among certain species of ants, one pheromone (oleic acid) indicates death. Its effect provides a good example of how rigidly the behavior of animals can be structured by stereotyped signals. As a dead ant decomposes, the death pheromone is produced in its body. Other worker ants react to it by picking up the carcass and carrying it to a refuse pile outside the nest. One behaviorist wondered what would happen if living ants were painted with death pheromone. It turns out that the painted ants know they're still alive, but their fellow workers do not. Thrashing about wildly, the victims are carried to the refuse pile, and as often as they scramble back to the nest, they are dutifully returned to the grave.

Ants are bound by a hereditary injunction which demands that they react to the death pheromone by heaving the body on the refuse pile, even if it is very much alive and kicking. Ethologists refer to such chemicals as **releasers,** for they signal (or release) a specific pattern of innate behavior in those that sense them. The messages touched off by releasers supersede all other stimuli. The workers can surely see and feel that the object they are carrying is struggling in a most undead manner, but the command of the pheromone overrides the contradicting physical reality.

Pheromones aren't the only behavioral releasers (sometimes called **sign stimuli**). Other kinds of signals also release innate sequences of behavior. Among them are visual displays (such as the brilliant plumage of the peacock) and auditory signals (such as bird songs).

The problem of anthropomorphism

Ethology is a science full of intellectual pitfalls. The basic problem is that humans are tempted to make *anthropomorphic* assumptions—that is, to assume that other animals have similar sensations and motivations as humans. The simple (but easily forgotten) fact is that other animals don't experience life in the same way we do. We don't really know, for example, how the death pheromone affects an ant. It is impossible for us to hear the song of a Chaffinch in the same way that a young bird of this species hears it, and we have no way of knowing how the silken sphere presented by a male Empis fly to his mate appeals to her. Nevertheless, since at least certain aspects of behavior are part of the evolutionary heritage of a species, and since we exist in the flow of evolution, it would seem that we may have some things to learn about ourselves from observing the behavior of other animals.

GENES AND BEHAVIOR— THE QUESTION OF SOCIOBIOLOGY

Because certain behavioral patterns are obviously inherited, it is generally assumed that they are somehow coded for by genes. How these codes are translated from the structure of DNA into the actual behavior, however, is unknown. Obviously this is a huge gap in knowledge. Nevertheless, certain contemporary biologists portray genes as the central motivators in the evolutionary struggle for survival. According to biologist Richard Dawkins, genes "swarm in huge colonies safe inside lumbering robots, sealed off from the outside world, manipulating it by remote control. They are in you and me; they created us body and mind; and their preservation is the ultimate rationale for our existence . . . we are their survival machines."

Dawkins could be accused of making overly anthropomorphic assumptions about the workings of genes. But, certain examples from nature support his view. Soldier ants, for example, will unflinchingly lay down their lives for the preservation of their colony, and individual birds will risk death for the good of the flock by calling out warnings when predators approach. These seemingly unselfish acts (altruism) cannot be accounted for by Charles Darwin's view of the evolutionary struggle for survival, in which every individual is out for itself. But, if one assumes that genes rather than individuals are the "combatants" in the evolutionary process, then "strategies" of survival which serve to preserve near relatives (even at the cost of the individual) make sense. The view that genes are somehow self-preserving entities is a central tenet of a controversial new field of biology, **sociobiology.** Sociobiology is the study of the biological basis of social behavior.

Apart from accounting for altruistic acts, sociobiological theory also accounts for the differences between male and female behavior of certain species. For example, females which nurture their offspring within their bodies for a period before giving birth are usually more protective of their young than are the males of the species. Sociobiologists point out that the "wisest" genetic strategy for the males of these species is to spread their genes widely by impregnating many females. Females, on the other hand, have an investment in time as well as genes with any given offspring. Thus, their best strategy is to be protective.

The views of sociobiology have been criticized as portraying existence (particularly human existence) as being too predetermined and mechanistic. Indeed, the most respected proponents of sociobiology caution against taking its theory too far and argue that culture shapes our behavior far more than genes do. We must not give up hope and believe that all our social problems are rooted in genetics.

Genes, society, and free will

Assuming that only a proportion of human behavior is genetically predetermined, several questions arise: Which behavioral characteristics are genetically preprogrammed, and which are not? What determines behavioral traits that are not genetically predetermined? Does free will figure into behavior?

For the question "Which behavioral characteristics are genetically preprogrammed?" there is no simple answer. However, in general, less and less behavior is predictable (hence, presumably, genetically determined) as people mature from infancy through childhood to adulthood. The innate sucking response with which all infants are born, for example, is eventually replaced by eating habits unique to the individual.

There are widely varying theories as to how innate drives govern adult behavior. Theorists who stress the power of genes believe that the act of copulation is an automatic response at the age of sexual maturity. Learning theorists, on the other hand, assume that patterns of sexual behavior are largely learned through a variety of psychological and social mechanisms, involving relationships between parents and offspring as well as relationships between peers.

According to learning theorists, experience accounts for behavioral traits that are not genetically predetermined. Learning theorists see the actions of an individual at any given moment as primarily determined by the sum of what the individual has learned up to that moment. How does one get to be a homosexual, for example? Is it the way in which we are brought up? Is it a free choice? Do we acquire genes for it? Or, do intrauterine hormone levels affect adult sexual preferences? Modern biological theory has presented many alternative answers but not the definitive one yet.

Obviously, learning experiences are significant to the development of human existence. Languages, for example, are acquired only by exposure to them. Moreover, there is an abundance of psychological evidence that learning can occur on a subconscious level. Emotional responses in the present are to a great extent reactions to deeply ingrained emotional experiences in the past that have been sensed but not thought about.

But, experience isn't the only determinant of emotional response. There is also a biochemical level. Aggressive tendencies in males, for example, seem to be directly related to levels of testosterone in the bloodstream. Studies of prison inmates have revealed higher than average levels of testosterone (as indicated by concentrations in urine) for those imprisoned for violent crimes. Thus, a genetic predisposition to high levels of testosterone may imply a predisposition to violence.

Hypothesis on the hormonal link to aggression is complicated, however, by the fact that genes aren't the only influences on testosterone levels. Social influences also come into play. Evidence for this has come from studies of various animals. In groups of monkeys, for instance, the predominant male almost always has the highest level of testosterone. When he is removed from the group, the male which succeeds to power usually experiences a rise in production of testosterone.

Apparently experience, social position, and genes are intertwined in the behavioral matrix of an individual; and one wonders if there is room in all of it for freedom of choice. To the extent that one can rationally appraise one's position in life and make logical decisions on courses of action, humans are capable of freedom of choice. But, words like rational and logical are highly subjective; what makes sense to one person may be nonsense to another. Moreover, what may seem to be a logical choice at one point, may later prove to have been illogical.

The fact that we are a thinking species complicates the matter of free will. At times one's brain may not be working for the benefit of one's self. In extreme cases insanity results when the brain (perhaps through a genetic hormonal imbalance, or perhaps through overwhelmingly negative experiences) totally turns against itself. The brain is at once our most powerful tool and our most potentially dangerous enemy.

Since we are a highly intercommunicating species, we possess a sort of collective consciousness, which has made choices (admittedly often through violence) for the species as a whole. Whether or not this collective consciousness is sane is subject to debate. The authors hope that the study of biology may help us to wisely use our undoubtedly powerful but questionably directed brilliance.

CHAPTER 16: SUMMARY

1   Night-day alternations (and to a lesser extent) seasonal cycles exert a significant effect upon behavior of both plants and animals.

2   Accurate migration and direction-finding by animals utilizes one or more of the following cues: sun, moon, stars, landmarks, and magnetic fields.

3   Plants can move in response to light, gravity, and touch.

4   It is often difficult to attribute a behavior only either to instinct or learning— both usually are involved.

5   Behavior is subject to the forces of natural selection.

6   Animals communicate with each other by sounds, visual displays, movements, and chemical scents.

7   Sociobiology is the study of the biology of social behavior in animals.

Suggested Readings

Barash, David P.   *Sociobiology and Behavior,* Elsevier, New York, 1977—An easy-to-read introduction to the field of sociobiology.

Bertram, Brian C. R.   "The Social System of Lions," *Scientific American,* May, 1975—A fascinating account of the organization of the social system of lions.

Callahan, Phillip   *Insect Behavior,* Four Winds Press, New York, 1970—An account of many aspects of insect behavior with a special section on possible projects and experiments.

Cohen, Daniel   *Human Nature-Animal Nature: The Biology of Human Behavior,* McGraw-Hill, New York, 1975—An examination of phylogenetic parallels between human and animal behavior.

Holldobler, Bert   "Communication Between Ants and Their Guests," *Scientific American,* March 1971—How ants and the animals they "domesticate" communicate.

Levine, Seymour   "Stress and Behavior," *Scientific American,* January, 1971—The effects of stress on the behavior of mammals.

Palmer, John   *An Introduction to Biological Rhythms,* Academic Press, New York, 1976—Discussion of rhythms in a variety of organisms, including humans.

ETHICS AND VALUES AS GENETIC TRAITS

According to E. O. Wilson, the preeminent spokesman for sociobiology and author of a book entitled *Sociobiology,* the study of human ethics must be "biologicized."

Sociobiology teems with . . . provocative suggestions about human social behavior. . . . Wilson speaks, for example, of the evolution of ethics. The emotional control centers of the brain, which flood our consciousness with hate, love, guilt, and fear, have evolved by natural selection, he says. Thus when ethical philosophers try to intuit the binding canons of morality, they are consulting the survival values programmed into their own brains by natural selection.

"Only by interpreting the activity of the emotive centers as a biological adaptation can the meaning of the canons be deciphered," says Wilson. The time has come "for ethics to be removed temporarily from the hands of the philosophers and biologicized." In a pending article in *BioScience* Wilson states: "The question that science is now in a position to answer is the very origin and meaning of human values, from which all ethical pronouncements and much of political practice flow."

Some of what we regard as our noblest sentiments may derive from behavior selected because of its basic survival value. Forms of altruism, for example, such as sacrificing one's life for the sake of one's family or group, may be programmed into us by natural selection because they favor the representation of the hero's genes in the next generation, which is all that nature cares about. The explorer's sense of exhilaration in breaking into virgin territory, or even the scientist's excitement at a new discovery, may be founded simply on the reward that nature pays for inquisitiveness. With a fuller understanding of the human brain, Wilson told *Science,* we may arrive at a new level of disillusionment: "In completing the Darwinian revolution we are likely to see some of our most exalted feelings explained in terms of traits which evolved. Human beings see themselves in transcendental terms. But we may find out that there is an overestimate of the nature of our deepest yearnings. We tend to be very respectful of these emotions, as we should be, because they are very human qualities, but we may discover that they have very humble origins."

Because of materials shortages and the threat to the environment, the creation of a planned society "seems inevitable in the coming century," Wilson says in *Sociobiology.* Yet the planners who try to discourage the beast in man may find that they lose the angel too because the one is the palimpsest of the other. If the planned society, Wilson says, "were to deliberately steer its members past those stresses and conflicts that once gave the destructive phenotypes their Darwinian edge, the other phenotypes might dwindle with them. In this, the ultimate genetic sense, social control would rob man of his humanity." Or as Horace puts it, you can't drive nature out with a pitchfork.

From N. Wade, "Ethics and Values as Genetic Traits," *Science* 191:1153, 1976. Copyright © 1976 by the American Association for the Advancement of Science.

GENES AND HUMAN BEHAVIOR: WHAT WILSON HAS NOT PROVED
Richard L. Currier

How applicable are the ideas of sociobiologists to human behavior? According to anthropologist Richard Currier, we know too little about the influences of genes on behavior to know yet, but it will be interesting to follow the dialogue between social scientists and biologists that will develop during the coming years.

© Copyright 1977 American Heritage Publishing Co., Inc. Reprinted by permission from *Horizon* (March 1977).

Sociobiology is both repugnant—and at the same time attractive—to the social sciences. Anthropology, psychology, and sociology are built on the assumption that human behavior is *learned,* and that except for a few primitive drives (sleep, hunger, sex), people do what their cultures require them to do. Accepting the notion that there are important biological factors in human social behavior means that most of the current theories of human behavior will have to be overhauled drastically or even abandoned. This is naturally unsettling to social scientists with careers and reputations to protect.

In addition, sociobiology poses a serious political problem. If important components of human behavior are natural and inborn, then liberal programs of social betterment can never really do away with the selfishness, oppression, and conflict that have been a part of human history as far back as we choose to look. That irrepressible "animal" component, which can be nasty as well as noble, will always be surfacing to throw an all-too-human monkey wrench into the gears of the utopians' social machinery.

To make matters worse, any human characteristic that has a genetic basis can—theoretically—vary from one racial group to another. And while we are willing to accept the idea that skin color has a genetic basis and varies from one racial group to another, are we equally willing to accept the proposition that intellectual ability, artistic talent, aggressiveness, and the intensity of sexual drive have genetic bases and also vary from one racial group to another? These are the kinds of ideas that can cause the humanitarian social scientist to wake up in the middle of the night in a cold sweat.

On the other hand, sociobiology is seductive, for—with the possible exception of Marxism or psychoanalysis—it offers something that theories of behavior never offered before: a system so vast, elegant, and all-embracing that it can relieve the intellectual self-doubt plaguing the social sciences today. Its adherents seem to believe in what they are doing, and this feeling is sure to be the object of envy and . . . imitation.

Most social scientists are still ambivalent about sociobiology, and, for the most part, they are not at all certain what it really is. Adding to the confusion are two prevalent misconceptions. The first is that genes control behavior. Wilson writes about "altruist genes" and "genes favoring homosexuality" in human beings, and from this it is easy to see human genes as producing certain kinds of behavior, just as genes in sweet peas produce blossoms of a certain color. But genes do not and cannot control behavior. Genes are only plans or blueprints according to which each living organism is constructed; genes control only the way in which atoms and molecules are arranged into living tissue. In the realm of behavior, the most important kind of living tissue is the central nervous system, especially the brain. We know *that* the brain controls behavior, but we do not know *how* the brain controls behavior, and—most importantly—we know very little about the relationship between the structure of the brain (how it is "wired up") and the behavior of each animal species.

The second misconception is that Wilson has demonstrated the existence of a biological basis to human behavior. This is not true. Rather, Wilson has synthesized a vast amount of research done in genetics, evolution, and animal behavior during the past twenty years. Students of animal behavior have been *assuming* that there is a biological basis to it; now the accumulated work has reached the stage at which Wilson and others can show how it all fits together. And, when one looks at this large picture, a general pattern of behavior for social animals, especially primates, emerges.

In all likelihood, it is more than simple coincidence that we share with many other animal species such things as territorial defense, possessiveness, and competition in sexual life. But the fact remains that zoologists have *not* studied human behavior—they have left the field to the social scientists, who obviously do not use the methods of biology. For this reason the meticulous gathering of data that the sociobiologists engage in as part of their studies of insects or monkeys in their natural surroundings has never really been done in the study of humans. Human behavior is studied almost entirely in rigidly controlled experimental situations, so that normal human social behavior ceases to exist and therefore cannot be properly studied! There is no real proof either way.

Wilson has given genuine coherence and respectability to the biological study of social behavior, but he has *not* proven that it applies to humans. My own guess is that once the groundwork has been done, others will succeed in demonstrating that the "biological component" in human behavior really exists. Whether this will be done by biologist-invaders who boldly pitch their tents in the fields of social science or by renegade social scientists who take up biology, it is too early to tell.

ART ACKNOWLEDGMENTS

Chapter 1 Chapter opening photograph, Ward's Natural Science Establishment; 1-2 (left), The Granger Collection; 1-2 (right), The Bettman Archive; 1-4, Paolo Koch/Photo Researchers; 1-5, Martin Trailer; 1-6, U.S. Forest Service; 1-8, Martin Trailer.

Chapter 2 Chapter opening photograph, American Museum of Natural History; 2-1, Hale Observatories; 2-3, *The Journal of the American Chemical Society* 77: 2351, 1955; 2-6, Pierre Berger/Photo Researchers; 2-11, from *Evolution*, Third Edition, by J.M. Savage. Copyright © 1963, 1969, 1977 by Holt Rinehart and Winston; 2-12 (top), Russ Kinne/Photo Researchers; 2-12 (bottom), Allan Roberts; 2-13, map copyright of Hammond Incorporated, Maplewood, New Jersey; 2-14 (bottom left), Paolo Koch/Rapho, Photo Researchers; 2-14 (bottom right), Toni Angermayer/Photo Researchers; 2-15, H.B. Kettlewell; 2-16, Jack Dermid/National Audubon Society Collection, Photo Researchers; 2-17 (top left), Erika Stone/Photo Researchers; 2-17 (bottom left), Paolo Koch/Rapho, Photo Researchers; 2-17 (right), Carl Purcell/Peace Corps; 2-18, Allan D. Cruickshank/National Audubon Society Collection, Photo Researchers; 2-19 (top), NOAA; 2-19 (bottom), A.W. Ambler/National Audubon Society Collection, Photo Researchers; 2-20, M.A. Geissinger/Van Cleve; 2-23 (left), Russ Kinne/Photo Researchers; 2-23 (right), Kenneth Fink/National Audubon Society Collection, Photo Researchers; 2-25, Camera Pix/Photo Researchers.

Chapter 3 Chapter opening photograph and 3-1 (top), Jen and Des Bartlett/Photo Researchers; 3-1 (bottom), Russ Kinne/Photo Researchers; 3-2, Walter Dawn/National Audubon Society Collection, Photo Researchers; 3-4, Houwink, Van Iterson and *Biochimica et Biophysica Acta*, Elsevier Publishing Co.; 3-6, T.F. Anderson, E.L. Wollman, F. Jacob; 3-22, John H. Gerard/National Audubon Society Collection, Photo Researchers; 3-23, Photri; 3-33, Eric V. Gravé/Photo Researchers; 3-34, Ron Church/Photo Researchers; 3-45 (left), Russ Kinne/Photo Researchers; 3-45 (right), Ron and Valerie Taylor/Bruce Coleman; 3-46 (left), Jen and Des Bartlett/Photo Researchers; 3-46 (right), and 3-47 (left), Robert C. Hermes/National Audubon Society Collection/Photo Researchers; 3-47 (right), Douglas Faulkner; 3-48 (left and right), Jen and Des Bartlett/Photo Researchers; 3-48 (center), Russ Kinne/Photo Researchers; 3-54 and 3-55, Jerome Wexler/National Audubon Society Collection, Photo Researchers; 3-56, C.H. Brett, North Carolina State University; 3-57 (top left), V.E. Ward/National Audubon Society Collection, Photo Researchers; 3-57 (top center), Jeanne White/National Audubon Society Collection, Photo Researchers; 3-57 (top right), Louis Quitt/Photo Researchers; 3-57 (bottom left), Harry Brevoort/National Audubon Society Collection, Photo Researchers; 3-57 (bottom right), Louis Quitt/National Audubon Society Collection, Photo Researchers; 3-61 (inset), photo by Hugh Spencer.

Chapter 4 Chapter opening photograph, Stanley Flegler/Van Cleve; 4-1, from *Life and Health*, New York: CRM Books, a division of Random House, Inc., 1972. Art by John Dawson; 4-2 and 4-3, © Douglas P. Wilson; 4-4, Lee Jenkins/National Audubon Society Collections, Photo Researchers; 4-6, U.S. Department of Agriculture Soil Conservation Service; 4-10 (photo), Stanley Flegler/Van Cleve; 4-11, from M.E. Hale, Jr., *The Biology of Lichens*, London: Edward Arnold, Ltd., 1967; 4-12, NOAA; 4-13, Walter Dawn/National Audubon Society Collection, Photo Researchers; 4-14, Fritz Goro; 4-15, from Lorus J. Milne and Margery Milne, *The Biotic World and Man*, 3rd ed., © 1965, p. 504. Reprinted by permission of Prentice-Hall, Inc., Englewood Cliffs, New Jersey; 4-17, B.C. McLean, U.S. Department of Agriculture; 4-18, Phoenix Chamber of Commerce Photo; 4-25, R. Schuster and J.H. Crowe from J.H. Crowe and A.F. Cooper, *Scientific American*, December 1971; 4-27, U.S. Department of Agriculture; 4-28, L. Nielsen from Peter Arnold; 4-29, Frederick Ayer/Photo Researchers; 4-30, Richard E. Ferguson, © William E. Ferguson; 4-33 (left), Russ Kinne/Photo Researchers; 4-33 (right), Howard Hall/Photophile.

Chapter 6 Chapter opening photograph, J. Arthur Herrick; 6-2, Parke, Davis, and Co.; 6-3, from F.P. Ottensmeyer et al., *Science* 179: 176, 1973: Copyright © 1973 by the American Association for the Advancement of Science; 6-5, W.G. Haley, Cell Research Institute, University of Texas at Austin; 6-7, Don W. Fawcett; 6-8, from S.J. Singer and G.L. Nicolson, *Science* 175: 723, 1972; Copyright © 1972 by the American Association for the Advancement of Science; 6-9, Don W. Fawcett; 6-10, George E. Palade; 6-11, Carlo Bruni, 6-12, Myron C. Ledbetter, Brookhaven National Laboratory; 6-18, J. Arthur Herrick; 6-19, from C. Avres, *Cell Biology*, New York: D. Van Nostrand, 1976; 6-20, George E. Palade; 6-21, Fisher Scientific Company, Educational Materials Division and S.T.E.M. Laboratories;

6-22, Phillip A. Harrington; 6-24, from William Bloom, and Don W. Fawcett, *A Textbook of Histology*, 10th edition, © 1975 by the W.B. Saunders Co., Philadelphia.

Chapter 7 Chapter opening photograph, Ward's Natural Science Establishment; 7-2, from R. Resnick and D. Halliday, *Physics*, New York: John Wiley and Sons, 1966; 7-3, from W. McElroy and C. Swanson, eds., *Foundations of Biology*, Englewood Cliffs, New Jersey: Prentice-Hall, 1968. Reprinted by permission; 7-6, L.K. Shumway; 7-10, Melvin Calvin; 7-17, from Nelson, Robinson and Boolootian, *Fundamental Concepts of Biology*, New York: John Wiley and Sons, 1970; 7-23, Stanley Flegler/Van Cleve; 7-25, Ward's Natural Science Establishment.

Chapter 8 8-1, Don W. Fawcett; 8-2, Renato Baserga; 8-8, from *Cell Structure and Function*, Second Edition, by Ariel G. Loewy and Philip Siekevitz. Copyright © 1963, 1969 by Holt, Rinehart and Winston; 8-14, from O.L. Miller et al., *Science* 1969: 392-395, 1970. Copyright 1970 by the American Association for the Advancement of Science; 8-16, from L. Pauling and R.B. Corey, *Proc. Int. Wool Textile Research Conf. B*, p. 249, 1955, as redrawn in C.B. Anfinsen, *The Molecular Basis of Evolution*, New York: John Wiley and Sons, 1959; 8-20, M.M.K. Nass, from M.M.K. Nass, *Science* 1965: 25, 1969. Copyright © 1969 by the American Association for the Advancement of Science; 8-21, R. Warner and R. Fishel; 8-22, from B. Lewin, *Gene Expression*, Vol. I, New York: John Wiley and Sons, 1974; 8-24, Louis Koster; 8-24 (insets), reproduced from *The Journal of Experimental Medicine* by copyright permission of The Rockefeller University Press and R. Austrian; 8-27, Photri.

Chapter 9 9-1, The Bettmann Archive; 9-12, W.R. Grace Co.; 9-15, Jerry Cooke/Sports Illustrated © Time Inc.; 9-16 (top), U.S. Department of Agriculture; 9-16 (bottom), American Hereford Association.

Chapter 10 Chapter opening photograph, Philips Electronic Instruments, Inc.; 10-1, Kenneth Karp; 10-2, from J.H. Tjio and T.T. Puck, *Proceedings of the National Academy of Sciences* 44: 1229-1237, 1958; 10-5 (top), Russ Kinne/Photo Researchers; 10-5 (bottom), Roche/National Audubon Society Collection, Photo Researchers; 10-6, Philips Electronic Instruments, Inc.; 10-8, D.H. Carr and M.L. Barr; 10-9, Bruce Roberts/Rapho, Photo Researchers; 10-10 and 10-11, Dr. Lillian Y. Hsu, Mount Sinai School of Medicine, N.Y.

Chapter 11 Chapter opening photograph, Photo Researchers; 11-1 from *Microbial Life*, Second Edition by William R. Sistrom. Copyright © 1962, 1969 by Holt, Rinehart and Winston. 11-2, Photo Researchers; 11-3, from W.K. Gregory, *Annals of the New York Academy of Sciences* 22, 1912; 11-6, from C.M. Dienhart, *Basic Human Anatomy and Physiology*, 2nd edition, © 1973 by the W.B. Saunders Company, Philadelphia; 11-10, M. Tegner, from M. Tegner and T. Epel, *Science*, 179: 685-688, 1973. Copyright © 1973 by the American Association for the Advancement of Science; 11-15, Jerome Wexler/National Audubon Society Collection, Photo Researchers; 11-23, U.S. Department of Agriculture; 11-24, A.W. Ambler/National Audubon Society, Photo Researchers.

Chapter 12 12-2, from C.M. Dienhart, *Basic Human Anatomy and Physiology*, 2nd edition, © 1973 by the W.B. Saunders Company, Philadelphia; Table 12-2, reproduced by permission from *Human Sexuality* by Wilson, Strong, Clarke and Johns, copyright © 1977, West Publishing Company. All rights reserved; 12-9b, from A.L. Colwin and L. H. Colwin, *Cellular Membranes and Development*, ed. M. Locke, New York: Academic Press, 1964; 12-11 and 12-12, reproduced by permission from *Elements of Human Anatomy and Physiology* by R. Boolootian, copyright © 1976, West Publishing Company. All rights reserved; 12-13, Landrum B. Shettles; 12-16, from N.M. Jessop, *Biosphere: A Study of Life*, Englewood Cliffs, New Jersey: Prentice-Hall, Inc., 1970. Reprinted by permission; 12-17, from W. Masters and V. Johnson, *Human Sexual Response*, Boston: Little, Brown and Company, 1966; 12-18, 12-19, and 12-20, reproduced by permission from *Human Sexuality* by Wilson, Strong, Clarke and Johns, copyright © 1977, West Publishing Company. All rights reserved; 12-24, from *International Planned Parenthood Federation Medical Handbook*, page 62. Reprinted by permission of the International Planned Parenthood Federation; 18-20 Lower Regent Street, London SW1, England; 12-27, from E. Jawetz et al., *Review of Medical Microbiology*, 12th ed., Los Altos, CA.: Lange Medical Publications, 1976; 12-28 and 12-29, Center for Disease Control, Atlanta.

Chapter 13 13-1, HRW photos by Russell Dian; 13-3, Percy W. Brooks; 13-7 (textual matter), from J.H. Otto and A. Towle, *Modern Biology*, New York: Holt, Rinehart and Winston, 1977; 13-7 (diagram), designed by W.A. Osburn, artwork by E. Cole, R. Demarest, G. Lashbrook and W. Osburn, Philadelphia: W.B. Saunders Co., 1966; 13-8, Percy W. Brooks; 13-11, Lester V. Bergman and Associates; 13-19, from E.O. Bernstein, cover of *Science*, vol. 166, 1969; 13-22 (top), Brookhaven National Laboratory; 13-22 (bottom), Percy W. Brooks; 13-28, U.S. Air Force from UPI; 13-29, M. BuKovac, Michigan State University; 13-30, and 13-31, U.S. Department of Agriculture; 13-32, reprinted from *Nature* 220: 498-499, 1968.

Chapter 14 Chapter opening photograph and 14-2, Center for Disease Control, Atlanta; 14-3, James G. Hirsch; 14-5 (top), U.S. Department of Agriculture; 14-5 (bottom), Virus Laboratory, University of California, Berkeley; 14-6 (photo), R.W. Horne; 14-7, Photo Trends; 14-8 (photo), T.F. Anderson, 14-9, A.K. Kleinschmidt and *Biochimica et Biophysica Acta*, Elsevier Publishing Co.; 14-10, Lewis Koster; 14-12, D.T. Brown and T.F. Anderson, Institute for Cancer Research, Philadelphia; 14-13, Photri; 14-14, Paolo Koch/Photo Researchers; 14-15, Pfizer, Inc.; 14-17, D. Greenwood, from D. Greenwood and F. O'Grady, *Science* 163: 1076-1077, 1969. Copyright © 1969 by the American Association for the Advancement of Science. 14-18, Eli Lilly and Co.; 14-19, from R. Ross and J.A. Glomset, *Science* 180: 1333, 1973. Copyright © 1973 by the American Association for the Advancement of Science.

Chapter 15 Chapter opening photograph and 15-3 (photo), Turtox/Cambosco, Macmillan Science Company, Inc., Chicago; 15-8, Sanford L. Palay from A. Peters et al., *The Fine Structure of the Nervous System*, New York: Harper and Row, 1970; 15-9 (photos), Lewis Koster; 15-11 (textual matter), from J.H. Otto and A. Towle, *Modern Biology*, New York: Holt, Rinehart and Winston, 1977; 15-11 (diagram), designed by W.A. Osburn, artwork by E. Cole, R. Demarest, G. Lashbrook, and W. Osburn, Philadelphia: W.B. Saunders Co., 1966; 15-12, U.S. Department of Agriculture; 15-18, from W. McElroy and C. Swanson, *Foundations of Biology*, Englewood Cliffs, New Jersey: Prentice-Hall, Inc., 1968. Reprinted by permission; 15-22, from R. Resnick and D. Halliday, *Physics*, New York: John Wiley and Sons, 1966; 15-29, John Loengard, *Life Magazine* © 1963 Time Inc.; 15-30, Van Bucher/Photo Researchers.

Chapter 16 Chapter opening photograph, Carolina Biological Supply Co.; 16-1, Douglas P. Wilson; 16-2, from *Biology* by James D. Ebert, Ariel G. Loewy, Richard S. Miller, and Howard A. Schneiderman. Copyright © 1973 by Holt, Rinehart and Winston; 16-3, reproduced from *The Journal of General Physiology* 62: 709, 1973 by copyright permission of The Rockefeller University Press; 16-5 (top and bottom left, bottom right), Lynwood M. Chace/National Audubon Society Collection, Photo Researchers; 16-5 (top right), Hugh Spencer/National Audubon Society Collection, Photo Researchers; 16-6, Dorothy M. Compton/National Audubon Society Collection, Photo Researchers; 16-7, Karl W. Kenyon/National Audubon Society Collection, Photo Researchers; 16-8, Hugh M. Halliday/National Audubon Society, Photo Researchers; 16-10, George Whiteley/Photo Researchers; 16-11, Tom McHugh/Photo Researchers; 16-12, J. Adler and S.W. Ramsey, from J. Adler, *Science* 166: 1588-1597, 1969. Copyright © 1969 by the American Association for the Advancement of Science; 16-13, photo by Hugh Spencer; 16-15, from Andreas Sievers, *Gravity and the Organisms*, eds. S.A. Gordon and M.J. Cohen, Chicago: University of Chicago Press, 1971; 16-16, Jack Dermid/National Audubon Society Collection, Photo Researchers; 16-17, William Harlow/National Audubon Society Collection, Photo Researchers; 16-18, Jerome Wexler/Photo Researchers; 16-21, Thomas McAvoy, *Life Magazine* © 1955 Time Inc.

Appendix from J.H. Otto and A. Towle, *Modern Biology*, New York: Holt, Rinehart and Winston, 1977.

Abdomen   in arthropods, the body region posterior to the thorax.

Abortion   termination of a pregnancy.

Absorption   the process by which water and dissolved substances pass into cells.

Acetylcholine   a chemical substance released at the motor end plate, causing muscle contraction.

Acid   any substance that gives off free hydrogen ions into solution.

Active transport   requiring the expenditure of energy, the passage of a substance through a cell membrane.

Acupuncture   a Chinese practice of puncturing the body with needles to relieve pain.

Adaptive radiation   a branching out of a population through variation and adaptation to many environments.

Addiction   the body's need for a drug that results from the use of the drug.

Adenoid   a mass of lymph tissue that grows from the back wall of the nasopharynx.

ADP (adenosine diphosphate)   a low-energy compound found in cells that functions in energy storage and transfer.

Adrenal glands   two ductless glands located one above each kidney.

Adventitious root   one that develops from the node of a stem or from a leaf.

Aerobic   requiring free atmospheric oxygen for normal activity.

Agglutinate   to clump together.

Agglutinogen   a protein substance on a blood cell's surface that is responsible for blood types.

Air sacs   in insects, the enlarged spaces in which the tracheae terminate; in birds, cavities extending from the lungs; in man, thin-walled divisions of the lungs.

Alimentary canal   those organs that compose the food tube in animals and man.

Allele   one of a pair of genes responsible for contrasting traits.

Alveoli   microscopic protrusions in the lungs in which the exchange of gases takes place.

Amber   a resin within which fossil organisms are preserved.

Ameoboytes   ameobalike cells in sponges that function in circulation and excretion.

Amino acids   substances from which organisms build protein; the end products of protein digestion.

Ammonification   the release of ammonia from decaying protein by means of bacterial action.

Amniocentesis   a procedure entailing extracting fetal cells from a pregnant woman to look for genetic abnormalities.

Amnionic fluid   secreted by the amnion and filling the cavity in which the embryo lies.

Ameoboid movement   a movement of cells involving extensions of protoplasm.

Amphetamines   a class of stimulant drugs.

Amphibian   a cold-blooded vertebrate, intermediate in characteristics between fish and reptiles, that is a gilled larva and an air-breathing adult.

Ampullae   storage vessels for sperm after they leave the epididymises.

anabolism   the process in a plant or animal by which food is changed into living tissue.

Anaerobic   deriving oxygen for life activity from chemical changes and, in some organisms, being unable to live actively in free oxygen.

Analogous organs   those that are similar in function.

Anaphase   a stage of mitosis during which the chromosomes migrate to opposite poles.

Angiosperm   any of a class of plants, including all the flowering plants, characterized by having the seeds enclosed in an ovary.

Annual   a plant that lives for one season only.

Annulus   the ring on the stipe of a mushroom that marks the point where the rim of the cap and the stipe were joined.

Antenna   a large "feeler" in insects and certain other animals.

Anterior   the front end of an organism.

Anther   that part of the stamen which bears pollen grains.

Anthocyanin   a red, blue, or purple pigment dissolved in cell sap.

Anthropology   the study of the physical and cultural history of humans.

Antibiotic   a germ-killing substance produced by a bacterium, mold, or other fungus.

Antibody   an immune substance in the blood and body fluids.

Anti-codon   the triplet nucleotide base code in transfer RNA which is the opposite of the messenger RNA codon.

Antigen   a substance, usually a protein, which when introduced into the body stimulates the formation of antibodies.

Antipodals   three nuclei found in the embryo sac at the end farthest from the micropyle.

Antitoxin   a substance in the blood that counteracts a specific toxin.

Anus   the opening at the posterior end of the digestive tube or alimentary canal.

Aorta   the great artery leading from the heart to the body.

Aquatic   living in water.

Aqueous humor   the watery fluid filling the cavity between the cornea and lens of the eye.

Arteriole   a tiny artery that eventually branches to become capillaries.

Arteriosclerosis   hardening of the arteries.

Artery   a large, muscular vessel that carries blood away from the heart.

Ascus   In Ascomycetes, the saclike structure that contains the spores.

Asexual reproduction   reproduction without eggs and sperms.

Association areas   nerve processes connecting different parts of the cerebral cortex.

Atom   the smallest unit of an element that can exist alone.

ATP (adenosine triphosphate)   a high-energy compound found in cells that functions in energy storage and transfer.

Atrioventricular valves   the heart valves located between the atria and ventricles.

Atrium   a thin-walled upper chamber of the heart that receives venous blood.

Auditory canal   the passageway from the external ear to the eardrum.

Auditory nerve   the nerve leading from the inner ear to the brain.

Australopithecus   the earliest form of anthropoid ape, whose skull resembles that of a human, discovered in Africa and dating back nearly a million years.

Autonomic nervous system   a division of the nervous system that regulates the vital internal organs.

Autosome   any paired chromosome other than the sex chromosomes.

Autotrophs   organisms capable of organizing organic molecules from inorganic molecules.

Auxin   a plant hormone that regulates growth.

Axon   a nerve process that carries an impulse away from the nerve body.

Bacillus   a rod-shaped bacterium.

Bacteria   a group of microscopic, one-celled protists.

Bacteriophage   one of several kinds of viruses that can destroy bacteria.

Barbituates   a group of depressant drugs known as sleeping pills.

Basal disk   cells of Hydra that secrete a substance to hold it to an object.

Basal metabolism   the quantity of energy used by any organism at rest.

Base   a compound which liberates hydroxyl ions in aqueous solutions; as used in nucleic acid structure, the molecular subunits forming the codons.

Basidium   a club-shaped structure found in the club fungus which bears the spores.

Benign   not malignant.

Benthic zone   bottom part of the marine biome.

Biennial   a plant that lives two seasons.

Big bang theory   that universe started with a big explosion.

Bile   a brownish-green emulsifying fluid secreted by the liver and stored in the gall bladder.

Binary fission   the division of cells into two approximately equal parts.

Biochemical compounds   molecules found in living organisms.

Biogeography   the study of the distribution of plants and animals throughout the earth.

Bioluminescence   the light produced by living organisms, such as the firefly.

Biome   a large geographical region identified mainly by its climax vegetation.

Biosphere   the area in which life is possible on our planet.

Biotic community   all the living organisms in an ecosystem.

Bipinnaria larva   the larva of starfish.

Bivalve   a mollusk possessing a shell of two valves hinged together.

Blastocyst   a group of cells resulting from fertilization that become the embryo.

Blastula   an early stage in the development of an embryo, in which cells have divided to produce a hollow sphere.

Blind spot   the small area, insensitive to light, in the retina of the eye where the optic nerve enters.

Blood   specialized fluid, circulating in the heart, arteries and veins of vertebrates, that transports nutrients to, and waste material from, body tissues.

Bowman's capsule   the cup-shaped structure forming one end of the tubule and surrounding a glomerulus in the nephron of a kidney.

Brain stem   an enlargement at the base of the brain where it connects with the spinal cord.

Breathing   the mechanical process of getting air into and out of the body.

Bronchial tube   a subdivision of a bronchus within a lung.

Bronchiole   one of numerous subdivisions of the bronchial tubes within a lung.

Budding   the uniting of a bud with a stock; also, a form of asexual reproduction in yeast and hydra.

Caeca   small pouches adjacent to the digestive organs.

Callus   a mass of cells, hardened and thickened on the skin.

Calorie   (food) used to measure food energy, the amount of heat required to raise the temperature of 1,000 grams (1 kilogram), or 1 liter, of water one centigrade degree.

Calorimeter   an instrument used to measure heat given off during metabolism.

Cambium   a ring of meristematic cells in roots and stems that forms secondary xylem and phloem.

Capillary   small, thin-walled vessel through which exchanges occur between blood and tissue fluid.

Capsule   a thick slime layer surrounding some bacteria; also, the spore case of mosses.

Carbon-oxygen cycle   the cycle that combines the processes of photosynthesis and respiration.

Carboxyl group   one which dissociates to yield hydrogen ions in water since it contains a COOH attached to a molecule.

Carcinogen   cancer-causing agent.

Cardiac muscle   muscle composing the heart wall.

Cardiovascular   pertaining to the circulatory system in higher animals.

Carnivore   a meat-eating organism.

Carpopedal spasms   twitching and clutching movements during sexual excitement.

Catabolism   the process in a plant or animal by which living tissue is changed into waste products of a simpler chemical composition.

Catalyst   a substance that accelerates a chemical reaction without itself being altered chemically.

Cell   a unit of structure and function of an organism.

Cell body   that part of a nerve cell containing the nucleus.

Cell cycle   the process of a cell's engaging in rest, DNA replication, and mitosis.

Cell sap   the material inside plant cell vacuoles.

Cell theory   a belief that the cell is the unit of structure and function of life and that cells come only from preexisting cells by reproduction.

Cellulose   a polysaccharide composing the cell walls or fibers of all plant tissue.

Cell wall   the outer, nonliving cellulose wall secreted around plant cells.

Central nervous system   the brain and spinal cord and the nerves arising from each.

Centriole   a cytoplasmic body lying just outside the nucleus in animal cells.

Centromere   a single granule that, during cell division, attaches a pair of chromatids after replication.

Cerebellum   the brain region between the cerebrum and medulla, concerned with equilibrium and muscular coordination.

Cerebrospinal fluid   a clear fluid in the brain ventricles and surrounding the spinal cord.

Cerebrum   the largest region of the human brain, considered to be the seat of emotions, intelligence, and voluntary nervous activities; also present in other vertebrates.

Cervix   the neck of the uterus.

Chemical bond   the linkage by which atoms or groups of atoms are combined in molecules.

Chemoreceptors sensory receptors that sense the presence of a variety of chemicals in the environment.

Chemosynthesis the organization of organic compounds by organisms by means of energy from inorganic chemical reactions instead of energy from light.

Chemotherapy the chemical treatment of disease.

Chemotropism the response of an organ or an organism to chemicals.

Chitin a material present in the exoskeleton of insects and other arthropods.

Chlorophyll green pigments essential to food manufacture in plants.

Chloroplast a cell plastid containing chlorophyll.

Cholinesterase a chemical substance released at the motor end plate that neutralizes acetylcholine.

Chordate any of a phylum of animals having at some stage of development a hollow dorsal nerve tract, a notochord, and gill slits. Includes the vertebrates, tunicates, and lancelets.

Chorion a membrane that forms early during development and attaches, in mammals, to the uterine wall. In birds and reptiles, this membrane is found under the shell.

Choroid layer the second and innermost layer of the eyeball.

Chromatid during cell division, each part of a double-stranded chromosome, joined by a centromere, after replication.

Chromatin material fine strands in the nucleoplasm believed to be forms of the chromosomes.

Chromoplasts pigment-containing bodies, other than chloroplasts, in certain plant cells.

Chromosomal aberration an alteration in the structure or number of a chromosome.

Chromosome a rod-shaped gene-bearing body in the cell nucleus, composed of DNA joined to protein molecules.

Chrysalis the hard covering of the pupa of a butterfly.

Cilia tiny, hairlike projections of cytoplasm.

Ciliary muscles those that control the shape of the lens.

Circadian rhythm rhythms associated with the 24-hour cycles of the earth's rotation as, in man, the regular metabolic, glandular, and sleep rhythms.

Circulation the flow of fluids through an organism.

Citric acid cycle (Krebs cycle) the chemical cycle of aerobic respiration.

Class the category of biological classification comprising closely related orders of organisms.

Climax plant a species that assumes final prominence in a region.

Clitoris a small, sensitive, erectile sexual organ in the female at the upper end of the vulva that is the center of sexual pleasure.

Cloning the development of identical organisms from single cells.

Cochlea the hearing apparatus of the inner ear.

Cocoon a silken case containing the pupa of a moth.

Codon one or more groups of base triplets of messenger RNA that code a specific amino acid.

Coelom the space between the mesodermal layers that forms the body cavity of an animal.

Coleoptile a protective sheath encasing the primary leaf of the oat and other grasses.

Collagen a fibrous protein found in connective tissue, bone, and cartilage in animals.

Collar cells flagellated cells in sponges that set up water currents.

Colloid a gelatinous substance, such as protoplasm or egg albumen, in which one or more solids are dispersed through a liquid.

Colonial very similar cells living together but performing different functions.

Colorblindness a genetic disease in which certain colors are perceived as shades of gray.

Commensalism one organism living in or on another, with only one of the two benefiting.

Competitive inhibition the competition by two different chemicals for the same place on a molecule.

Condom a male contraceptive device, a thin protective sheath for the penis, generally made of rubber.

Cones cells in the retina that are involved in color vision.

Conifer a cone-bearing gymnosperm.

Conjugation a primitive form of sexual reproduction in Spirogyra and certain other algae and fungi in which the content of two cells unites; also, an exchange of nuclear substance in the paramecium.

Contractile vacuole a large cavity in protozoans associated with the discharge of water from the cell and the regulating of osmotic pressure.

Contraception the intentional prevention of conception, of fertilization of the human ovum.

Contraction (muscle) the shortening of muscle fibers.

Convergent evolution the type in which organisms of entirely different origin evolve in a manner that results in certain similarities.

Convolution one of many irregular, rounded ridges on the surface of the brain.

Cork a tissue formed by the cork cambium that replaces the epidermis in woody stems and roots.

Cornea a transparent bulge of the sclerotic layer of the eye in front of the iris, through which light rays pass.

Corpora cavernosa the two tubes of tissue above the corpus spongiosum in the penis.

Corpus luteum refers to the follicle after the ovum is discharged.

Corpus spongiosum a tube of tissue that surrounds the male urethra.

Cortex in roots and stems, a storage tissue; in organs such as the kidneys and brain, the outer region.

Cotyledon a seed leaf present in the embryo plant that serves as a food reservoir.

Covalent compound atoms joined together by sharing electrons.

Countershading a form of protective coloration in which darker colors on the upper side of the animal fade into lighter colors on the lower side.

Cowper's gland located near the upper end of the male urethra. It secretes a fluid which is added to the sperms.

Cranial nerves the twelve pairs of nerves communicating directly with the human brain.

Cretinism a condition caused by a defective thyroid gland leading to stunted development.

Cristae membranous infolded partitions in mitochondria.

Crop an organ of the alimentary canal of the earthworm, bird, and certain other animals that serves for food storage.

Cultural evolution the continual change in a society's way of life.

Culture the ideas, customs, skills, arts, etc. of a given people in a given period.

Cuticle a waxy, transparent layer covering the upper epidermis of certain leaves; also, the outer covering of an earthworm.

Cyclosis flowing, or streaming, of cell cytoplasm.

Cytochromes any of several iron-containing enzymes found in almost all animal and plant cells; proteins in the electron transport chain.

Cytokinins a group of plant hormones that influence cell division.

Cytolysis the swelling and bursting of cells when put into a hypotonic medium.

Cytoplasm the protoplasmic materials lying outside the nucleus and inside the cell membrane.

Dark reactions the second stage of photosynthesis whereby carbon dioxide is transformed to glucose.

Daughter cells newly formed cells resulting from the division of a previously existing cell, called a mother cell. The two daughter cells receive identical nuclear materials.

Deamination the breaking down of amino acids into simpler substances.

Decomposers organisms that break down the tissues and excretions of other organisms into simpler substances through the process of decay.

Dehydration loss of water from body tissues.

Dendrite a branching nerve process that carries an impulse toward the nerve cell body.

Denitrification the process carried on by denitrifying bacteria in breaking down ammonia, nitrites, and nitrates and liberating free nitrogen.

Depressant a drug having an anesthetic effect on the nervous system.

Dermis the thick, active layer of tissue lying beneath the epidermis.

Diaphragm a muscular partition separating the thoracic cavity from the abdominal cavity or a contraceptive device.

Diastole part of the cycle of the heart during which the ventricles relax and receive blood.

Dicotyledon a seed plant with two seed leaves, or cotyledons.

Diffusion the spreading out of molecules in a given space from a region of greater concentration to one of lesser concentration.

Digestion the process during which foods are chemically simplified and made soluble for use.

Dihybrid an offspring resulting from parents differing from one another in two pairs of alleles.

Diploid term used to indicate a cell that contains (or an organism whose cells contain) a full set of homologous pairs of chromosomes.

Disaccharide any of a group of sugars with a common formula which on hydrolysis yield two monosaccharides.

Diurnal active during the day.

DNA (deoxyribonucleic acid) a supermolecule consisting of alternating units of nucleotides, composed of deoxyribose sugar, phosphates, and nitrogen bases.

Dominance principle first observed by Mendel, that one gene may prevent the expression of an allele.

Dormancy a period of inactivity.

Dorsal the top of an organism.

Down's Syndrome a genetic disease characterized by mental deficiency, a broad face, slanting eyes, a short fifth finger caused by an additional autosome.

Eardrum the membrane in the ear that vibrates in response to sound waves.

Ecology the study of the relationship of living things to their surroundings.

Ecosystem a unit of the biosphere in which living and nonliving things interact, and in which materials are used over and over again.

Ectoderm the outer layer of cells of a simple animal body; in vertebrates, the layer of cells from which the skin and nervous system develop.

Egg a female reproductive cell.

Electromagnetic spectrum the complete range of frequencies of electromagnetic waves from the lowest to the highest frequencies including radio, infrared, visible light, ultraviolet, X ray, gamma ray, and cosmic ray waves.

Electron a negatively charged particle that moves around an atomic nucleus.

Electron microscope a high resolution microscope.

Electron transport chain the chain of events in the respiration system after the Krebs cycle that produces ATP.

Electrostatic forces the interaction of positive and negative charges.

Element one of nearly 100 types of matter that singly or in combination constitute all substances.

Embryo a developing organism.

Embryo sac the tissue in a plant ovule that contains the egg, the antipodals, the polar nuclei, and the synergids.

Empirical formula a chemical formula which gives the composition of elements in a molecule in their lowest relative proportions but does not specify the structural arrangement or true molecular weight.

Emulsification the breakdown of substances into small droplets.

Endocrine gland a ductless gland that secretes hormones directly into the blood.

Endoderm the inner layer of cells of a simple animal body; in vertebrates, the layer of cells from which the linings of the digestive system, liver, lungs, and so on develop.

Endogenous a biological rhythm that persists in the absence of external cues.

Endometrium the lining of the uterus.

Endoplasmic reticulum a complex system of parallel membranes that extend from the plasma membrane to the nuclear membrane of a cell and that function as a system of canals.

Endorphins naturally occurring proteins found in the brain that have opiate-like properties.

Endoskeleton internal framework of vertebrates composed of bone and/or cartilage.

Endosperm the tissue in some seeds containing stored food.

Endosperm nucleus the body formed by the fusion of a sperm with the polar nuclei during the double fertilization of a spermatophyte.

Endospore a resting stage of some bacteria that is resistant to adverse environmental conditions.

Endotoxins insoluble poisons formed by certain bacteria within the cells.

Energy level a stable state in which the energy of a physical system remains constant.

Enzyme an organic catalyst.

Epicotyl in a seed, the part of the embryo plant that lies above the attachment of the cotyledons and from which the stem and leaves will develop.

Epidermis the outer tissue of a young root or stem, a leaf, other plant parts, or skin.

Epididymis the sperm storage area in the testicles, a long oval-shaped structure attached to the rear upper surface of each testicle.

Epiphyte a plant that grows on a tree or other plant.

Epiglottis a cartilaginous flap at the upper end of the trachea.

Esophagus the food tube, or gullet, that connects the mouth and the stomach.

Ethology the study of behavior patterns in animals.

Euglena a small alga.

Eukaryotic cell a cell possessing a true nucleus separated from the rest of the cell interior by a special membrane.

Eustachian tube   a tube connecting the pharynx with the middle ear.

Eutrophication   an increase in nutrients in fresh waters.

Evaporation   movement of water in the form of water vapor from the earth to the atmosphere.

Evolution   the slow process of change by which organisms have acquired their distinguishing characteristics.

Excretion   the process by which metabolic waste materials are removed from living cells or from the body.

Excurrent pore   in sponges, the osculum.

Excurrent siphon   the structure in the clam through which water passes out of the body.

Exoskeleton   the hard outer covering or skeleton of certain animals, especially arthropods.

Exotoxin   a soluble toxin excreted by certain bacteria and absorbed by the tissues of the host.

Expiration   the discharge of air from the lungs.

External ear   the outside of the entire ear.

Eyespot   the sensory structure in the euglena and planarian that is believed to perceive light and dark.

Fallopian tube   oviduct in the mammal.

Family   the category of biological classification comprising closed related genera of organisms.

Feces   intestinal solid waste material.

Feedback   an operating mechanism in the body that produces a delicate check-and-balance system.

Fermentation   glucose oxidation that is anaerobic and in which lactic acid or alcohol is formed.

Fertilization   the union of sperm and egg.

Fertilization membrane   the membrane that forms around an egg immediately after fertilization.

Fetus   mammalian embryo after the main body features are apparent.

Fibrin   a substance formed during blood clotting by the union of thrombin and fibrinogen.

Fibrinogen   a blood protein present in the plasma, involved in clotting.

Filament   a stalk of a stamen, bearing the anther at its tip; in algae, a threadlike group of cells.

Fission   the splitting of the nuclei of atoms into two fragments of approximately equal mass, accompanied by conversion of part of the mass into energy.

Flagellate   an organism bearing one or more whiplike appendages, or flagella.

Flagellum   a whiplike projection of cytoplasm used in locomotion by certain simple organisms and by the sperms of many multicellular organisms.

Follicle   an indentation in the skin from which hair grows; also, a mass of ovarian cells that produces an ovum.

Food   any substance absorbed into the cells of the body that yields material for energy, growth, and repair of tissue and regulation of the life processes, without harming the organism.

Food chain   the transfer of the sun's energy from producers to consumers as organisms feed on one another.

Food pyramid   a quantitative representation of a food chain, with the food producers forming the base and the top carnivore at the apex.

Food vacuole   a cavity associated with the intake and expulsion of food in cells.

Food web   complex food chains existing within an ecosystem.

Fossil   the imprint or preserved remains of an organism that once lived.

Fovea   a small, sensitive spot on the retina of the eye where cones are specially abundant.

Fraternal twins   twins developing from two eggs and two sperms.

Frequency   The number of vibrations per second.

Fruit   The ripe ovary of a seed plant.

Fungi   a kingdom of thallophytes, including molds, mildews, mushrooms, rusts, and smuts, which are parasites on living organisms or feed upon dead organic material, lack chlorophyll, true roots, stems, and leaves, and reproduce by means of spores.

Fusion   the merging of lightweight atomic nuclei into a nucleus of heavier mass with a resultant loss in the combined mass which is converted into energy.

Gall bladder   a sac in which bile from the liver is stored and concentrated.

Gamete   a male or female reproductive cell, or germ cell.

Gametophyte   the stage that produces gametes in an organism having alternation of generations.

Ganglion   a mass of nerve cells lying outside the central nervous system.

Gastric fluid   glandular secretions of the stomach.

Gastrula   a stage in embryonic development during which the primary germ layers are formed.

Germmule   a coated cell mass produced by the parent sponge and capable of growing into an adult sponge.

Gene   that portion of a DNA molecule that is genetically active and capable of replication and mutation.

Gene pool   all the genes present in a given population.

Generative nucleus   the nucleus in a pollen grain that divides to form two sperm nuclei.

Genetic code   the sequential arrangement of the bases in the DNA molecule, which controls traits of an organism.

Genetics   the science of heredity.

Genitalia   sex organs.

Genotype   the hereditary makeup of an organism.

Genus   the category of biological classification comprising closely related species of organisms.

Geotropism   the response of plants to gravity.

Gibberellins   growth regulating substances promoting cell elongation in plants.

Gill   an organ modified for absorbing dissolved oxygen from water; in mushrooms, a platelike structure bearing the reproductive hyphae and spores.

Gizzard   an organ in the digestive system of the earthworm and birds modified for grinding food.

Glomerulus   the knob of capillaries in a Bowman's capsule.

Glycogen   a partially soluble, starchlike substance produced in animal tissues and changed into a simple sugar as the body needs it.

Glycolysis   the conversion of glucose to pyrmic acid by an organism when ample oxygen is not available; that is, under anaerobic conditions.

Goiter   a condition caused by iodine deficiency.

Golgi complex   small groups of parallel membranes in the cytoplasm near the nucleus.

Gonadotropins   group of hormones secreted by the pituitary which supports and stimulates the function and growth of the gonads. Hormones that affect male and female sexual development by supporting and stimulating the function and growth of the gonads.

Gonads   the male and female reproductive organs in which the gametes are produced.

Grafting   the union of the cambium layers of two woody stems, one of the stock and the other of the scion.

Grana   disklike bodies in chloroplasts.

Gravitation   the attraction between matter in the universe.

Guard cell   one of the two epidermal cells surrounding a stoma.

Gullet the passageway to a food vacuole in paramecia; the food tube or esophagus.

Guttation the forcing of water from the leaves of plants, usually when the stomata are closed.

Gymnosperm any of a large class of seed plants having the ovules born on open scales, usually in cones, and usually lacking true vessels in the woody tissue. Includes the pines, spruces, cedars and others.

Habitat place where an organism lives.

Half-life the time it takes for half of a radioactive substance to disintegrate.

Haploid a term used to indicate a cell, such as a gamete, that contains only one chromosome of each homologous pair; also, an organism having cells of this type.

Hemoglobin an iron-containing protein compound giving red corpuscles their color.

Hemophilia a hereditary blood disease in which the normal blood-clotting factor is absent.

Hemorrhage heavy bleeding.

Herbivores plant-eating animals.

Heredity the transmission of traits from parents to offspring.

Hermaphroditic having the organs of both sexes.

Heterogametes male and female gametes that are unlike in appearance and structure.

Heterotrophs organisms that are unable to synthesize organic molecules from inorganic molecules; that is, nutritionally dependent on other organisms or their products.

Heterozygous refers to an organism in which the paired genes for a particular trait are different.

Hibernate to spend the winter months in an inactive condition.

Hilum the scar on a seed where it was attached to the ovary wall.

Holdfast the special cell at the base of certain algae that anchors them to the substrate.

Homeostasis a steady state that an organism maintains by self-regulating adjustments.

Homeothermic warm-blooded, as applied to birds and mammals.

Homo habilis an advanced hominid, generally recognized as a precursor of modern man.

Homologous organs those similar in origin and structure but not necessarily in function.

Homo sapiens the species name for modern man.

Homozygous refers to an organism in which the paired genes for a particular trait are identical.

Hormone the chemical secretion of a ductless gland producing a definite physiological effect.

Humus organic matter in the soil formed by the decomposition of plant and animal remains.

Hybrid an offspring from a cross between parents differing in one or more traits.

Hybridization (or outbreeding) the crossing of different strains, varieties, or species to establish new genetic characteristics.

Hydrogen bonding the linking together of certain atoms with hydrogen atoms.

Hydrolysis the chemical breakdown of a substance by combination with water.

Hydronium ion the H_3O^+ ion.

Hydrophytes plants that grow in water or partially submerged in water in very wet surroundings.

Hydroxyl ion the OH^- ion.

Hypertonic solution a solution containing a higher concentration of solutes and a lower concentration of water molecules than another solution.

Hypha a threadlike filament of the vegetative body of a fungus.

Hypocotyl that part of a plant embryo from whose lower end the root develops.

Hypothalamus the part of the brain that forms the third ventricle and regulates many body functions including hormone secretion.

Hypothesis a scientific idea or working theory.

Hypotonic solution a solution containing a lower concentration of solutes and a higher concentration of water molecules than another solution.

Identical twins twins developing from a single egg and sperm.

Immune therapy the assistance and stimulation of the natural body defenses in preventing disease.

Immunity the ability of the body to resist a disease by natural or artificial means.

Immunological memory the process whereby antibodies are produced faster when exposed to an antigen for the second time.

Imprinting a learning mechanism operating very early in the life of an animal in which a particular stimulus immediately establishes an irreversible behavior pattern with reference to the same stimulus in the future.

Incomplete dominance a blend of two traits, resulting from a cross of these characteristics.

Incurrent pore one of many holes in the sponge through which water passes into the animal.

Independent assortment, law of a law based on Mendel's hypothesis that the separation of gene pairs on a given pair of chromosomes, and the distribution of the genes to gametes during meiosis, are entirely independent of the distribution of other gene pairs on other pairs of chromosomes.

Industrial melanism the process of color change in an organism as a result of environmental changes caused by industry.

Inert nonreacting.

Innate behavior inborn behavior.

Inner ear that part of the ear containing the cochlea and semicircular canals.

Instinct a natural urge, or drive, not depending on experience or intelligence.

Intelligence quotient I.Q., a supposed measure of human intelligence.

Interferon a cellular chemical defense against a virus.

Interphase the period of growth of a cell that occurs between mitotic divisions.

Interstitial cells the site of testosterone production in the testes.

Intertidal zone the area of the beach that is periodically covered and uncovered by water.

Intestinal fluid a digestive secretion of the intestinal glands.

Intrauterine device (I.U.D.) a female contraceptive device inserted into the uterus.

Intromission insertion of a penis into a vagina.

Invertebrate an animal lacking a backbone.

Ion a charged atom.

Ionic compound atoms attracted together because of charge differences.

Iris the muscular, colored portion of the eye, behind the cornea and surrounding the pupil.

Irritability the ability to respond to a stimulus.

Islets of Langerhans groups of cells in the pancreas that secrete insulin.

Isogametes male and female gametes that are structurally similar.

Isotonic solution a solution containing the same concentration of solutes and the same concentration of water molecules as another solution.

442

Isotopes different forms of an element resulting from varying numbers of neutrons.

Kidney an excretory organ that excretes urine.

Kingdom the broadest division of organisms in biological classification.

Kinin (also cytokinin) a plant hormone, but one not transported out of the cell.

Lacteal a lymph vessel that absorbs the end products of fat digestion from the intestinal wall.

Large intestine the part of the intestine that stores wastes and leads to the anus.

Larva an immature stage in the life of an animal.

Larynx the voice box; also called the "Adam's apple."

Layering propagation by stimulating the growth of roots on a stem, as in burying a stem in the ground and then cutting it when roots form.

Legume any of a large family of herbs, shrubs, and trees that usually have compound leaves, flowers having a single carpel, and fruit that is a dry pod splitting along two sutres. Many are nitrogen-fixing.

Leucoplast a colorless plastid serving as a food reservoir in certain plant cells.

Light reactions the first stage of photosynthesis where light energy is used to form ATP from ADP and to reduce electron carrier molecules.

Lignin an amorphous, cellulose-like, organic substance which acts as a binder for the cellulose fibers in wood and certain plants and adds strength and stiffness to the cell walls.

Limiting factor any factor that is essential to organisms and for which there is competition.

Littoral zone the area above the sea floor.

Liver the largest gland in the human body, associated with digestion and sugar metabolism.

Lobe a division of the brain.

Lunar rhythms biological rhythms influenced by the position and/or phases of the moon.

Lung an organ for air breathing and external respiration in higher animals.

Lymph the clear, liquid part of blood that enters the tissue spaces and lymph vessels.

Lymphocyte a variety of leukocyte formed in lymphatic tissue.

Lysis the dissolution or destruction of cells.

Lysosome a spherical body within the cytoplasm of most cells.

Lysozyme an enzyme that dissolves the cell walls of many bacteria.

Lytic cycle the cycle of a virulent phage resulting in destruction of a bacterial cell.

Macronucleus the large nucleus of the paramecium and certain other protozoans.

Malpighian tubules long excretory tubules attached to the junction of the stomach and intestine of the grasshopper, and that collect nitrogenous wastes from the blood.

Mammal a class of higher vertebrates characterized by hairiness, mammary glands to nurse their young, and constant body temperature.

Mantle a thin membrane covering the visceral hump of a mollusk; in some, it secretes a shell.

Marine inhabitants of salt water.

Medulla in the kidney, the inner portion containing pyramids that, in turn, contain numerous tubules; in the adrenal gland, the inner portion that secretes epinephrine.

Medulla oblongata the enlargement at the upper end of the spinal cord, at the base of the brain.

Medusa the bell-shaped, free-swimming form in the jellyfish.

Megaspores four cells formed from the megaspore mother cell, three of which disintegrate and one of which develops into the embryo sac.

Meiosis the type of cell division in which during oogenesis and spermatogenesis there is reduction of chromosomes to the haploid number.

Meninges the three membranes covering the brain and spinal cord.

Menstruation the periodic breakdown and discharge of uterine tissues that occur in the absence of fertilization.

Meristematic tissue small, actively dividing cells that produce growth in plants.

Mesentery a folded membrane that connects to the intestines and the dorsal body wall.

Mesoderm the middle layer of cells in an embryo.

Mesoglea a jellylike material between the two cell layers composing the body of a coelenterate.

Mesophytes plants that occupy neither extremely wet nor extremely dry surroundings.

Messenger RNA the type of RNA that is thought to receive a code for a specific protein from the DNA in the nucleus and to act as a template for protein synthesis on the ribosome.

Metabolic water water produced within an organism.

Metabolism the sum of the chemical processes of the body.

Metamorphosis a marked change in structure of an animal during its growth.

Metaphase the stage of mitosis in which the chromosomes line up at the equator.

Metathorax the posterior portion of the thorax of an insect, bearing the third pair of legs and the second pair of wings.

Microfilaments 50-Å diameter protein filaments found in the cytoplasm.

Micronucleus a small nucleus found in the paramecium and certain other protozoans.

Micropyle the opening in the ovule wall through which the pollen tube enters.

Microspore mother cells diploid cells in the anther that divide twice, forming four haploid microspores.

Microspores four cells, formed from the microspore mother cell, that develop into pollen grains.

Microtubules 300-Å diameter tubes found in the cytoplasm.

Middle ear the part of the ear behind the eardrum and containing the Eustachian tube.

Middle lamella a thin plate, composed largely of pectin, forming the middle portion of a wall between two adjacent plant cells.

Migration seasonal movement of animals from one place to another.

Mimicry a form of protective coloration in which an animal closely resembles another kind of animal or an object in its envrionment.

Mitochondria rod-shaped bodies in the cytoplasm known to be centers of cellular respiration.

Mitosis the division of chromosomes preceding the division of cytoplasm.

Molecule the smallest unit of a compound substance consisting of two or more atoms.

Mollusc invertebrate animal characterized by a soft, unsegmented body usually enclosed in a shell.

Molting shedding of the outer layer of exoskeleton of arthropods, or of a scale layer of reptiles.

Monera a kingdom of organisms that is made up of bacteria and blue-green algae.

Monocotyledon a flowering plant that develops a single seed leaf, or cotyledon.

Monoecious bearing staminate and pistillate flowers on different parts of the same plant.

Monohybrid an offspring from a cross between parents differing in one trait.

Monosaccharide a carbohydrate not decomposable by hydrolysis; simple sugar.

Mosaic an irregular pattern.

Mother cell a cell that has undergone growth and is ready to divide.

Motor area the area of the brain that controls motor (movement) events.

Motor end plate the terminus of the axon of a motor nerve in a muscle.

Motor neuron one that carries impulses from the brain or spinal cord to a muscle or gland.

Motor unit the nerve cell and the individual muscle fibers it stimulates to contract.

Mucous membrane a form of epithelial tissue that lines the body openings and digestive tract and secretes mucus.

Mucus a slimy secretion of mucous glands.

Mullerian ducts forerunners of Fallopian tubes.

Multiple alleles one of two or more pairs of genes that act together to produce a specific trait.

Muscle tissue cells that are specialized to contract and cause movement.

Mutagen any agent or substance that is capable of noticeably increasing the frequency of mutation in DNA.

Mutation a change in genetic makeup resulting in a new characteristic that can be inherited.

Mutualism a form of symbiosis in which two organisms live together to the advantage of both.

Myofibrils fine, parallel threads arranged in groups to form a muscle fiber.

Myosin a form of thick filament that together with actin filaments composes a myofibril.

Myotonia prolonged muscular spasm, often a manifestation of certain diseases of muscles.

Narcotics a group of drugs that have a pronounced effect on the nervous system and that are addictive with continued use.

Nastic movement turgor movement in plants, such as the daily opening and closing of flowers.

Natural selection the result of survival in the struggle for existence among organisms possessing those characteristics that give them an advantage.

Nematocyst a stinging cell in coelenterates.

Nephridia the excretory structures in worms, mollusks, and certain arthropods.

Nephron one of the numerous excretory structures in the kidney, including Bowman's capsule, the glomerulus, and the tubules.

Nerve impulse an electrochemical stimulus causing change in a nerve fiber.

Nerve net a group of nerve cells.

Nervous tissue specialized cells for transmitting impulses for coordination, perception, or automatic body functions.

Neuron a nerve cell body and its processes.

Neurotoxin a poison that affects the parts of the nervous system that control breathing and heart action.

Neutron a particle in the atomic nucleus with neither a positive nor negative charge.

Niche the particular role played by organisms of a species in relation to those of other species in a community. No two species occupy the same niche.

Nitrification the action of a group of soil bacteria on ammonia, producing nitrates.

Nitrogen cycle a series of chemical reactions in which nitrogen compounds change form.

Nitrogen fixation the process by which certain bacteria in soil or on the roots of leguminous plants convert free nitrogen into nitrogen compounds that the plant can use.

Nocturnal an organism active at night.

Notochord a rod of cartilage running longitudinally along the dorsal side of lower chordates and always present in the early embryological stages of vertebrates.

Nuclear membrane a living membrane surrounding the nucleus.

Nucleic acid an organic acid composed of nucleotides; principally DNA (dioxyribonucleic acid) and RNA (ribonucleic acid).

Nucleolus a small, spherical body within the nucleus.

Nucleoplasm the dense, gelationous living content of the nucleus.

Nucleoprotein a complex biomolecule consisting of nucleic acids and proteins.

Nucleotide a unit composed of a ribose or deoxyribose sugar, a phosphate, and an organic base. Many such units make up an RNA or DNA molecule.

Nucleus the part of the cell that contains chromosomes; also, the central mass of an atom, containing protons and neutrons.

nymph an early stage in the development of an insect.

Olfactory nerve the nerve leading from the olfactory receptor endings to the olfactory lobe.

Omnivore animals that eat both plants and animals.

Oocyte an egg that has not yet undergone maturation.

Oogenesis the process of the development of female reproductive cells whereby the diploid chromosome number is reduced to the haploid.

Oogonium an egg-producing cell in certain thallophytes.

Ootid a cell that matures into an egg.

Operon a group of adjacent genes with related functions acting as a coordinated unit controlled by a regulatory gene.

Optic lobe the region of the brain that registers sight.

Optic nerve the nerve leading from the retina of the eye to the optic lobe of the brain.

Oral groove a deep cavity along one side of the paramecium and similar protozoans.

Orbital the path of an electron around an atom.

Order the category of biological classification comprising related families of organisms.

Organ different tissues grouped together to perform a function or functions.

Organelles specialized structures present in the cell.

Organism a complete and entire living thing.

Orgasmic platform the first third of the vagina.

Os the connection of the vagina with the uterus.

Osmosis the diffusion of water through a semipermeable membrane from a region of greater concentration of water to a region of lesser concentration.

Osmotic system the separation of two different solutions by a selectively permeable membrane.

Ovary the basal part of the pistil containing the ovules; a female reproductive organ.

Ovipositor an egg-laying organ in insects.

Ovule a structure in the ovary of a flower that can become a seed when the egg is fertilized.

Ovum   egg.

Oxidative reaction   a chemical reaction involving the addition of oxygen to a chemical.

Oxygen debt   the buildup of lactic acid in muscles when oxygen is lacking.

Paleontologist   one who studies ancient life forms.

Pancreas   a gland located near the stomach and duodenum that is both endocrine and digestive.

Pancreatic fluid   a digestive secretion of the pancreas.

Paramecium   any of a genus of one-celled, elongated, slipper-shaped protozoans moving by means of cilia and having a backward-curving oral groove that ends with the mouth.

Parasite   an organism that takes nourishment from a living host.

Parasympathetic nervous system   a division of the autonomic nervous system.

Parathyroid   one of the four small ductless glands embedded in the thyroid.

Parenchyma   the thin-walled, soft tissue in plants forming cortex and pith.

Parthenogenesis   the development of an egg without fertilization.

Passive transport   the movement of molecules by their own energy during diffusion.

Pathogenic   disease-causing.

Peat   the organic remains of aquatic plants.

Pectin   water-soluble carbohydrate, obtained from certain ripe fruit, which yields a gel that is the basis of jellies and jams. A polysaccharide found intercellularly in plants.

Pelagic zone   the deep part of the marine biome.

Pellicle   a thickened membrane surrounding the cell of a paramecium.

Pelvis   the hip girdle; in man consisting of the ilium, ischium, and pubis bones; also, the central portion of a kidney.

Peptide   another name for a protein; any of a group of compounds formed from two or more amino acids by the linkage of amino groups or some of the acids with carboxyl groups of others, or by the hydrolysis of proteins.

Peptide bond   the bond formed when a carboxyl group of one molecule of an amino acid is condensed with an amino group of a second molecule to form proteins.

Peptones   initial breakdown products of proteins.

Perennials   plants that grow through more than two growing seasons.

Pericardium   the membrane around the heart.

Periodicity   alternating periods of activity.

Peripheral nervous system   the nerves communicating with the central nervous system and other parts of the body.

Peristalsis   the rhythmic, wavelike motion of the walls of the hollow organs consisting of alternate contractions and dilations of transverse and longitudinal muscles that move the contents of the tube onward.

Permeable membrane   one that allows substances to pass through it.

Perspiration   the excretion of waste products through the skin.

Petal   one of the colored parts of the flower. (In some flowers the sepals are also colored.)

pH   the measure of acidity and alkalinity.

Phage   a bacteriophage, or virus, that reproduces in a bacterium.

Phagocytic cells   those that engulf bacteria and digest them by means of enzymes.

Pharynx   the muscular throat cavity extending up over the soft palate and to the nasal cavity in vertebrates.

Phenotype   the outward appearance of an organism as the result of gene action.

Pheromone   a chemical signal used between members of the same species to convey information and produce specific responses.

Phloem   the tissue in leaves, stems, and roots that conducts dissolved food substances.

Photon   a quantum of electromagnetic energy having both particle and wave behavior: the energy of light, X-rays, gamma rays is carried by photons.

Photoperiodism   the dependence of some plants on the relation between the length of light and the length of darkness in a given day.

Photoreceptor   an organ that is sensitive to light.

Photosensitive   responsive to radiant energy, especially to light.

Photosynthesis   the process by which certain living plant cells combine carbon dioxide and water in the presence of chlorophyll and light energy, to form carbohydrates and release oxygen.

Phototropism   the response of plants to light.

Phylogenetic tree   a scheme used to show evolutionary stages of development.

Phylum   the category of biological classification comprising closely related classes of organisms.

Pinocytosis   the engulfing of large particles in pockets in a cell membrane.

Pistil   the part of a flower bearing the ovary at its base.

Pith   a storage tissue of roots and stems consisting of thin-walled parenchyma cells.

Pithecanthropus   Java man, an extinct humanlike primate.

Pituitary gland   a ductless gland composed of two lobes, located beneath the cerebrum.

Placenta   a large, thin membrane in the uterus, in the area of the chorionic villi, that transports substances between the mother and developing young by means of the umbilical cord.

Planarian   any of a family or order of related small, soft-bodied, free-living, turbellarian flatworms moving by means of cilia.

Plankton   minute floating organisms suspended near the surface in a body of water that serve as food for larger animals.

Plaques   holes in a colony of bacteria resulting from destruction of cells by a bacteriophage or clusters of cells.

Plasma   the liquid portion of blood tissue.

Plasma cell   a type of white blood cell involved in immunological memory.

Plasma membrane   a thin, living membrane located at the outer edge of the cytoplasm.

Plasmodium   name given to the vegetative body of a typical slime mold.

Plasmolysis   the collapse of cell protoplasm due to loss of water.

Plastids   living bodies in the cytoplasm of plant cells.

Platelet   the smallest of the solid components in the blood, releasing thromboplastin in clotting.

Pleiotropic   a gene's affecting more than one trait in an organism.

Pleural membrane   one of two membranes surrounding each lung.

Poikilothermic   cold-blooded, in reference to certain animals.

Polar body   a cell that decomposes during egg development.

Polar compound   a molecule with uneven charge distribution.

Polarization   a charge difference across cells.

Polar nuclei   the two nuclei in the embryo sac in flowers that fuse with one of the sperm nuclei to form the endosperm nucleus.

Pollen   the microgametophyte produced in the anther of a spermatophyte.

Pollen sacs   structures in the anther containing pollen grains.

Pollen tube   the tube formed by a pollen grain when it grows down the style of a pistil.

Pollination   the transfer of pollen from anther to stigma.

Polymerization   the process of joining two or more like molecules to form a more complex one.

Polyp   one of the stages in the life cycle of coelenterates.

Polyssaccharide   any of a group of complex carbohydrates that decompose by hydrolysis into a large number of monosaccharide units.

Population   a group of individuals of any one kind of organism in a given ecosystem.

Posterior   the back end of an organism.

Predator   any animal that preys on other animals.

Primary germ layers   the ectoderm, endoderm, and mesoderm.

Primate   member of the order of mammals comprising humans, apes, monkeys and related forms.

Proglottid   a segment of a tapeworm's body.

Prokaryotic cell   a very simple cell having no truly developed nucleus or separate membrane.

Prophase   the stage of mitosis in which chromosomes shorten and appear distinctly double and the nuclear membrane disappears.

Protective coloration   an animal's coloring that blends with its surroundings to enable it to escape the notice of predators.

Protein synthesis   a universal phase of cell anabolism whereby protein molecules are built up from amino acid molecules.

Protista   a kingdom of one-celled organisms having characteristics found in both plants and animals. Includes the algae, yeasts, bacteria, and protozoans.

Proton   a positively charged particle in an atomic nucleus.

Protoplasm   organized complex system of substances found in living organisms.

Pseudocoel   a cavity or false coelom that is not lined with specialized covering cells.

Psychedelic drugs   a class of drugs, including LSD, that affect the mind by changing the perception of all the senses, the interpretation of space and time, and the rate and content of thought.

Ptyalin   a digestive enzyme secreted into the mouth that converts starch into various dextrins and maltose.

Puberty   the age at which the secondary sex characteristics appear.

Pulmonary   pertaining to the lungs.

Punnett square   a grid system used in computing possible combinations of genes resulting from random fertilization.

Pupa   the stage in an insect having complete metamorphosis that follows the larva stage.

Pupil   the opening in the front of the eyeball, the size of which is controlled by the iris.

Purine   a nitrogenous base that is one of the components of nucleic acids.

Pyloric valve   a sphincter valve regulating the passing of substances from the stomach to the duodenum.

Race   a division of mankind into those possessing similar characteristics which are distinct from those of other human types.

Radicle   embryonic root in a seed.

Radioactive   refers to an element that spontaneously gives off radiations.

Radioautography   a method of labeling cells with an isotope, and then exposing a photographic plate to show the location of the radioactive material.

Receptor   a cell, or group of cells, that receives a stimulus.

Recessive   refers to a gene or character that is masked when a dominant allele is present.

Recombinant DNA   DNA that has recombined with the DNA of another organism.

Red corpuscles (or erythrocytes)   disk-shaped blood cells containing hemoglobin.

Red marrow   found in flat bones and the ends of long bones; forms red corpuscles and certain white corpuscles.

Reflex action   a nervous reaction in which a stimulus causes the passage of a sensory nerve impulse to the spinal cord or brain, from which, involuntarily, a motor impulse is transmitted to a muscle or gland.

Refractory period   a time during which a previous response cannot be repeated.

Releasers   anything that releases a specific pattern of innate behavior.

Renal   relating to the kidneys.

Replication   self-duplication, or the process whereby a DNA molecule makes an exact duplicate of itself.

Reptiles   air-breathing vertebrate usually characterized by scales or bony plates.

Respiration   the exchange of oxygen and carbon dioxide between cells and their surroundings, accompanied by oxidation and energy release.

Response   the reaction to a stimulus.

Retina   the inner layer of the eyeball, formed from the expanded end of the optic nerve.

Rh factor   any one of six or more protein substances found on the surface of the red blood cells of most people.

Rhizoid   a rootlike structure that carries on absorption.

Rhizome   horizontal underground stem, often enlarged for storage, that can carry on vegetative propagation.

Rhythmic   regular periodicity in organisms.

Ribosomal RNA   the RNA that forms part of the ribosome.

Ribosomes   tiny, dense granules attached to the endoplasmic reticulum and lying between its folds. They contain RNA and protein-synthesizing enzymes. Also found free.

RNA (ribonucleic acid)   a nucleic acid in which the sugar is ribose. A product of DNA, it serves in controlling certain cell activities, including protein synthesis.

Rod   a cell of the retina of the eye that receives impulses from light rays and that is sensitive to shades but not to colors.

Saliva   a fluid secreted into the mouth by the salivary glands.

Salivary gland   a group of secretory cells producing saliva.

saprophyte   an organism that lives on dead or nonliving organic matter.

Sarcoplasmic reticulum   specialized area within a muscle cell from which calcium ions move.

Savannah   a treeless plain or a grassland characterized by scattered trees, especially in tropical or subtropical regions having seasonal rains.

Scanning electron microscope   a type of electron microscope designed to observe cell surfaces.

Scavengers   animals that feed on dead organisms.

Scion   the portion of a twig grafted onto a rooted stock.

Sclerenchyma   a plant-strengthening tissue, including fibers, mechanical tissue, and stone cells.

Sclerotic layer   the outer layer of the wall of the eyeball.

Scolex   knob-shaped head with hooks or suckers, as seen on some parasitic flatworms.

Scrotum   the pouch outside the body that contains the testes.

Second law of thermodynamics   states that all natural processes tend to proceed in such a way that disorder increases.

Secretion   formation of essential chemical substances by cells.

Sediment   deposits of debris of dead organisms mixed with soil settling to the bottom of a liquid.

Seed a complete embryo plant surrounded by an endosperm and protected by seed coats.

Segregation, law of Mendel's first law, based on his third hypothesis, stating that a pair of factors (genes) is segregated, or separated, during the formation of gametes (spores in lower plants).

Selectively permeable membrane one that lets substances pass through more readily than other substances.

Semen fertilizing fluid consisting of sperms and fluids from the seminal vesicle, prostate gland, and Cowper's gland.

Semicircular canals the three curved passages in the inner ear that are associated with balance.

Semilunar valves cup-shaped valves at the base of the aorta and the pulmonary artery that prevent backflow into the heart ventricles.

Seminal receptacles structures that receive sperm cells in certain animals.

Seminal vesicles structures that store sperm cells in certain animals.

Seminiferous tubules a mass of coiled tubes in which the sperms are formed within the testes.

Sensory area the area of the brain that controls sensations.

Sensory neurons those that carry impulses from a receptor to the spinal cord or brain.

Sepal the outermost part of a flower, usually green and not involved in the reproductive process.

Septum a wall separating two cavities or masses of tissues; as, the nasal or heart septum.

Serum a substance (usually an extract of blood containing antibodies) used in treating disease and to produce immediate passive immunity.

Sessile a leaf lacking a petiole.

Setae bristles on the earthworm used in locomotion.

Sex chromosomes the two kinds of chromosomes (X and Y) that determine the sex of an offspring.

Sex-limited character a characteristic that develops only in the presence of sex hormones.

Sex-linked character a recessive characteristic carried on the X type of sex chromosome.

Sexual reproduction that involving the union of a female gamete, or egg, and a male gamete, or sperm.

Sieve plate in the starfish, the opening of the water-vascular system to the outside.

Sign stimuli stimuli that release innate behaviors.

Silk scar the point at which the stalk of a pistil is attached to a kernel.

Skeletal muscle that which is striated and voluntary.

Slime layer that which surrounds a bacterium.

Small intestine the digestive tube, about twenty-three feet long in man, that begins at the pylorus.

Smooth muscle that which is involuntary and that is found lining the walls of the intestine, stomach, and arteries.

Sociobiology the study of the biological basis of social behavior.

Solute the dissolved substance in a solution.

Solvent the dissolving component of a solution.

Somatic any of the cells of an organism that become differentiated into the tissues, organs, etc. of the body; not a reproductive or germ cell.

Speciation the development of a species.

Species a group of related organisms capable of interbreeding and distinct reproductively from other similar groups.

Specific heat the heat necessary to raise 1 gram of water 1°C.

Sperm a male reproductive cell.

Spermatid a structure formed from a secondary spermatocyte that matures into a sperm.

Spermatocyte (primary) a structure formed by meiosis from a spermatogonial cell; (secondary) a structure formed by division of a primary spermatocyte.

Spermatogenesis the process of the development of male reproductive cells whereby the diploid chromosome number is reduced to the haploid.

Spicule the material forming the skeleton of certain sponges.

Spinal cord the main dorsal nerve of the central nervous system in vertebrates, extending down the back from the medulla.

Spinal nerves large nerves connecting the spinal cord with various parts of the body.

Spindle the numerous fine threads formed between the poles of the nucleus during mitosis.

Spiracles external openings of the insect's tracheal tubes on the thorax and abdomen.

Spontaneous generation a disproved belief that certain nonliving or dead materials could be transformed into living organisms.

Sporangium a structure that produces spores.

Spore an asexual reproductive cell.

Sporophyte the stage that produces spores in an organism having alternation of generations.

Stamen the male reproductive part of the flower bearing an anther at its tip.

Statolith starch granules in plants that respond to gravity.

Steady-state theory states that the universe always existed and maintains itself with no net change.

Sterile incapable of producing others of its kind.

Sterilization elimination of the ability to reproduce, either by removing the organs of reproduction or by preventing them from functioning effectively.

Stigma the part of the pistil that receives pollen grains.

Stimulant an agent that increases or elevates body activity.

Stimulus a factor or environmental change capable of producing activity in protoplasm.

Stipe the stalk portion of a fruiting body of a mushroom.

Stock the plant on which a scion has been grafted; a line of descent.

Stomach an organ that receives ingested food, prepares it for digestion, and begins protein digestion.

Stomata pores regulating the passage of air and water vapor to and from the leaf.

Strata layers of rock and soil.

Streaming the movement of cytoplasm within cells.

Substrate a layer or substance an organism takes root in or adheres to; what an enzyme acts on.

Succession the changing plant and animal populations of a given area.

Suspension a mixture formed by particles that are larger than ions or molecules.

Swimmerets appendages on the abdomen of a crustacean.

Symbiosis the relationship in which two organisms live together for the advantage of each.

Symmetry similarity of form or arrangement on either side of a dividing line or plane.

Sympathetic nervous system a division of the autonomic nervous system.

Synapse the space between nerve endings.

Synapsis the coming together of homologous pairs of chromosomes during meiosis.

Synaptic vesicles small bodies on the presynaptic side of a neuron containing the neurotransmitters.

Synergid one of two structures formed on either side of the egg in the embryo sac of flowers.

Systemic relating to the body.

Systole part of the cycle of the heart during which the ventricles contract and force blood into the arteries.

Systolic blood pressure arterial pressure produced when the ventricles contract.

Taste buds flask-shaped structures in the tongue containing nerve endings that are stimulated by flavors.

Taxonomy the science of classification of living things.

Telophase the last stage of mitosis, during which two daughter cells are formed.

Tentacle a long appendage, or "feeler," of certain invertebrates.

Tenting the expansion of the vagina during the excitement sexual phase.

Terrestrial land-living.

Testa the outer seed coat.

Testes the male reproductive organs of higher animals.

Thallus a plant body, such as an alga, that lacks differentiation into stems, leaves, and roots and that does not grow from an apical point.

Theory a hypothesis that is continually supported by experimental evidence.

Thigmotropism a response of an organ or an organism to touching an object; for example, the wrapping of the tendrils of a climbing plant around an object.

Thorax the middle region of the body of an insect between the head and abdomen; the chest region of mammals.

Thrombin a substance formed in blood clotting as a result of the reaction of prothrombin, thromboplastin, and calcium.

Thromoplastin a substance essential to blood clotting formed by disintegration of blood platelets.

Thyroid the ductless gland, located in the neck on either side of the larynx, that regulates metabolism.

Tissue a group of cells that are similar in structure and function.

Tissue culture the process or technique of growing cells in the laboratory in a special, sterile culture medium.

Tolerance adaptation to a drug from regular use.

Tonsil a mass of lymphatic tissue in the throat of higher animals.

Toxin a poisonous substance produced by bacteria and other organisms that acts in the body or on foods.

Toxin-antitoxin a mixture of diptheria antitoxin and toxin, formerly used to develop active immunity.

Toxoid toxin weakened by mixing with formaldehyde or salt solution, used extensively to develop immunity to diphtheria, scarlet fever, and tetanus.

Trachea an air tube in insects and spiders; the windpipe in air-breathing vertebrates.

Tranquilizers a depressant group of drugs that reduces nervous activity.

Transcription copying of DNA by RNA.

Transfer RNA a form of RNA thought to deliver amino acids to the template formed by messenger RNA on the ribosomes.

Transformation in pneumococcus, the change from a noncapsulated to a capsulated form, brought about by the transfer of DNA.

Translation the synthesis of protein from mRNA on RNA.

Transmitter substance the chemical contained in synaptic vesicles. It changes the permeability of the postsynaptic membrane.

Transpiration the loss of water from plants.

Tree line the elevation above which trees do not grow.

Trichocysts sensitive protoplasmic threads in the paramecium, concerned with protection.

Triplet in DNA, a group of three bases.

Trochophore a larval form of mollusks.

Tropism an involuntary growth response of an organism to a stimulus.

Tubal ligation female sterilization involving the cutting of the Fallopian tubes.

Tube feet movable suction disks on the rays of most echinoderms.

Tube nucleus one of the three nuclei present in a pollen tube.

Tubule a tiny collecting tube extending from a Bowman's capsule of a kidney.

Tumor a mass of new abnormal tissue growth that is either benign or cancerous.

Tundra the treeless plains of arctic regions.

Turgor the stiffness of plant cells due to the presence of water.

Two-messenger model a model of hormone action that involves the hormone and cyclic AMP.

Tympanum a membrane in certain arthropods, serving a vibratory function.

Umbilical cord found in female mammals, leading from the placenta to the embryo.

Unit character, principle of Mendel's concept that the various hereditary characteristics are controlled by factors (genes), and that these factors occur in pairs.

Urea a nitrogenous waste substance found chiefly in the urine of mammals, but also formed in the liver from broken-down proteins.

Ureter a tube leading from a kidney to the bladder or cloaca.

Urethra the tube leading from the urinary bladder to an external opening of the body.

Uric acid a nitrogenous waste product of cell activity.

Urinary bladder the sac at the base of the ureters that stores urine.

Urine the liquid waste filtered from the blood in the kidney and voided by the bladder.

Uterus the organ in which young mammals are nourished until they are ready for birth.

Vaccination method of producing immunity by inoculating with a vaccine.

Vaccine a substance used to produce immunity.

Vacuolar membrane a membrane surrounding a vacuole in a cell and regulating the movement of materials in and out of the vacuole.

Vacuole one of the spaces scattered through the cytoplasm of a cell and containing fluid.

Vagina cavity of the female immediately outside and surrounding the cervix of the uterus.

Valence electrons the orbital electrons in the outermost shell of an atom.

Vasa deferentia ducts that transport sperms from the testes.

Vascular tissue fluid-conducting tissues characteristic of the tracheophytes.

Vasectomy male sterilization involving the surgical removal of all or part of the vas deferens.

Vasocongestion the accumulation of blood in blood vessels.

Vegetative reproduction reproduction of an organism not involving sexual union.

Veins strengthening and conducting structures in leaves; vessels carrying blood to the heart.

Vena cava a large collecting vein found in many vertebrates.

Venereal disease a contagious disease transmitted through sexual intercourse.

Ventral   the bottom of an organism.

Ventricle   a muscular chamber of the heart; also, a space in the brain.

Vertebrate   an animal with a backbone.

Vessel   a large, tubular cell of xylem through which water and minerals are conducted; in animals, what blood flows through.

Vestigal organs   those that are poorly developed and not functioning.

Villi   microscopic projections of the wall of the small intestine that increase the absorbing surface.

Viruses   particles that are noncellular and have no nucleus, no cytoplasm, and no surrounding membrane. They may reproduce in living tissue.

Vitamin   an organic substance that, though not a food, is essential for normal body activity.

Vitreous humor   a transparent substance that fills the interior of the eyeball.

Vulva   the external genital organs of the female.

Water cycle   the continuous movement of water from the atmosphere to the earth and from the earth to the atmosphere.

Water table   the level at which water is standing underground.

Wavelength   the distance between peaks of a wave.

White corpuscles   colorless, nucleated blood cells.

Withdrawal symptoms   nervous reactions and hallucinations resulting from the lack of a drug to which the victim is addicted.

Wolffian ducts   forerunners of sperm ducts.

X chromosome   a sex chromosome present singly in human males and as a pair in females.

Xerophyte   a plant that requires very little water to live.

Xylem   the woody tissue of a root or stem that conducts water and dissolved minerals upward.

Y chromosome   a sex chromosome found only in males.

Yolk sac   an extraembryonic membrane providing food for the embryo.

Zinjanthropus   an anthropoid ape with massive molars and premolars discovered in fossils nearly two million years old in East Africa.

Zygote   a fusion body formed when two gametes unite.

METRIC TABLES

I   THE METRIC SYSTEM

Most scientists all over the world use the **metric system** of measurement. Most other people of the world also use the metric system in their daily lives. In the United States, use of the metric system is rapidly growing.

The units of the metric system of measurement are the **meter** (for length), the **gram** (for weight), and the **liter** (for volume). Prefixes placed before the units of *meter, gram,* or *liter,* will tell you the multiple of the unit. Therefore, once you have an understanding of each basic unit, this system is easy to use.

PREFIXES OF THE METRIC SYSTEM		
Prefix	Scientific Notation	Decimal
KILO-	10^3	or 1000 times the unit
HECTO-	10^2	or 100 times the unit
DECA-	10	or 10 times the unit
the UNIT		
DECI-	10^{-1}	or 1/10 the unit
CENTI-	10^{-2}	or 1/100 the unit
MILLI-	10^{-3}	or 1/1000 the unit

Linear Measures

The unit of length in the metric system is the *meter* (abbreviation *m*), which is equal to 39.37 inches.

KILOmeter (km)	=	1000 meters
HECTOmeter	=	100 meters
DECAmeter	=	10 meters
meter		
DECImeter	=	1/10 meter
CENTImeter	=	1/100 meter
MILLImeter (mm)	=	1/1000 meter
MICROmeter (μm)—unit of measurement used in microscopic work	=	1/1,000,000 meter (10^{-6} m)
NANOmeter (nm)	=	1/1,000,000,000 meter (10^{-9} m)
Angstrom (Å)	=	1/10,000,000,000 meter (10^{-10} m)

Weight Measures

The unit of weight in the metric system is the *gram* (abbreviation g). One gram is the weight of one cubic centimeter of distilled water at 4°C.

KILOgram (kilo or kg)	=	1000 grams
HECTOgram	=	100 grams
DECAgram	=	10 grams
gram		
DECIgram	=	1/10 gram
CENTIgram (cg)	=	1/100 gram
MILLIgram (mg)	=	1/1,000 gram
MICROgram or gamma (γ)	=	1/1,000,000 gram (10^{-6} g)

Volume Measures

The unit of volume in the metric system is the *liter* (abbreviation l). One *liter* of distilled water weighs one kilogram. The most commonly used division is:

MILLIliter (ml)   =   1/1000 liter.

II   METRIC-ENGLISH EQUIVALENTS

1 meter = 39.37 inches
1 millimeter = approximately 1/25 inch
1 micron = approximately 1/25,000 inch
2.54 centimeters = 1 inch
1 kilogram = approximately 2.2 pounds
1 liter = approximately 1.06 quarts

III   CELSIUS AND FAHRENHEIT TEMPERATURE SCALES

Zero on the Celsius (also called centigrade) scale marks the freezing temperature of water. The equivalent on the Fahrenheit scale is 32°. Zero on the Fahrenheit scale is an arbitrary point: it marks the lowest temperature observed by the German scientist Fahrenheit during the winter of 1709. Zero degrees F corresponds to -17.77°C.

The temperature of boiling water, at sea level, is marked 100° on the Celsius scale. This is 212° on the Fahrenheit scale. There are 100 degrees between the melting point of ice and the boiling point of water on the Celsius scale. But there are 180 degrees between these two temperatures on the Fahrenheit scale. Therefore, one Fahrenheit degree is equal to 5/9 (100/180) of one Celsius degree.

The following procedure may be used to convert the temperatures of one scale to those of the other:

°F to °C: subtract 32, multiply by 5, divide by 9.

°C to °F: multiply by 9, divide by 5, add 32.

Expressed as formulae:

$$°C = 5/9 \times (°F - 32°)$$
$$°F = (9/5 \times C°) + 32°$$

INDEX

Saline injection, 306
Saliva, 324, 326
Salivary amylase, 141
Salivary glands, 84, 324, 326
Salmon, 33
Salvarsan, 367
Sand dollars, 77
Saprophytes, 55
Sarcoplasmic reticulum, 384
Sardines, 116
Saturated fatty acids, 138
Savannah, 113
Scallops, 75
Scanning electron microscope, 146–147
Scavengers, 95
Schistosoma japonicum, 99
Schistosomiasis, 99
Schizophrenia, 383, 404
Schleiden, Matthias, 146
Schwann, Theodor, 146
Scientific method, 4–5
Sclerenchyma, 64
Sclerotic layer, 389
Scolex, 72
Scorpions, 79, 80
Sea, 18
Sea anemones, 70
Sea lettuce, 266
Sea urchins, 77
Seagulls, 115
Seasonal change, effect on behavior, 416–420
Second polar body, 263
Secondary oocyte, 263
Secondary sex characteristics, 244–245, 284, 344–345
Secondary spermatocytes, 262
Secondary syphillis, 309
Secondary walls, 154
Sedges, 272
Sediment, 20
Sedimentary rock, 20–21
Seed plants, 62–66
Seedcoats, 275, 276
Seeds, 32, 62–63, 99, 272
  dispersal of, 64, 272–274
  dormancy of, 274
  germination of, 274–275
Segmented body, 80
Segmented worms, 73–75, 79
Segregation, law of, 220
Selective breeding, 227–232
Selectively permeable membrane, 156
Self-pollination, 218, 228, 270
Semen, 285, 300
Semicircular canals, 388, 389
Semideserts, 105
Semilunar valves, 333
Seminal receptacles, 75, 85
Seminal vesicles, 75, 284, 300
Seminiferous tubules, 261, 284
Semipermeability, 155
Sensation, 386–394

Sensory areas, 393
Sensory neurons, 380–381
Septum, 333
Sequoia gigantea, 63
Serine, 198
Serum globulin, 327
Sessile animals, 67
Setae, 73
Sewage disposal plants, 96
Sex chromosomes:
  abnormal numbers of, 248
  of Drosophila, 225–226
  human, 226, 243–244
Sex flush, 297, 299
Sex hormones, 245, 284, 287, 293, 304–306, 344–345, 372, 374
Sex-limited genes, 244–245
Sex-linked genes, 224–226, 243–244
Sexual development, 295
Sexual reproduction, 7, 263–264
  by bacteria, 53
  conjugation, 54–56, 59
  defined, 263
  by earthworms, 75
  evolution and, 53
  by ferns, 62
  fertilization, see Fertilization
  by flowering plants, 269–277
  by green algae, 59
  human, 284–296
  by hydra, 69
  by insects, 85
  by molds, 56
  by mosses, 61
  by paramecium, 54–55
  by planarians, 71
  by sponges, 67
  by starfish, 78
  by tapeworms, 72
Sharks, 86, 100
Sheep, 95, 113, 416
Shells, 75
Shorter, Frank, 183
Shrubs, 103, 109
Sickle-cell anemia, 245–246
Sieve plate, 77
Sight, 388–392
Sign stimuli, 427
Silicon, 135
Silk ear, 277
Silt, 104
Silver maple, 35
Silverfish, 95
Simple goiter, 342
Skates, 86
Skeletal muscles, 384, 385
Skeleton, 37, 79, 80
Skin:
  as barrier to bacteria, 357
  cancer of, 371
  coloration, genetics of, 242, 243
  functions of, 339–340
  sensations of, 386–387
Skulls, fossil, 37, 38–39

Skunks, 413, 417
Sleep, 415, 416
Slime layer, 50
Small intestine, 325, 326
Smallpox, 359, 362, 364
Smell, sense of, 387–388
Smooth muscles, 384, 385
Snails, 36, 75, 76, 95
Snakes, 87, 113, 413
Sociobiology, 427–429
Sodium:
  concentration in human body, 135
  nerve impulse and, 382–383
Sodium bicarbonate, 134
Sodium chloride, 130–131, 319, 321
Sodium hydroxide, 133
Soil, 104–105
Solar energy, 6, 94, 105
Somatic cells, 158–160, 221, 259–261
Somatic mutations, 207
Somatotrophic hormones, 342
Sorenson, Jonenen, 133
South Pole, 110
Southern slopes, 102
Sow bugs, 80
Speciation, 34–35
Species, 48
  characteristics, 227
  competition between, 109–110
  defined, 34
  diurnal, 413
  extinction of, 119
  habitats of, 109
  immunity, 363
  interactions between, 10
  natural selection and, 27–30, 101, 119, 412
  niches of, 109
  nocturnal, 413
  succession and, 108–109
Specific heat, 132
Sperm:
  of earthworms, 75
  division of, 158–160
  of ferns, 62
  in fertilization, 289–290
  formation of, 261–262
  human, 7, 158–160, 261–262, 284–286, 289–290
  of insects, 85
  of plants, 272
  of sponges, 67
  of starfish, 78
Sperm nuclei, 272
Spermatids, 262
Spermatocytes, 261, 262
Spermatogenesis, 261–262
Spermicidal foams, 301, 305
Spherical symmetry, 70
Spicules, 67
Spiders, 79, 80
Spinal cord, 85, 396
Spinal nerves, 396
Spindle, 300
Spines, 113

Spiracles, 84
Spirogyra, 59
Spleen, 350
Sponges, 66–67, 327, 337
Spongin, 67
Spontaneous generation, 2–4
Sporangia, 56
Spores, 62, 265
Sporophytes, 61, 62
Spruce, 104, 112
Squid, 36, 75, 76, 116
Squirrels, 33, 112, 113, 412, 417
Stamens, 218, 269, 271
Stanley, Wendell, 357–358
Staphylococci, 367–368
Starches:
  animal, 137, 320
  in diet, 320
  digestion of, 141, 324, 325
  penetration of plasma membranes by,
    156
  plant, 137
  stored in seeds, 275
Starfish, 77–78, 115, 116
Statoliths, 422
Steady-state theory, 17
Stems, 65, 422
Sterility, 244
Sterilization, 306–307
Steroids, 138–139
Stickleback fish, 412
Stigma, 269
Stimulant drugs, 400–403
Stimulus, 9
Stips, 56
Stirrup, 388–389
Stock, 268
Stomach:
  cancer of, 371
  digestion in, 324–326
  of grasshoppers, 84
Stomata, 185–186
Strata, 20
Strawberry plants, 266–267
Streaming, 150
Streptomyces griseus, 368
Streptomycin, 368
Striped maple, 35
Stroke, 369–370
Sub-species, 48
Subclavian veins, 336
Substrates, 140–141
Succession, 108–109
Sucrase, 325, 326
Sucrose, 136, 156, 320
Sugar maple, 35
Sugarcane, 170
Sugars, 132
  in diet, 320
  digestion of, 177–178, 325
  structure of, 136
Sulfa drugs, 367, 397–398
Sulfur, 18, 135
Sulfur bacteria, 171, 180
Sundew, 386

Superior vena cava, 334
Survival of the fittest, 29
Suspensions, 134
Swamps, 104
Sycamore, 112
Symbiosis, 99–100
Symmetry, 70
Sympathetic nervous system, 396
Synapse, 381, 382–383
Synapsis, 261
Synaptic vesicles, 382–383
Syphillis, 309–310, 368
Systemic circulation, 335
Systolic pressure, 334

Tap roots, 113
Tapeworms, 71–72, 99
Tarantula hawks, 82
Tardigrade, 111
Taste, sense of, 387
Taste buds, 387
Tatum, Edward, 209
Taxonomy, 48–49
Tay-Sachs disease, 204, 245
Tears, 357, 390
Telophase, 260, 261
Temperate regions, 103, 112
Temperature:
  effect on ecosystems, 102–103
  nastic movement and, 423
  plant growth and, 350
  seasonal, 416
Tentacles, 67–68
Tenting, 297
Termites, 82, 98, 99–100, 114
Testes:
  human, 261, 284, 342–343, 370
  of hydra, 69
  of insects, 85
Testosterone, 138–139, 148, 284, 342,
    429
Tetracyclines, 368
Tetrahydrocannabinols (THC), 403
Thalli, 58, 59
Theories, 5
Thermodynamically stable mixture,
    18
Thermodynamics, second law of, 6
Thiamin, 323
Thigmotropism, 422
Thoracic duct, 336
Thorax, 79, 80
Thorpe, W. H., 425
Threonine, 196
Thrombin, 330
Thromboplastin, 330
Thymine, 196–197
Thymus gland, 358
Thyroid gland, 341–342
Thyroid stimulating hormone, 342
Thyroxine, 341
Ticks, 80
Tidal rhythms, 420
Tides, 115

Tigers, 100–101, 115
Tissue, 160
  connective, 85
  cultures, 161
  of seed plants, 64–65
Toads, 87
Tobacco (plant), 170
Tobacco (smoking), 400–402
Tocopherol, 323
Tolerance, 398, 402
Tongue, 401
Tongue rolling, 240–241
Tonsils, 336
Tools, use of, 37–38
Topsoil, 104
Touch, sense of, 386–387
Tracheae, 80, 81, 84, 181
Tracheophyta, 62–66
Tranquilizers, 404
Transcription, 197–200
Transfer RNA, 199–200
Transformation, 206
Transfusions, blood, 331
Translation, 198
Transmission electron microscope,
    146–147
Transmitter substances, 382–383
Transpiration, 185
Transverse binary fission, 51
Transverse nerves, 71
Tree frogs, 101
Tree line, 111
Trees, 103
  See also Forests; specific species
Treponema pallidum, 309
Trichocysts, 54
Triplets, 195
Triploid chromosome number, 272
Trisomy-X, 248
Trochophre, 75, 79
Tropical grasslands, 113
Tropical rain forests, 114
Tropisms, 421–422
Trypsin, 325, 326
Tryptophan, 198
Tubal ligation, 305, 307
Tube feet, 77
Tube nucleus, 269
Tulips, 269
Tumors, 370
Tuna, 116
Tundra, 111, 413
Turgor, 156–157, 185–186, 422–424
Turkeys, 231
Turtles, 36, 87
Twins, studies of, 250–251
Two-messenger model, 345–346
Tympanum, 84–85
Tyrosine, 198

Ulva, 266
Umbilical cord, 293, 296
Undercrowding, 98
Unit characters, concept of, 220